Modern Developments in Transcranial Magnetic Stimulation (TMS) –Applications and Perspectives in Clinical Neuroscience

Modern Developments in Transcranial Magnetic Stimulation (TMS) –Applications and Perspectives in Clinical Neuroscience

Editors

Nico Sollmann
Petro Julkunen

MDPI • Basel • Beijing • Wuhan • Barcelona • Belgrade • Manchester • Tokyo • Cluj • Tianjin

Editors
Nico Sollmann
Department of Diagnostic
and Interventional
Neuroradiology
Klinikum rechts der Isar
Technical University
of Munich
Munich
Germany

Petro Julkunen
Department of Clinical
Neurophysiology
Kuopio University Hospital
Kuopio
Finland

Editorial Office
MDPI
St. Alban-Anlage 66
4052 Basel, Switzerland

This is a reprint of articles from the Special Issue published online in the open access journal *Brain Sciences* (ISSN 2076-3425) (available at: www.mdpi.com/journal/brainsci/special_issues/Transcranial_Magnetic_Stimulation).

For citation purposes, cite each article independently as indicated on the article page online and as indicated below:

LastName, A.A.; LastName, B.B.; LastName, C.C. Article Title. *Journal Name* **Year**, *Volume Number*, Page Range.

ISBN 978-3-0365-4398-7 (Hbk)
ISBN 978-3-0365-4397-0 (PDF)

© 2022 by the authors. Articles in this book are Open Access and distributed under the Creative Commons Attribution (CC BY) license, which allows users to download, copy and build upon published articles, as long as the author and publisher are properly credited, which ensures maximum dissemination and a wider impact of our publications.

The book as a whole is distributed by MDPI under the terms and conditions of the Creative Commons license CC BY-NC-ND.

Contents

Nico Sollmann and Petro Julkunen
Modern Developments in Transcranial Magnetic Stimulation: The Editorial
Reprinted from: *Brain Sci.* **2022**, *12*, 628, doi:10.3390/brainsci12050628 1

Nico Sollmann, Sandro M. Krieg, Laura Säisänen and Petro Julkunen
Mapping of Motor Function with Neuronavigated Transcranial Magnetic Stimulation: A Review on Clinical Application in Brain Tumors and Methods for Ensuring Feasible Accuracy
Reprinted from: *Brain Sci.* **2021**, *11*, 897, doi:10.3390/brainsci11070897 5

Haosu Zhang, Severin Schramm, Axel Schröder, Claus Zimmer, Bernhard Meyer and Sandro M. Krieg et al.
Function-Based Tractography of the Language Network Correlates with Aphasia in Patients with Language-Eloquent Glioblastoma
Reprinted from: *Brain Sci.* **2020**, *10*, 412, doi:10.3390/brainsci10070412 57

Severin Schramm, Alexander F. Haddad, Lawrence Chyall, Sandro M. Krieg, Nico Sollmann and Phiroz E. Tarapore
Navigated TMS in the ICU: Introducing Motor Mapping to the Critical Care Setting
Reprinted from: *Brain Sci.* **2020**, *10*, 1005, doi:10.3390/brainsci10121005 75

Valentina Baro, Samuel Caliri, Luca Sartori, Silvia Facchini, Brando Guarrera and Pietro Zangrossi et al.
Preoperative Repetitive Navigated TMS and Functional White Matter Tractography in a Bilingual Patient with a Brain Tumor in Wernike Area
Reprinted from: *Brain Sci.* **2021**, *11*, 557, doi:10.3390/brainsci11050557 93

Ann-Katrin Ohlerth, Roelien Bastiaanse, Chiara Negwer, Nico Sollmann, Severin Schramm and Axel Schröder et al.
Bihemispheric Navigated Transcranial Magnetic Stimulation Mapping for Action Naming Compared to Object Naming in Sentence Context
Reprinted from: *Brain Sci.* **2021**, *11*, 1190, doi:10.3390/brainsci11091190 105

Shohreh Kariminezhad, Jari Karhu, Laura Säisänen, Jusa Reijonen, Mervi Könönen and Petro Julkunen
Brain Response Induced with Paired Associative Stimulation Is Related to Repetition Suppression of Motor Evoked Potential
Reprinted from: *Brain Sci.* **2020**, *10*, 674, doi:10.3390/brainsci10100674 135

Sascha Freigang, Christian Lehner, Shane M. Fresnoza, Kariem Mahdy Ali, Elisabeth Hlavka and Annika Eitler et al.
Comparing the Impact of Multi-Session Left Dorsolateral Prefrontal and Primary Motor Cortex Neuronavigated Repetitive Transcranial Magnetic Stimulation (nrTMS) on Chronic Pain Patients
Reprinted from: *Brain Sci.* **2021**, *11*, 961, doi:10.3390/brainsci11080961 147

Shuo Xu, Qing Yang, Mengye Chen, Panmo Deng, Ren Zhuang and Zengchun Sun et al.
Capturing Neuroplastic Changes after iTBS in Patients with Post-Stroke Aphasia: A Pilot fMRI Study
Reprinted from: *Brain Sci.* **2021**, *11*, 1451, doi:10.3390/brainsci11111451 165

Davide Tabarelli, Arianna Brancaccio, Christoph Zrenner and Paolo Belardinelli
Functional Connectivity States of Alpha Rhythm Sources in the Human Cortex at Rest: Implications for Real-Time Brain State Dependent EEG-TMS
Reprinted from: *Brain Sci.* **2022**, *12*, 348, doi:10.3390/brainsci12030348 . . . . **179**

Manuel F. Casanova, Mohamed Shaban, Mohammed Ghazal, Ayman S. El-Baz, Emily L. Casanova and Ioan Opris et al.
Effects of Transcranial Magnetic Stimulation Therapy on Evoked and Induced Gamma Oscillations in Children with Autism Spectrum Disorder
Reprinted from: *Brain Sci.* **2020**, *10*, 423, doi:10.3390/brainsci10070423 . . . . **197**

Milorad Dragić, Milica Zeljković, Ivana Stevanović, Marija Adžić, Andjela Stekić and Katarina Mihajlović et al.
Downregulation of CD73/A_{2A}R-Mediated Adenosine Signaling as a Potential Mechanism of Neuroprotective Effects of Theta-Burst Transcranial Magnetic Stimulation in Acute Experimental Autoimmune Encephalomyelitis
Reprinted from: *Brain Sci.* **2021**, *11*, 736, doi:10.3390/brainsci11060736 . . . . **217**

Andrea Guerra, Lorenzo Rocchi, Alberto Grego, Francesca Berardi, Concetta Luisi and Florinda Ferreri
Contribution of TMS and TMS-EEG to the Understanding of Mechanisms Underlying Physiological Brain Aging
Reprinted from: *Brain Sci.* **2021**, *11*, 405, doi:10.3390/brainsci11030405 . . . . **237**

Connor J. Phipps, Daniel L. Murman and David E. Warren
Stimulating Memory: Reviewing Interventions Using Repetitive Transcranial Magnetic Stimulation to Enhance or Restore Memory Abilities
Reprinted from: *Brain Sci.* **2021**, *11*, 1283, doi:10.3390/brainsci11101283 . . . . **255**

Editorial

Modern Developments in Transcranial Magnetic Stimulation: The Editorial

Nico Sollmann [1,2,3,*] and Petro Julkunen [4,5]

1. Department of Diagnostic and Interventional Radiology, University Hospital Ulm, Albert-Einstein-Allee 23, 89081 Ulm, Germany
2. Department of Diagnostic and Interventional Neuroradiology, School of Medicine, Klinikum rechts der Isar, Technical University of Munich, Ismaninger Str. 22, 81675 Munich, Germany
3. TUM-Neuroimaging Center, Klinikum rechts der Isar, Technical University of Munich, 81675 Munich, Germany
4. Department of Clinical Neurophysiology, Kuopio University Hospital, P.O. Box 100, 70029 KYS Kuopio, Finland; petro.julkunen@kuh.fi
5. Department of Applied Physics, University of Eastern Finland, 70210 Kuopio, Finland
* Correspondence: nico.sollmann@tum.de

Transcranial magnetic stimulation (TMS) is being increasingly applied in neuroscience and the clinical setup [1–5]. Applications of TMS are focused on treatment and diagnostics. Modern advances include, but are not limited to, the combination of TMS with precise neuronavigation, as well as the integration of TMS into a multimodal environment, mainly by guiding TMS applications using complementary techniques such as functional magnetic resonance imaging (fMRI), electroencephalography (EEG), diffusion tensor imaging (DTI), or magnetoencephalography (MEG). The impact of stimulation can be identified and characterized by such multimodal approaches, thus helping to shed light on basic neurophysiology and TMS effects in the human brain [6–9].

This Special Issue entitled "Modern Developments in Transcranial Magnetic Stimulation (TMS)–Applications and Perspectives in Clinical Neuroscience" in *Brain Sciences* received studies covering various applications of TMS, with focuses on neuronavigated TMS (nTMS) for mapping of cortical functions [10–14], treatment and modulatory effects [15–19], and basic neuromechanisms [20–22], all clinically relevant and supporting the aims set for the Special Issue.

Sollmann et al. comprehensively reviewed nTMS motor mapping in clinical application with accompanying validating evidence, and provided additional and new evidence on the parametric prerequisites crucial to ensuring feasible mapping accuracy [13]. In addition, the accompanied multimodal information used (e.g., for fiber tracking of the descending motor tracts) was concluded to provide potential aid and improvements in nTMS applicability, particularly in the critical cohort of patients harboring motor-eloquent brain tumors [13]. Highlighted applications of nTMS in clinical motor mapping include, in addition to the conventionally acknowledged imaging information, the mode of risk stratification and prediction of potential surgical outcomes, as well as observations of neural plasticity related to adjustment and relocation of motor functions within the brains of such patients [13]. Sollmann et al. also highlighted the role of methodological integration into clinical routines and accompanied systems for achieving the full potential of nTMS, while considering that successful application requires comprehensive knowledge of the application and its methodological constraints [13]. The authors see great potential in nTMS and its multimodal applications, not only as a surgical planning tool, but also for providing longitudinal information applicable to prognostics and follow-up examinations [13].

Applying motor mapping in critically ill patients, Schramm et al. demonstrated the use of nTMS as a safe and reliable method for motor mapping in the intensive care unit (ICU) setting, and outlined its possible benefits [12]. The authors demonstrated that in the

ICU environment, where imaging with computed tomography (CT) is more applicable than magnetic resonance imaging (MRI) that is conventionally used with nTMS, the post-processed CT images provided a feasible alternative to MRI for neuronavigation [12]. The ICU environment is notoriously challenging, given that electromyography (EMG) for motor mapping with nTMS is highly sensitive to noise coupled to weak signals. The noise in the measured EMG signal has been successfully reduced to a feasible level [12]. Schramm et al. considered the potential applications in the ICU to involve prognostics and monitoring of certain transient complications related to the nTMS motor mapping procedure [12].

By mapping language-related areas, Zhang et al. provided evidence for structural differences on cortical and subcortical levels between language-positive (i.e., locations where upon stimulation, a modulatory effect on language performance was detected) and language-negative areas (i.e., locations where no effect of nTMS was detected) during nTMS language mapping among patients with language-eloquent brain tumors [14]. Their results provided additional and new evidence in patients with glioblastoma multiforme regarding the connection of speech and language function and brain anatomy, with nTMS demonstrating that responsiveness to stimulation is critically related to cortical and subcortical interplay and the rate of speech impairment [14]. They considered that the results further increase confidence in nTMS language mapping and nTMS-based tractography in the clinical setting [14]. In their study, Baro et al. reported a case of nTMS-based tractography application in neurosurgery in a bilingual patient affected by a brain tumor in the left temporal lobe, who underwent nTMS mapping for both languages (Romanian and Italian) [10]. This procedure was considered to disclose the true eloquence of the anterior part of the lesion in both language-related tests [10]. The outcome was verified after surgery, with language abilities remaining intact in both languages [10]. To further develop the protocol of language mapping with nTMS, Ohlerth et al. compared a conventional noun-naming task to an action-naming task, and reported that action naming may be more favorable in nTMS mapping in terms of error rates and may hence improve the accuracy of nTMS-aided preoperative planning [11]. Their findings may have distinct impact on nTMS language mapping routines in clinical setups, where an object-naming task is routinely used despite limited specificity.

In their study, Phipps et al. reviewed the current heterogeneous literature-based evidence for using repetitive TMS (rTMS) to enhance or restore memory, e.g., for applications in Alzheimer's disease treatment, and offered several recommendations for the design of future investigations using rTMS to modulate human memory performance [17]. Regarding the potential analgesic effects of rTMS for treatment of chronic refractory pain, Freigang et al. compared two treatment targets and sequences against a sham setting in multi-session therapy for lower back pain, and found indications that treatment on the left dorsolateral prefrontal cortex with 5-Hz rTMS may induce greater pain and stress relief than treatment on the primary motor cortex (M1) with 20 Hz [16]. In addition, Xu et al. provided evidence that an intermittent theta-burst stimulation (TBS) protocol on the ipsilesional M1 could induce immediate neural activity and functional connectivity changes in motor, language, and other brain regions in patients with post-stroke aphasia as observed through fMRI, which could promote functional recovery [18]. In an experimental setting in rats, to shed light on the efficacy of rTMS in multiple sclerosis (MS) treatment, Dragić et al. reported that continuous TBS counteracted with experimental autoimmune encephalomyelitis (EAE)-induced effects on adenosine signaling [19]. Furthermore, it attenuated the reactive state of microglia and astrocytes, thus suggesting a potential TBS-induced reduction in the neuroinflammatory process known to be related to MS [19].

In their exploratory study combining EEG and rTMS modulation, Casanova et al. found that in autism spectrum disorder (ASD) subjects, visually evoked and induced gamma oscillations were evident at higher magnitudes of gamma oscillations before rTMS modulation than in neurotypical controls [15]. Recordings after rTMS treatment in ASD revealed a significant reduction in gamma responses to task-irrelevant stimuli, and participants made fewer errors after rTMS neuromodulation [15]. In addition, behavioral questionnaires conducted after treatment revealed decreased irritability, hyperactivity, and repetitive behavior scores [15].

Guerra et al. provide a status update and summarizing review on previously reported findings regarding the potential contribution of TMS in combination with EEG to the understanding of the mechanisms underlying normal brain aging [20]. Continuing with combined TMS and EEG, and to demonstrate the functional phase-dependent relationships between frontal, parietal, and occipital areas of the brain to support EEG-state-dependent TMS, Tabarelli et al. studied a large open dataset [22]. They found a consistent connectivity between parietal and prefrontal regions, whereas occipito-prefrontal connectivity was less marked and occipito-parietal connectivity was comparatively low [22]. The authors consider their results a relevant add-on feature for individualized brain-state-dependent TMS, with possible contributions to personalized therapeutic nTMS applications [22].

Kariminezhad et al. reported that the individual paired associate stimulation (PAS) response, whether expressed as long-term depression (LTD)-like or long-term potentiation (LTP)-like effects on motor-evoked potentials (MEPs), were related to the individual repetition suppression (RS) responses observed in the MEP amplitudes [21]. Kariminezhad et al. considered this finding a promising step for predicting TMS neuromodulation outcome based on the individual RS effect [21].

Overall, the original articles, reviews, and case studies included in this Special Issue provide interesting reading, solid evidence, important indicative findings, and summarizing conclusions based on recent literature, hence being of great interest for all those working with clinical applications of nTMS or neuroscience research. We would like to thank the authors for their contributions and wish the readers happy and fruitful reading.

Funding: P.J. received funding from the Academy of Finland (grant no: 322423).

Conflicts of Interest: P.J. received consultation fees and shares a patent with Nexstim Oyj (Helsinki, Finland). N.S. received honoraria from Nexstim Oyj (Helsinki, Finland).

References

1. Rossi, S.; Hallett, M.; Rossini, P.M.; Pascual-Leone, A.; Safety of TMS Consensus Group. Safety, ethical considerations, and application guidelines for the use of transcranial magnetic stimulation in clinical practice and research. *Clin. Neurophysiol. Off. J. Int. Fed. Clin. Neurophysiol.* **2009**, *120*, 2008–2039. [CrossRef] [PubMed]
2. Rossi, S.; Antal, A.; Bestmann, S.; Bikson, M.; Brewer, C.; Brockmoller, J.; Carpenter, L.L.; Cincotta, M.; Chen, R.; Daskalakis, J.D.; et al. Safety and recommendations for TMS use in healthy subjects and patient populations, with updates on training, ethical and regulatory issues: Expert Guidelines. *Clin. Neurophysiol. Off. J. Int. Fed. Clin. Neurophysiol.* **2021**, *132*, 269–306. [CrossRef] [PubMed]
3. Hallett, M. Transcranial magnetic stimulation: A primer. *Neuron* **2007**, *55*, 187–199. [CrossRef] [PubMed]
4. Hallett, M. Transcranial magnetic stimulation and the human brain. *Nature* **2000**, *406*, 147–150. [CrossRef] [PubMed]
5. Sollmann, N.; Meyer, B.; Krieg, S.M. Implementing Functional Preoperative Mapping in the Clinical Routine of a Neurosurgical Department: Technical Note. *World Neurosurg.* **2017**, *103*, 94–105. [CrossRef]
6. Bergmann, T.O.; Varatheeswaran, R.; Hanlon, C.A.; Madsen, K.H.; Thielscher, A.; Siebner, H.R. Concurrent TMS-fMRI for causal network perturbation and proof of target engagement. *NeuroImage* **2021**, *237*, 118093. [CrossRef]
7. Kimiskidis, V.K. Transcranial magnetic stimulation (TMS) coupled with electroencephalography (EEG): Biomarker of the future. *Rev. Neurol.* **2016**, *172*, 123–126. [CrossRef]
8. Bestmann, S.; Ruff, C.C.; Blankenburg, F.; Weiskopf, N.; Driver, J.; Rothwell, J.C. Mapping causal interregional influences with concurrent TMS-fMRI. *Exp. Brain Res.* **2008**, *191*, 383–402. [CrossRef]
9. Esposito, R.; Bortoletto, M.; Miniussi, C. Integrating TMS, EEG, and MRI as an Approach for Studying Brain Connectivity. *Neuroscientist* **2020**, *26*, 471–486. [CrossRef]
10. Baro, V.; Caliri, S.; Sartori, L.; Facchini, S.; Guarrera, B.; Zangrossi, P.; Anglani, M.; Denaro, L.; d'Avella, D.; Ferreri, F.; et al. Preoperative Repetitive Navigated TMS and Functional White Matter Tractography in a Bilingual Patient with a Brain Tumor in Wernike Area. *Brain Sci.* **2021**, *11*, 557. [CrossRef]
11. Ohlerth, A.K.; Bastiaanse, R.; Negwer, C.; Sollmann, N.; Schramm, S.; Schroder, A.; Krieg, S.M. Bihemispheric Navigated Transcranial Magnetic Stimulation Mapping for Action Naming Compared to Object Naming in Sentence Context. *Brain Sci.* **2021**, *11*, 1190. [CrossRef] [PubMed]
12. Schramm, S.; Haddad, A.F.; Chyall, L.; Krieg, S.M.; Sollmann, N.; Tarapore, P.E. Navigated TMS in the ICU: Introducing Motor Mapping to the Critical Care Setting. *Brain Sci.* **2020**, *10*, 1005. [CrossRef] [PubMed]

13. Sollmann, N.; Krieg, S.M.; Saisanen, L.; Julkunen, P. Mapping of Motor Function with Neuronavigated Transcranial Magnetic Stimulation: A Review on Clinical Application in Brain Tumors and Methods for Ensuring Feasible Accuracy. *Brain Sci.* **2021**, *11*, 897. [CrossRef] [PubMed]
14. Zhang, H.; Schramm, S.; Schroder, A.; Zimmer, C.; Meyer, B.; Krieg, S.M.; Sollmann, N. Function-Based Tractography of the Language Network Correlates with Aphasia in Patients with Language-Eloquent Glioblastoma. *Brain Sci.* **2020**, *10*, 412. [CrossRef] [PubMed]
15. Casanova, M.F.; Shaban, M.; Ghazal, M.; El-Baz, A.S.; Casanova, E.L.; Opris, I.; Sokhadze, E.M. Effects of Transcranial Magnetic Stimulation Therapy on Evoked and Induced Gamma Oscillations in Children with Autism Spectrum Disorder. *Brain Sci.* **2020**, *10*, 423. [CrossRef]
16. Freigang, S.; Lehner, C.; Fresnoza, S.M.; Mahdy Ali, K.; Hlavka, E.; Eitler, A.; Szilagyi, I.; Bornemann-Cimenti, H.; Deutschmann, H.; Reishofer, G.; et al. Comparing the Impact of Multi-Session Left Dorsolateral Prefrontal and Primary Motor Cortex Neuronavigated Repetitive Transcranial Magnetic Stimulation (nrTMS) on Chronic Pain Patients. *Brain Sci.* **2021**, *11*, 961. [CrossRef]
17. Phipps, C.J.; Murman, D.L.; Warren, D.E. Stimulating Memory: Reviewing Interventions Using Repetitive Transcranial Magnetic Stimulation to Enhance or Restore Memory Abilities. *Brain Sci.* **2021**, *11*, 283. [CrossRef]
18. Xu, S.; Yang, Q.; Chen, M.; Deng, P.; Zhuang, R.; Sun, Z.; Li, C.; Yan, Z.; Zhang, Y.; Jia, J. Capturing Neuroplastic Changes after iTBS in Patients with Post-Stroke Aphasia: A Pilot fMRI Study. *Brain Sci.* **2021**, *11*, 1451. [CrossRef]
19. Dragic, M.; Zeljkovic, M.; Stevanovic, I.; Adzic, M.; Stekic, A.; Mihajlovic, K.; Grkovic, I.; Ilic, N.; Ilic, T.V.; Nedeljkovic, N.; et al. Downregulation of CD73/A2AR-Mediated Adenosine Signaling as a Potential Mechanism of Neuroprotective Effects of Theta-Burst Transcranial Magnetic Stimulation in Acute Experimental Autoimmune Encephalomyelitis. *Brain Sci.* **2021**, *11*, 736. [CrossRef]
20. Guerra, A.; Rocchi, L.; Grego, A.; Berardi, F.; Luisi, C.; Ferreri, F. Contribution of TMS and TMS-EEG to the Understanding of Mechanisms Underlying Physiological Brain Aging. *Brain Sci.* **2021**, *11*, 405. [CrossRef]
21. Kariminezhad, S.; Karhu, J.; Säisänen, L.; Reijonen, J.; Könönen, M.; Julkunen, P. Brain Response Induced with Paired Associative Stimulation Is Related to Repetition Suppression of Motor Evoked Potential. *Brain Sci.* **2020**, *10*, 674. [CrossRef] [PubMed]
22. Tabarelli, D.; Brancaccio, A.; Zrenner, C.; Belardinelli, P. Functional Connectivity States of Alpha Rhythm Sources in the Human Cortex at Rest: Implications for Real-Time Brain State Dependent EEG-TMS. *Brain Sci.* **2022**, *12*, 348. [CrossRef] [PubMed]

Review

Mapping of Motor Function with Neuronavigated Transcranial Magnetic Stimulation: A Review on Clinical Application in Brain Tumors and Methods for Ensuring Feasible Accuracy

Nico Sollmann [1,2,3,4,*], Sandro M. Krieg [3,5], Laura Säisänen [6,7] and Petro Julkunen [6,7]

1. Department of Diagnostic and Interventional Radiology, University Hospital Ulm, Albert-Einstein-Allee 23, 89081 Ulm, Germany
2. Department of Diagnostic and Interventional Neuroradiology, School of Medicine, Klinikum rechts der Isar, Technical University of Munich, Ismaninger Str. 22, 81675 Munich, Germany
3. TUM-Neuroimaging Center, Klinikum rechts der Isar, Technical University of Munich, 81675 Munich, Germany; sandro.krieg@tum.de
4. Department of Radiology and Biomedical Imaging, University of California San Francisco, 185 Berry Street, San Francisco, CA 94143, USA
5. Department of Neurosurgery, School of Medicine, Klinikum rechts der Isar, Technical University of Munich, Ismaninger Str. 22, 81675 Munich, Germany
6. Department of Clinical Neurophysiology, Kuopio University Hospital, 70029 Kuopio, Finland; laura.saisanen@kuh.fi (L.S.); Petro.Julkunen@kuh.fi (P.J.)
7. Department of Applied Physics, University of Eastern Finland, 70211 Kuopio, Finland
* Correspondence: nico.sollmann@tum.de

Abstract: Navigated transcranial magnetic stimulation (nTMS) has developed into a reliable non-invasive clinical and scientific tool over the past decade. Specifically, it has undergone several validating clinical trials that demonstrated high agreement with intraoperative direct electrical stimulation (DES), which paved the way for increasing application for the purpose of motor mapping in patients harboring motor-eloquent intracranial neoplasms. Based on this clinical use case of the technique, in this article we review the evidence for the feasibility of motor mapping and derived models (risk stratification and prediction, nTMS-based fiber tracking, improvement of clinical outcome, and assessment of functional plasticity), and provide collected sets of evidence for the applicability of quantitative mapping with nTMS. In addition, we provide evidence-based demonstrations on factors that ensure methodological feasibility and accuracy of the motor mapping procedure. We demonstrate that selection of the stimulation intensity (SI) for nTMS and spatial density of stimuli are crucial factors for applying motor mapping accurately, while also demonstrating the effect on the motor maps. We conclude that while the application of nTMS motor mapping has been impressively spread over the past decade, there are still variations in the applied protocols and parameters, which could be optimized for the purpose of reliable quantitative mapping.

Keywords: brain stimulation; brain tumor; electric field; eloquent cortex; functional mapping; motor mapping; motor threshold; navigated transcranial magnetic stimulation; neuronavigation; presurgical evaluation

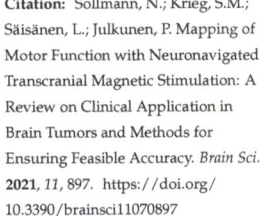

Citation: Sollmann, N.; Krieg, S.M.; Säisänen, L.; Julkunen, P. Mapping of Motor Function with Neuronavigated Transcranial Magnetic Stimulation: A Review on Clinical Application in Brain Tumors and Methods for Ensuring Feasible Accuracy. *Brain Sci.* 2021, 11, 897. https://doi.org/10.3390/brainsci11070897

Academic Editor: Lorenzo Rocchi

Received: 4 June 2021
Accepted: 2 July 2021
Published: 7 July 2021

Publisher's Note: MDPI stays neutral with regard to jurisdictional claims in published maps and institutional affiliations.

Copyright: © 2021 by the authors. Licensee MDPI, Basel, Switzerland. This article is an open access article distributed under the terms and conditions of the Creative Commons Attribution (CC BY) license (https://creativecommons.org/licenses/by/4.0/).

1. Introduction

Navigated transcranial magnetic stimulation (nTMS) combines the use of neuronavigation with TMS to target neurostimulation inductively to the brain cortex, utilizing views of the brain anatomy with sub-centimeter precision and enabling tracking of the coil during stimulation (e.g., using an infrared (IR) tracking system combined with the stimulator; Figure 1) [1]. For over a decade, nTMS has been used in diagnostic setups (e.g., to perform non-invasive clinical mapping of motor or other brain functions) or for therapeutic purpose (e.g., to treat major depression or other psychiatric diseases as well as chronic pain). Preceding nTMS as we know it today, early approaches conducted mapping based on anatomical landmarks placed over the scalp or used frameless stereotactic systems to localize and track stimulations [2–4]. However, such approaches seemed limited for clinical use where high precision and reliability is warranted in relation to individual brain anatomy that can be both spatially and functionally deranged due to pathology. Technological advancements have enabled the estimation of the spatial extent and geometry of the electric field (EF) induced by stimulation [1,5–7]. Specifically, the development of nTMS using EF-based neuronavigation acknowledges that the EF induced by stimulation strongly depends on several factors such as skull thickness and coil tilting, amongst others. In this context, accurate and efficient modelling of the EF is essential to pinpoint the impact of stimulation and to understand its effects [1,5–7]. Since nTMS using such EF-based neuronavigation is currently considered the most accurate method for targeting stimulation, this article focusses on this technique.

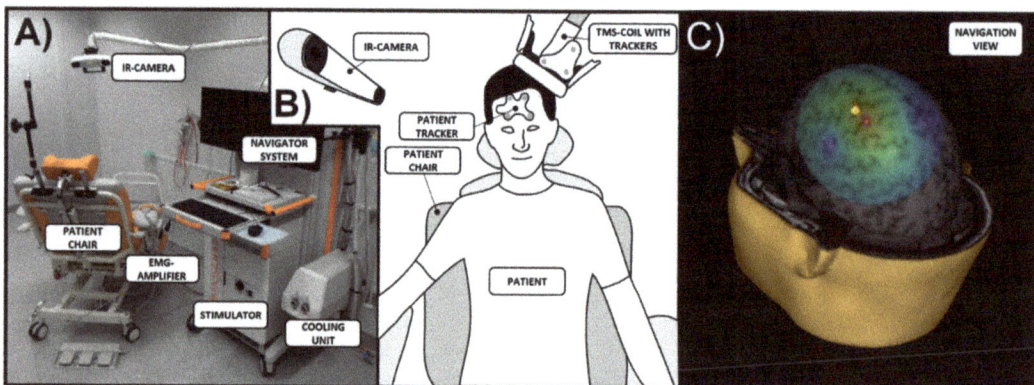

Figure 1. Setup for nTMS mapping. (**A**) Hardware of the nTMS system, including an infrared camera, EMG amplifier, neuronavigation monitor, and the stimulator with a cooling unit as the central components. (**B**) Patient positioning during mapping procedures, requiring initial co-registration of the patient's head with cranial MRI (using a head tracker attached to the patient's forehead) to be able to track the stimulating coil (equipped with infrared trackers) in relation to individual brain anatomy. (**C**) Navigation view during the mapping procedure, showing the stimulating coil (*yellow marker*) with its orientation (*red arrow*) and modelled EF distribution.

Overall, interest towards combining neuronavigation with TMS has experienced more and more interest in the research community as indicated by the increasing trend in published studies (Figure 2). From a clinical perspective, nTMS has been gaining importance especially as a tool for preoperative mapping used for planning and intraoperative resection guidance in patients harboring functionally eloquent brain neoplasms [1,8–11]. As of today, mapping of motor function has become the mainstay of nTMS, making so-called nTMS motor mapping in patients with motor-eloquent lesions an ideal use case for the technique. In modern neuro-oncological surgery, achieving an optimal balance between the extent of resection (EOR) of a brain tumor and the individual functional status of the patient is the major principle for surgical resection [12]. Prognostically, a safely performed maximized tumor resection is of utmost importance as incomplete resection is correlated to lower survival rates and quality of life for patients with malignant glioma as the prominent entity of intra-axial brain tumors [13–18]. To maximize the EOR while keeping surgery-related decline of function (e.g., paresis in the context of the motor system) to the lowest level achievable, intraoperative direct electrical stimulation (DES) is performed as the gold-standard method for assessing subcortical and cortical functional representations [19–23]. Outside of the operating theater, nTMS motor mapping and later-developed nTMS-based tractography can be performed for this purpose. As the major field of current application, the first part of this article focusses on the clinical utility of nTMS-based motor mapping and derived tractography in patients with motor-eloquent brain neoplasms.

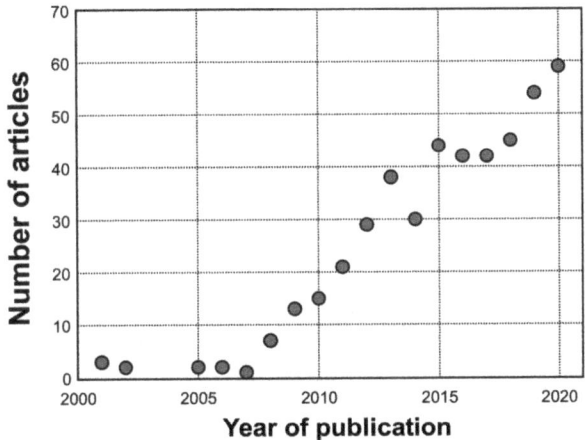

Figure 2. Number of articles on nTMS published annually between 2002 and 2020 as indexed in PubMed (on the 21st of May in 2021). The search query in PubMed was defined as follows: ("navigated transcranial magnetic stimulation" OR "navigated TMS" OR "neuronavigated TMS" OR "neuronavigated transcranial magnetic stimulation").

From a methodological perspective, the nTMS parameters selected for motor mapping, such as the stimulation intensity (SI), number of applied stimuli, and the density of the applied stimulation grid, affect the outcome and accuracy of the motor mapping (Figure 3) [24–27]. Specifically, optimal selection becomes especially important when quantitative parameters are drawn from the motor maps, e.g., to characterize the location of motor function or to define its extent on the cortex. Quantitative mapping refers to those quantitative parameters derived from nTMS motor mapping, based on the recorded responses and location information related to the responses. The quantitative parameters in nTMS motor mapping include the locations of the motor hotspot and center of gravity (CoG), and area and volume of the motor map, which are all based on recorded amplitudes of motor-evoked potentials (MEPs) elicited by stimulation (Figure 3) [24–27]. The motor hotspot location is often considered as the location of the maximum MEP amplitude response [28,29]. It not only characterizes the major location of the motor representation, but is also used as the location where the resting motor threshold (rMT) is determined and around which the motor map extends. The CoG is defined as the amplitude-weighted location representing the location in the motor representation area where the center of motor activation lies [28,30,31]. The volume of the motor map is usually calculated by summing the MEP amplitudes within a motor map, e.g., within a stimulus grid [30–33]. The area of the motor map on the other hand binarizes the MEP responses to positive and negative responses and represents the area covered by positive responses (i.e., spots where stimulation elicited an MEP with a certain amplitude) [27,34,35]. Importantly, wrong selection or wrong interpretation of these quantitative parameters affects the confidence in the motor map and, in the worst case, could lead to imprecise motor maps that may have harmful consequences to the patient when used for preoperative planning and intraoperative resection guidance in neuro-oncological surgery. The technical solutions and refinements realized in the nTMS systems by the manufacturers are also important for the overall performance of motor mapping, albeit representative comparisons between different manufacturer-specific approaches have not yet been made in the literature. Given the crucial role quantitative parameters play for nTMS motor mapping, the second part of this article demonstrates the significance of certain parameters used in motor mapping to apply quantitative measures on the resulting individual motor maps. Finally, we speculate on the future of nTMS motor mapping by the potential in advancing from current trends towards purely quantitative motor mapping, which would have direct clinical impact. Relevant studies were identified by PubMed search (http://www.ncbi.nlm.nih.gov/pubmed; accessed on 21 May 2021).

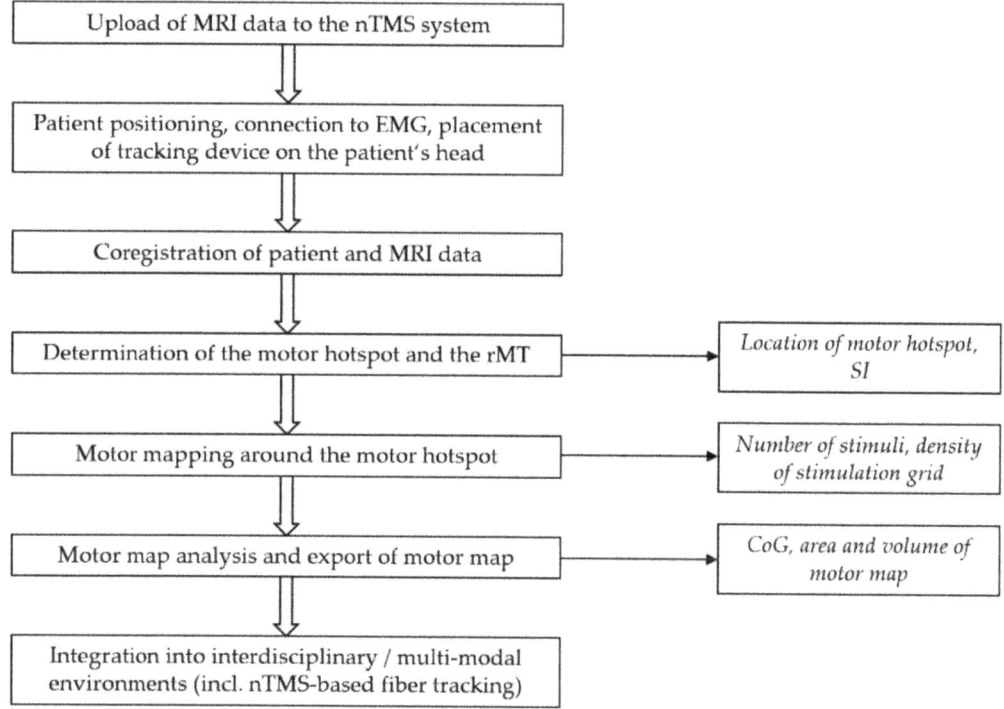

Figure 3. Overview of the main steps during the nTMS mapping procedure, starting with upload of the MRI data to the nTMS system and resulting in the generation of the individual motor map that can be further used for nTMS-based fiber tracking and other interdisciplinary or multi-modal applications. In addition, the main quantitative parameters that can be extracted from the single steps are shown (including the location of the motor hotspot, SI, number of stimuli, density of the stimulation grid, CoG, area, and volume of the motor map).

2. Clinical Application of nTMS for Mapping Motor Function

2.1. Feasibility, Reliability, and Comparison with Other Methods

The gold standard for functional brain mapping is defined by DES. Thus, any other technique for mapping purposes needs to be validated against it. First case reports in patients with intracranial metastasis (MET), meningioma (MEN), and low-grade glioma (LGG) have already indicated good agreement between preoperative motor mapping by nTMS and DES [32–34]. Specifically, in an early series of 20 patients with different entities of intracranial neoplasms, the motor hotspot was located on the same gyrus for nTMS and DES mappings in all enrolled cases, with distances of motor hotspots between techniques amounting to values <1 cm for specific cortical muscle representations [35]. In this regard, Table 1 provides an overview of studies on reliability of nTMS motor mapping and comparison with other methods, mainly including DES.

The initial impression of high concordance between motor maps provided by preoperatively acquired nTMS motor mapping and those derived from DES was confirmed by subsequent studies that aimed to enroll larger patient cohorts with different entities of brain lesions and performed comparisons with a third method. Specifically, comparative analyses were achieved for nTMS and task-based functional magnetic resonance imaging (fMRI) versus DES mappings since task-based fMRI appeared as a widely used alternative to nTMS, with all major studies consistently revealing closer distances for localizations of cortical motor function between preoperative nTMS and DES compared with task-based fMRI activation maps [36–41]. Exemplarily, in a series of 26 patients, distances between motor hotspots of nTMS and DES amounted to 4.4 ± 3.4 mm on average [38]. However, when comparing distances between activation in task-based fMRI and nTMS motor hotspots, larger distances of 9.8 ± 8.5 mm for upper extremity (uE) and 14.7 ± 12.4 mm for lower extremity (lE) muscle representations were observed [38]. More recently, considering motor representations of the uE, lE, and tongue in 36 patients with intracranial tumors, there were significantly smaller Euclidean distances as well as a better spatial overlap between DES and nTMS compared with task-based fMRI [41]. In this regard, one study suggested that nTMS motor mapping is as accurate in recurrent gliomas as it seems to be prior to the initial surgery, opening up the possibility to reuse nTMS motor mapping in the individual course of disease [39]. Similar to results for nTMS motor mapping versus fMRI, in an investigation comparing the accuracy of nTMS with magnetoencephalography (MEG) and DES mapping, distances between motor hotspots of nTMS and DES mappings were significantly smaller than those between the activation in MEG and nTMS motor hotspots [42].

The implementation of nTMS motor mapping confirmed the expected anatomy in 22%, added awareness of high-risk functional areas in 27%, modified the approach in 16%, changed the planned EOR in 8%, and changed the surgical indication in 3% of patients among a cohort of 73 patients with different entities of brain neoplasms [43]. Similarly, another study concluded that nTMS motor mapping enabled an exact localization of the motor cortex in 88.2%, provided the neurosurgeon with new unexpected information about functional anatomy in 70.6%, and facilitated a modification of the surgical approach to spare the motor cortex from damage in 29.4% of patients [44]. Further, the availability of nTMS data improved the neurosurgeons' confidence in identifying the motor strip or central region [38,40]. Importantly, valuable motor mapping data from nTMS may be generated irrespective of the distinct experience level of the operator (after a certain interval of training), with average intra-examiner distances between CoGs for an expert investigator amounting to 4.4 mm and 4.9 mm for the expert versus novice investigator, with comparable values in the investigated healthy subjects as well as in patients with brain tumors [45].

Regarding potential variability of MEP latencies or the SI applied during nTMS motor mapping as a function of the rMT, it has been revealed in large cohorts of patients with motor-eloquent tumors that sex and antiepileptic drug (AED) intake, amongst other factors, may considerably contribute to MEP latency variability as well as variability of the determined rMT [46,47]. Interestingly, also the tumor grading according to the World Health Organization (WHO) scheme may be held accountable for a certain degree of variability introduced to nTMS motor mappings because MEP latencies of lE muscles increased with the WHO grading, and correlations between the increase in WHO grading and a decreased rMT were observed for lE muscles [48]. From a more methodological perspective, efforts have been made to expand accurate nTMS-based motor mapping beyond the limits of the supposed primary motor cortex, and to increase robustness by introducing altered threshold values for MEP amplitudes [49,50]. Future research will determine the distinct relevance of such influencing factors for motor mapping by nTMS in patients with brain tumors, in whom such potential confounders of motor mapping results are still largely neglected in the clinical routine workflow.

Table 1. Feasibility, reliability, and comparison with other methods. This table outlines the studies published on feasibility and reliability of motor mapping by nTMS (using an electric-field-navigated system) in patients harboring brain neoplasms. Furthermore, studies comparing nTMS motor mapping with other modalities (intraoperative DES, preoperative fMRI, and preoperative MEG) are included.

Author	Year	Cohort	Tumor Entities	nTMS Method	Techniques for Comparison	Main Objective	Main Findings
Coburger et al. [32]	2011	75-year-old woman	MEN	Preoperative motor mapping (60–80 V/m)	Intraoperative DES	To correlate nTMS with DES	– The motor strip as determined by DES was located in the same cortical area where motor responses were elicited by nTMS. – nTMS motor mapping is a helpful tool for preoperative planning.
Picht et al. [33]	2011	63-year-old woman	MET	Preoperative and postoperative (4th postoperative day) motor mapping (110% rMT)	Intraoperative DES	To test nTMS in a patient with severe preoperative motor impairment and to correlate with DES	– nTMS motor mapping demonstrated intact corticospinal pathways in presence of hemiplegia. – nTMS motor mapping modified the surgical strategy. – nTMS and DES mappings agreed well.
Picht et al. [35]	2011	20 adult patients	HGG, MET, MEN, Other	Preoperative motor mapping (110% rMT)	Intraoperative DES	To compare the accuracy of nTMS with DES	– Motor hotspots were located on the same gyrus for nTMS and DES mappings in all cases. – Distances between motor hotspots of nTMS and DES mappings amounted to 7.8 ± 1.2 mm for the APB and to 7.1 ± 0.9 mm for the TA muscles. – nTMS and DES mappings agreed well.
Forster et al. [37]	2011	11 adult patients	LGG, HGG, MET, Other	Preoperative motor mapping (110% rMT)	Preoperative fMRI Intraoperative DES	To compare the accuracy of nTMS with fMRI and DES	– Distances between motor hotspots of nTMS and DES mappings (10.5 ± 5.7 mm) were significantly smaller than those between the activation in fMRI and DES motor hotspots (15.0 ± 7.6 mm). – nTMS motor mapping is more precise than fMRI when correlated to results of DES.
Picht et al. [43]	2012	73 adult patients	LGG, HGG, MET, MEN, Other	Preoperative motor mapping (110% rMT) and nTMS-based tractography (but not considered)	Intraoperative DES	To evaluate the influence, benefit, and impact of nTMS on surgery	– nTMS motor mapping confirmed the expected functional anatomy in 22%, added awareness of high-risk areas in 27%, modified the approach in 16%, changed the planned EOR in 8%, and changed the surgical indication in 3% of patients. – nTMS motor mapping has positive impact on surgical planning and on the surgery itself.

Table 1. *Cont.*

Author	Year	Cohort	Tumor Entities	nTMS Method	Techniques for Comparison	Main Objective	Main Findings
Krieg et al. [38]	2012	26 adult patients	LGG, HGG, MET, Other	Preoperative motor mapping (110% rMT for uE and 130% for lE muscles)	Preoperative fMRI Intraoperative DES	To compare the accuracy of nTMS with fMRI and DES	– Distances between motor hotspots of nTMS and DES mappings (4.4 ± 3.4 mm) were significantly smaller than those between the activation in fMRI and nTMS motor hotspots (9.8 ± 8.5 mm for uE and 14.7 ± 12.4 mm for lE muscles). – In most cases of tumors in the precentral gyrus the neurosurgeon admitted easier identification of the central region in awareness of nTMS motor maps. – nTMS and DES mappings agreed well.
Coburger et al. [34]	2012	3-year-old boy	LGG	Preoperative motor mapping (130 V/m for uE and 220 V/m for lE muscles)	Intraoperative DES	To test nTMS in a child and to correlate with DES	– DES verified the location of nTMS motor hotspots. – nTMS is a precise tool for preoperative motor mapping and is feasible even in young, pediatric patients.
Tarapore et al. [42]	2012	24 adult patients	LGG, HGG, Other	Preoperative motor mapping (110% rMT)	Preoperative MEG Intraoperative DES	To compare the accuracy of nTMS with MEG and DES	– Distances between motor hotspots of nTMS and DES mappings (2.1 ± 0.3 mm) were significantly smaller than those between the activation in MEG and nTMS motor hotspots (4.7 ± 1.1 mm). – DES did not reveal a motor site that was unrecognized by nTMS motor mapping in any patient. – nTMS motor mapping agreed well with DES as well as preoperative MEG.
Coburger et al. [36]	2013	28 adult patients and 2 patients <18 years	LGG, HGG, MET, MEN, Other	Preoperative motor mapping (110% rMT in adult patients and aMT in patients <18 years)	Preoperative fMRI Intraoperative DES	To compare the accuracy of nTMS with fMRI and DES	– Mean accuracy to localize the motor cortex was higher for nTMS motor mapping compared with fMRI. – In the subgroup of intrinsic tumors, nTMS motor mapping produced significantly higher accuracy for lE muscle representations than fMRI.
Krieg et al. [39]	2013	31 adult patients	LGG, HGG, MET, MEN, Other	Preoperative motor mapping (110% rMT for uE and 130% for lE muscles)	Preoperative fMRI Intraoperative DES	To compare the accuracy of nTMS with fMRI and DES in recurrent gliomas vs. initially operated tumors	– nTMS motor mapping correlated well with DES in recurrent gliomas (6.2 ± 6.0 mm) and newly diagnosed tumors (5.7 ± 4.6 mm), yet without a significant difference. – Compared with fMRI, the difference was larger for uE muscle representations (recurrent: 8.5 ± 7.2 mm; new: 9.8 ± 8.6 mm) and lE muscle representations (recurrent: 17.1 ± 10.6 mm; new: 13.8 ± 13.0 mm), yet without a significant difference. – nTMS motor mapping was as accurate in recurrent gliomas as it has been prior to the initial surgery.

Table 1. *Cont.*

Author	Year	Cohort	Tumor Entities	nTMS Method	Techniques for Comparison	Main Objective	Main Findings
Mangraviti et al. [40]	2013	8 adult patients	LGG, HGG, MET, Other	Preoperative motor mapping (110% rMT for uE and 130% for lE muscles)	Preoperative fMRI Intraoperative DES	To compare the accuracy of nTMS with fMRI and DES	– Distances between motor hotspots of nTMS and DES mappings (8.5 ± 4.6 mm) were significantly smaller than those between the activation in fMRI and DES motor hotspots (12.9 ± 5.7 mm). – Visualization of nTMS motor hotspots improved the neurosurgeons' confidence in identifying the motor strip as well as in planning of surgical strategies.
Zdunczyk et al. [45]	2013	10 adult patients (and 10 healthy volunteers)	LGG, HGG, MET, MEN, Other	Preoperative motor mapping (110% rMT) (and two motor mappings per subject in the healthy volunteers)	-	To assess the intra- and inter-examiner reliability of nTMS mapping	– Distances between CoGs for the expert examiner in healthy volunteers were 4.4 (1.9–7.7) mm and 4.9 (2.4–9.2) mm for the expert vs. novice examiner. – The reliability of nTMS motor mapping in tumor patients appeared to be comparable to those in healthy subjects.
Rizzo et al. [44]	2014	17 adult patients	LGG, HGG, MET, MEN, Other	Preoperative motor mapping (110–115% rMT for uE and 130% for lE muscles)	Intraoperative DES	To evaluate the influence, benefit, and impact of nTMS on surgery	– nTMS motor mapping exactly localized the motor cortex in 88.2%, provided the neurosurgeon with new unexpected information about functional anatomy of the motor area in 70.6%, and led to a change of the surgical strategy in 29.4% of patients. – nTMS motor mapping has objective and subjective benefits on surgical planning and on the surgery itself.
Sollmann et al. [46]	2017	100 adult patients	LGG, HGG, MET, Other	Preoperative motor mapping (110% rMT for uE and 130% for lE muscles)	-	To investigate factors that have impact on MEP latencies	– Common factors (relevant to APB, ADM, and FCR) for MEP latency variability were sex and AED intake. – Muscle-specific factors (relevant to APB, ADM, or FCR) for MEP latency variability were tumor side, tumor location, and rMT.
Sollmann et al. [47]	2017	100 adult patients	LGG, HGG, MET, Other	Preoperative motor mapping (110% rMT for uE and 130% for lE muscles)	-	To investigate factors that influence the determination of the rMT	– Edema and age at exam in the ADM model only jointly reduced the unexplained variance for rMT determination. – The other factors kept in the ADM model (sex, AED intake, and motor deficit) and each of the factors kept in the APB and FCR models independently and significantly reduced the unexplained variance for rMT determination.

Table 1. *Cont.*

Author	Year	Cohort	Tumor Entities	nTMS Method	Techniques for Comparison	Main Objective	Main Findings
Lam et al. [49]	2019	20 adult patients	n/a	Preoperative motor mapping (105% rMT)	-	To investigate the feasibility of increasing the MEP threshold (50 μV vs. 500 μV) to improve the robustness of motor mapping	- Both the standard (50 μV) MEP threshold as well as the experimental (500 μV) MEP threshold yielded motor maps in all patients. - No significant differences in motor area sizes were found between the conventional (500 μV) MEP threshold and the experimental (50 μV) MEP threshold. - MEP latency time was significantly reduced for recordings from 500 μV compared with recordings from 50 μV MEP thresholds.
Weiss-Lucas et al. [41]	2020	36 adult patients	LGG, HGG, MET, MEN, Others	Preoperative motor mapping (110% rMT)	Preoperative fMRI Intraoperative DES	To compare the accuracy of nTMS with fMRI and DES	- There were significantly smaller Euclidean distances (11.4 ± 8.3 vs. 16.8 ± 7.0 mm) and better spatial overlaps (64 ± 38% vs. 37 ± 37%) between DES and nTMS mappings compared with DES vs. fMRI. - Contrary to DES, fMRI and nTMS motor mappings were feasible for all regions and patients without complications (reliable and accurate DES was only obtained in 25 of the included patients).
Mirbagheri et al. [50]	2020	12 adult patients (and six healthy volunteers)	n/a	Preoperative motor mapping (105% rMT for primary motor areas, 120%/150% for non-primary motor areas)	-	To investigate whether nTMS reliably elicits MEPs outside of the primary motor cortex	- 88.8% of stimulations in suspected non-primary motor areas did not result in motor-positive spots with MEPs ≥50 μV. - Positive nTMS motor mapping in non-primary motor areas was associated with higher SI and larger primary motor areas. - Particularly when mapped with 150% rMT, more MEP artifacts occurred in patients than in healthy volunteers.
Lavrador et al. [48]	2020	45 adult patients	LGG, HGG	Preoperative motor mapping (105% rMT)	-	To assess the excitability of the motor system in relation to tumor grading	- MEP latencies of IE muscles increased with an increase in the WHO grading of the tumor. - An association between the increase in the WHO grading and a decreased rMT was observed for IE muscles. - Higher WHO grading of the tumor and isocitrate dehydrogenase wild-type tumors were associated with the number of abnormal interhemispheric rMT ratios.

Abbreviations: nTMS—navigated transcranial magnetic stimulation; LGG—low-grade glioma; HGG—high-grade glioma; MET—metastasis; MEN—meningioma; rMT—resting motor threshold; aMT—active motor threshold; uE—upper extremity; lE—lower extremity; DES—direct electrical stimulation; fMRI—functional magnetic resonance imaging; APB—abductor pollicis brevis; TA—tibialis anterior; EOR—extent of resection; MEG—magnetoencephalography; CoG—center of gravity; ADM—abductor digiti minimi; FCR—flexor carpi radialis; MEP—motor-evoked potential; AED—antiepileptic drug; SI—stimulation intensity; WHO—World Health Organization.

2.2. Fiber Tractography

The motor-positive nTMS points can be used for seeding to obtain regions of interest (ROIs) for subsequent delineation of the corticospinal tract (CST) in the context of fiber tracking. The combination of nTMS motor maps at the cortical surface with diffusion MRI (dMRI) and tractography, i.e., the delineation of the primary motor cortex and its subcortical connections to the peripheral nervous system with visualization of the course of the CST, can provide a more complete picture than one of the techniques alone (Figure 4). In this regard, Table 2 outlines studies using data from nTMS motor mapping for tractography of the CST in patients with different kinds of brain neoplasms.

In a first tractography study using nTMS data, 30 patients harboring motor-eloquent brain tumors underwent presurgical nTMS motor mapping followed by nTMS-based tractography, using the nTMS motor-positive spots and the ipsilateral cerebral peduncle as ROIs [51]. Compared with the conventionally used approach with manual delineation of the suspected primary motor cortex and the ipsilateral cerebral peduncle as ROIs, the novel setup led to a lower number of fibers displayed, a reduced volume of the CST, and, most importantly, lower fractions of aberrant tracts most likely not belonging to the CST [51]. The approach of nTMS-based tractography of the CST was subsequently refined by investigating individually adapted adjustments for the fractional anisotropy (FA) that had to be defined for the deterministic tractography algorithm: the FA was increased stepwise until no fibers were displayed, followed by reducing the FA value by 0.01, thus delineating only a thin fiber course; the obtained FA value was defined as 100% FA threshold (FAT), and nTMS-based tractography was then carried out with 50% and 75% FAT with motor-positive nTMS points and the manually delineated internal capsule or brainstem as ROIs [52]. This method influenced the surgical strategy in 46% of patients, in contrast to conventional tractography without nTMS data for ROI generation where an impact was only observed for 22% of patients [52].

Furthermore, a study among 20 patients with different entities of brain neoplasms achieved detailed somatotopic CST reconstruction derived from nTMS motor maps combined with dMRI for uE, lE, and face muscle representations, with a decreased number of fibers and a greater overlap between the motor cortex and the cortical end-region of the CST when compared with conventional tractography with only anatomical seeding [53]. Of note, the obtained CST course as well as the somatotopic organization were confirmed by DES mapping [53]. In another study on somatotopic reconstruction of the CST considering parts subserving uE, lE, and face muscles, a higher fraction of plausible fibers was observed for seeding at the anterior inferior pontine level when compared with seeding at the internal capsule, combined with nTMS-based seeding at the cortical level [54]. When setting somatotopic nTMS-based tractography in contrast to fMRI-based seeding, a higher plausibility was observed for the nTMS-based approach, and fMRI-originated tracts presented with a more posterior course relative to the nTMS-based reconstruction of tracts [55]. Recently, in a comprehensive study systematically comparing different setups for tractography (deterministic and probabilistic algorithms with variable ROI definitions) and correlating tractography with DES mapping and fMRI findings in 11 adult patients, highest accuracy of tractography was achieved when using seeding with a manually generated mask enclosing the precentral gyrus, but none of the applied setups showed clear superiority and nTMS- or fMRI-based tractography differed only slightly [56]. Yet, probabilistic tracking resulted in an optimized correlation with DES mapping when compared with the more commonly used deterministic tractography algorithm [56]. Upcoming work is needed to further investigate optimal settings and algorithms in representative cohort sizes to achieve results of CST tractography as close as possible to DES mapping results in order to assure accuracy and reliability of nTMS-based tractography.

Table 2. Fiber tractography. This table presents the studies published on tractography of the CST primarily using motor mapping by nTMS (using an electric-field-navigated system) for seeding in patients harboring brain neoplasms.

Author	Year	Cohort	Tumor Entities	dMRI Acquisition	Tractography Specifics	Main Objective	Main Findings
Krieg et al. [51]	2012	30 adult patients	LGG, HGG, MET, MEN, Other	6 diffusion directions, b-values: 0–800 s/mm² (3 Tesla)	Two ROIs: motor-positive nTMS points and ipsilateral brainstem (conventional: manually delineated motor cortex and ipsilateral brainstem) Deterministic tracking; FA <0.2, FL ~100 mm, angular threshold of 30°	To assess the feasibility of nTMS-based tractography in relation to conventional seeding without nTMS data	- nTMS-based tractography resulted in a lower number of aberrant tracts (i.e., tracts not belonging to the CST) when compared with conventional seeding without nTMS. - The proximity of the tracts to the tumor was not different between nTMS-based and conventional tractography for CST reconstruction. - Conventional seeding showed to be user-dependent, whereas nTMS-based tractography seemed to be less subjective.
Frey et al. [52]	2012	50 adult patients	LGG, HGG, MET, MEN	23 diffusion directions, b-values: 0–1000 s/mm² (3 Tesla)	One ROI: motor-positive nTMS points (conventional: manually delineated internal capsule or brainstem) Deterministic tracking; FA = 50% and 75% FAT, FL = 110 mm, angular threshold of 30°	To assess the feasibility and impact on surgery of nTMS-based tractography in relation to conventional seeding without nTMS data and to provide a new algorithm for FA determination	- nTMS-based tractography changed or modified the surgical strategy in 46% of patients, whereas conventional tractography would have changed the surgical strategy in only 22% of patients. - Tractography facilitated intraoperative situs orientation and application of DES in 56% of patients. - Tracking at 75% FAT was considered most beneficial by the neurosurgeons.
Conti et al. [53]	2014	20 adult patients	LGG, HGG, MET, Other	32 diffusion directions (3 Tesla)	Two ROIs: motor-positive nTMS points, subdivided for uE, lE, and face muscles, and ipsilateral brainstem (conventional: ipsilateral brainstem) Deterministic tracking; FA = 0.2, FL = 20 mm, angular threshold of 45°	To assess somatotopic organization by nTMS-based tractography in relation to conventional seeding without nTMS data and to verify nTMS-based tractography by intraoperative DES	- Detailed somatotopic CST reconstruction was possible by nTMS-based tractography with a greater overlap between the motor cortex and the cortical end-region of the CST (90.5 ± 8.8% vs. 58.3 ± 16.6%) when compared with conventional tractography. - DES mapping confirmed the CST location and the somatotopic reconstruction in all cases. - nTMS-based tractography of the CST appeared to be more accurate, less user-dependent, and capable of providing reliable CST delineation compared with conventional tractography.
Weiss et al. [54]	2015	32 adult patients	LGG, HGG, MET, MEN, Other	30 diffusion directions, b-values: 0–800 s/mm² (3 Tesla)	Two ROIs: motor-positive nTMS points, subdivided for uE, lE, and face muscles, and internal capsule and/or anterior inferior pontine level Deterministic tracking; FA = 75% and 100% FAT, FL = 1 mm, angular threshold of 30°	To assess the impact of subcortical seed regions and of somatotopic location of cortical seed regions on plausibility of tractography in relation to clinical factors	- A higher proportion of plausible tracts was observed for seeding at the anterior inferior pontine level when compared with seeding at the internal capsule. - Low FAT and the presence of peritumoral edema within the internal capsule led to less plausible tractography, and most plausible tractography was obtained when the FAT ranged above a cut-off of 0.105. - A strong effect of somatotopic location of the seed region was observed, with the best plausibility present for tractography of fibers subserving the bilateral uE muscle representations (>95%).

Table 2. Cont.

Author	Year	Cohort	Tumor Entities	dMRI Acquisition	Tractography Specifics	Main Objective	Main Findings
Weiss et al. [55]	2017	18 adult patients	LGG, HGG, MET, MEN, Other	30 diffusion directions, b-values: 0–800 s/mm^2 (3 Tesla)	Two ROIs: motor-positive nTMS points, subdivided for uE, lE, and face muscles, and anterior inferior pontine level (fMRI-based: task-derived activation map and anterior inferior pons) Deterministic tracking: FA = 100% FAT, FL = 1 mm, angular threshold of 30°	To assess the impact of the modality used for somatotopic location of cortical seed regions on plausibility of tractography	– A higher plausibility was observed for nTMS-based tractography compared with fMRI-based tractography, with fMRI-originated tracts showing a significantly more posterior course relative to the nTMS-based tracts. – nTMS motor mapping seems to be the method of choice to identify seed regions for tractography of the CST in patients with close vicinity of the primary motor cortex to a brain neoplasm.
Münnich et al. [56]	2019	11 adolescent or adult patients	HGG, MET, Other	20 diffusion directions, b-values: 0–700 s/mm^2 (3 Tesla)	Several ROIs for motor-positive nTMS points, task-derived fMRI activation maps, or conventional seeding Deterministic and probabilistic tracking: minimal FA threshold = 0.05/∈ [0.1;0.45], tracking step length = 1 mm/∈ [0.5;7] mm, maximum fiber curvature = 0.3/∈ [0.2; 0.65]	To compare different seeding setups and tracking algorithms, and to correlate tractography with intraoperative DES and MRI findings	– The best accuracy of tractography was achieved using the segmented precentral gyrus for seeding (marginal R^2 = 0.146); however, since the marginal R^2 of fMRI and nTMS motor mapping differed very little, none of the methods showed distinct superiority. – Both nTMS-based and fMRI-based tractography showed significant correlations between distances and the SI of DES for the CST, but only with respect to uE muscle representations. – The use of the probabilistic tracking algorithm led to a better correlation between DES mapping and tractography. – Tractography demands for careful interpretation of its results by considering all influencing variables (e.g., seeding approach and tracking algorithm used).

Abbreviations: nTMS—navigated transcranial magnetic stimulation; LGG—low-grade glioma; HGG—high-grade glioma; MET—metastasis; MEN—meningioma; CST—corticospinal tract; dMRI—diffusion magnetic resonance imaging; ROI—region of interest; FA—fractional anisotropy; FAT—fractional anisotropy threshold; FL—fiber length; fMRI—functional magnetic resonance imaging; uE—upper extremity; lE—lower extremity; DES—direct electrical stimulation; SI—stimulation intensity.

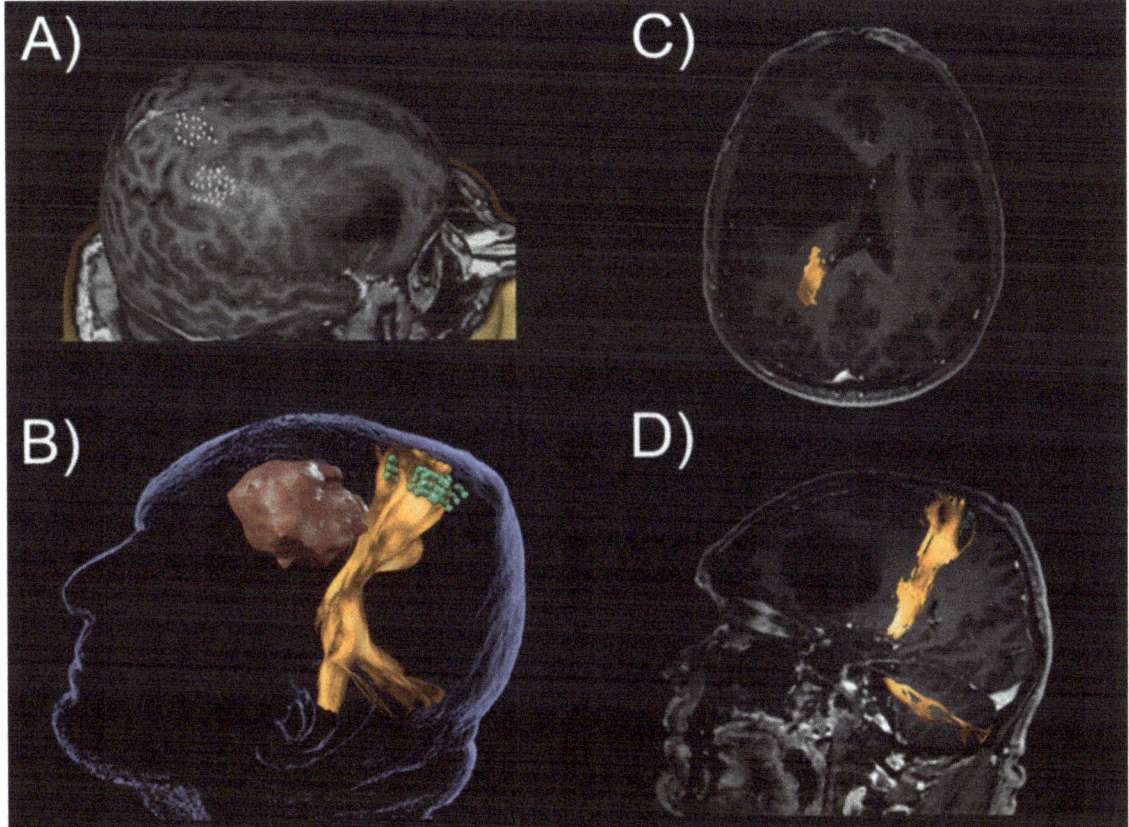

Figure 4. Exemplary patient case (right-hemispheric glioma in a 56-year-old male patient) for illustration of CST reconstruction using tractography based on motor maps derived from motor mapping with nTMS. (**A**) Motor map with binarization into motor-positive (*white*) and motor-negative (*grey*) stimulation points. (**B**) Tractography of the CST (*orange*) based on a ROI constituted of motor-positive nTMS points (*green*). (**C**) Fusion of T1-weighted imaging and tractography results (axial plane). (**D**) Fusion of T1-weighted imaging and tractography results (sagittal plane).

2.3. Improvement of Clinical Outcome

When added to the armamentarium of the preoperative workup of patients harboring brain neoplasms, the question arises whether nTMS motor mapping and derived nTMS-based tractography may be capable of improving the clinical outcome, as measured by an ideally increased EOR combined with lowered rates of functional perioperative decline. In this regard, Table 3 gives an overview of studies focused on clinical outcome.

An initial study compared 11 patients who underwent preoperative nTMS to 11 patients without nTMS motor mapping, revealing that preoperative nTMS motor mapping changed the treatment plan towards early and more extensive resection in 6 out of 11 patients [57]. Furthermore, one of four patients of the nTMS group with preoperative motor deficits improved by one year, whereas increased deficits were observed in three of the eight patients of the historical group not having surgery [57]. In two retrospective studies in considerably large cohorts of patients with different types of brain tumors, the utility of nTMS motor mapping and nTMS-based tractography of the CST may have facilitated a more extensive EOR, combined with tendencies towards better motor function after surgery [58,59].

Specifically, nTMS disproved suspected involvement of the primary motor cortex in 25.1% of the 250 enrolled patients, and it enabled expanding surgical indication in 14.8%, thus facilitating planning of more extensive resections in 35.2% of patients [58]. Furthermore, the distinct add-on value of nTMS-based tractography of the CST has been evaluated in a study including 70 adult patients with different brain tumor entities, revealing that patients having nTMS-based tractography available are characterized by an improved risk-benefit profile, showed an increased EOR, and demonstrated reduced rates of worsening in motor function in cases of already preexisting preoperative motor deficits [60].

In a follow-up study investigating the role of nTMS motor mapping and nTMS-based tractography in 70 patients presenting with high-grade glioma (HGG), residual tumor tissue and unexpected tumor residuals were less frequent in the nTMS group compared with historical control patients, with patients of the nTMS group being more frequently eligible for postoperative radiotherapy and showing prolonged 3-, 6-, and 9-month survival rates [61]. A significantly higher EOR was subsequently confirmed by another study for patients with a diagnosis of glioblastoma multiforme (GBM), in which patients of the nTMS group showed a gross total resection (GTR) rate of 61% versus 45% for the non-nTMS group [62]. Analogously, in studies focusing on patients with intracranial MET, patients of the nTMS group showed a lower rate of residuals combined with comparatively low rates of perioperative decline of motor function [63,64]. Specifically, in a retrospective comparative study pooling patients with intracranial MET from three different neurosurgical centers, surgery-related paresis was clearly less frequent in patients of the nTMS group [64]. In 47 patients with MEN located in the rolandic area, nTMS motor mapping and tractography facilitated a modification of the surgical strategy in 42.5% of cases, and a new permanent motor deficit (i.e., deficit that did not resolve to the preoperative status within the follow-up interval) occurred in 8.5% of cases, which is at the lower edge of the range for motor deficits known from the literature of the pre-nTMS era [65]. Furthermore, the combination of sodium fluorescein-guided resection (FGR) with preoperative nTMS motor mapping and tractography has been explored recently, revealing a higher GTR rate for patients operated on using nTMS and FGR as well as lower rates of new surgery-related permanent motor deficits when compared with controls [66,67].

One study in 43 patients with LGG and HGG showed that 72% of patients had motor-positive nTMS points in areas frontal of the rolandic area and, thus, outside of the expected spatial dimensions of the primary motor cortex [68]. Interestingly, 10 of the 13 patients who underwent resection of motor-positive nTMS points presented with postoperative paresis (8 patients with a new permanent surgery-related paresis), suggesting that even motor-positive nTMS points within the superior or middle frontal gyrus should be considered carefully for resection planning and guidance to avoid perioperative functional decline [68]. Hence, nTMS motor mapping and derived tractography may help to understand individual functional anatomy, allowing for optimized resection that provides a high EOR and low rates of surgically induced motor function decline.

Table 3. Improvement of clinical outcome. This table outlines the studies published on improvements of clinical outcome through the use of motor mapping by nTMS (using an electric-field-navigated system) with or without additional nTMS-based tractography in patients harboring brain neoplasms.

Author	Year	Cohort	Group for Comparison	Tumor Entities	nTMS Method	Outcome Parameters	Main Objective	Main Findings
Picht et al. [57]	2013	11 adult patients	11 historical controls	(suspected) LGG, HGG	Preoperative motor mapping	– Influence on surgery – EOR/tumor volume – Motor function (BMRC)	To assess the impact of nTMS on the treatment strategy and clinical outcome	– In 6 out of 11 patients, nTMS changed the treatment plan towards early and more extensive resection. – One of 4 patients of the nTMS group with preoperative motor deficits improved by one year, whereas increased motor deficits were observed in 3 of the 8 patients of the non-nTMS group not having surgery. – Median change of tumor volume from baseline to one year was −83% in the nTMS group and +12% in the non-nTMS group.
Frey et al. [58]	2014	250 adult patients	115 historical controls	LGG, HGG, MET, Other	Preoperative motor mapping and nTMS-based tractography	– Influence on surgery – EOR/tumor volume – Motor function (BMRC) – KPS – PFS	To assess the impact of nTMS on the treatment strategy, clinical outcome, and survival	– nTMS disproved suspected involvement of the primary motor cortex in 25.1%, expanded surgical indication in 14.8%, and led to planning of more extensive resections in 35.2% and more restrictive resections in 3.5% of patients. – The rate of GTR was significantly higher in the nTMS group (42% vs. 59%), and PFS for patients with LGG was better in the nTMS group (at 22.4 months) than in the non-nTMS group (at 15.4 months). – Integration of nTMS led to a non-significant change of postoperative deficits from 8.5% in the non-nTMS group to 6.1% in the nTMS group. – Expanding surgical indications and EOR based on nTMS might enable more patients to undergo surgery and could lead to better motor function outcome and improved PFS.

Table 3. *Cont.*

Author	Year	Cohort	Group for Comparison	Tumor Entities	nTMS Method	Outcome Parameters	Main Objective	Main Findings
Krieg et al. [59]	2014	100 adult patients	100 historical controls (matching criteria: tumor location, preoperative paresis, and histology)	LGG, HGG, MET, Other	Preoperative motor mapping (110% rMT for uE and 130% for lE muscles) and nTMS-based tractography	– Influence on surgery – EOR/tumor volume – Motor function (BMRC)	To assess the impact of nTMS on the treatment strategy, clinical outcome, and survival	– Patients of the nTMS group showed significantly smaller craniotomies. – 12% of patients of the non-nTMS group and 1% of patients of the nTMS group improved, while 13% and 18% of patients in the nTMS and non-nTMS groups, respectively, deteriorated in postoperative motor function on long-term follow-up. – Patients of the nTMS group showed a lower rate of residual tumor tissue according to postoperative MRI (odds ratio = 0.3828; 95% confidence interval = 0.2062–0.7107).
Krieg et al. [61]	2015	70 adult patients	70 historical controls (matching criteria: tumor location, preoperative paresis, and histology)	HGG	Preoperative motor mapping (110% rMT for uE and 130% for lE muscles) and nTMS-based tractography	– Influence on surgery – Perioperative complications – Adjuvant therapy – EOR/tumor volume – Motor function (BMRC) – KPS – PFS/overall survival	To assess the impact of nTMS on the treatment strategy, clinical outcome including direct perioperative complications, and survival	– Patients of the nTMS group showed significantly smaller craniotomies. – Residual tumor tissue (nTMS group: 34.3%, non-nTMS group: 54.3%) and unexpected tumor residuals (nTMS group: 15.7%, non-nTMS group: 32.9%) were significantly less frequent in the nTMS group. – Patients of the nTMS group were significantly more frequently eligible for postoperative radiotherapy (nTMS group: 67.1%, non-nTMS group: 48.6%). – 3-, 6-, and 9-month survival rates were significantly better in the nTMS group.
Picht et al. [62]	2016	93 adult patients	34 controls (with nTMS mapping not available)	HGG (only GBMs)	Preoperative motor mapping and nTMS-based tractography	– Influence on surgery – Perioperative complications – EOR/tumor volume – Motor function (BMRC)	To assess the impact of nTMS on the treatment strategy and clinical outcome including direct perioperative complications (at two campuses)	– In 10% of patients of the nTMS group the initial recommendation for biopsy or a "wait and see" approach was changed to resection because nTMS disproved the suspected invasion of motor structures. – Patients of the nTMS group showed a significantly higher rate of GTR. – A higher impact from nTMS was found in patients with tumors located subcortically when compared with tumors restricted to the cortex.

Table 3. Cont.

Author	Year	Cohort	Group for Comparison	Tumor Entities	nTMS Method	Outcome Parameters	Main Objective	Main Findings
Hendrix et al. [63]	2016	61 adult patients	-	LGG, HGG, MET, MEN, Other	Preoperative motor mapping (110% rMT)	– Influence on surgery – EOR/tumor volume – Motor function (BMRC)	To assess the impact of nTMS on the treatment strategy and clinical outcome	– Paresis resolved or improved in 56.7% of patients one week after surgery, and 89.5% of patients with postoperative paresis improved during the follow-up interval. – Only 4.3% of patients with a metastatic lesion, but 26.3% of patients with a non-metastatic lesion experienced deterioration of motor function after surgery. – All metastatic lesions were completely resected compared with 78.9% of non-metastatic lesions.
Krieg et al. [64]	2016	120 adult patients	130 historical controls	MET	Preoperative motor mapping (110% rMT for uE and 130% for lE muscles) and nTMS-based tractography	– Influence on surgery – EOR/tumor volume – Motor function (BMRC)	To assess the impact of nTMS on the treatment strategy and clinical outcome (multi-centric with three sites)	– Patients of the nTMS group showed significantly smaller craniotomies. – Patients of the nTMS group showed a lower rate of residual tumor tissue after surgery (odds ratio: 0.3025, 95% confidence interval: 0.1356–0.6749). – Surgery-related paresis was significantly less frequent in patients of the nTMS group (nTMS group: improved: 30.8%, unchanged: 65.8%, worse: 3.4%, non-nTMS group: improved: 13.1%, unchanged: 73.8%, worse: 13.1% of patients).
Moser et al. [68]	2017	43 adult patients	-	LGG, HGG	Preoperative motor mapping (110% rMT for uE and 130% for lE muscles)	– Latency analyses – Motor function (BMRC)	To assess the impact of resection of motor-positive prerolandic nTMS points on clinical outcome	– 72% of patients showed motor-positive nTMS points in the prerolandic gyri and, thus, outside of the anatomically suspected extent of the primary motor cortex. – Out of the 13 patients who underwent resection of motor-positive nTMS points, 10 patients showed postoperative paresis (2 patients with transient and 8 patients with permanent surgery-related paresis). – Motor-positive nTMS points within the superior or middle frontal gyrus should be considered carefully and can result in motor deficits when affected during resection.

Table 3. Cont.

Author	Year	Cohort	Group for Comparison	Tumor Entities	nTMS Method	Outcome Parameters	Main Objective	Main Findings
Raffa et al. [60]	2018	70 adult patients (50% also having nTMS-based fiber tracking)	35 historical controls	LGG, HGG, MET, Other	Preoperative motor mapping (120% rMT) and nTMS-based tractography	– Influence on surgery – EOR/tumor volume – Motor function (BMRC) – KPS	To assess the impact of nTMS with or without nTMS-based tractography on the treatment strategy and clinical outcome	– Patients of the nTMS and nTMS + nTMS-based tractography groups received significantly smaller craniotomies and had better postoperative motor performance and KPS scores than patients of the non-nTMS group. – Patients of the nTMS-based tractography group exhibited an improved risk-benefit analysis, a significantly increased EOR in absence of preoperative motor deficits, and significantly less motor and KPS score worsening (in case of preoperative motor deficits when compared with the nTMS group). – Risk-benefit analysis, EOR, and outcome could be improved when nTMS-based tractography is added to nTMS motor mapping.
Raffa et al. [66]	2019	79 adult patients	55 historical controls	HGG	Preoperative motor mapping (120% rMT) and nTMS-based tractography	– EOR/tumor volume – Motor function (BMRC)	To assess the impact of nTMS and nTMS-based tractography with sodium-fluorescein guidance on the treatment strategy and clinical outcome	– In patients operated on considering nTMS + FGR, the GTR rate was significantly higher compared with controls (64.5% vs. 47.2%). – Surgery-related permanent motor deficits were reduced in the nTMS + FGR group compared with controls (11.4% vs. 20%).
Raffa et al. [67]	2019	41 adult patients	41 historical controls	HGG	Preoperative motor mapping (120% rMT) and nTMS-based tractography	– Influence on surgery – EOR/tumor volume – Motor function (BMRC) – KPS	To assess the impact of nTMS and nTMS-based tractography with sodium-fluorescein guidance on the treatment strategy and clinical outcome	– Use of nTMS motor mapping and nTMS-based tractography reliably identified the spatial tumor-to-function relationship with an accuracy of 92.7%. – Patients of the nTMS group showed an increased EOR and higher rate of GTR (73.2% vs. 51.2%). – The number of cases with new surgery-related permanent motor deficits was lower in the nTMS group compared with controls (9.8% vs. 29.3%). – The number of cases with KPS worsening was lower in the nTMS group compared with controls (12.2% vs. 31.7%).

Table 3. Cont.

Author	Year	Cohort	Group for Comparison	Tumor Entities	nTMS Method	Outcome Parameters	Main Objective	Main Findings
Raffa et al. [65]	2019	47 adult patients	-	MEN	Preoperative motor mapping and nTMS-based tractography	– Influence on surgery – EOR/tumor volume – Motor function (BMRC) – Arachnoidal cleavage plane	To analyze the role of nTMS motor mapping for planning resection of rolandic meningiomas and predicting arachnoidal cleavage plane.	– Use of nTMS motor mapping and nTMS-based tractography was considered useful in 89.3% of patients and changed the surgical strategy in 42.5% of patients. – A new permanent motor deficit occurred in 8.5% patients. – A higher rMT and the lack of an intraoperative arachnoidal cleavage plane were independent predictors of poor motor function outcome. – A higher rMT and perilesional edema predicted the lack of an arachnoidal cleavage plane.

Abbreviations: nTMS—navigated transcranial magnetic stimulation; LGG—low-grade glioma; HGG—high-grade glioma; MET—metastasis; MEN—meningioma; rMT—resting motor threshold; uE—upper extremity; lE—lower extremity; EOR—extent of resection; GTR—gross total resection; BMRC—British Medical Research Council; KPS—Karnofsky performance status; PFS—progression-free survival; OS—overall survival; FGR—fluorescein-guided resection; GBM—glioblastoma multiforme.

2.4. Risk Stratification and Prediction

Besides the role for preoperative planning and intraoperative resection guidance, nTMS motor mapping and tractography could also be efficiently used for risk stratification and prediction of the motor status in patients with brain neoplasms. This has already been acknowledged by a growing body of studies, which are summarized in Table 4.

An early study characterized the neurophysiological status as derived from nTMS motor mapping in 100 patients, and already suggested that interhemispheric differences for MEP latencies may be considered as potential warning signs for the motor system at risk as comparatively similar latencies are commonly observed between the two hemispheres [69]. On a similar note, a high interhemispheric rMT ratio (i.e., the ratio between the two hemispheres regarding the rMT, which is commonly higher in a tumor-affected hemisphere) could suggest immanent deterioration of the functional motor status [69]. Furthermore, two studies in patients with various tumor entities investigated the role of nTMS-based tractography of the CST for risk stratification, evaluating the cut-off value for the lesion-to-CST distance that amounted to 8 mm and 12 mm to avoid new surgery-related permanent motor deficits, respectively [70,71]. Hence, patients that showed a lesion-to-CST distance above this cut-off value based on preoperative nTMS-based tractography were unlikely to suffer from surgery-related postoperative permanent paresis [70,71]. Moreover, statistically significant negative correlations were observed between the rMT value and lesion-to-CST distances in patients with a new surgery-related paresis, emphasizing the interplay between the SI used during motor mapping and results of nTMS-based tractography [71]. Correspondingly, motor function did not improve in cases with the rMT being significantly higher in the tumor-affected hemisphere than in the contralateral hemisphere, as expressed by an interhemispheric rMT ratio of >110% [70]. In a study investigating patients harboring HGG, lower FA values within the tumor-affected CST and higher average apparent diffusion coefficient (ADC) values were significantly correlated to worsened postoperative motor function, thus further exploring the contribution of dMRI-derived metrics to risk modelling [72].

In an innovative approach investigating postoperative nTMS motor mapping—instead of standardly used presurgical mapping—compared with intraoperative neuromonitoring (IONM) for predicting recovery of motor function, it was revealed that IONM and postoperative nTMS motor mapping were equally predictive for long-term motor recovery [73]. Specifically, when postoperative motor mapping was able to elicit MEPs, motor strength recovered to a score of at least 4/5 on the British Medical Research Council (BMRC) scale within one month after surgery, whereas when postoperative nTMS motor mapping did not elicit MEPs, the patient did not recover [73]. Furthermore, when implementing presurgical nTMS motor mapping and tractography in multi-modal neuroimaging with multi-sequence MRI and dedicated positron emission tomography (PET) protocols, it has been demonstrated that PET may be superior to contrast-enhanced T1-weighted MRI for proposing a motor deficit prior to surgery, and that the highest association with clinical impairment was revealed for the T2-weighted lesion overlap with functional brain tissue (i.e., the spatial overlap between the lesion volume on T2-weighted images of MRI and the functional primary motor cortex and/or CST volumes as derived from nTMS motor mapping and nTMS-based tractography) [74]. Future research may further explore the role of nTMS in multi-modal environments, given that data from various methods are frequently available for clinical needs prior to surgery. Opportunistic use of data from adjunct modalities (e.g., PET) as well as performance of dedicated longitudinal motor mapping (e.g., during the immediate postoperative course and during long-term follow-up examinations) could pave the way for a more efficient use of nTMS motor mapping and related tractography.

Table 4. Risk stratification and prediction. This table provides an overview of the studies published on risk stratification and prediction using motor mapping by nTMS (using an electric-field-navigated system) with or without additional nTMS-based tractography in patients harboring brain neoplasms.

Author	Year	Cohort	Tumor Entities	nTMS Method	Tractography Specifics	Main Objective	Main Findings
Picht et al. [69]	2012	100 adult patients	LGG, HGG, MET, MEN, Other	Preoperative motor mapping (110% rMT)	-	To provide reference values for parameters of the functional status and neurophysiological measurements	– The MEP latency was almost never different in the tumor-affected hemisphere compared with the healthy hemispheres; thus, interhemispheric differences for MEP latencies may reflect a warning sign for functional decline. – A high interhemispheric rMT ratio or a low interhemispheric MEP amplitude ratio may suggest immanent deterioration of the motor status.
Neuschmelting et al. [74]	2016	30 adult patients	HGG, MET, Other	Preoperative motor mapping (110% rMT) and nTMS-based tractography	One ROI: motor-positive nTMS points Deterministic tracking; FA = 100% FAT, FL = 1 mm, angular threshold of 30°	To better understand motor deficits in patients with brain tumors using structural, functional, and metabolic neuroimaging and to determine the predictive value of this approach	– Motor deficits were detected in almost all patients in whom the contrast-enhanced T1-weighted or O-(2-[18F]fluoroethyl)-L-tyrosine PET lesion area overlapped with functional tissue. – All patients who declined in motor function perioperatively showed such overlap on presurgical maps, while the absence of overlap predicted a favorable motor function outcome. – O-(2-[18F]fluoroethyl)-L-tyrosine-PET was superior to contrast-enhanced T1-weighted imaging for proposing a motor deficit before surgery. – The highest association with clinical impairment was revealed for the T2-weighted lesion area overlap with functional tissue.
Rosenstock et al. [70]	2017	113 adult patients	LGG, HGG	Preoperative motor mapping (105% rMT for uE and ~130% for lE muscles) and nTMS-based tractography	Two ROIs: motor-positive nTMS points and ipsilateral brainstem Deterministic tracking; FA = 75% FAT, FL = 110 mm, angular threshold of 30°	To establish risk stratification by examining whether the results of nTMS motor mapping and its neurophysiological data predict postoperative motor function outcome	– No new surgery-related permanent motor deficit was observed when the lesion-to-CST distance was >8 mm and the precentral gyrus was not infiltrated by the tumor mass. – New postoperative motor deficits were associated with a pathological excitability of the motor cortices (as indicated by an interhemispheric rMT ratio of <90% or >110%). – Motor function did not improve in any patient when the rMT was significantly higher in the tumor-affected hemisphere than in the healthy hemisphere (related to an interhemispheric rMT ratio of >110%).

Table 4. Cont.

Author	Year	Cohort	Tumor Entities	nTMS Method	Tractography Specifics	Main Objective	Main Findings
Rosenstock et al. [72]	2017	30 adult patients	HGG	Preoperative motor mapping (105% rMT for uE and 130–150% for lE muscles) and nTMS-based tractography	– Two ROIs: motor-positive nTMS points and cubic ROI in the pons – Deterministic tracking; FA = 75% FAT, FL = 110 mm, angular threshold of 30°	To analyze FA and ADC within the CST in different locations and their usefulness for predicting motor function outcome	– Lower FA within the tumor-affected CST as well as higher average ADC values were significantly correlated to worsened postoperative motor function. – Segmental analyses within the CST indicated that the extent of impairment of diffusion metrics correlates with motor function deficits.
Sollmann et al. [71]	2018	86 adult patients	LGG, HGG, MET	Preoperative motor mapping (110% rMT for uE and 130% for lE muscles) and nTMS-based tractography	– Two ROIs: motor-positive nTMS points and ipsilateral brainstem – Deterministic tracking; FA = 50% FAT/75% FAT/100% FAT, FL = 110 mm, angular threshold of 30°	To explore whether nTMS-based tractography can be used for individual preoperative risk evaluation for surgery-related motor impairment	– For tractography with certain FATs, a significant difference in lesion-to-CST distances was observed between patients with HGGs who had no impairment and those who developed surgery-related transient or permanent motor function deficits. – As a cut-off value, no patient with a lesion-to-CST distance ≥12 mm showed a new surgery-related permanent paresis. – Significant negative associations were observed between the rMT and lesion-to-CST distances of patients with surgery-related transient paresis or surgery-related permanent paresis.
Seidel et al. [73]	2019	13 adult patients	LGG, HGG, MET	Postoperative motor mapping (within 14 days after surgery; "MEP loss" if not 5/10 stimulations could be elicited with 70–100% of the MSO)	–	To investigate the value of postoperative nTMS motor mapping compared with intraoperative MEP monitoring for predicting recovery of motor function	– Motor strength recovered to a score of at least 4/5 of the BMRC scale within one month after surgery if postoperative nTMS motor mapping elicited MEPs (PPV = 90.9%). – When postoperative nTMS motor mapping did not elicit MEPs, the patient was unlikely to recover in terms of motor function. – Intraoperative MEP monitoring and postoperative nTMS motor mapping were equally predictive for long-term motor recovery. – ~2/3 of patients with intraoperative MEP alterations or signal loss but positive postoperative nTMS motor mapping demonstrated motor function recovery.

Abbreviations: nTMS—navigated transcranial magnetic stimulation; LGG—low-grade glioma; HGG—high-grade glioma; MET—metastasis; MEN—meningioma; CST—corticospinal tract; ROI—region of interest; FA—fractional anisotropy; FAT—fractional anisotropy threshold; FL—fiber length; uE—upper extremity; lE—lower extremity; rMT—resting motor threshold; MEP—motor-evoked potential; PET—positron emission tomography; MSO—maximum stimulator output.

2.5. Plasticity and Reallocation of Motor Function

Repeated application of nTMS motor mapping and tractography has potential to provide insights into brain plasticity that is likely to occur to a certain degree due to the presence and growth of a brain tumor. Few studies have already tried to investigate the role of nTMS motor mapping in this regard, and these studies are outlined in Table 5.

The non-invasive character of nTMS makes possible the acquisition of data from multiple time points, ideally spanning from the preoperative to the postoperative and follow-up interval. Correspondingly, an early explorative study in five patients and five controls used preoperative motor mapping by nTMS as well as mapping during follow-up examinations, revealing a shift of CoGs over a mean interval of 18 months of 6.8 ± 3.4 mm and a shift of motor hotspots of 8.7 ± 5.1 mm for the dominant hemispheres [75]. In a case report on a patient with a LGG that was situated within the frontal lobe and affected the suspected primary motor cortex, motor representation shifted from the precentral to the postcentral gyrus over an interval of 18 months according to serial nTMS motor mappings, which was confirmed by DES mapping during re-resection [76].

In general, a connection between the distinct location of the motor map as enclosed by nTMS as well as its extent and tumor location has been demonstrated in the sense of tumor location-dependent changes in the distribution of polysynaptic MEP latencies and spread of motor maps, especially along the anterior-posterior direction [77]. In the further course, it was revealed that in a majority of patients with mixed tumor entities, MEP counts, when elicited by nTMS to the precentral gyrus, were higher than average, potentially reflecting robust and less variable motor representations within the primary motor cortex [78]. Additionally, patients with tumors affecting the postcentral gyrus and other parietal areas primarily showed high MEP counts when stimulation by nTMS was delivered to the postcentral gyrus [78]. Hence, functional reorganization patterns seem to be reflected by a reorganization within anatomical constraints, such as of the postcentral gyrus [78]. Using again serial nTMS motor mappings from presurgical and follow-up sessions, the initial observation of CoG or motor hotspot shifts have been confirmed in further series including 22 and 20 patients with different tumor entities, respectively [79,80]. Additionally, motor representations appeared to shift more clearly toward the tumor mass if the lesion was anterior to the rolandic region than if it was located posterior to the rolandic region, and a preferential regrowth pattern of tumor recurrence towards the primary motor cortex and/or CST as defined by nTMS-based motor mapping and tractography has been suggested by exploratory approaches [80,81].

Table 5. Plasticity and reallocation of motor function. This table outlines the studies published on plasticity and reallocation of motor function as revealed by motor mapping using nTMS (using an electric-field-navigated system) with or without additional nTMS-based tractography in patients harboring brain neoplasms.

Author	Year	Cohort	Tumor Entities	nTMS Method	Main Objective	Main Findings
Forster et al. [75]	2012	5 adult patients (and 5 healthy volunteers)	LGG, HGG	Preoperative and follow-up motor mapping (110% rMT for uE and lE muscles; interval between mappings: 18 months on average) (two motor mappings in healthy volunteers)	To investigate cortical motor representation after resection of perirolandic WHO grade II and III gliomas	– Shift of CoGs over time was 0.7 ± 0.3 cm in the dominant and 0.8 ± 0.4 cm in the non-dominant hemisphere. – Shift of motor hotspots amounted to 0.9 ± 0.5 cm for the dominant and 0.8 ± 0.5 cm in the non-dominant hemisphere. – In one patient CoG and motor hotspot shifts significantly differed from the control group of healthy volunteers.
Takahashi et al. [76]	2013	20-year-old man	LGG	Preoperative and follow-up motor mapping (110% for uE muscles; interval between mappings: 18 months)	To confirm induced brain plasticity by nTMS motor mapping	– Primary motor representation as determined by nTMS motor mapping shifted from the precentral to the postcentral gyrus over time, which was confirmed by DES mapping. – Plastic changes in primary motor representations permitted complete tumor removal without neurological decline.
Bulubas et al. [77]	2016	100 adult patients	LGG, HGG, MET, Other	Preoperative motor mapping (110% rMT for uE and 130% for lE muscles)	To investigate whether brain lesions induce a change in motor cortex representation depending on tumor localization	– Motor areas according to nTMS motor mapping were not restricted to the precentral gyrus. – The dominant hemisphere showed a significantly greater number of longer MEP latencies than the non-dominant hemisphere. – Tumor location-dependent changes in the distribution of polysynaptic MEP latencies were observed.

Table 5. *Cont.*

Author	Year	Cohort	Tumor Entities	nTMS Method	Main Objective	Main Findings
Conway et al. [80]	2017	22 adult patients	LGG, HGG	Preoperative and follow-up motor mapping (110% rMT for uE and ≥130% for lE muscles; interval between mappings: 3–42 months)	To demonstrate the frequency of plastic reshaping and reveal clues to the patterns of reorganization	– Motor hotspots showed an average shift of 5.1 ± 0.9 mm on the medio-lateral axis, and a shift of 10.7 ± 1.6 mm on the antero-posterior axis. – CoGs shifted by 4.6 ± 0.8 mm on the medio-lateral axis and by 8.7 ± 1.5 mm on the antero-posterior axis. – Motor-positive nTMS points tended to shift more clearly toward the tumor if the lesion was anterior to the rolandic area than if it was located posterior to the rolandic area.
Bulubas et al. [78]	2018	100 adult patients	LGG, HGG, MET, Other	Preoperative motor mapping (110% rMT for uE and 130% for lE muscles)	To investigate the spatial distributions of motor sites to reveal tumor-induced brain plasticity in patients with brain tumors	– High MEP counts were elicited less frequently by stimulating the precentral gyrus in patients with tumors directly affecting this gyrus. – In more than 50% of these patients, the MEP counts elicited by stimulating the precentral gyrus were higher than average, indicating robust motor representations within the primary motor cortex. – Patients with parietal tumors (and specifically tumors within the postcentral gyrus) showed high MEP counts when stimulating the postcentral gyrus. – The functional reorganization seemed to be reflected by a reorganization within anatomical constraints, such as of the postcentral gyrus.

Table 5. *Cont.*

Author	Year	Cohort	Tumor Entities	nTMS Method	Main Objective	Main Findings
Barz et al. [79]	2018	20 adolescent to adult patients (and 12 healthy volunteers)	LGG, HGG, Other	Preoperative and follow-up motor mapping (110% rMT for uE and IE muscles; interval between mappings: 26.1 ± 24.8 months and 46.3 ± 25.4 months) (two motor mappings in healthy volunteers)	To evaluate motor cortex reorganization in patients after perirolandic glioma surgery	— Pre- and postoperatively pooled CoGs from the areas of the dominant APB muscle and non-dominant IE representation area differed significantly from those of healthy individuals. — During the follow-up period, reorganization of all muscle representation areas occurred in 3 patients, and significant shifts of uE muscle representations were detected in another 3 patients.
Sollmann et al. [81]	2018	60 adult patients	HGG	Preoperative motor mapping (110% rMT for uE and ≥130% for IE muscles) and nTMS-based tractography (and fMRI for a subsample of the cohort)	To evaluate whether brain tumor relapse has a preference to grow towards motor-eloquent brain structures	— 69.0% of patients without residual tumor, 64.3% with residual tumor away from motor areas, and 66.7% with residual tumor facing motor areas showed tumor recurrence that was directed towards motor eloquence. — Average growth towards was highest for patients with residual tumor already facing motor areas, suggesting a preference in growth patterns towards (reshaping) motor areas.

Abbreviations: nTMS—navigated transcranial magnetic stimulation; LGG—low-grade glioma; HGG—high-grade glioma; MET—metastasis; rMT—resting motor threshold; uE—upper extremity; lE—lower extremity; DES—direct electrical stimulation; fMRI—functional magnetic resonance imaging; APB—abductor pollicis brevis; CoG—center of gravity; ADM—abductor digiti minimi; MEP—motor-evoked potential; WHO—World Health Organization.

2.6. Integration into the Clinical Environment

For broad application of nTMS motor mapping and derived nTMS-based tractography in neuro-surgical oncology, seamless integration into existing hospital infrastructure and processes is key for acceptance and optimal use of generated data. In this regard, a structured workflow has already been proposed [82]. It starts with admission of the patient and when the indication for mapping is made and includes, amongst other steps, transfer of nTMS data to a hospital-intern picture archiving and communication system (PACS) as well as reporting within dedicated masks for the hospital-intern electronic patient charts [82].

An example of inter-disciplinary integration into different systems requiring nTMS data transfer is represented by the versatile use of the motor maps for planning and treatment purposes in radiosurgery and radiotherapy [83–87]. The first published approach achieved easy and reliable integration of nTMS, fMRI, and tractography data for radiosurgery treatment planning, which led to an average radiation dose reduction of 17% to functional brain areas in a cohort with mixed entities of pathologies [83]. Another study approved flawless integration of specifically nTMS data for radiosurgery, which influenced the radiosurgical planning procedure by improving risk-benefit balancing in all cases, achieved dose plan modifications in 81.9%, facilitated treatment indication in 63.7%, and reduced radiation doses in 72.7% of cases [86]. Compared with radiosurgery plans without nTMS data, treatment plans with integration of nTMS data demonstrated a significant decrease in dose to eloquent cortex volume, which was achievable without a reduction of the dose applied to intracranial MET [87]. Moreover, integration of nTMS motor maps with radiotherapy planning software for hypofractioned stereotactic treatment regimens for patients diagnosed with intracranial MET has been proposed, and by constraining the dose applied to the nTMS motor maps outside the planning target volume (PTV) to 15 Gy, the mean dose was significantly reduced from 23.0 Gy to 18.9 Gy, while the mean dose of the PTV increased [85]. Analogously, in patients with HGG, mean dose to the nTMS motor maps was significantly reduced by 14.3% when constraining the dose to nTMS motor areas, while the dose to the PTV was not compromised [84].

Furthermore, integrating nTMS motor mapping in clinical workflows has provided initial evidence for the usefulness of the method for planning of a stereotactic tumor biopsy, performing endoscopic cystoventriculostomy, or facilitating a transparietal approach to the trigone of the lateral ventricle in patients with brain neoplasms [88–90]. In a special environment such as the intensive care unit with critically ill patients, an approach for safe and reliable use of nTMS motor mapping has been described recently, yet preliminarily in patients suffering from other diseases than brain tumors (e.g., central cord syndrome after trauma, ischemic or hemorrhagic stroke) [91]. In particular, the use of computed tomography (CT) instead of MRI data may help to establish nTMS motor mapping also in special environments with patients who may only be eligible to undergo CT due to specific infrastructural constraints (e.g., non-availability of timely imaging by MRI) or medical conditions (e.g., specific implanted devices as contraindications for MRI) [91,92]. While this underlines the broad applicability of nTMS motor mapping, which requires little patient interaction while creating valuable data on the motor system in a non-invasive way, high accuracy has to be ensured and other imaging sources than MRI have to be regarded as second-line alternatives in selected cases.

3. Methodological Considerations on Application of nTMS Motor Mapping

3.1. Current Practices and Protocols

Previous large-scale studies have converged to feasible routines for nTMS motor mapping in clinical practice [93]. As an example, the usual SI is normalized to the rMT, and 110% rMT has become the standard, though also 105% rMT is often used. Alternative methods, such as using Mills–Nithi upper threshold (UT), exist as well [93].

There is large variance in the motor mapping protocols since the studies intend to answer specific research questions. On a similar note, the terminology used is not always clear and uniform. New methods and analysis tools have emerged. In addition to formerly used metrics, new and parallel map measures (e.g., area and volume in defining the extent of motor maps) are increasingly used and reported, which brings variance to the studies and makes them more difficult to compare with each other. For instance, motor map topography based on counting the number of discrete peaks, which in turn was based on MEP amplitudes, was introduced in 2017 [94]. Another measure that seems to be interesting and reliable and may also have clinical relevance is the overlap of the motor representation of muscles, though its potential meaning is not yet fully understood [95–97].

When it comes to quantitative mapping, a lot of research on quantitative parameters derived from nTMS motor mapping is ongoing. Despite many motor mapping results in patients (such as greater or smaller area of motor maps, closeness of CoGs, or location and shift of CoGs) have already been published, there is clear need of comparison data on healthy volunteers as a basis against which to evaluate cortical reorganization in clinical populations [98,99]. In addition, the normativity of hemispheric side-to-side differences needs to be ensured before comparison between potentially affected and non-affected hemispheres takes place. Fine-scale topography seems to be complex and variable between subjects. To understand it better, multi-modal approaches would be important to better track and understand nTMS-induced effects across the brain [100,101]. Furthermore, instead of mapping single muscles, the importance of groups of muscles, their synergy, and the role of movements and their relation to posture and biomechanics have been pointed out [102].

It should not be forgotten that the steps during initial mapping to locate a motor hotspot, sometimes called coarse mapping or technical mapping, are an important part of the examination to define the precise and correct hotspot. Regarding motor areas of lE muscle representations, a double-cone coil is recommended to reach deeper [103]. Different coil types hinder direct comparison between studies. The coil orientation may also have impact, which partly depends on the area of interest (uE, lE, or face muscle representations) [104–106]. The need for preactivation of muscles is a special issue that needs to be taken into account, particularly for lE and face muscle representations [106]. Another important issue is that when mapping the extent of several muscle representations within a limb, the rMT is usually only determined for one specific small hand muscle (abductor pollicis brevis (APB) or first dorsal interosseous (FDI) muscle), and this is used as a reference for motor mapping of representations of other muscles as well. This could perhaps be tackled by targeted post-hoc analysis [95]. For the lE, there is much more variance when choosing the muscle of primary interest, which should preferably also have the lowest rMT. The clinical importance is, however, mostly unknown.

In the analysis of motor mapping data, large variability in MEP amplitudes is a challenge. Another challenge is that the amplitudes are often small. The usual response criterion for an accepted MEP in rest is often 50 µV [26], but lower amplitude criteria have been successfully applied [107]. Some of the measures need to be normalized to the maximum recorded MEP amplitude. Mapping-related biomarkers of sensorimotor plasticity could be used in the study of pathophysiology of different diseases and an important application is rehabilitation. Based on these reflections on the current status and routine procedures for nTMS motor mapping, in the following we outline the most important quantitative parameters and methods for ensuring feasible accuracy of nTMS motor mapping.

3.2. Selecting Stimulation Intensity

The proper SI used in quantitative mapping is conventionally and fundamentally dependent on the rMT, which can be determined in several ways [47,108–112]. These days, the most convenient and most widespread method is adaptive threshold hunting (ATH), which estimates the threshold SI in an iterative fashion with excellent confidence [109,110,113–115].

Selection of the SI used in motor mapping overall is a crucial part of the experimental design, as the SI defines the amplitude and spread of the responses, and the spatial accuracy of individual stimuli [25]. In general, the map size increases with the SI [24,25]. A common practice is to use a SI that is related to the individual rMT by percentage increase, i.e., 110% of the rMT. A workshop report including recommendations for nTMS motor mapping in patients harboring brain tumors suggest the use of 105% rMT when mapping the uE muscle representations and 110% (with additional 20 V/m) to map the representations of lE muscle representations [93]. While these are practical solutions easily applied for clinical practice, they are likely to cause protocol-induced variation to the results. The individual input–output characteristics vary with factors such as age [116,117]. Furthermore, they are then also affecting the outcome of the mapping, as the used supra-threshold SI is dependent on those characteristics [24,25,107,118,119].

From a risk-benefit perspective, the risks in selection of the SI are the following: (1) too low SI will not activate the cortex that contains motor functions in mapped areas (and, as a result, false-negative responses are gained), and (2) too high SI that excites neurons outside the stimulated target region (resulting in a response that is falsely positive, i.e., positive responses not associated with the stimulated target). Provided that the fluctuation in cortical excitability is normal, and muscles are maintained in rest, the benefits corresponding with the above-mentioned risks are that with (1) low SI if the stimulation of a cortical target is producing a response, it can be assumed with high confidence as a true-positive response, and with (2) high SI the probability of inducing a response is greater (Figure 5), and it is likely that if there is no response the stimulated target is a true-negative response. Therefore, in selecting the SI, the risks and benefits should be weighted regarding the information that is needed to be acquired. In preoperative motor mapping in patients with brain tumors, a motor map that only shows a minimum number of false-positive stimulation points is warranted in order to avoid overestimation of the extent of the motor map. Such overestimation could lead to incomplete tumor resection given that false-positive spots are unnecessarily spared from resection due to anticipated, but faulty "true" motor function representations. However, high fractions of false-negative points would put the patient at a theoretically higher risk of functional deterioration in cases in which such points are included in the surgical resection. Correspondingly, Thordstein et al. speculated that using a low SI could include an additional risk since the activation area of a muscle could be distributed non-continuously along the motor cortex [107,120].

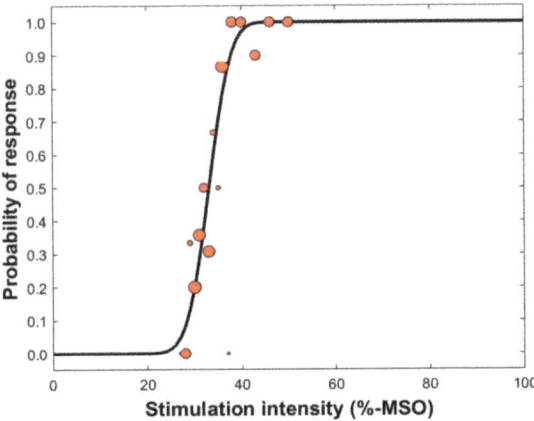

Figure 5. Example of the cumulative distribution function (*black line*) fitted to the experimental data of one subject. The *red dots* indicate the probability of response based on repeated MEP trials. The number of repetitions at each SI (given as %-MSO) is reflected in the marker size, which has been used to weight the fitting of the cumulative distribution function.

When using a SI that is at supra-threshold level normalized to the rMT (e.g., 110% rMT), it is unclear what the individual likelihood for induction of a response is (Figure 6).

However, it may largely avoid false-negative stimulation points within the motor map by arbitrarily creating some sort of a "safety margin" around unequivocally motor-positive stimulation spots. Certainly, the likelihood for producing a response at the motor hotspot where the rMT was determined is greater than 50%—but is it 60% or 95%? This dilemma makes it difficult to estimate the real accuracy of individual motor mapping if the input–output characteristics are unknown. Alternative techniques for determining the mapping intensity based on a threshold value that relies on greater probability of responses than 50% has been suggested as an alternative way to determine the SI [25]. Specifically, Kallioniemi and Julkunen proposed that the use of the so-called Mills–Nithi UT could be used directly as the SI for mapping, and demonstrated that it indeed reduces the inter-individual variation in the quantified motor map size [25,108]. The core principle in using the UT instead of a SI normalized to the rMT seems justified as with UT the probability of a response is ~90%, hence reducing the dependence of mapping outcome from the individual input–output characteristics. However, there are drawbacks in that methodology: (1) the confidence in the estimated UT is likely lower than for the rMT that is determined using ATH [109,110], and (2) it may take a few minutes more time to determine the UT [25].

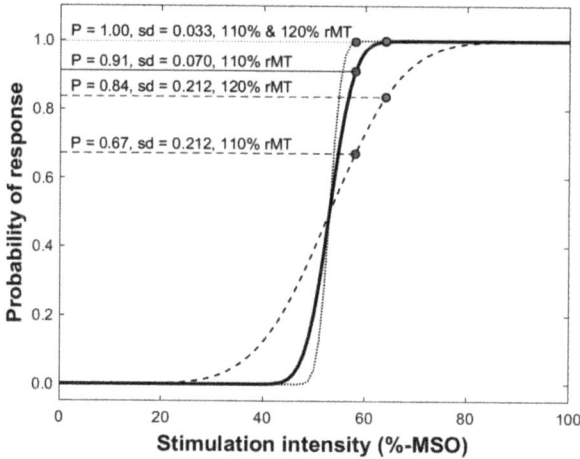

Figure 6. Demonstration of the cumulative distribution function representing the probability of inducing a response at different SIs. As the slope, represented as relative spread [110], can differ between subjects, the use of the SI related to the rMT can induce differences to the absolute size of the motor map [25]. In the plot, mean relative spread (*solid black line*) and the minimum (*dotted black line*) and maximum (*dashed black line*) found in the study population were used to compare probabilities of inducing a response at 110% rMT (*blue dots*) and 120% rMT (*red dots*). At a low relative spread, 110% and 120% rMT produce an MEP at the stimulation target with a probability close to 100%, but with the high relative spread, the probabilities of MEPs at the target at 110% and 120% rMT are 68% and 83%, respectively.

To demonstrate the variability of motor maps due to uncertainty in response occurrence probability, we simulated motor maps based on experimental motor mapping data by assuming that 10% of the responses were falsely negative to estimate the general confidence of motor mapping based on uncertainties related to response occurrence (Figure 7). The simulation demonstrated that despite the uncertainty, the SI-dependent differences in the motor map area were still apparent, and the shape and extent of the motor maps were maintained from simulation to simulation. Unlike the area of the motor map, the location of CoGs does not appear to depend on the SI [24]. It was demonstrated by Thickbroom et al. that when moving the coil from one cortical location to another, the shape of the input–output curve does not change significantly, only the offset that is the crucial part being represented as the rMT [121].

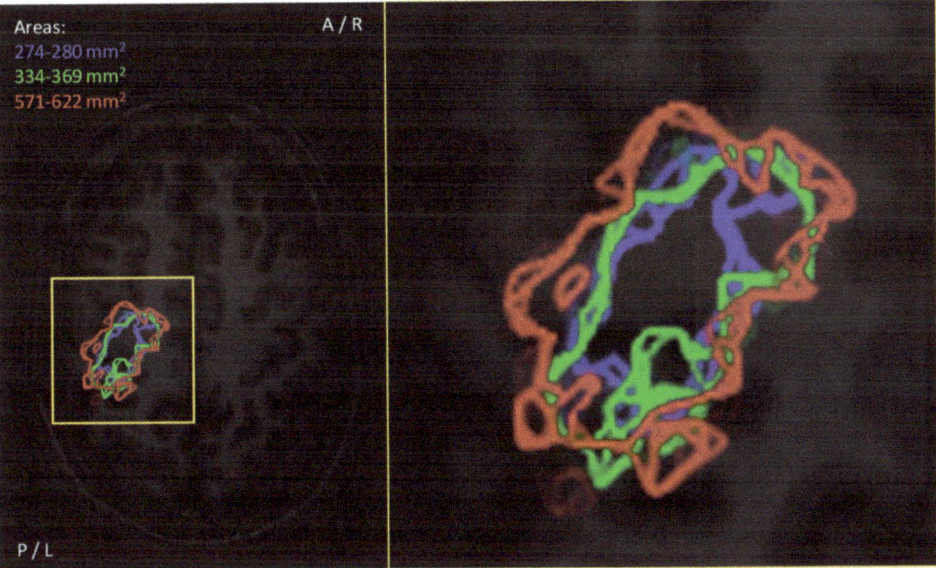

Figure 7. Visualization of the motor map resulting from mapping with three SIs (*blue*: 110% of the rMT, *green*: Mills–Nithi UT method [108], *red*: 120% rMT). Data from each of the experiments were bootstrapped 1000 times, assuming that 10% of the responses observed were false negative. The image is visualized in neurological projection. The stimulations were placed on average 0.4 mm apart. The 95% confidence limits are indicated for quantified areas in the images.

The SI is commonly represented as a percentage of the maximum stimulator output (%-MSO), by definition making it dependent on the maximal stimulator performance that is highly dependent on the used instrumentation including the characteristics of stimulation coils and the stimulating pulse [122–128]. This means that when comparing the used SIs between individuals, one has to consider the characteristics of the instrumentation. In addition, the SI, when represented as %-MSO, is not considering the individual distance of the stimulated cortex from the stimulating coil that is placed on the scalp [129–133]. To account for individual coil-to-cortex distance by estimating the cortically-induced EF by stimulation, an EF estimate could be used [134–136]. However, the EF estimate may not account for differences in pulse characteristics [137]; yet it could reduce the difference in representing SIs as EFs between stimulator manufactures differ [126]. Nevertheless, when applying EFs in different individuals, there exists a challenge to determine the anatomic location where the EF should be estimated, as the exact location of response induction has not been unambiguously determined since the activation by stimulation is not limited to the crown of specific layers of the cortex, but, instead, the coil distance-dependent EF affects a large part of the cortex [29,105,138].

Recently, Nazarova et al. mapped the representations of multiple muscles simultaneously to distinguish between muscle representations of the individual muscles [98]. As previously suggested, they observed that the use of a single SI may be a possible limitation as the different muscles could potentially have different rMTs [98]. Previously, it has been observed that the different somatotopically adjacent muscle representations could have different excitability profiles [95,107,118]. Albeit, at the group level the effect size may be minor or acceptable, at the individual level the clinical significance for such different profiles may be crucial [95,119,139–141]. This means that if a muscle has a lower rMT for activation and a steeper rise for the input–output curve than the other mapped muscles, the motor map will be biased due to the responses of that muscle, and will mostly present the representation area of that specific muscle over the other mapped muscles. Therefore, when determining quantitative characteristics for a group of muscles, the mapped outcome may be biased with certain muscles due to the differences in the individual muscle rMTs, and perhaps also due to individual motor hotspots.

Furthermore, the coil-to-cortex distance varies with stimulated cortical regions, which may require adjustment of the applied SI [130,131,135,136,142–144]. Because of the differences between target sites, the SI needs to be adjusted by taking into account the differences in coil-to-cortex distances, the secondary field caused by charge accumulation at conductivity discontinuities, and the coil orientation, and adjustment based only on the SI or primary EF is not sufficient [135].

3.3. Stimulation Grid

To enable quantitative mapping and to set the spatial density for stimulation targets and, thus, spatial accuracy of the quantitative mapping, stimulation grids are frequently used [27,145–147]. The grids are especially crucial for non-navigated estimation of motor maps [145,146]. However, they are also used in nTMS approaches where the underlying anatomy is visible [27,147]. The use of the grids enables straightforward calculation of the motor map size, i.e., by calculation of the number of active squares (producing acceptable responses when that square is stimulated) within the grid [118]. The definition of the active grid square varies in terms of interpreting a response (e.g., response/no response, maximum MEP amplitude, mean MEP amplitude, MEP count, etc.). In nTMS, the size of the grid squares affects the accuracy of the motor map that can be related to anatomical structures. The accuracy is limited also by the resolution of the underlying structural MRI and the accuracy of the neuronavigation system [1]. If using a grid as aid for enabling homogeneous spacing between the stimuli, the selection of the grid size should consider the required spatial resolution of the motor map. Figure 8 demonstrates how the size of a stimulation grid square could affect the appearance of a motor map. The larger grid squares result in bulk-shaped motor maps with the grid potentially overestimating the true map size, while reduction of the grid square size converges towards the true motor map size. However, the smallest grid squares produce lower map size than the reference value because the original data were gathered with inferior density and there might be no data for all grid squares (Figure 9).

It may soon be obsolete to consider motor mapping in terms of grids and grid targets, as modern nTMS motor mapping could potentially be performed in a quantitative fashion without the use of grids as long as quality criteria are set [148]. This means that the placement of stimuli is anatomically guided with denser placement of stimuli at specific regions, e.g., in the vicinity of anatomically interesting landmarks or at the edges of a motor map. This also means that the spatial and regional accuracy within the motor map may vary, while the extent of the motor map could be more accurately defined. The effects could be inverse for the accuracy of the CoGs, and for calculation of CoGs the coverage of each individual stimulus target in the motor map needs to be accounted for and be weighted in the calculation [27].

3.4. Number of Stimuli Required (per Target Location)

The accuracy of the motor mapping has been shown to relate to the number of stimuli used [26]. The number of stimulated responses within each grid square varies, as does the size of the grid squares [26,118,140,145,147,149–153]. Previous studies have investigated the required number of stimuli in a stimulation grid, having observed that two or more stimuli should be used to improve the confidence in the resulting motor map parameters [26,154]. Specifically, Cavaleri et al. reported that at least two responses induced by targeting each grid square were required for reliable calculation of the CoG and motor map volume in non-navigated TMS with a grid square size of 10 mm [154]. The used stimulus grid, or the density of the stimulus target spacing plays an essential role when defining the parameters of the motor map. In essence, the CoG location is dependent on the grid square size, as it is the case with the cortical area (Figure 9). The larger the grid square is, the less accurate is the CoG location or the motor area (Figure 9).

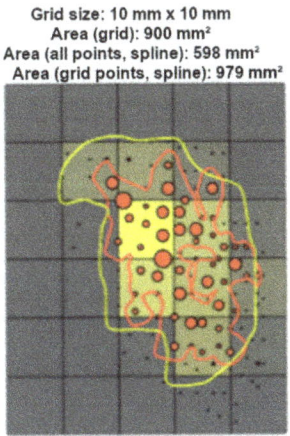
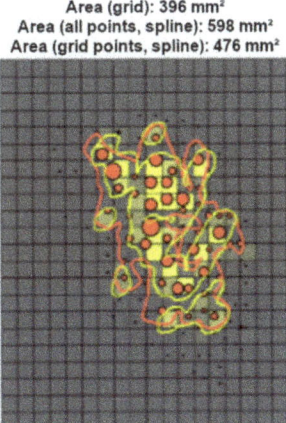

Figure 8. Demonstration of the use of a grid in calculating the area of the motor map. The original individual responses are presented as *red-filled dots*, with the dot size reflecting the MEP amplitude. The *black dots* are 0-amplitude responses. Average MEP amplitudes were calculated within grid squares in different size grids. The resulting average MEP amplitude size is reflected in the *yellow color* of the grid squares. The area of the motor map was evaluated based on the sum of the grid square areas with average MEP amplitudes of at least 50 µV, and by using spline interpolation (*yellow line*). The corresponding motor map areas are displayed above the grids with the grid sizes. For comparison, the motor map area is displayed for the original responses with spline interpolation, indicated by the *red line* in each plot.

When using nTMS, the conventional grid squares (e.g., 10×10 mm or 15×15 mm in size) are likely to include more stimuli. This is demonstrated in Figure 10, placing a stimulus grid of typical size over the mapped region and demonstrating that several stimuli are placed within the grid squares [146,149]. In fact, due to the spatial averaging caused by the large number of MEPs recorded during the mapping, the inherent variability effect in MEPs may be reduced, and placing multiple stimuli per location is compensated by closer spacing of the stimulus locations [95,155]. Chernyavskiy et al. showed that with an increasing number of stimuli included in the motor map, the accuracy is improved in nTMS mapping without application of a grid, as the coverage of a single response in a motor map and, hence, the contribution a response for the total motor map is decreased [155].

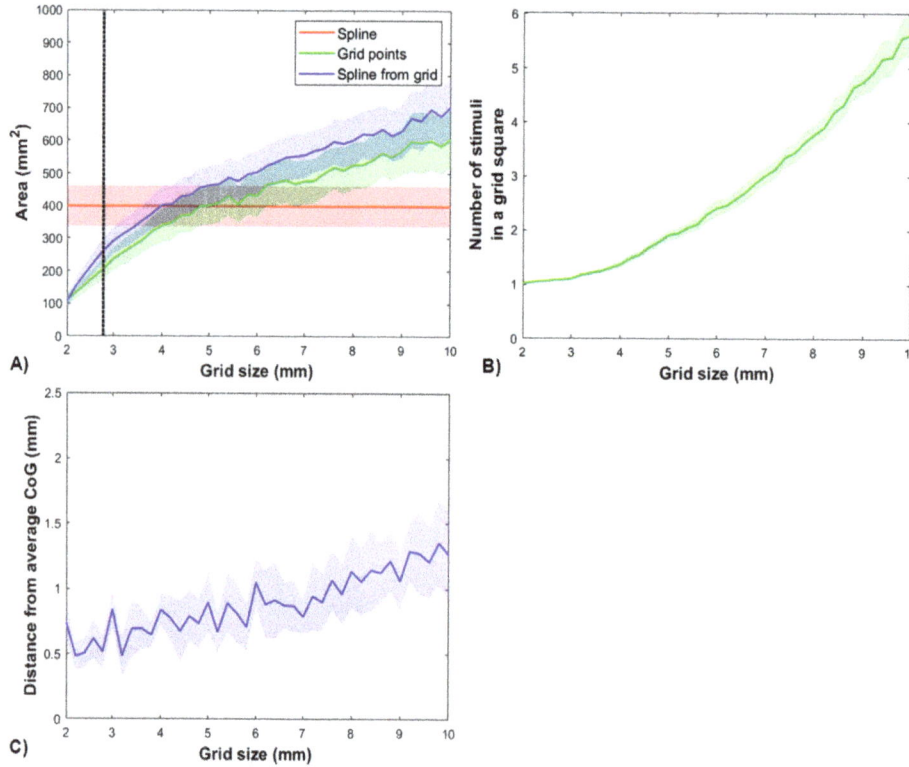

Figure 9. (**A**) The effect of grid size on the resulting motor areas with different methods of calculation. As a comparison, there is the area calculated with the spline interpolation (*red*), which is independent of the grid size, but is instead dependent on the local spacing of the stimuli. The area calculated from active grid squares is presented in *green* and, for ease of comparison, the spline interpolation area calculated from the active grid sites is shown in *blue*. The shaded areas indicate the 95% confidence interval within the study population, which was 24 mapping experiments at 110% of rMT. (**B**) The average number of stimuli falling within each grid square is shown as a function of grid size, with the shaded area indicating the 95% confidence interval within the study population. (**C**) The effect of grid size on the CoG location is shown, with the shaded area indicating the 95% confidence interval within the study population.

Often, when utilizing stimulation grids, a few stimuli may be repeated per grid to reduce the effect of MEP amplitude variation. However, then the number of stimuli is in a different scale than the number of required stimuli needed to obtain the average MEP amplitude confidently. To reach for a reliable and stable value of the MEP amplitude, previous studies have found that at least 20 repeated trials should be averaged [156,157]. However, these confidences are not fully comparable, since the overall data on MEP amplitudes within the motor maps are more extensive than in studies assessing single-target MEP amplitudes. Thus, the effect of individual grid square MEP amplitudes variability affects less the binarized map parameters such as the area, and, hence, the "spatial filtering" reduces the apparent MEP variability. Obviously, parameters utilizing absolute MEP amplitudes such as the map volume could be more affected.

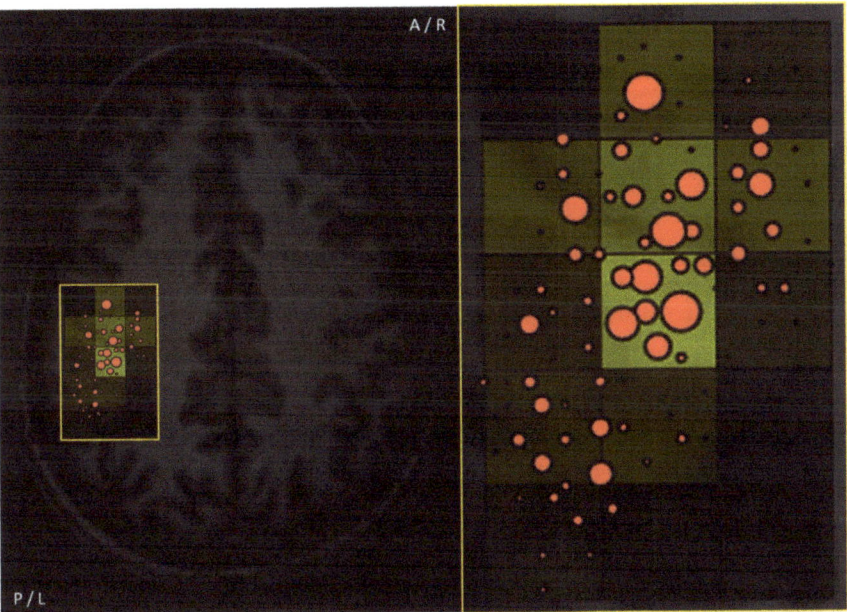

Figure 10. The visualized grid square size used was 10 mm × 10 mm, which is a quite commonly used size [146,149]. The stimulations were placed on average 2.6 mm apart. The individual stimulus locations are shown with *red dots*, the size of which is indicative of the associated MEP amplitude (the larger the amplitude, the larger the dot). The *yellow colors* in the grid squares indicate the size of the mean MEP amplitude for the MEPs induced by stimulating points within the grid square (the brighter the color, the higher the mean MEP amplitude).

3.5. Coil Orientation with Respect to Anatomy

The degrees of freedom (DOF) of the stimulation coil include the coil location, orientation and tilt, while additionally, one system-dependent DOF is also the previously discussed SI. The coil tilt affects the efficiency of dose delivery on the cortex [158,159]. With respect to the underlying cortical anatomy, the coil orientation affects the observed response (i.e., suboptimal coil orientation may result in MEPs with low amplitude or a non-response) [105,160–162]. Balslev et al. showed with non-navigated TMS that a 45° angle with respect to the interhemispheric midline is generally the optimal coil orientation [163]. Further, the optimal coil orientation has been observed from experiments and simulations to be perpendicular to the gyral wall [104,105,135,138,160,164]. While there may not be group-level differences in the optimal coil orientation, the individually quantified parameters, such as the motor area, may depend on it [106].

The microanatomy of the gyrus may also affect the efficacy of nTMS, i.e., how aligned are the stimulations and the activated neuronal structures, and how organized or anisotropic are the activated individual neurons in their population [138,165]. While these factors cannot be directly visualized during application of nTMS, they may take relevant effect on the mapping outcome.

3.6. Other Quantitative Parameters in Motor Maps

Other quantitative parameters commonly used to characterize motor maps based on MEP amplitudes include the motor hotspot, CoG, motor area, and motor map volume. The motor hotspot location typically is defined by the location of the maximum MEP amplitude response (x_{max}; y_{max}) or the "optimal location" for stimulation [28,29,166–168]. The motor hotspot is not only used to characterize the location of the motor representation, but is also used as the location where the rMT is determined and around which the motor mapping is extended [169]. Considering the definition of the motor hotspot, Reijonen et al. characterized a hotspot as a region instead of a unique target and found that if the definition is based on MEP amplitudes within individualized motor maps, the hotspot is on average 13 mm^2 in size, and if the hotspot is defined on the basis of the stimulation-induced EF, the size is on average 26 mm^2 [29]. These hotspot definitions consider the accuracy of the definition of the hotspot, including within-session neuronavigation system accuracy. Specifically, with nTMS, it has been shown that intra- and inter-observer variability for motor hotspot determination are on average ≤ 1 cm, with values ranging within the calculated precision of the used system [170].

The CoG is defined as the amplitude-weighted location in coordinates representing the location in the motor representation area where the center of motor activation lies, and is represented by the following equation:

$$x_{CoG} = \sum M_i x_i / \sum M_i ; y_{CoG} = \sum M_i y_i / \sum M_i \qquad (1)$$

where CoG location is defined in two dimensions with the Cartesian coordinates x_{CoG} and y_{CoG}, individual MEP amplitudes, and M_i corresponding to single stimulation targets $(x_i; y_i)$ [171]. The CoG locations are dependent on the muscle representation mapped [95,118,172]. The inter-session repeatability of the CoG has been demonstrated to be good to excellent [28,31,98,118]. In our simulations, we found that the SI has a minor effect on the CoG location, as does the chance of false-negative responses (Figure 11).

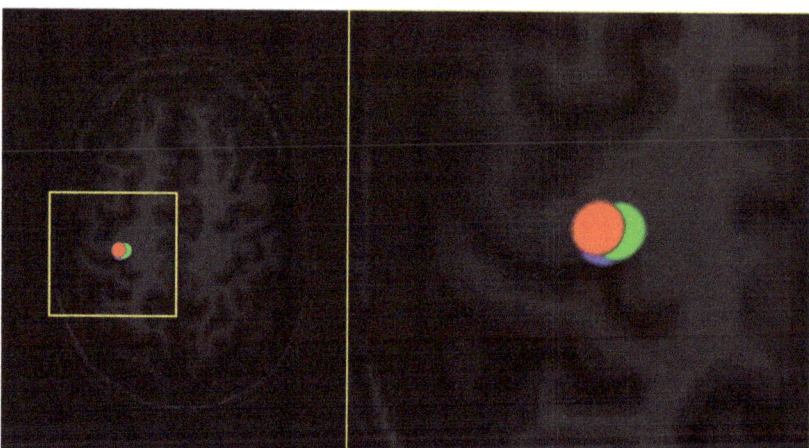

Figure 11. Visualization of CoGs based on motor maps with three SIs (blue: 110% of the rMT, green: Mills–Nithi UT method [108], red: 120% rMT). Data from each of the experiments were bootstrapped 1000 times, assuming that 10% of the responses observed were false negative. The dots indicate a product of 1000 CoG locations each. The image is visualized in neurological projection.

If the spacing of the stimulus targets is not homogeneous within the motor map (i.e., when a stimulus grid is not used), then the response amplitudes M_i need to be weighted by their stimulus targets coverage in the motor map A_i [27]:

$$x_{CoG} = \sum M_i A_i x_i / \sum M_i A_i ; y_{CoG} = \sum M_i A_i y_i / \sum M_i A_i \qquad (2)$$

The volume of a motor map is usually defined in a grid by summing the MEP amplitudes associated with the grid squares that exceed a given response threshold [24,30,31]:

$$Volume = \sum M_i \qquad (3)$$

where the index *i* refers to each grid element. Alternatively, the volume of the motor map has been determined as the volume of an interpolated amplitude surface on the cortex [96]. The repeatability of the motor map volumes has been demonstrated to be between poor and good [30,31].

The area has been defined in different ways. When using a simulation grid, previous studies have multiplied the grid square area with the number of active stimulation sites within the grid [118,147,173]. With nTMS, recent studies have utilized different means for calculating the area, such as spline interpolation or Voronoi tessellation [25,27,95,153,155,173–176]. These analysis techniques of the representation are not directly comparable, as they include systematic differences [173]. Previous studies have shown that the motor map area may suffer from poor to excellent session-to-session/within-session repeatability [27,30,31,98,118]. Here, we simulated the session-to-session repeatability of a motor map by assuming the potential of false-negative motor responses to find that the quantitative size is well-preserved within the sessions as are the shapes and locations, while minor session-to-session differences were observed in the motor map area (Figure 12). The set amplitude criterion also affects the motor area, and commonly the amplitude criterion of 50 µV is used (Figure 13).

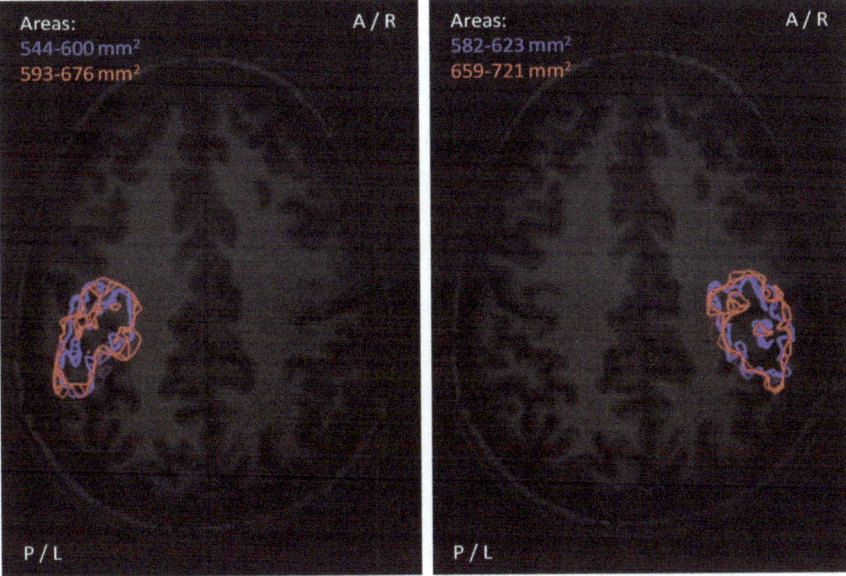

Figure 12. Visualization of motor map outlines of the FDI muscle representation in repeated measurements. Data from each of the experiments were bootstrapped 1000 times, assuming that 10% of the responses observed were false negative. The image is visualized in neurological projection. The *blue maps* were mapped a bit more densely (mean distance between stimulus locations was 2.6 mm on both hemispheres, as a 3.0 mm grid was used as aid) than the *red maps* (mean distance between stimulus locations was 2.7 mm and 3.0 mm on the left and right hemisphere, respectively, without a stimulation grid). The 95% confidence limits are indicated for quantified areas in the images.

Sinitsyn et al. found, based on their experiments comparing multiple techniques and multiple stimuli/grid squares in calculating the motor map areas, that area weighted by the probability of an MEP within a grid square appeared overall best in terms of accuracy [26]. As the SI applied with respect to the rMT determines the probability of an MEP in each grid square, the reliable determination of the probability would likely require multiple repetitions per grid square, or for one stimulus per grid square the probability would be 0 or 1.

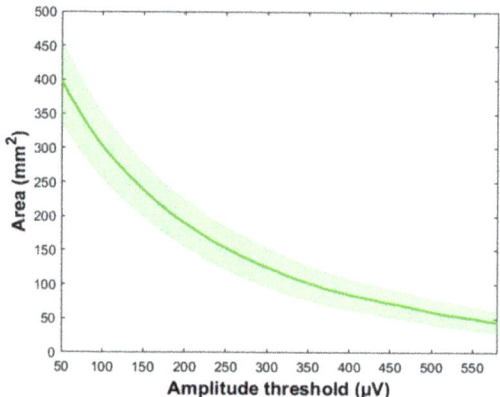

Figure 13. Effect of amplitude criterion for accepted MEPs measured on the resulting motor map size as determined using spline interpolation [27]. In the figure, 24 cortical mapping experiments for locating and outlining the FDI muscle representation were used to calculate the mean (*green line*) and 95% confidence interval (*green area*) for the motor area at different amplitude threshold criteria for accepted MEPs.

4. Perspectives and Future Directions

Over the past decade, nTMS has found its way into clinical routine, particularly for motor mapping among patients with motor-eloquent brain neoplasms since the method combines spatially resolved identification of brain function with high accuracy even in cases with deranged anatomo-functional architecture. The good agreement between preoperative nTMS motor mapping and intraoperative DES mapping as the reference method seems one of the key factors contributing to the current role of nTMS motor mapping in neuro-oncological surgery [35–42]. Furthermore, motor maps derived from nTMS can be used for seeding to achieve function-based tractography, which enables the identification of the spatial course of critical subcortical structures such as the CST [51–56]. However, the initial application for preoperative planning and intraoperative resection guidance in patients harboring functionally eloquent brain neoplasms has been greatly expanded over the years, thus enabling researchers to also address basic research questions in the context of a brain tumor as the use case, spanning from risk stratification for motor function to plasticity and reshaping of functional anatomy [69–73,77–81].

The key to successful integration of nTMS motor mapping and derived tractography into the clinical workflow is closely related to feasibility aspects and potential difficulties when embedding these approaches in existing environments. In this context, seamless integration into pre-existing infrastructure (e.g., hospital information system or PACS) can be achieved by standardized data export and transfer from the nTMS system [82]. However, it has to be noted that performing the mapping procedure in the most accurate fashion requires some training and time (~60 to 90 min per patient, excluding preparation and depending on factors such as the extent of motor mapping and patient cooperation) [6]. Thus, trained personnel and dedicated time slots for mapping purposes may be required in the clinical setting, which are not always granted under economic and time constraints. However, alternatives to nTMS motor mapping for the preoperative workup in patients harboring brain neoplasms (e.g., fMRI or MEG) also come with expenses and may only be available in specialized centers. Given that an overall better agreement between nTMS motor mapping and DES mapping has been observed in comparative studies with fMRI or MEG, efforts during the preoperative setup with nTMS seem justified [35–42].

Further potential of nTMS motor mapping lies in the interdisciplinary use of derived data. Particularly, radiotherapists may take advantage of such maps to modify treatment plans with the aim of limiting dose exposure to motor-eloquent cortex, as shown in first studies on the matter [83–87]. Integration of functional data derived from nTMS motor mapping could also be achieved for case discussions of interdisciplinary tumor boards and forwarded to follow-up treatment, which may make use of such information for rehabilitation strategies. In this context, a first example in patients suffering from acute surgery-related paresis of uE muscles after glioma resection provides evidence for the beneficial use of nTMS as a therapeutic tool in neuro-oncology, with the exact site of stimulation being determined based on nTMS motor maps [177]. Specifically, the combination of low-frequency nTMS with physical therapy for seven consecutive days after surgery improved motor function outcome according to the Fugl–Meyer assessment performed postoperatively and until the 3-months follow-up examinations [177].

Additionally, nTMS data could also be efficiently used in multi-modal scenarios. For instance, the combination of motor mapping with mapping of other functions such as language and derived tractography of motor- and language-related subcortical pathways has already been achieved in a few studies, which may help to gain a broader picture of functional anatomy in patients harboring large or critically situated neoplasms that most likely do not solely affect the motor system [178,179]. Further, to understand better the distinct effects of nTMS on the brain's connectivity profile, combinations of nTMS application with pre- and post-stimulation fMRI acquisitions and functional connectivity analyses are possible and have shown promising results in healthy volunteers [100,101]. Most notably, it has been revealed that modulation by nTMS critically depends on the connectivity profile of the target region, with imaging biomarkers derived from fMRI possibly playing a role to improve sensitivity of nTMS for research and clinical applications [100]. Based on such data, it seems likely that in patients with brain neoplasms, the impact and effects of nTMS also depend on a connectivity profile that may fluctuate over time or due to yet unidentified parameters, possibly interfering with the mapping outcome. Yet, multi-modal scenarios specifically in patients with motor-eloquent brain tumors may also exert difficulties on data acquisition, processing, and interpretation of data, which need to be addressed prior to routine application. For fMRI, presence of particularly high-grade tumors with increased cerebral blood flow characteristics can negatively interfere with signal interpretation [180,181]. Thus, combinations of fMRI with nTMS in such patients need to be considered carefully to avoid errors in calculations of connectivity characteristics.

Regarding dMRI, images can suffer from geometric image distortions compared with anatomical MRI, which may introduce spatial inaccuracies when dMRI data is linearly projected on conventional T1- or T2-weighted sequences and used for tractography. This may be retrospectively corrected for by non-linear, semi-elastic image fusion, thus potentially enabling tractography with improved accuracy and clinical feasibility [182]. Expanding on this, intraoperative MRI-based elastic image fusion for anatomically accurate tractography of the CST using nTMS motor maps has been achieved, correlating well with IONM and disproving the severity of brain shift in selected cases [183,184]. Furthermore, most work on nTMS-based fiber tracking of the CST has used standard deterministic algorithms implemented in commercially available packages for clinical use (e.g., fiber assignment by continuous tracking (FACT) algorithms) [51,52,54,55,70–72]. Using a FACT algorithm, fiber bundles are reconstructed in a voxel-by-voxel fashion with respect to the direction of the main eigenvector, which works purely data-based (no interpolation function) and needs only comparatively low computation time [185,186]. However, on the other hand, FACT algorithms create some predictable inherent errors, which limit the accuracy of the method and could lead to error-prone or incomplete tractography results [186–189]. Probabilistic approaches disperse trajectories more than deterministic methods and may delineate a greater proportion of white matter tracts, particularly when combined with more advanced dMRI sequences [189]. Additionally, the potential value of diffusion measures besides mere delineation of the spatial course of the CST by nTMS-based tractography may be of merit. One study has shown that the extent of impairment of diffusion metrics (such as FA and ADC) correlates with motor function deficits according to segmental analyses within the CST [72]. Hence, supplementing nTMS-based tractography of the CST with diffusion metrics may improve the predictive power for postoperative motor impairment, but other parameters such as mean diffusivity (MD) have not yet been routinely considered. As such, MD is a quantitative measure of the mean motion of water and reflects the rotationally invariant magnitude of water diffusion, which could also be representative of structural integrity of white matter [190]. Future work may explore the potential benefits of further, even more elaborate quantitative markers of white matter structure, composition, and integrity, which can be derived from advanced dMRI techniques such as high angular resolution diffusion imaging, multi-shell imaging, diffusion kurtosis imaging, neurite orientation dispersion, and density imaging [191–193].

The continuous optimization of the nTMS system technique has recently enabled a novel paired-pulse nTMS (pp-nTMS) paradigm for biphasic pulse wave application to induce short-interval intracortical facilitation [194–196]. Use of pp-nTMS may increase efficacy of motor mapping in patients with brain tumors as accurate motor maps are achieved even in cases where conventional single-pulse mapping fails (e.g., due to tumor-affected motor structures or edema) [174,175]. Further, pp-nTMS would result in a lower rMT, thus allowing motor mapping with lower SI but without clinically relevant constraints for motor map extent or location [174,175]. Particularly in patients with brain tumors, the rMT can be frequently high in the tumor-affected hemisphere, and accurate mapping with a lower SI related to a lower determined rMT could permit successful use of nTMS even in the most demanding cases. This may also be highly relevant for other applications of nTMS in patients with brain tumors, such as language mapping [197–201]. In such applications, a higher SI is often used and stimulation is more widespread and, thus, can entail discomfort that negatively interferes with the mapping outcome. However, studies using pp-nTMS for other purposes than motor mapping are currently lacking.

From a methodological perspective, the presently used motor mapping approach is focused on determining the volume, area, coil locations, or corresponding cortical EF maximum locations associated with motor responses. However, as the EF induced by nTMS is not ideally focused, the spread of the EF also stimulates adjacent areas and, thus, when stimulating an area in the cortex adjacent to the target muscle representation area, the target muscle may activate even though the representation area was not targeted. Hence, quantitative mapping at present cannot precisely capture the true representation size of the motor representation area, and may over- or underestimate it. Previous studies have shown that inclusion of the EF information in generating the motor maps may aid in localizing the motor representation [202–204]. Specifically, minimum norm estimation (MNE) has recently been employed for estimating the true representation area of muscles by accounting for the EF spread and the input–output characteristics of the MEP values [204]. Figure 14 demonstrates the application of MNE in a single subject in comparison with the outlined cortical maximum EF locations with associated positive responses. A quite similar method to MNE was presented by Weise et al., also utilizing the input–output characteristics of the MEPs [203]. The practical difference between these methods arises from the number of coil locations applied and how the input–output characteristics are estimated spatially.

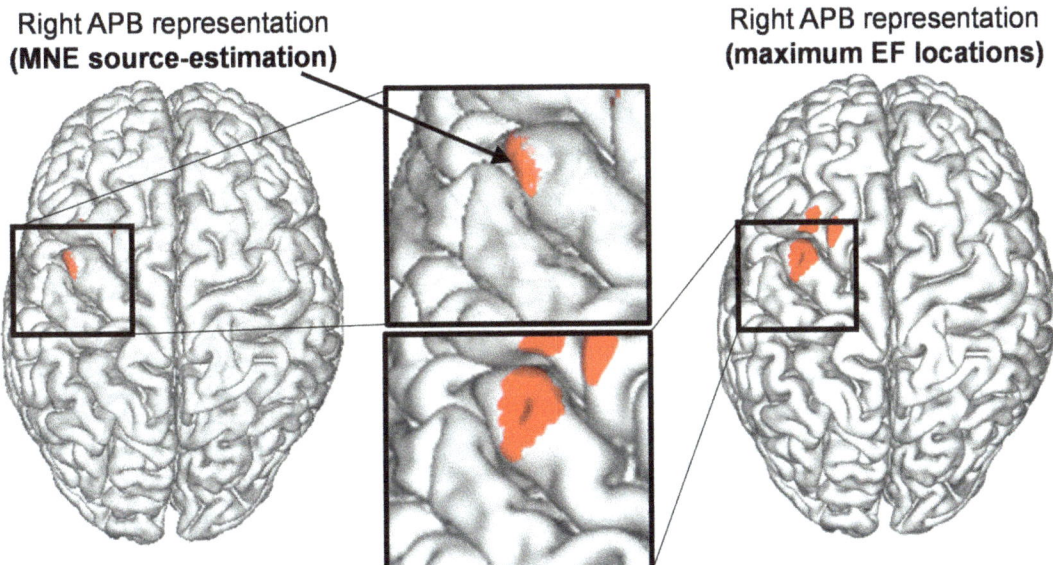

Figure 14. The MNE method (**left**) utilizes the spread of the EF at each stimulation point in addition to the MEP amplitude to estimate the source of motor activation, as opposed to the more conventional motor mapping approach not considering the spread of the EF associated with each stimulus (**right**) [204]. The active stimulation sites in the conventional motor map were outlined using the spline interpolation method [27].

In essence, one challenge for more accurate clinical motor mapping with nTMS in the future may be the accurate inclusion of the EF information to estimate the source of the induced MEPs. Currently, there are no clinically validated tools classified as medical devices available for source estimation utilizing EF spread, and the tools that are currently available require special skills and are not feasible for clinical routines. However, the development seems to be heading in the right direction. With tools such as the SimNIBS, EF simulations have been made quite straightforward, albeit the pipelines feasible for clinical mapping applications are still lacking [205,206]. Furthermore, the shape of the motor map may be evaluated by the aspect ratio (i.e., the ratio between map extension along the EF direction and perpendicular to it) [195]. This means that if the aspect ratio is 1, the shape of the motor map is approximated as circular, if the ratio is >1, the map is elongated along the EF direction, and with an aspect ratio <1, the motor map is elongated to the direction perpendicular to the EF direction. With single-pulse nTMS, the aspect ratio tends to be >1 [174,175,195].

Despite increasing acceptance of nTMS motor mapping in clinical routine, it cannot be emphasized enough that the value of the method stands and falls with its accuracy. In this regard, many parameters such as the location of the motor hotspot and CoG, and area and volume of the motor map are associated with the applied SI, which needs to be determined with highest diligence. Therefore, with the increase in applications of nTMS, methods for ensuring feasible accuracy become more and more important. In the context of quantitative mapping, awareness of the relevant parameters and control over them is warranted to assure best practices and reliable mapping outcomes. Specifically, quantitative mapping has the potential to derive parameters related to the motor maps of the patients that could impact diagnostics, prognostics, and follow-up examinations by enabling spatial and spatio-temporal metrics related to cortical motor function. Major future development efforts should be put to understanding the correlations between motor mapping, interpretation of stimulation targets, and resulting responses in relation to their origin in the cortex by consideration of the physical effects of induced EFs.

5. Conclusions

The technique of nTMS is increasingly used particularly for preoperative motor mapping in patients harboring brain tumors, which is due to its sufficient accuracy and reliability in a clinical setup. The combination of nTMS motor mapping with tractography as well as the option of serial mapping over time profoundly expands its role beyond a mere surgical planning tool. Development of quantitative motor mapping can include further applications while the accuracy of current mapping modalities can be improved by standardized protocols and increased consideration of EF information.

Author Contributions: All authors declare that they have made substantial contributions to the conception and design of the work and have drafted the work or substantively revised it. All authors approved the submitted version. All authors have read and agreed to the published version of the manuscript.

Funding: This research was supported by the Academy of Finland (grant no: 322423). L.S. was funded by UEF in a research project funded by Business Finland (grant no: 2956/31/2018) and State Research Funding (grant no: 1689/2020).

Institutional Review Board Statement: Not applicable.

Informed Consent Statement: Not applicable.

Acknowledgments: The authors would like to thank Axel Schroeder for his support during the generation of figures to illustrate the motor mapping procedure by nTMS and fiber tracking.

Conflicts of Interest: P.J. has a shared patent with, and has received consulting fees from Nexstim Plc (Helsinki, Finland), a manufacturer of navigated TMS systems. N.S. received honoraria from Nexstim Plc (Helsinki, Finland). S.M.K. is a consultant for Nexstim Plc (Helsinki, Finland) and Spineart Deutschland GmbH (Frankfurt, Germany) and received honoraria from Medtronic (Meerbusch, Germany) and Carl Zeiss Meditec (Oberkochen, Germany). S.M.K. received research grants and is a consultant for Brainlab AG (Munich, Germany). The funders had no role in the design of the study; in the collection, analyses, or interpretation of data; in the writing of the manuscript, or in the decision to publish the results.

Abbreviations

ADC	Apparent diffusion coefficient
ADM	Abductor digiti minimi
AED	Antiepileptic drug
aMT	Active motor threshold
APB	Abductor pollicis brevis
ATH	Adaptive threshold hunting
BMRC	British Medical Research Council
CoG	Center of gravity
CST	Corticospinal tract
CT	Computed tomography
DES	Direct electrical stimulation
DOF	Degrees of freedom
dMRI	Diffusion magnetic resonance imaging
EF	Electric field
EOR	Extent of resection
FA	Fractional anisotropy
FACT	Fiber assignment by continuous tracking
FAT	Fractional anisotropy threshold
FCR	Flexor carpi radialis
FDI	First dorsal interosseous
FGR	Fluorescein-guided resection

FL	Fiber length
fMRI	Functional MRI
GBM	Glioblastoma multiforme
GTR	Gross total resection
HGG	High-grade glioma
IONM	Intraoperative neuromonitoring
IR	Infrared
KPS	Karnofsky performance status
lE	Lower extremity
LGG	Low-grade glioma
MD	Mean diffusivity
MEG	Magnetoencephalography
MEN	Meningioma
MEP	Motor-evoked potential
MET	Metastasis
MNE	Minimum norm estimation
MSO	Maximum stimulator output
nTMS	Navigated transcranial magnetic stimulation
PACS	Picture archiving and communication system
PET	Positron emission tomography
PFS	Progression-free survival
pp-nTMS	Paired-pulse navigated transcranial magnetic stimulation
PTV	Planning target volume
rMT	Resting motor threshold
ROI	Region of interest
SI	Stimulation intensity
TA	Tibialis anterior
uE	Upper extremity
UT	Upper threshold
WHO	World Health Organization

References

1. Ruohonen, J.; Karhu, J. Navigated transcranial magnetic stimulation. *Clin. Neurophysiol.* **2010**, *40*, 7–17. [CrossRef]
2. Krings, T.; Buchbinder, B.R.; Butler, W.E.; Chiappa, K.H.; Jiang, H.J.; Rosen, B.R.; Cosgrove, G.R. Stereotactic transcranial magnetic stimulation: Correlation with direct electrical cortical stimulation. *Neurosurgery* **1997**, *41*, 1319–1326. [CrossRef]
3. Krings, T.; Naujokat, C.; von Keyserlingk, D.G. Representation of cortical motor function as revealed by stereotactic transcranial magnetic stimulation. *Electroencephalogr. Clin. Neurophysiol.* **1998**, *109*, 85–93. [CrossRef]
4. Krings, T.; Chiappa, K.H.; Foltys, H.; Reinges, M.H.; Cosgrove, G.R.; Thron, A. Introducing navigated transcranial magnetic stimulation as a refined brain mapping methodology. *Neurosurg. Rev.* **2001**, *24*, 171–179. [CrossRef]
5. Ruohonen, J.; Ilmoniemi, R.J. Modeling of the stimulating field generation in TMS. *Electroencephalogr. Clin. Neurophysiol. Suppl.* **1999**, *51*, 30–40.
6. Sollmann, N.; Goblirsch-Kolb, M.F.; Ille, S.; Butenschoen, V.M.; Boeckh-Behrens, T.; Meyer, B.; Ringel, F.; Krieg, S.M. Comparison between electric-field-navigated and line-navigated TMS for cortical motor mapping in patients with brain tumors. *Acta Neurochir.* **2016**, *158*, 2277–2289. [CrossRef]
7. Bijsterbosch, J.D.; Barker, A.T.; Lee, K.H.; Woodruff, P.W. Where does transcranial magnetic stimulation (TMS) stimulate? Modelling of induced field maps for some common cortical and cerebellar targets. *Med. Biol. Eng. Comput.* **2012**, *50*, 671–681. [CrossRef]
8. Lefaucheur, J.P.; Picht, T. The value of preoperative functional cortical mapping using navigated TMS. *Clin. Neurophysiol.* **2016**, *46*, 125–133. [CrossRef]
9. Haddad, A.F.; Young, J.S.; Berger, M.S.; Tarapore, P.E. Preoperative applications of navigated transcranial magnetic stimulation. *Front. Neurol.* **2020**, *11*, 628903. [CrossRef]
10. Raffa, G.; Scibilia, A.; Conti, A.; Ricciardo, G.; Rizzo, V.; Morelli, A.; Angileri, F.F.; Cardali, S.M.; Germano, A. The role of navigated transcranial magnetic stimulation for surgery of motor-eloquent brain tumors: A systematic review and meta-analysis. *Clin. Neurol. Neurosurg.* **2019**, *180*, 7–17. [CrossRef]
11. Picht, T. Current and potential utility of transcranial magnetic stimulation in the diagnostics before brain tumor surgery. *CNS Oncol.* **2014**, *3*, 299–310. [CrossRef]
12. Duffau, H.; Mandonnet, E. The "onco-functional balance" in surgery for diffuse low-grade glioma: Integrating the extent of resection with quality of life. *Acta Neurochir.* **2013**, *155*, 951–957. [CrossRef]

13. Pope, W.B.; Sayre, J.; Perlina, A.; Villablanca, J.P.; Mischel, P.S.; Cloughesy, T.F. MR imaging correlates of survival in patients with high-grade gliomas. *AJNR Am. J. Neuroradiol.* **2005**, *26*, 2466–2474.
14. Molinaro, A.M.; Hervey-Jumper, S.; Morshed, R.A.; Young, J.; Han, S.J.; Chunduru, P.; Zhang, Y.; Phillips, J.J.; Shai, A.; Lafontaine, M.; et al. Association of maximal extent of resection of contrast-enhanced and non-contrast-enhanced tumor with survival within molecular subgroups of patients with newly diagnosed glioblastoma. *JAMA Oncol.* **2020**. [CrossRef]
15. Brown, P.D.; Maurer, M.J.; Rummans, T.A.; Pollock, B.E.; Ballman, K.V.; Sloan, J.A.; Boeve, B.F.; Arusell, R.M.; Clark, M.M.; Buckner, J.C. A prospective study of quality of life in adults with newly diagnosed high-grade gliomas: The impact of the extent of resection on quality of life and survival. *Neurosurgery* **2005**, *57*, 495–504. [CrossRef]
16. Haj, A.; Doenitz, C.; Schebesch, K.M.; Ehrensberger, D.; Hau, P.; Putnik, K.; Riemenschneider, M.J.; Wendl, C.; Gerken, M.; Pukrop, T.; et al. Extent of resection in newly diagnosed glioblastoma: Impact of a specialized neuro-oncology care center. *Brain Sci.* **2017**, *8*, 5. [CrossRef]
17. Ammirati, M.; Vick, N.; Liao, Y.L.; Ciric, I.; Mikhael, M. Effect of the extent of surgical resection on survival and quality of life in patients with supratentorial glioblastomas and anaplastic astrocytomas. *Neurosurgery* **1987**, *21*, 201–206. [CrossRef]
18. Hervey-Jumper, S.L.; Berger, M.S. Evidence for Improving Outcome Through Extent of Resection. *Neurosurg. Clin. N. Am.* **2019**, *30*, 85–93. [CrossRef]
19. Szelenyi, A.; Bello, L.; Duffau, H.; Fava, E.; Feigl, G.C.; Galanda, M.; Neuloh, G.; Signorelli, F.; Sala, F.; Workgroup for intraoperative management in low-grade glioma surgery within the european low-grade glioma. Intraoperative electrical stimulation in awake craniotomy: Methodological aspects of current practice. *Neurosurg. Focus* **2010**, *28*, E7. [CrossRef]
20. Duffau, H. Lessons from brain mapping in surgery for low-grade glioma: Insights into associations between tumour and brain plasticity. *Lancet. Neurol.* **2005**, *4*, 476–486. [CrossRef]
21. De Witt Hamer, P.C.; Robles, S.G.; Zwinderman, A.H.; Duffau, H.; Berger, M.S. Impact of intraoperative stimulation brain mapping on glioma surgery outcome: A meta-analysis. *J. Clin. Oncol.* **2012**, *30*, 2559–2565. [CrossRef]
22. Mandonnet, E.; Winkler, P.A.; Duffau, H. Direct electrical stimulation as an input gate into brain functional networks: Principles, advantages and limitations. *Acta Neurochir.* **2010**, *152*, 185–193. [CrossRef]
23. Hervey-Jumper, S.L.; Berger, M.S. Maximizing safe resection of low- and high-grade glioma. *J. Neurooncol.* **2016**, *130*, 269–282. [CrossRef]
24. Van de Ruit, M.; Grey, M.J. The TMS map scales with increased stimulation intensity and muscle activation. *Brain Topogr.* **2016**, *29*, 56–66. [CrossRef]
25. Kallioniemi, E.; Julkunen, P. Alternative stimulation intensities for mapping cortical motor area with navigated TMS. *Brain Topogr.* **2016**, *29*, 395–404. [CrossRef]
26. Sinitsyn, D.O.; Chernyavskiy, A.Y.; Poydasheva, A.G.; Bakulin, I.S.; Suponeva, N.A.; Piradov, M.A. Optimization of the navigated TMS mapping algorithm for accurate estimation of cortical muscle representation characteristics. *Brain Sci.* **2019**, *9*, 88. [CrossRef]
27. Julkunen, P. Methods for estimating cortical motor representation size and location in navigated transcranial magnetic stimulation. *J. Neurosci. Methods* **2014**, *232*, 125–133. [CrossRef]
28. Wolf, S.L.; Butler, A.J.; Campana, G.I.; Parris, T.A.; Struys, D.M.; Weinstein, S.R.; Weiss, P. Intra-subject reliability of parameters contributing to maps generated by transcranial magnetic stimulation in able-bodied adults. *Clin. Neurophysiol.* **2004**, *115*, 1740–1747. [CrossRef]
29. Reijonen, J.; Pitkanen, M.; Kallioniemi, E.; Mohammadi, A.; Ilmoniemi, R.J.; Julkunen, P. Spatial extent of cortical motor hotspot in navigated transcranial magnetic stimulation. *J. Neurosci. Methods* **2020**, *346*, 108893. [CrossRef]
30. Ngomo, S.; Leonard, G.; Moffet, H.; Mercier, C. Comparison of transcranial magnetic stimulation measures obtained at rest and under active conditions and their reliability. *J. Neurosci. Methods* **2012**, *205*, 65–71. [CrossRef]
31. Kraus, D.; Gharabaghi, A. Neuromuscular plasticity: Disentangling stable and variable motor maps in the human sensorimotor cortex. *Neural. Plast.* **2016**, *2016*, 7365609. [CrossRef] [PubMed]
32. Coburger, J.; Musahl, C.; Weissbach, C.; Bittl, M. Navigated transcranial magnetic stimulation-guided resection of a left parietal tumor: Case report. *Minim. Invasive Neurosurg.* **2011**, *54*, 38–40. [CrossRef] [PubMed]
33. Picht, T.; Schmidt, S.; Woitzik, J.; Suess, O. Navigated brain stimulation for preoperative cortical mapping in paretic patients: Case report of a hemiplegic patient. *Neurosurgery* **2011**, *68*, E1475–E1480. [CrossRef] [PubMed]
34. Coburger, J.; Karhu, J.; Bittl, M.; Hopf, N.J. First preoperative functional mapping via navigated transcranial magnetic stimulation in a 3-year-old boy. *J. Neurosurg. Pediatrics* **2012**, *9*, 660–664. [CrossRef]
35. Picht, T.; Schmidt, S.; Brandt, S.; Frey, D.; Hannula, H.; Neuvonen, T.; Karhu, J.; Vajkoczy, P.; Suess, O. Preoperative functional mapping for rolandic brain tumor surgery: Comparison of navigated transcranial magnetic stimulation to direct cortical stimulation. *Neurosurgery* **2011**, *69*, 581–588. [CrossRef]
36. Coburger, J.; Musahl, C.; Henkes, H.; Horvath-Rizea, D.; Bittl, M.; Weissbach, C.; Hopf, N. Comparison of navigated transcranial magnetic stimulation and functional magnetic resonance imaging for preoperative mapping in rolandic tumor surgery. *Neurosurg. Rev.* **2013**, *36*, 65–76. [CrossRef]
37. Forster, M.T.; Hattingen, E.; Senft, C.; Gasser, T.; Seifert, V.; Szelenyi, A. Navigated transcranial magnetic stimulation and functional magnetic resonance imaging: Advanced adjuncts in preoperative planning for central region tumors. *Neurosurgery* **2011**, *68*, 1317–1325. [CrossRef]

38. Krieg, S.M.; Shiban, E.; Buchmann, N.; Gempt, J.; Foerschler, A.; Meyer, B.; Ringel, F. Utility of presurgical navigated transcranial magnetic brain stimulation for the resection of tumors in eloquent motor areas. *J. Neurosurg.* **2012**, *116*, 994–1001. [CrossRef]
39. Krieg, S.M.; Shiban, E.; Buchmann, N.; Meyer, B.; Ringel, F. Presurgical navigated transcranial magnetic brain stimulation for recurrent gliomas in motor eloquent areas. *Clin. Neurophysiol.* **2013**, *124*, 522–527. [CrossRef]
40. Mangraviti, A.; Casali, C.; Cordella, R.; Legnani, F.G.; Mattei, L.; Prada, F.; Saladino, A.; Contarino, V.E.; Perin, A.; DiMeco, F. Practical assessment of preoperative functional mapping techniques: Navigated transcranial magnetic stimulation and functional magnetic resonance imaging. *Neurol. Sci.* **2013**, *34*, 1551–1557. [CrossRef]
41. Weiss Lucas, C.; Nettekoven, C.; Neuschmelting, V.; Oros-Peusquens, A.M.; Stoffels, G.; Viswanathan, S.; Rehme, A.K.; Faymonville, A.M.; Shah, N.J.; Langen, K.J.; et al. Invasive versus non-invasive mapping of the motor cortex. *Hum. Brain Mapp.* **2020**, *41*, 3970–3983. [CrossRef]
42. Tarapore, P.E.; Tate, M.C.; Findlay, A.M.; Honma, S.M.; Mizuiri, D.; Berger, M.S.; Nagarajan, S.S. Preoperative multimodal motor mapping: A comparison of magnetoencephalography imaging, navigated transcranial magnetic stimulation, and direct cortical stimulation. *J. Neurosurg.* **2012**, *117*, 354–362. [CrossRef]
43. Picht, T.; Schulz, J.; Hanna, M.; Schmidt, S.; Suess, O.; Vajkoczy, P. Assessment of the influence of navigated transcranial magnetic stimulation on surgical planning for tumors in or near the motor cortex. *Neurosurgery* **2012**, *70*, 1248–1257. [CrossRef]
44. Rizzo, V.; Terranova, C.; Conti, A.; Germano, A.; Alafaci, C.; Raffa, G.; Girlanda, P.; Tomasello, F.; Quartarone, A. Preoperative functional mapping for rolandic brain tumor surgery. *Neurosci. Lett.* **2014**, *583*, 136–141. [CrossRef]
45. Zdunczyk, A.; Fleischmann, R.; Schulz, J.; Vajkoczy, P.; Picht, T. The reliability of topographic measurements from navigated transcranial magnetic stimulation in healthy volunteers and tumor patients. *Acta Neurochir.* **2013**, *155*, 1309–1317. [CrossRef] [PubMed]
46. Sollmann, N.; Bulubas, L.; Tanigawa, N.; Zimmer, C.; Meyer, B.; Krieg, S.M. The variability of motor evoked potential latencies in neurosurgical motor mapping by preoperative navigated transcranial magnetic stimulation. *BMC Neurosci.* **2017**, *18*, 5. [CrossRef] [PubMed]
47. Sollmann, N.; Tanigawa, N.; Bulubas, L.; Sabih, J.; Zimmer, C.; Ringel, F.; Meyer, B.; Krieg, S.M. Clinical factors underlying the inter-individual variability of the resting motor threshold in navigated transcranial magnetic stimulation motor mapping. *Brain Topogr.* **2017**, *30*, 98–121. [CrossRef]
48. Lavrador, J.P.; Gioti, I.; Hoppe, S.; Jung, J.; Patel, S.; Gullan, R.; Ashkan, K.; Bhangoo, R.; Vergani, F. Altered motor excitability in patients with diffuse gliomas involving motor eloquent areas: The impact of tumor grading. *Neurosurgery* **2020**, *88*, 183–192. [CrossRef] [PubMed]
49. Lam, S.; Lucente, G.; Schneider, H.; Picht, T. TMS motor mapping in brain tumor patients: More robust maps with an increased resting motor threshold. *Acta Neurochir.* **2019**, *161*, 995–1002. [CrossRef]
50. Mirbagheri, A.; Schneider, H.; Zdunczyk, A.; Vajkoczy, P.; Picht, T. NTMS mapping of non-primary motor areas in brain tumour patients and healthy volunteers. *Acta Neurochir.* **2020**, *162*, 407–416. [CrossRef]
51. Krieg, S.M.; Buchmann, N.H.; Gempt, J.; Shiban, E.; Meyer, B.; Ringel, F. Diffusion tensor imaging fiber tracking using navigated brain stimulation—A feasibility study. *Acta Neurochir.* **2012**, *154*, 555–563. [CrossRef] [PubMed]
52. Frey, D.; Strack, V.; Wiener, E.; Jussen, D.; Vajkoczy, P.; Picht, T. A new approach for corticospinal tract reconstruction based on navigated transcranial stimulation and standardized fractional anisotropy values. *NeuroImage* **2012**, *62*, 1600–1609. [CrossRef] [PubMed]
53. Conti, A.; Raffa, G.; Granata, F.; Rizzo, V.; Germano, A.; Tomasello, F. Navigated transcranial magnetic stimulation for "somatotopic" tractography of the corticospinal tract. *Neurosurgery* **2014**, *10* (Suppl. 4), 542–554. [CrossRef] [PubMed]
54. Weiss, C.; Tursunova, I.; Neuschmelting, V.; Lockau, H.; Nettekoven, C.; Oros-Peusquens, A.M.; Stoffels, G.; Rehme, A.K.; Faymonville, A.M.; Shah, N.J.; et al. Improved nTMS- and DTI-derived CST tractography through anatomical ROI seeding on anterior pontine level compared to internal capsule. *NeuroImage. Clin.* **2015**, *7*, 424–437. [CrossRef]
55. Weiss Lucas, C.; Tursunova, I.; Neuschmelting, V.; Nettekoven, C.; Oros-Peusquens, A.M.; Stoffels, G.; Faymonville, A.M.; Jon, S.N.; Langen, K.J.; Lockau, H.; et al. Functional MRI vs. navigated TMS to optimize M1 seed volume delineation for DTI tractography. A prospective study in patients with brain tumours adjacent to the corticospinal tract. *NeuroImage. Clin.* **2017**, *13*, 297–309. [CrossRef] [PubMed]
56. Munnich, T.; Klein, J.; Hattingen, E.; Noack, A.; Herrmann, E.; Seifert, V.; Senft, C.; Forster, M.T. Tractography verified by intraoperative magnetic resonance imaging and subcortical stimulation during tumor resection near the corticospinal tract. *Oper. Neurosurg. (Hagerstown)* **2019**, *16*, 197–210. [CrossRef] [PubMed]
57. Picht, T.; Schulz, J.; Vajkoczy, P. The preoperative use of navigated transcranial magnetic stimulation facilitates early resection of suspected low-grade gliomas in the motor cortex. *Acta Neurochir.* **2013**, *155*, 1813–1821. [CrossRef]
58. Frey, D.; Schilt, S.; Strack, V.; Zdunczyk, A.; Rosler, J.; Niraula, B.; Vajkoczy, P.; Picht, T. Navigated transcranial magnetic stimulation improves the treatment outcome in patients with brain tumors in motor eloquent locations. *Neuro-oncol.* **2014**, *16*, 1365–1372. [CrossRef]
59. Krieg, S.M.; Sabih, J.; Bulubasova, L.; Obermueller, T.; Negwer, C.; Janssen, I.; Shiban, E.; Meyer, B.; Ringel, F. Preoperative motor mapping by navigated transcranial magnetic brain stimulation improves outcome for motor eloquent lesions. *Neuro-oncol.* **2014**, *16*, 1274–1282. [CrossRef]

60. Raffa, G.; Conti, A.; Scibilia, A.; Cardali, S.M.; Esposito, F.; Angileri, F.F.; La Torre, D.; Sindorio, C.; Abbritti, R.V.; Germano, A.; et al. The impact of diffusion tensor imaging fiber tracking of the corticospinal tract based on navigated transcranial magnetic stimulation on surgery of motor-eloquent brain lesions. *Neurosurgery* **2018**, *83*, 768–782. [CrossRef]
61. Krieg, S.M.; Sollmann, N.; Obermueller, T.; Sabih, J.; Bulubas, L.; Negwer, C.; Moser, T.; Droese, D.; Boeckh-Behrens, T.; Ringel, F.; et al. Changing the clinical course of glioma patients by preoperative motor mapping with navigated transcranial magnetic brain stimulation. *BMC Cancer* **2015**, *15*, 231. [CrossRef]
62. Picht, T.; Frey, D.; Thieme, S.; Kliesch, S.; Vajkoczy, P. Presurgical navigated TMS motor cortex mapping improves outcome in glioblastoma surgery: A controlled observational study. *J. Neurooncol.* **2016**, *126*, 535–543. [CrossRef] [PubMed]
63. Hendrix, P.; Senger, S.; Griessenauer, C.J.; Simgen, A.; Schwerdtfeger, K.; Oertel, J. Preoperative navigated transcranial magnetic stimulation in patients with motor eloquent lesions with emphasis on metastasis. *Clin. Anat.* **2016**, *29*, 925–931. [CrossRef] [PubMed]
64. Krieg, S.M.; Picht, T.; Sollmann, N.; Bahrend, I.; Ringel, F.; Nagarajan, S.S.; Meyer, B.; Tarapore, P.E. Resection of motor eloquent metastases aided by preoperative nTMS-based motor maps-comparison of two observational cohorts. *Front. Oncol.* **2016**, *6*, 261. [CrossRef] [PubMed]
65. Raffa, G.; Picht, T.; Scibilia, A.; Rosler, J.; Rein, J.; Conti, A.; Ricciardo, G.; Cardali, S.M.; Vajkoczy, P.; Germano, A. Surgical treatment of meningiomas located in the rolandic area: The role of navigated transcranial magnetic stimulation for preoperative planning, surgical strategy, and prediction of arachnoidal cleavage and motor outcome. *J. Neurosurg.* **2019**, 1–12. [CrossRef]
66. Raffa, G.; Picht, T.; Angileri, F.F.; Youssef, M.; Conti, A.; Esposito, F.; Cardali, S.M.; Vajkoczy, P.; Germano, A. Surgery of malignant motor-eloquent gliomas guided by sodium-fluorescein and navigated transcranial magnetic stimulation: A novel technique to increase the maximal safe resection. *J. Neurosurg. Sci.* **2019**, *63*, 670–678. [CrossRef]
67. Raffa, G.; Scibilia, A.; Conti, A.; Cardali, S.M.; Rizzo, V.; Terranova, C.; Quattropani, M.C.; Marzano, G.; Ricciardo, G.; Vinci, S.L.; et al. Multimodal Surgical Treatment of High-Grade Gliomas in the Motor Area: The impact of the combination of navigated transcranial magnetic stimulation and fluorescein-guided resection. *World Neurosurg.* **2019**, *128*, e378–e390. [CrossRef]
68. Moser, T.; Bulubas, L.; Sabih, J.; Conway, N.; Wildschutz, N.; Sollmann, N.; Meyer, B.; Ringel, F.; Krieg, S.M. Resection of navigated transcranial magnetic stimulation-positive prerolandic motor areas causes permanent impairment of motor function. *Neurosurgery* **2017**, *81*, 99–110. [CrossRef]
69. Picht, T.; Strack, V.; Schulz, J.; Zdunczyk, A.; Frey, D.; Schmidt, S.; Vajkoczy, P. Assessing the functional status of the motor system in brain tumor patients using transcranial magnetic stimulation. *Acta Neurochir.* **2012**, *154*, 2075–2081. [CrossRef]
70. Rosenstock, T.; Grittner, U.; Acker, G.; Schwarzer, V.; Kulchytska, N.; Vajkoczy, P.; Picht, T. Risk stratification in motor area-related glioma surgery based on navigated transcranial magnetic stimulation data. *J. Neurosurg.* **2017**, *126*, 1227–1237. [CrossRef]
71. Sollmann, N.; Wildschuetz, N.; Kelm, A.; Conway, N.; Moser, T.; Bulubas, L.; Kirschke, J.S.; Meyer, B.; Krieg, S.M. Associations between clinical outcome and navigated transcranial magnetic stimulation characteristics in patients with motor-eloquent brain lesions: A combined navigated transcranial magnetic stimulation-diffusion tensor imaging fiber tracking approach. *J. Neurosurg.* **2018**, *128*, 800–810. [CrossRef] [PubMed]
72. Rosenstock, T.; Giampiccolo, D.; Schneider, H.; Runge, S.J.; Bahrend, I.; Vajkoczy, P.; Picht, T. Specific DTI seeding and diffusivity-analysis improve the quality and prognostic value of TMS-based deterministic DTI of the pyramidal tract. *NeuroImage. Clin.* **2017**, *16*, 276–285. [CrossRef] [PubMed]
73. Seidel, K.; Hani, L.; Lutz, K.; Zbinden, C.; Redmann, A.; Consuegra, A.; Raabe, A.; Schucht, P. Postoperative navigated transcranial magnetic stimulation to predict motor recovery after surgery of tumors in motor eloquent areas. *Clin. Neurophysiol.* **2019**, *130*, 952–959. [CrossRef]
74. Neuschmelting, V.; Weiss Lucas, C.; Stoffels, G.; Oros-Peusquens, A.M.; Lockau, H.; Shah, N.J.; Langen, K.J.; Goldbrunner, R.; Grefkes, C. Multimodal imaging in malignant brain tumors: Enhancing the preoperative risk evaluation for motor deficits with a combined hybrid MRI-PET and navigated transcranial magnetic stimulation approach. *AJNR Am. J. Neuroradiol.* **2016**, *37*, 266–273. [CrossRef] [PubMed]
75. Forster, M.T.; Senft, C.; Hattingen, E.; Lorei, M.; Seifert, V.; Szelenyi, A. Motor cortex evaluation by nTMS after surgery of central region tumors: A feasibility study. *Acta Neurochir.* **2012**, *154*, 1351–1359. [CrossRef] [PubMed]
76. Takahashi, S.; Vajkoczy, P.; Picht, T. Navigated transcranial magnetic stimulation for mapping the motor cortex in patients with rolandic brain tumors. *Neurosur. Focus* **2013**, *34*, E3. [CrossRef] [PubMed]
77. Bulubas, L.; Sabih, J.; Wohlschlaeger, A.; Sollmann, N.; Hauck, T.; Ille, S.; Ringel, F.; Meyer, B.; Krieg, S.M. Motor areas of the frontal cortex in patients with motor eloquent brain lesions. *J. Neurosurg.* **2016**, 1–12. [CrossRef]
78. Bulubas, L.; Sollmann, N.; Tanigawa, N.; Zimmer, C.; Meyer, B.; Krieg, S.M. Reorganization of motor representations in patients with brain lesions: A navigated transcranial magnetic stimulation study. *Brain Topogr.* **2018**, *31*, 288–299. [CrossRef] [PubMed]
79. Barz, A.; Noack, A.; Baumgarten, P.; Seifert, V.; Forster, M.T. Motor cortex reorganization in patients with glioma assessed by repeated navigated transcranial magnetic stimulation- A longitudinal study. *World Neurosurg.* **2018**, *112*, e442–e453. [CrossRef] [PubMed]
80. Conway, N.; Wildschuetz, N.; Moser, T.; Bulubas, L.; Sollmann, N.; Tanigawa, N.; Meyer, B.; Krieg, S.M. Cortical plasticity of motor-eloquent areas measured by navigated transcranial magnetic stimulation in patients with glioma. *J. Neurosurg.* **2017**, 1–11. [CrossRef]

81. Sollmann, N.; Laub, T.; Kelm, A.; Albers, L.; Kirschke, J.S.; Combs, S.E.; Meyer, B.; Krieg, S.M. Predicting brain tumor regrowth in relation to motor areas by functional brain mapping. *Neurooncol. Pract.* **2018**, *5*, 82–95. [CrossRef] [PubMed]
82. Sollmann, N.; Meyer, B.; Krieg, S.M. Implementing functional preoperative mapping in the clinical routine of a neurosurgical department: Technical note. *World Neurosurg.* **2017**, *103*, 94–105. [CrossRef]
83. Conti, A.; Pontoriero, A.; Ricciardi, G.K.; Granata, F.; Vinci, S.; Angileri, F.F.; Pergolizzi, S.; Alafaci, C.; Rizzo, V.; Quartarone, A.; et al. Integration of functional neuroimaging in CyberKnife radiosurgery: Feasibility and dosimetric results. *Neurosur. Focus* **2013**, *34*, E5. [CrossRef] [PubMed]
84. Diehl, C.D.; Schwendner, M.J.; Sollmann, N.; Oechsner, M.; Meyer, B.; Combs, S.E.; Krieg, S.M. Application of presurgical navigated transcranial magnetic stimulation motor mapping for adjuvant radiotherapy planning in patients with high-grade gliomas. *Radiother. Oncol.* **2019**, *138*, 30–37. [CrossRef]
85. Schwendner, M.J.; Sollmann, N.; Diehl, C.D.; Oechsner, M.; Meyer, B.; Krieg, S.M.; Combs, S.E. The role of navigated transcranial magnetic stimulation motor mapping in adjuvant radiotherapy planning in patients with supratentorial brain metastases. *Front. Oncol.* **2018**, *8*, 424. [CrossRef]
86. Picht, T.; Schilt, S.; Frey, D.; Vajkoczy, P.; Kufeld, M. Integration of navigated brain stimulation data into radiosurgical planning: Potential benefits and dangers. *Acta Neurochir.* **2014**, *156*, 1125–1133. [CrossRef]
87. Tokarev, A.S.; Rak, V.A.; Sinkin, M.V.; Evdokimova, O.L.; Stepanov, V.N.; Koynash, G.V.; Krieg, S.M.; Krylov, V.V. Appliance of navigated transcranial magnetic stimulation in radiosurgery for brain metastases. *J. Clin. Neurophysiol.* **2020**, *37*, 50–55. [CrossRef]
88. Hendrix, P.; Senger, S.; Griessenauer, C.J.; Simgen, A.; Linsler, S.; Oertel, J. Preoperative navigated transcranial magnetic stimulation and tractography in transparietal approach to the trigone of the lateral ventricle. *J. Clin. Neurosci.* **2017**, *41*, 154–161. [CrossRef] [PubMed]
89. Hendrix, P.; Senger, S.; Griessenauer, C.J.; Simgen, A.; Linsler, S.; Oertel, J. Preoperative navigated transcranial magnetic stimulation and tractography to guide endoscopic cystoventriculostomy: A technical note and case report. *World Neurosurg.* **2018**, *109*, 209–217. [CrossRef] [PubMed]
90. Bartek, J., Jr.; Cooray, G.; Islam, M.; Jensdottir, M. Stereotactic brain biopsy in eloquent areas assisted by navigated transcranial magnetic stimulation: A technical case report. *Oper. Neurosurg.* **2019**, *17*, E124–E129. [CrossRef]
91. Schramm, S.; Haddad, A.F.; Chyall, L.; Krieg, S.M.; Sollmann, N.; Tarapore, P.E. Navigated TMS in the ICU: Introducing motor mapping to the critical care setting. *Brain Sci.* **2020**, *10*, 1005. [CrossRef] [PubMed]
92. Ferreira Pinto, P.; Nigri, F.; Caparelli-Daquer, E.M.; Viana, J.D.S. Computed tomography-guided navigated transcranial magnetic stimulation for preoperative brain motor mapping in brain lesion resection: A case report. *Surg. Neurol. Int.* **2019**, *10*, 134. [CrossRef]
93. Krieg, S.M.; Lioumis, P.; Makela, J.P.; Wilenius, J.; Karhu, J.; Hannula, H.; Savolainen, P.; Lucas, C.W.; Seidel, K.; Laakso, A.; et al. Protocol for motor and language mapping by navigated TMS in patients and healthy volunteers; workshop report. *Acta Neurochir.* **2017**, *159*, 1187–1195. [CrossRef] [PubMed]
94. Te, M.; Baptista, A.F.; Chipchase, L.S.; Schabrun, S.M. Primary motor cortex organization is altered in persistent patellofemoral pain. *Pain Med.* **2017**, *18*, 2224–2234. [CrossRef]
95. Saisanen, L.; Kononen, M.; Niskanen, E.; Lakka, T.; Lintu, N.; Vanninen, R.; Julkunen, P.; Maatta, S. Primary hand motor representation areas in healthy children, preadolescents, adolescents, and adults. *Neuroimage* **2021**, *228*. [CrossRef] [PubMed]
96. Novikov, P.A.; Nazarova, M.A.; Nikulin, V.V. TMSmap—Software for quantitative analysis of TMS mapping results. *Front. Hum. Neurosci.* **2018**, *12*, 239. [CrossRef] [PubMed]
97. Melgari, J.M.; Pasqualetti, P.; Pauri, F.; Rossini, P.M. Muscles in "concert": Study of primary motor cortex upper limb functional topography. *PLoS ONE* **2008**, *3*, e3069. [CrossRef] [PubMed]
98. Nazarova, M.; Novikov, P.; Ivanina, E.; Kozlova, K.; Dobrynina, L.; Nikulin, V.V. Mapping of multiple muscles with transcranial magnetic stimulation: Absolute and relative test-retest reliability. *Hum. Brain Mapp.* **2021**, *42*, 2508–2528. [CrossRef] [PubMed]
99. Davies, J.L. Using transcranial magnetic stimulation to map the cortical representation of lower-limb muscles. *Clin. Neurophysiol. Pract.* **2020**, *5*, 87–99. [CrossRef]
100. Castrillon, G.; Sollmann, N.; Kurcyus, K.; Razi, A.; Krieg, S.M.; Riedl, V. The physiological effects of noninvasive brain stimulation fundamentally differ across the human cortex. *Sci. Adv.* **2020**, *6*, eaay2739. [CrossRef]
101. Zhang, H.; Sollmann, N.; Castrillon, G.; Kurcyus, K.; Meyer, B.; Zimmer, C.; Krieg, S.M. Intranetwork and internetwork effects of navigated transcranial magnetic stimulation using low- and high-frequency pulse application to the dorsolateral prefrontal cortex: A combined rTMS-fMRI approach. *J. Clin. Neurophysiol.* **2020**, *37*, 131–139. [CrossRef] [PubMed]
102. Tsao, H.; Galea, M.P.; Hodges, P.W. Reorganization of the motor cortex is associated with postural control deficits in recurrent low back pain. *Brain* **2008**, *131*, 2161–2171. [CrossRef] [PubMed]
103. Groppa, S.; Oliviero, A.; Eisen, A.; Quartarone, A.; Cohen, L.G.; Mall, V.; Kaelin-Lang, A.; Mima, T.; Rossi, S.; Thickbroom, G.W.; et al. A practical guide to diagnostic transcranial magnetic stimulation: Report of an IFCN committee. *Clin. Neurophysiol.* **2012**, *123*, 858–882. [CrossRef] [PubMed]
104. Richter, L.; Neumann, G.; Oung, S.; Schweikard, A.; Trillenberg, P. Optimal coil orientation for transcranial magnetic stimulation. *PLoS ONE* **2013**, *8*, e60358. [CrossRef] [PubMed]

105. Reijonen, J.; Saisanen, L.; Kononen, M.; Mohammadi, A.; Julkunen, P. The effect of coil placement and orientation on the assessment of focal excitability in motor mapping with navigated transcranial magnetic stimulation. *J. Neurosci. Methods* **2020**, *331*, 108521. [CrossRef]
106. Saisanen, L.; Julkunen, P.; Kemppainen, S.; Danner, N.; Immonen, A.; Mervaala, E.; Maatta, S.; Muraja-Murro, A.; Kononen, M. Locating and outlining the cortical motor representation areas of facial muscles with navigated transcranial magnetic stimulation. *Neurosurgery* **2015**, *77*, 394–405. [CrossRef]
107. Thordstein, M.; Saar, K.; Pegenius, G.; Elam, M. Individual effects of varying stimulation intensity and response criteria on area of activation for different muscles in humans. A study using navigated transcranial magnetic stimulation. *Brain Stimul.* **2013**, *6*, 49–53. [CrossRef]
108. Mills, K.R.; Nithi, K.A. Corticomotor threshold to magnetic stimulation: Normal values and repeatability. *Muscle Nerve* **1997**, *20*, 570–576. [CrossRef]
109. Awiszus, F. TMS and threshold hunting. *Suppl. Clin. Neurophysiol.* **2003**, *56*, 13–23.
110. Julkunen, P. Mobile application for adaptive threshold hunting in transcranial magnetic stimulation. *IEEE Trans. Neural. Syst. Rehabil. Eng.* **2019**, *27*, 1504–1510. [CrossRef]
111. Engelhardt, M.; Schneider, H.; Gast, T.; Picht, T. Estimation of the resting motor threshold (RMT) in transcranial magnetic stimulation using relative-frequency and threshold-hunting methods in brain tumor patients. *Acta Neurochir.* **2019**, *161*, 1845–1851. [CrossRef]
112. Tranulis, C.; Gueguen, B.; Pham-Scottez, A.; Vacheron, M.N.; Cabelguen, G.; Costantini, A.; Valero, G.; Galinovski, A. Motor threshold in transcranial magnetic stimulation: Comparison of three estimation methods. *Neurophysiol. Clin.* **2006**, *36*, 1–7. [CrossRef]
113. Dissanayaka, T.; Zoghi, M.; Farrell, M.; Egan, G.; Jaberzadeh, S. Comparison of Rossini-Rothwell and adaptive threshold-hunting methods on the stability of TMS induced motor evoked potentials amplitudes. *J. Neurosci. Res.* **2018**, *96*, 1758–1765. [CrossRef] [PubMed]
114. Ah Sen, C.B.; Fassett, H.J.; El-Sayes, J.; Turco, C.V.; Hameer, M.M.; Nelson, A.J. Active and resting motor threshold are efficiently obtained with adaptive threshold hunting. *PLoS ONE* **2017**, *12*, e0186007. [CrossRef] [PubMed]
115. Silbert, B.I.; Patterson, H.I.; Pevcic, D.D.; Windnagel, K.A.; Thickbroom, G.W. A comparison of relative-frequency and threshold-hunting methods to determine stimulus intensity in transcranial magnetic stimulation. *Clin. Neurophysiol.* **2013**, *124*, 708–712. [CrossRef] [PubMed]
116. Saisanen, L.; Julkunen, P.; Lakka, T.; Lindi, V.; Kononen, M.; Maatta, S. Development of corticospinal motor excitability and cortical silent period from mid-childhood to adulthood—A navigated TMS study. *Neurophysiol. Clin.* **2018**, *48*, 65–75. [CrossRef] [PubMed]
117. Pitcher, J.B.; Doeltgen, S.H.; Goldsworthy, M.R.; Schneider, L.A.; Vallence, A.M.; Smith, A.E.; Semmler, J.G.; McDonnell, M.N.; Ridding, M.C. A comparison of two methods for estimating 50% of the maximal motor evoked potential. *Clin. Neurophysiol.* **2015**, *126*, 2337–2341. [CrossRef]
118. Malcolm, M.P.; Triggs, W.J.; Light, K.E.; Shechtman, O.; Khandekar, G.; Gonzalez Rothi, L.J. Reliability of motor cortex transcranial magnetic stimulation in four muscle representations. *Clin. Neurophysiol.* **2006**, *117*, 1037–1046. [CrossRef]
119. Saisanen, L.; Kononen, M.; Julkunen, P.; Maatta, S.; Vanninen, R.; Immonen, A.; Jutila, L.; Kalviainen, R.; Jaaskelainen, J.E.; Mervaala, E. Non-invasive preoperative localization of primary motor cortex in epilepsy surgery by navigated transcranial magnetic stimulation. *Epilepsy Res.* **2010**, *92*, 134–144. [CrossRef]
120. Schieber, M.H. Constraints on somatotopic organization in the primary motor cortex. *J. Neurophysiol.* **2001**, *86*, 2125–2143. [CrossRef]
121. Thickbroom, G.W.; Sammut, R.; Mastaglia, F.L. Magnetic stimulation mapping of motor cortex: Factors contributing to map area. *Electroencephalogr. Clin. Neurophysiol.* **1998**, *109*, 79–84. [CrossRef]
122. Thielscher, A.; Kammer, T. Electric field properties of two commercial figure-8 coils in TMS: Calculation of focality and efficiency. *Clin. Neurophysiol.* **2004**, *115*, 1697–1708. [CrossRef] [PubMed]
123. Sommer, M.; Alfaro, A.; Rummel, M.; Speck, S.; Lang, N.; Tings, T.; Paulus, W. Half sine, monophasic and biphasic transcranial magnetic stimulation of the human motor cortex. *Clin. Neurophysiol.* **2006**, *117*, 838–844. [CrossRef]
124. Hannah, R.; Rothwell, J.C. Pulse duration as well as current direction determines the specificity of transcranial magnetic stimulation of motor cortex during contraction. *Brain Stimul.* **2017**, *10*, 106–115. [CrossRef]
125. Saisanen, L.; Julkunen, P.; Niskanen, E.; Danner, N.; Hukkanen, T.; Lohioja, T.; Nurkkala, J.; Mervaala, E.; Karhu, J.; Kononen, M. Motor potentials evoked by navigated transcranial magnetic stimulation in healthy subjects. *J. Clin. Neurophysiol.* **2008**, *25*, 367–372. [CrossRef] [PubMed]
126. Danner, N.; Julkunen, P.; Kononen, M.; Saisanen, L.; Nurkkala, J.; Karhu, J. Navigated transcranial magnetic stimulation and computed electric field strength reduce stimulator-dependent differences in the motor threshold. *J. Neurosci. Methods* **2008**, *174*, 116–122. [CrossRef]
127. Nieminen, J.O.; Koponen, L.M.; Ilmoniemi, R.J. Experimental characterization of the electric field distribution induced by TMS devices. *Brain Stimul.* **2015**, *8*, 582–589. [CrossRef]
128. Zacharias, L.R.; Peres, A.S.C.; Souza, V.H.; Conforto, A.B.; Baffa, O. Method to assess the mismatch between the measured and nominal parameters of transcranial magnetic stimulation devices. *J. Neurosci. Methods* **2019**, *322*, 83–87. [CrossRef] [PubMed]

129. McConnell, K.A.; Nahas, Z.; Shastri, A.; Lorberbaum, J.P.; Kozel, F.A.; Bohning, D.E.; George, M.S. The transcranial magnetic stimulation motor threshold depends on the distance from coil to underlying cortex: A replication in healthy adults comparing two methods of assessing the distance to cortex. *Biol. Psychiatry* **2001**, *49*, 454–459. [CrossRef]
130. Stokes, M.G.; Chambers, C.D.; Gould, I.C.; Henderson, T.R.; Janko, N.E.; Allen, N.B.; Mattingley, J.B. Simple metric for scaling motor threshold based on scalp-cortex distance: Application to studies using transcranial magnetic stimulation. *J. Neurophysiol.* **2005**, *94*, 4520–4527. [CrossRef] [PubMed]
131. Stokes, M.G.; Chambers, C.D.; Gould, I.C.; English, T.; McNaught, E.; McDonald, O.; Mattingley, J.B. Distance-adjusted motor threshold for transcranial magnetic stimulation. *Clin. Neurophysiol.* **2007**, *118*, 1617–1625. [CrossRef]
132. Cai, W.; George, J.S.; Chambers, C.D.; Stokes, M.G.; Verbruggen, F.; Aron, A.R. Stimulating deep cortical structures with the batwing coil: How to determine the intensity for transcranial magnetic stimulation using coil-cortex distance. *J. Neurosci. Methods* **2012**, *204*, 238–241. [CrossRef] [PubMed]
133. Herbsman, T.; Forster, L.; Molnar, C.; Dougherty, R.; Christie, D.; Koola, J.; Ramsey, D.; Morgan, P.S.; Bohning, D.E.; George, M.S.; et al. Motor threshold in transcranial magnetic stimulation: The impact of white matter fiber orientation and skull-to-cortex distance. *Hum. Brain Mapp.* **2009**, *30*, 2044–2055. [CrossRef] [PubMed]
134. Rosanova, M.; Casali, A.; Bellina, V.; Resta, F.; Mariotti, M.; Massimini, M. Natural frequencies of human corticothalamic circuits. *J. Neurosci.* **2009**, *29*, 7679–7685. [CrossRef] [PubMed]
135. Janssen, A.M.; Oostendorp, T.F.; Stegeman, D.F. The effect of local anatomy on the electric field induced by TMS: Evaluation at 14 different target sites. *Med. Biol. Eng. Comput.* **2014**, *52*, 873–883. [CrossRef]
136. Julkunen, P.; Saisanen, L.; Danner, N.; Awiszus, F.; Kononen, M. Within-subject effect of coil-to-cortex distance on cortical electric field threshold and motor evoked potentials in transcranial magnetic stimulation. *J. Neurosci. Methods* **2012**, *206*, 158–164. [CrossRef]
137. Danner, N.; Kononen, M.; Saisanen, L.; Laitinen, R.; Mervaala, E.; Julkunen, P. Effect of individual anatomy on resting motor threshold-computed electric field as a measure of cortical excitability. *J. Neurosci. Methods* **2012**, *203*, 298–304. [CrossRef] [PubMed]
138. Laakso, I.; Hirata, A.; Ugawa, Y. Effects of coil orientation on the electric field induced by TMS over the hand motor area. *Phys. Med. Biol.* **2014**, *59*, 203–218. [CrossRef]
139. Nazarova, M.; Kulikova, S.; Piradov, M.A.; Limonova, A.S.; Dobrynina, L.A.; Konovalov, R.N.; Novikov, P.A.; Sehm, B.; Villringer, A.; Saltykova, A.; et al. Multimodal assessment of the motor system in patients with chronic ischemic stroke. *Stroke* **2021**, *52*, 241–249. [CrossRef]
140. Schabrun, S.M.; Ridding, M.C. The influence of correlated afferent input on motor cortical representations in humans. *Exp. Brain Res.* **2007**, *183*, 41–49. [CrossRef]
141. Ziemann, U.; Ilic, T.V.; Alle, H.; Meintzschel, F. Cortico-motoneuronal excitation of three hand muscles determined by a novel penta-stimulation technique. *Brain* **2004**, *127*, 1887–1898. [CrossRef]
142. Knecht, S.; Sommer, J.; Deppe, M.; Steinstrater, O. Scalp position and efficacy of transcranial magnetic stimulation. *Clin. Neurophysiol* **2005**, *116*, 1988–1993. [CrossRef] [PubMed]
143. Davis, N.J. Variance in cortical depth across the brain surface: Implications for transcranial stimulation of the brain. *Eur. J. Neurosci.* **2021**, *53*, 996–1007. [CrossRef] [PubMed]
144. Trillenberg, P.; Bremer, S.; Oung, S.; Erdmann, C.; Schweikard, A.; Richter, L. Variation of stimulation intensity in transcranial magnetic stimulation with depth. *J. Neurosci. Methods* **2012**, *211*, 185–190. [CrossRef]
145. Foltys, H.; Krings, T.; Meister, I.G.; Sparing, R.; Boroojerdi, B.; Thron, A.; Topper, R. Motor representation in patients rapidly recovering after stroke: A functional magnetic resonance imaging and transcranial magnetic stimulation study. *Clin. Neurophysiol.* **2003**, *114*, 2404–2415. [CrossRef]
146. Schabrun, S.M.; Hodges, P.W.; Vicenzino, B.; Jones, E.; Chipchase, L.S. Novel adaptations in motor cortical maps: The relation to persistent elbow pain. *Med. Sci. Sports Exerc.* **2015**, *47*, 681–690. [CrossRef]
147. Nicolini, C.; Harasym, D.; Turco, C.V.; Nelson, A.J. Human motor cortical organization is influenced by handedness. *Cortex* **2019**, *115*, 172–183. [CrossRef]
148. Pitkanen, M.; Kallioniemi, E.; Julkunen, P. Extent and Location of the excitatory and inhibitory cortical hand representation maps: A navigated transcranial magnetic stimulation study. *Brain Topogr.* **2015**, *28*, 657–665. [CrossRef]
149. Uy, J.; Ridding, M.C.; Miles, T.S. Stability of maps of human motor cortex made with transcranial magnetic stimulation. *Brain Topogr.* **2002**, *14*, 293–297. [CrossRef] [PubMed]
150. Marconi, B.; Pecchioli, C.; Koch, G.; Caltagirone, C. Functional overlap between hand and forearm motor cortical representations during motor cognitive tasks. *Clin. Neurophysiol.* **2007**, *118*, 1767–1775. [CrossRef]
151. Kleim, J.A.; Kleim, E.D.; Cramer, S.C. Systematic assessment of training-induced changes in corticospinal output to hand using frameless stereotaxic transcranial magnetic stimulation. *Nat. Protoc.* **2007**, *2*, 1675–1684. [CrossRef]
152. Schabrun, S.M.; Stinear, C.M.; Byblow, W.D.; Ridding, M.C. Normalizing motor cortex representations in focal hand dystonia. *Cereb. Cortex* **2009**, *19*, 1968–1977. [CrossRef] [PubMed]
153. Borghetti, D.; Sartucci, F.; Petacchi, E.; Guzzetta, A.; Piras, M.F.; Murri, L.; Cioni, G. Transcranial magnetic stimulation mapping: A model based on spline interpolation. *Brain Res. Bull.* **2008**, *77*, 143–148. [CrossRef]
154. Cavaleri, R.; Schabrun, S.M.; Chipchase, L.S. Determining the optimal number of stimuli per cranial site during transcranial magnetic stimulation mapping. *Neurosci. J.* **2017**, *2017*, 6328569. [CrossRef]

155. Chernyavskiy, A.Y.; Sinitsyn, D.O.; Poydasheva, A.G.; Bakulin, I.S.; Suponeva, N.A.; Piradov, M.A. Accuracy of Estimating the Area of Cortical Muscle Representations from TMS Mapping Data Using Voronoi Diagrams. *Brain Topogr.* **2019**, *32*, 859–872. [CrossRef]
156. Goldsworthy, M.R.; Hordacre, B.; Ridding, M.C. Minimum number of trials required for within- and between-session reliability of TMS measures of corticospinal excitability. *Neuroscience* **2016**, *320*, 205–209. [CrossRef]
157. Chang, W.H.; Fried, P.J.; Saxena, S.; Jannati, A.; Gomes-Osman, J.; Kim, Y.H.; Pascual-Leone, A. Optimal number of pulses as outcome measures of neuronavigated transcranial magnetic stimulation. *Clin. Neurophysiol.* **2016**, *127*, 2892–2897. [CrossRef]
158. Schmidt, S.; Bathe-Peters, R.; Fleischmann, R.; Ronnefarth, M.; Scholz, M.; Brandt, S.A. Nonphysiological factors in navigated TMS studies; confounding covariates and valid intracortical estimates. *Hum. Brain Mapp.* **2015**, *36*, 40–49. [CrossRef] [PubMed]
159. De Santo, G.; Tanner, P.; Witt, T.N.; Gebrecht, V.; Bötzel, K. Investigation of coil tilt by transcranial magnetic stimulation (TMS) of the motor cortex using a navigation system. *Klin. Neurophysiol.* **2006**, *37*, A38. [CrossRef]
160. Raffin, E.; Pellegrino, G.; Di Lazzaro, V.; Thielscher, A.; Siebner, H.R. Bringing transcranial mapping into shape: Sulcus-aligned mapping captures motor somatotopy in human primary motor hand area. *Neuroimage* **2015**, *120*, 164–175. [CrossRef] [PubMed]
161. Kallioniemi, E.; Kononen, M.; Julkunen, P. Repeatability of functional anisotropy in navigated transcranial magnetic stimulation–coil-orientation versus response. *Neuroreport* **2015**, *26*, 515–521. [CrossRef]
162. Kallioniemi, E.; Kononen, M.; Saisanen, L.; Grohn, H.; Julkunen, P. Functional neuronal anisotropy assessed with neuronavigated transcranial magnetic stimulation. *J. Neurosci. Methods* **2015**, *256*, 82–90. [CrossRef] [PubMed]
163. Balslev, D.; Braet, W.; McAllister, C.; Miall, R.C. Inter-individual variability in optimal current direction for transcranial magnetic stimulation of the motor cortex. *J. Neurosci. Methods* **2007**, *162*, 309–313. [CrossRef] [PubMed]
164. Gomez-Tames, J.; Hamasaka, A.; Laakso, I.; Hirata, A.; Ugawa, Y. Atlas of optimal coil orientation and position for TMS: A computational study. *Brain Stimul.* **2018**, *11*, 839–848. [CrossRef]
165. Opitz, A.; Windhoff, M.; Heidemann, R.M.; Turner, R.; Thielscher, A. How the brain tissue shapes the electric field induced by transcranial magnetic stimulation. *Neuroimage* **2011**, *58*, 849–859. [CrossRef]
166. Niskanen, E.; Julkunen, P.; Saisanen, L.; Vanninen, R.; Karjalainen, P.; Kononen, M. Group-level variations in motor representation areas of thenar and anterior tibial muscles: Navigated Transcranial Magnetic Stimulation Study. *Hum. Brain Mapp.* **2010**, *31*, 1272–1280. [CrossRef] [PubMed]
167. Ahdab, R.; Ayache, S.S.; Brugieres, P.; Farhat, W.H.; Lefaucheur, J.P. The hand motor hotspot is not always located in the hand knob: A neuronavigated transcranial magnetic stimulation study. *Brain Topogr.* **2016**, *29*, 590–597. [CrossRef] [PubMed]
168. Julkunen, P.; Saisanen, L.; Danner, N.; Niskanen, E.; Hukkanen, T.; Mervaala, E.; Kononen, M. Comparison of navigated and non-navigated transcranial magnetic stimulation for motor cortex mapping, motor threshold and motor evoked potentials. *Neuroimage* **2009**, *44*, 790–795. [CrossRef]
169. Rossini, P.M.; Burke, D.; Chen, R.; Cohen, L.G.; Daskalakis, Z.; Di Iorio, R.; Di Lazzaro, V.; Ferreri, F.; Fitzgerald, P.B.; George, M.S.; et al. Non-invasive electrical and magnetic stimulation of the brain, spinal cord, roots and peripheral nerves: Basic principles and procedures for routine clinical and research application. An updated report from an I.F.C.N. Committee. *Clin. Neurophysiol.* **2015**, *126*, 1071–1107. [CrossRef]
170. Sollmann, N.; Hauck, T.; Obermuller, T.; Hapfelmeier, A.; Meyer, B.; Ringel, F.; Krieg, S.M. Inter- and intraobserver variability in motor mapping of the hotspot for the abductor policis brevis muscle. *BMC Neurosci.* **2013**, *14*, 1–7. [CrossRef]
171. Wassermann, E.M.; McShane, L.M.; Hallett, M.; Cohen, L.G. Noninvasive mapping of muscle representations in human motor cortex. *Electroencephalogr. Clin. Neurophysiol.* **1992**, *85*, 1–8. [CrossRef]
172. Saisanen, L.; Maatta, S.; Julkunen, P.; Niskanen, E.; Kallioniemi, E.; Grohn, H.; Kemppainen, S.; Lakka, T.A.; Lintu, N.; Eloranta, A.M.; et al. Functional and structural asymmetry in primary motor cortex in Asperger syndrome: A navigated TMS and imaging study. *Brain Topogr.* **2019**, *32*, 504–518. [CrossRef]
173. Jonker, Z.D.; van der Vliet, R.; Hauwert, C.M.; Gaiser, C.; Tulen, J.H.M.; van der Geest, J.N.; Donchin, O.; Ribbers, G.M.; Frens, M.A.; Selles, R.W. TMS motor mapping: Comparing the absolute reliability of digital reconstruction methods to the golden standard. *Brain Stimul.* **2019**, *12*, 309–313. [CrossRef]
174. Sollmann, N.; Zhang, H.; Kelm, A.; Schroder, A.; Meyer, B.; Pitkanen, M.; Julkunen, P.; Krieg, S.M. Paired-pulse navigated TMS is more effective than single-pulse navigated TMS for mapping upper extremity muscles in brain tumor patients. *Clinical Neuropsychol.* **2020**, *131*, 2887–2898. [CrossRef]
175. Zhang, H.; Julkunen, P.; Schroder, A.; Kelm, A.; Ille, S.; Zimmer, C.; Pitkanen, M.; Meyer, B.; Krieg, S.M.; Sollmann, N. Short-interval intracortical facilitation improves efficacy in nTMS motor mapping of lower extremity muscle representations in patients with supra-tentorial brain tumors. *Cancers* **2020**, *12*, 3233. [CrossRef] [PubMed]
176. Jussen, D.; Zdunczyk, A.; Schmidt, S.; Rosler, J.; Buchert, R.; Julkunen, P.; Karhu, J.; Brandt, S.; Picht, T.; Vajkoczy, P. Motor plasticity after extra-intracranial bypass surgery in occlusive cerebrovascular disease. *Neurology* **2016**, *87*, 27–35. [CrossRef] [PubMed]
177. Ille, S.; Kelm, A.; Schroeder, A.; Albers, L.E.; Negwer, C.; Butenschoen, V.M.; Sollmann, N.; Picht, T.; Vajkoczy, P.; Meyer, B.; et al. Navigated repetitive transcranial magnetic stimulation improves the outcome of postsurgical paresis in glioma patients—A randomized, double-blinded trial. *Brain Stimul.* **2021**. [CrossRef]

178. Sollmann, N.; Zhang, H.; Fratini, A.; Wildschuetz, N.; Ille, S.; Schroder, A.; Zimmer, C.; Meyer, B.; Krieg, S.M. Risk assessment by presurgical tractography using navigated TMS maps in patients with highly motor- or language-eloquent brain tumors. *Cancers* **2020**, *12*, 1264. [CrossRef]
179. Sollmann, N.; Giglhuber, K.; Tussis, L.; Meyer, B.; Ringel, F.; Krieg, S.M. nTMS-based DTI fiber tracking for language pathways correlates with language function and aphasia—A case report. *Clin. Neurol. Neurosurg.* **2015**, *136*, 25–28. [CrossRef] [PubMed]
180. Holodny, A.I.; Schulder, M.; Liu, W.C.; Maldjian, J.A.; Kalnin, A.J. Decreased BOLD functional MR activation of the motor and sensory cortices adjacent to a glioblastoma multiforme: Implications for image-guided neurosurgery. *AJNR Am. J. Neuroradiol.* **1999**, *20*, 609–612.
181. Holodny, A.I.; Schulder, M.; Liu, W.C.; Wolko, J.; Maldjian, J.A.; Kalnin, A.J. The effect of brain tumors on BOLD functional MR imaging activation in the adjacent motor cortex: Implications for image-guided neurosurgery. *AJNR Am. J. Neuroradiol.* **2000**, *21*, 1415–1422. [PubMed]
182. Gerhardt, J.; Sollmann, N.; Hiepe, P.; Kirschke, J.S.; Meyer, B.; Krieg, S.M.; Ringel, F. Retrospective distortion correction of diffusion tensor imaging data by semi-elastic image fusion—Evaluation by means of anatomical landmarks. *Clin. Neurol. Neurosurg.* **2019**, *183*, 105387. [CrossRef]
183. Ille, S.; Schwendner, M.; Zhang, W.; Schroeder, A.; Meyer, B.; Krieg, S.M. Tractography for Subcortical Resection of Gliomas Is Highly Accurate for Motor and Language Function: ioMRI-Based Elastic Fusion Disproves the Severity of Brain Shift. *Cancers* **2021**, *13*, 1787. [CrossRef] [PubMed]
184. Ille, S.; Schroeder, A.; Wagner, A.; Negwer, C.; Kreiser, K.; Meyer, B.; Krieg, S.M. Intraoperative MRI-based elastic fusion for anatomically accurate tractography of the corticospinal tract: Correlation with intraoperative neuromonitoring and clinical status. *Neurosurg. Focus* **2021**, *50*, E9. [CrossRef] [PubMed]
185. Mori, S.; Crain, B.J.; Chacko, V.P.; van Zijl, P.C. Three-dimensional tracking of axonal projections in the brain by magnetic resonance imaging. *Ann. Neurol.* **1999**, *45*, 265–269. [CrossRef]
186. Mori, S.; van Zijl, P.C. Fiber tracking: Principles and strategies—A technical review. *NMR Biomed.* **2002**, *15*, 468–480. [CrossRef]
187. Lazar, M.; Alexander, A.L. An error analysis of white matter tractography methods: Synthetic diffusion tensor field simulations. *Neuroimage* **2003**, *20*, 1140–1153. [CrossRef]
188. Anderson, A.W. Theoretical analysis of the effects of noise on diffusion tensor imaging. *Magn. Reson. Med. Off. J. Soc. Magn. Reson. Med.* **2001**, *46*, 1174–1188. [CrossRef] [PubMed]
189. Mukherjee, P.; Berman, J.I.; Chung, S.W.; Hess, C.P.; Henry, R.G. Diffusion tensor MR imaging and fiber tractography: Theoretic underpinnings. *AJNR Am. J. Neuroradiol.* **2008**, *29*, 632–641. [CrossRef]
190. Beaulieu, C. The basis of anisotropic water diffusion in the nervous system—A technical review. *NMR Biomed.* **2002**, *15*, 435–455. [CrossRef]
191. Michailovich, O.; Rathi, Y.; Dolui, S. Spatially regularized compressed sensing for high angular resolution diffusion imaging. *IEEE Trans. Med. Imaging* **2011**, *30*, 1100–1115. [CrossRef]
192. Rathi, Y.; Michailovich, O.; Setsompop, K.; Bouix, S.; Shenton, M.E.; Westin, C.F. Sparse multi-shell diffusion imaging. *Med. Image Comput. Comput. Assist. Interv.* **2011**, *14*, 58–65. [CrossRef]
193. Zhang, H.; Schneider, T.; Wheeler-Kingshott, C.A.; Alexander, D.C. NODDI: Practical in vivo neurite orientation dispersion and density imaging of the human brain. *Neuroimage* **2012**, *61*, 1000–1016. [CrossRef] [PubMed]
194. Julkunen, P.; Jarnefelt, G.; Savolainen, P.; Laine, J.; Karhu, J. Facilitatory effect of paired-pulse stimulation by transcranial magnetic stimulation with biphasic wave-form. *Med. Eng. Phys.* **2016**, *38*, 813–817. [CrossRef] [PubMed]
195. Pitkanen, M.; Kallioniemi, E.; Jarnefelt, G.; Karhu, J.; Julkunen, P. Efficient mapping of the motor cortex with navigated biphasic paired-pulse transcranial magnetic stimulation. *Brain Topogr.* **2018**, *31*, 963–971. [CrossRef]
196. Mohammadi, A.; Ebrahimi, M.; Kaartinen, S.; Jarnefelt, G.; Karhu, J.; Julkunen, P. Individual characterization of fast intracortical facilitation with paired biphasic-wave transcranial magnetic stimulation. *IEEE Trans. Neural Syst. Rehabil. Eng.* **2018**. [CrossRef] [PubMed]
197. Picht, T.; Krieg, S.M.; Sollmann, N.; Rosler, J.; Niraula, B.; Neuvonen, T.; Savolainen, P.; Lioumis, P.; Makela, J.P.; Deletis, V.; et al. A comparison of language mapping by preoperative navigated transcranial magnetic stimulation and direct cortical stimulation during awake surgery. *Neurosurgery* **2013**, *72*, 808–819. [CrossRef] [PubMed]
198. Tarapore, P.E.; Findlay, A.M.; Honma, S.M.; Mizuiri, D.; Houde, J.F.; Berger, M.S.; Nagarajan, S.S. Language mapping with navigated repetitive TMS: Proof of technique and validation. *Neuroimage* **2013**, *82*, 260–272. [CrossRef] [PubMed]
199. Sollmann, N.; Ille, S.; Hauck, T.; Maurer, S.; Negwer, C.; Zimmer, C.; Ringel, F.; Meyer, B.; Krieg, S.M. The impact of preoperative language mapping by repetitive navigated transcranial magnetic stimulation on the clinical course of brain tumor patients. *BMC Cancer* **2015**, *15*, 1–8. [CrossRef]
200. Sollmann, N.; Kubitscheck, A.; Maurer, S.; Ille, S.; Hauck, T.; Kirschke, J.S.; Ringel, F.; Meyer, B.; Krieg, S.M. Preoperative language mapping by repetitive navigated transcranial magnetic stimulation and diffusion tensor imaging fiber tracking and their comparison to intraoperative stimulation. *Neuroradiology* **2016**, *58*, 807–818. [CrossRef]
201. Babajani-Feremi, A.; Narayana, S.; Rezaie, R.; Choudhri, A.F.; Fulton, S.P.; Boop, F.A.; Wheless, J.W.; Papanicolaou, A.C. Language mapping using high gamma electrocorticography, fMRI, and TMS versus electrocortical stimulation. *Clin. Neurophysiol.* **2016**, *127*, 1822–1836. [CrossRef] [PubMed]

202. Laakso, I.; Murakami, T.; Hirata, A.; Ugawa, Y. Where and what TMS activates: Experiments and modeling. *Brain Stimul.* **2018**, *11*, 166–174. [CrossRef]
203. Weise, K.; Numssen, O.; Thielscher, A.; Hartwigsen, G.; Knosche, T.R. A novel approach to localize cortical TMS effects. *Neuroimage* **2020**, *209*, 116486. [CrossRef] [PubMed]
204. Pitkanen, M.; Kallioniemi, E.; Julkunen, P.; Nazarova, M.; Nieminen, J.O.; Ilmoniemi, R.J. Minimum-norm estimation of motor representations in navigated TMS mappings. *Brain Topogr.* **2017**, *30*, 711–722. [CrossRef] [PubMed]
205. Thielscher, A.; Antunes, A.; Saturnino, G.B. Field modeling for transcranial magnetic stimulation: A useful tool to understand the physiological effects of TMS? In Proceedings of the 2015 37th Annual International Conference of the IEEE Engineering in Medicine and Biology Society, Milan, Italy, 25–29 August 2015; pp. 222–225.
206. Saturnino, G.B.; Madsen, K.H.; Thielscher, A. Electric field simulations for transcranial brain stimulation using FEM: An efficient implementation and error analysis. *J. Neural. Eng.* **2019**, *16*, 066032. [CrossRef]

Article

Function-Based Tractography of the Language Network Correlates with Aphasia in Patients with Language-Eloquent Glioblastoma

Haosu Zhang [1], Severin Schramm [1], Axel Schröder [1], Claus Zimmer [2], Bernhard Meyer [1], Sandro M. Krieg [1,3] and Nico Sollmann [2,3,*]

[1] Department of Neurosurgery, Klinikum rechts der Isar, Technische Universität München, Ismaninger Str. 22, 81675 Munich, Germany; Haosu.Zhang@gmail.com (H.Z.); Schrammseverin@gmail.com (S.S.); Axel.Schroeder@tum.de (A.S.); Bernhard.Meyer@tum.de (B.M.); Sandro.Krieg@tum.de (S.M.K.)
[2] Department of Diagnostic and Interventional Neuroradiology, Klinikum rechts der Isar, Technische Universität München, Ismaninger Str. 22, 81675 Munich, Germany; Claus.Zimmer@tum.de
[3] TUM-Neuroimaging Center, Klinikum rechts der Isar, Technische Universität München, 81675 Munich, Germany
* Correspondence: nico.sollmann@tum.de

Received: 5 June 2020; Accepted: 23 June 2020; Published: 1 July 2020

Abstract: To date, the structural characteristics that distinguish language-involved from non-involved cortical areas are largely unclear. Particularly in patients suffering from language-eloquent brain tumors, reliable mapping of the cortico-subcortical language network is of high clinical importance to prepare and guide safe tumor resection. To investigate differences in structural characteristics between language-positive and language-negative areas, 20 patients (mean age: 63.2 ± 12.9 years, 16 males) diagnosed with language-eloquent left-hemispheric glioblastoma multiforme (GBM) underwent preoperative language mapping by navigated transcranial magnetic stimulation (nTMS) and nTMS-based diffusion tensor imaging fiber tracking (DTI FT). The number of language-positive and language-negative points as well as the gray matter intensity (GMI), normalized volumes of U-fibers, interhemispheric fibers, and fibers projecting to the cerebellum were assessed and compared between language-positive and language-negative nTMS mappings and set in correlation with aphasia grades. We found significantly lower GMI for language-positive nTMS points (5.7 ± 1.7 versus 7.1 ± 1.6, $p = 0.0121$). Furthermore, language-positive nTMS points were characterized by an enhanced connectivity profile, i.e., these points showed a significantly higher ratio in volumes for U-fibers ($p \leq 0.0056$), interhemispheric fibers ($p = 0.0494$), and fibers projecting to the cerebellum ($p = 0.0094$). The number of language-positive nTMS points ($R \geq 0.4854$, $p \leq 0.0300$) as well as the ratio in volumes for U-fibers ($R \leq -0.4899$, $p \leq 0.0283$) were significantly associated with aphasia grades, as assessed pre- or postoperatively and during follow-up examinations. In conclusion, this study provides evidence for structural differences on cortical and subcortical levels between language-positive and language-negative areas, as detected by nTMS language mapping. The results may further increase confidence in the technique of nTMS language mapping and nTMS-based tractography in the direct clinical setting. Future studies may confirm our results in larger cohorts and may expand the findings to patients with other tumor entities than GBM.

Keywords: brain stimulation; fiber tractography; glioblastoma multiforme; gray matter; language mapping; navigated transcranial magnetic stimulation

1. Introduction

Resection of intracranial glioma aims at a maximum extent of resection, which should ideally be achieved without causing surgery-related functional deficits that could severely reduce the patients' quality of life [1–4]. To establish maximum resection whilst avoiding functional decline as far as possible, several pre- and intraoperative techniques have been developed to assist in neurosurgical planning and resection guidance [4–7]. For the intraoperative setting, cortical and subcortical direct electrical stimulation (DES) serves as the current gold-standard method [8–10]. Regarding the preoperative setting, navigated transcranial magnetic stimulation (nTMS) has found its way into neurosurgery over the last decade [6,11,12].

The technique of nTMS has lately been used to conduct language mappings in patients suffering from language-eloquent glioma or other entities of brain tumors [13–16]. Furthermore, it has been combined with diffusion tensor imaging (DTI) derived from preoperative magnetic resonance imaging (MRI) to provide spatially resolved maps that visualize language-related structures [17–25]. Integration into clinical routine and the perioperative workflow is seamless, and the approach is currently regarded as a valuable adjunct to intraoperative DES [14,15]. However, in direct comparison to intraoperative DES, nTMS language mapping has shown a rather low specificity of 23.8% and a positive predictive value of 35.6% [13]. The mere additional use of nTMS-based DTI fiber tracking (DTI FT) did not improve the identification of DES-positive language areas during awake surgery [24]. Nevertheless, the visualization of the subcortical language network becomes possible purely based on functional data by using nTMS-based DTI FT, which has shown high potential for surgical planning, resection guidance, and risk assessment in patients with language-eloquent lesions [17–25]. However, explorations of the differences between language-positive and language-negative nTMS mappings are still largely missing, which is one reason contributing to the lack of understanding of the comparatively low specificity of nTMS language mapping in relation to intraoperative DES. Further insights may lead to a better definition of the role of nTMS in the neurosurgical setting and to improved understanding of nTMS characteristics.

The cortico-subcortical network behind human language function is complex [26–29]. On the cortical level, differences in gray matter (GM) distribution could probably differentiate between language-related and non-related areas or, at least, between highly and less involved areas. In this regard, previous research has demonstrated that the degree to which language is lateralized to one of the hemispheres is positively predicted by the degree to which GM is lateralized on a voxel-by-voxel basis [30]. In subjects with dyslexia, a GM deficit involving a fronto-temporal network important for phonological processing was revealed, whereas region-specific increases in GM volume are possible for developmental language disorders as a result of compensatory mechanisms [31,32]. On the subcortical level, differences in connectivity profiles between areas depending on language involvement seem likely. Besides major language-related white matter tracts known to be involved in language processing, various short or long interconnecting fibers, such as short association fibers (commonly referred to as arcuate fibers or U-fibers), interhemispheric transcallosal fibers, and fibers projecting to the cerebellum play a role [28,29,33,34].

Against this background, this study's objective is to systematically explore characteristics of language-positive structures according to nTMS language mapping and nTMS-based DTI FT in direct comparison to language-negative counterparts. We hypothesize that (1) language-positive cortical areas may show different focal GM intensity (GMI, as a potential expression of higher GM density as a correlate of increased functional involvement), and that (2) language-positive cortical areas show a different connectivity profile (higher volumes of U-fibers, transcallosal fibers, and fibers projecting to the cerebellum) when compared to language-negative cortical areas.

2. Materials and Methods

2.1. Ethics

The current study is in accordance with the Declaration of Helsinki and its later amendments and has been approved by the local institutional review board (registration numbers: 2793/10, 5811/13, 223/14, and 336/17). Written informed consent was obtained from all patients enrolled in this study.

2.2. Patients and Study Inclusion

This study is a post-hoc analysis including patients of our prospectively enrolled cohort that underwent language mapping by nTMS and nTMS-based DTI FT of language-related fiber tracts prior to resection of a brain tumor. The following inclusion criteria were defined for the present study:

(1) Written informed consent,
(2) Age above 18 years,
(3) German as first language,
(4) Left-hemispheric perisylvian tumor location (MRI suggesting infiltration and/or compression of anatomically suspected cortical language-eloquent areas and/or suspected close proximity to subcortical language-related pathways),
(5) Availability of preoperative 3-Tesla MRI, including a DTI sequence with 32 diffusion directions,
(6) Clinical indication for preoperative nTMS language mapping and nTMS-based DTI FT,
(7) Surgery for tumor resection and final diagnosis of a glioblastoma multiforme (GBM) according to histopathological examination (based on tumor tissue probes taken during resection), and
(8) Follow-up time of at least 3 months after surgery.

The exclusion criteria were defined as follows:

(1) Multilingual background (regular input in more than one language between birth and adolescence),
(2) Neurological or psychiatric diseases (except for the diagnosis of a GBM), and
(3) Aphasia to a degree not allowing for preoperative language mapping by nTMS.

2.3. Clinical Examination

A standardized assessment of sensory function, coordination, muscle strength, and cranial nerve function was performed as part of the initial clinical examination. In particular, the language status was evaluated by a neuropsychologist using the Aachen Aphasia Test and by categorizing language deficits on a four-point scale [14,35–37]. In detail, four grades were established, which were no deficit (grade 0), mild deficit (grade 1: normal language comprehension and/or conversational language with slight amnesic aphasia, adequate communication ability), medium deficit (grade 2: minor disruption of language comprehension and/or conversational language, adequate communication ability), and severe deficit (grade 3: major disruption of language comprehension and/or conversational language, clear impairment of communication ability) [14,35–37]. Furthermore, handedness was assessed by the Edinburgh Handedness Inventory (EHI) [38]. The clinical examinations including detailed assessments of language function were repeated postoperatively and during the routine follow-up examinations.

2.4. Cranial Magnetic Resonance Imaging

Imaging was performed on a 3-Tesla MRI scanner using a 32-channel head coil (Achieva dStream or Ingenia; Philips Healthcare, Best, The Netherlands). As part of a standardized, multi-sequence imaging protocol for brain tumors, a three-dimensional fluid attenuated inversion recovery (FLAIR) sequence (repetition time (TR)/echo time (TE): 4800/277 ms, 1 mm^3 isovoxel covering the whole head), three-dimensional T1-weighted gradient echo sequence (TR/TE: 9/4 ms, 1 mm^3 isovoxel covering the whole head) without and with application of an intravenous contrast agent (Dotagraf, Jenapharm GmbH & Co. KG, Jena, Germany), and a DTI sequence (TR/TE 5000/78 ms, voxel size of $2 \times 2 \times 2$ mm^3, 32 diffusion gradient directions) were acquired preoperatively and were used for nTMS language mapping (contrast-enhanced T1-weighted sequences) and nTMS-based DTI FT (contrast-enhanced T1-weighted, DTI, and FLAIR sequences).

Postoperative (within the first 48 h after surgery) and follow-up imaging were performed using the same imaging protocol. In postoperative MRI, special attention was paid to the assessment of residual tumor tissue or achieved gross total resection (GTR) as well as perioperative bleeding or ischemia.

2.5. Language Mapping by Navigated Transcranial Magnetic Stimulation

2.5.1. Mapping Procedure

Preoperative language mapping by nTMS of the tumor-affected left hemisphere (LH) was carried out during the days before scheduled surgery for tumor resection using a Nexstim eXimia NBS system (version 4.3; Nexstim Plc, Helsinki, Finland) [11,14,15,39,40]. Language mappings were conducted in German as the native language of all enrolled patients.

After initial co-registration of the patient's head and the contrast-enhanced T1-weighted image dataset, pictures of common objects were presented to the patient in the context of two baseline trials (presentation of objects without simultaneous stimulation) [13,40–44]. The objects shown were part of a standardized object-naming task, and the purpose of baseline assessment was to systematically discard objects that did not elicit a quick and fluent response and to familiarize the patients with the task and setup. The remaining stack of objects that were named correctly and fluently was then shown under stimulation of up to 46 target points that had been placed on the LH, with each point being stimulated six times in total (stimulation intensity: 100% of the individual resting motor threshold, stimulation frequency: 5 Hz/5 pulses) [13,40,42,43].

The 46 target points for stimulation were tagged with close respect to the cortical parcellation system (CPS), thus covering almost the whole LH except for polar, occipital, and inferior temporal regions, similar to previous studies on nTMS language mapping (Figure 1) [13,42,43]. Exclusion was due to considerable muscle activation and/or discomfort that can be observed following stimulation of these particular regions. Furthermore, in patients showing large tumor masses, the number of 46 target points for stimulation needed to be reduced according to individual cortical architecture given destructions and derangements caused by such lesions. The picture-to-trigger interval during stimulation was set as 0 ms by default [45,46]. The patients' performance during object naming was video-recorded for later evaluation to detect and categorize errors elicited by targeted nTMS [41,44].

2.5.2. Mapping Evaluation

Video data of the nTMS language mappings were systematically searched for naming errors of different categories after the mapping procedure. No responses, performance errors, hesitations, neologisms, phonological paraphasias, and semantic paraphasias were defined and considered during evaluation [13,41–44]. The stimulation points that elicited any naming error except hesitations (no responses, performance errors, neologisms, phonological paraphasias, and semantic paraphasias together) were considered as language-positive nTMS points. Correspondingly, stimulation points that did not elicit a naming error were defined as language-negative nTMS points.

Per patient, two export datasets were then generated, which included the language-positive nTMS points of the LH and the language-negative nTMS points of the LH, respectively. Each export file was in Digital Imaging and Communications in Medicine (DICOM) format and included the nose and ears as anatomical landmarks together with the stimulation points. Furthermore, in each patient, CPS regions were counted as language-positive CPS regions when they contained at least one language-positive nTMS point, while CPS regions without language-positive nTMS points were considered as language-negative CPS regions. CPS regions that were not stimulated (polar, occipital, and inferior temporal regions) were not considered. The total number of language-positive CPS regions was recorded as N_p, and the number of language-negative CPS regions as N_n.

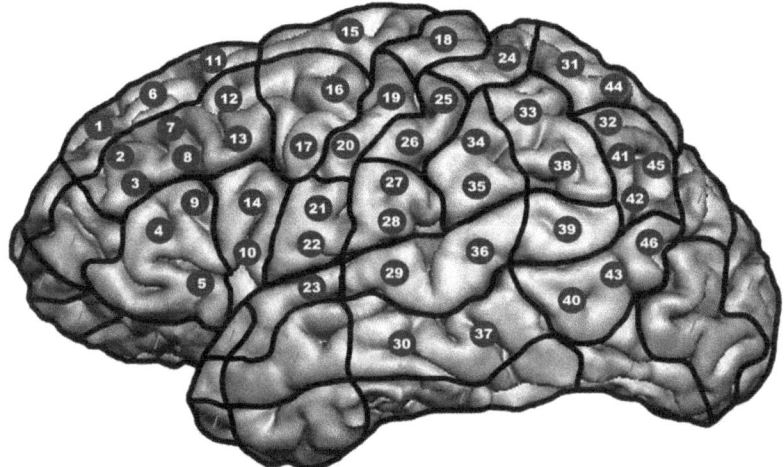

Figure 1. Cortical parcellation system (CPS). The tumor-affected left hemisphere (LH) was mapped by navigated transcranial magnetic stimulation (nTMS) with respect to 46 target points that were placed in relation to the CPS (reduced numbers of target points were accepted in patients with large tumor masses, which hampered placement of the total amount of 46 target points). The points were stimulated six times in total each, with a stimulation intensity of 100% of the individual resting motor threshold and a stimulation frequency of 5 Hz/5 pulses. Numbers in circles schematically represent the stimulation targets on a standardized brain template, red dashes mark regions that were not subject to nTMS.

2.6. Tractography Based on Navigated Transcranial Magnetic Stimulation

Deterministic fiber tractography was based on nTMS language mapping data without additional anatomical seeding (Brainlab Elements, version 3.1.0; Brainlab AG, Munich, Germany). Both export datasets were first transferred to an external server and then auto-fused with the MRI sequences of the respective patient, using manual correction in case of registration misalignments and eddy current correction for DTI data. The datasets containing language-positive or language-negative nTMS points were separately defined as three-dimensional objects, followed by generation of regions of interest (ROIs) out of these objects by adding a rim of 5 mm to each stimulation point [14,15,17–19].

An individual fractional anisotropy (FA) value, the fractional anisotropy threshold (FAT), was defined separately in each patient for nTMS-based DTI FT using the ROI constituted of language-positive or language-negative nTMS points, which was determined by setting angulation to 90°, the minimum fiber length (FL) to 30 mm, and increasing the FA stepwise until no fibers were displayed, followed by decreasing the FA by 0.01, thus visualizing a minimum fiber course. The corresponding FA value was defined as 100% FAT. A similar approach using a higher minimum FL for FAT determination has been used previously for nTMS-based DTI FT [17,19,47].

After FAT definition, nTMS-based DTI FT was carried out separately with the ROIs of language-positive and language-negative nTMS points using the respective FA values for 100% FAT, 75% FAT, 50% FAT, and 25% FAT, and a minimum FL of 30 mm as well as 3 mm, respectively. The values of 30 and 3 mm were considered for the minimum FL due to the characteristic length of U-fibers that have been described to be typically in this range [48].

2.7. Data Analyses

The data analyses were divided into two parts (Figure 2). The first part focused on the cortical level to investigate differences in GMI between language-positive and language-negative nTMS points (Figure 2). The second part focused on the subcortical level to assess differences in fiber tractography between nTMS-based DTI FT using language-positive or language-negative nTMS points as ROIs by means of investigating short fibers for short-distance connections (U-fibers), long fibers projecting to the contralateral hemisphere (Cross-F), and fibers projecting to the cerebellum (Cereb-F; Figure 2).

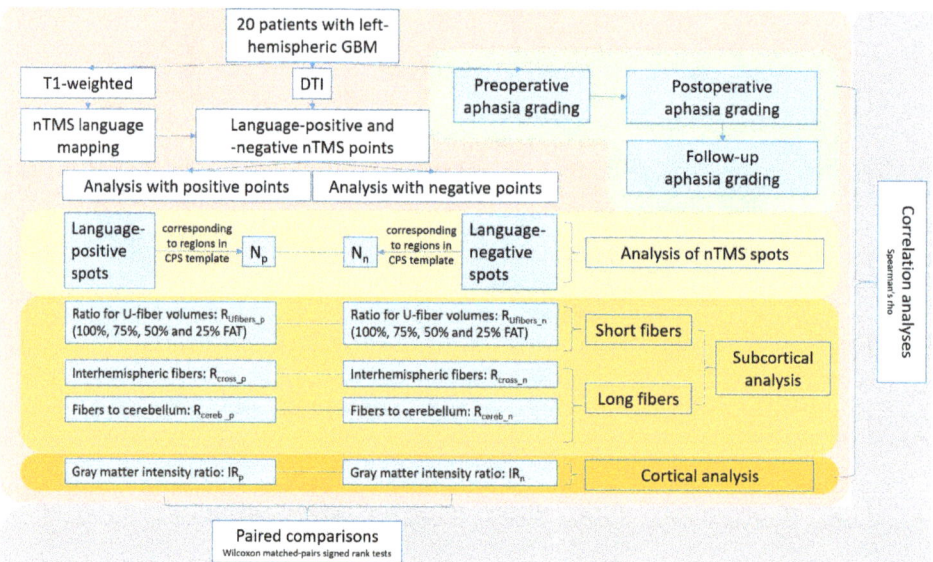

Figure 2. Overview of procedures and analyses. This scheme summarizes the study's steps of analyses, which were performed based on data derived from language mapping by navigated transcranial magnetic stimulation (nTMS) and nTMS-based diffusion tensor imaging fiber tracking (DTI FT). Language-positive and language-negative nTMS points were used separately for the different analyses. The study cohort included 20 patients with left-hemispheric glioblastoma multiforme (GBM).

2.7.1. Gray Matter Intensity

Using the T1-weighted sequences co-registered to language-positive or language-negative nTMS points, the GMI was measured per stimulation point (iPlan Net server, version 3.0.1; Brainlab AG, Munich, Germany). Three pixels were randomly selected within a stimulation point and used for GMI measurements to represent the GMI per stimulation point by averaging the three values obtained (Figure 3). The mean GMI of language-positive nTMS points was recorded as $mGMI_p$, and the mean GMI of language-negative nTMS points as $mGMI_n$ in each patient (Figure 3). Then, intensity extraction was also conducted for the cerebrospinal fluid (CSF) using three randomly selected pixels in the lateral ventricles, and the mean was recorded as mI_{csf} to obtain an internal control value (Figure 3). The signal intensity ratio (IR) was calculated as follows [49]:

$$IR_p = \frac{mGMI_p}{mI_{csf}} \qquad (1)$$

$$IR_n = \frac{mGMI_n}{mI_{csf}} \qquad (2)$$

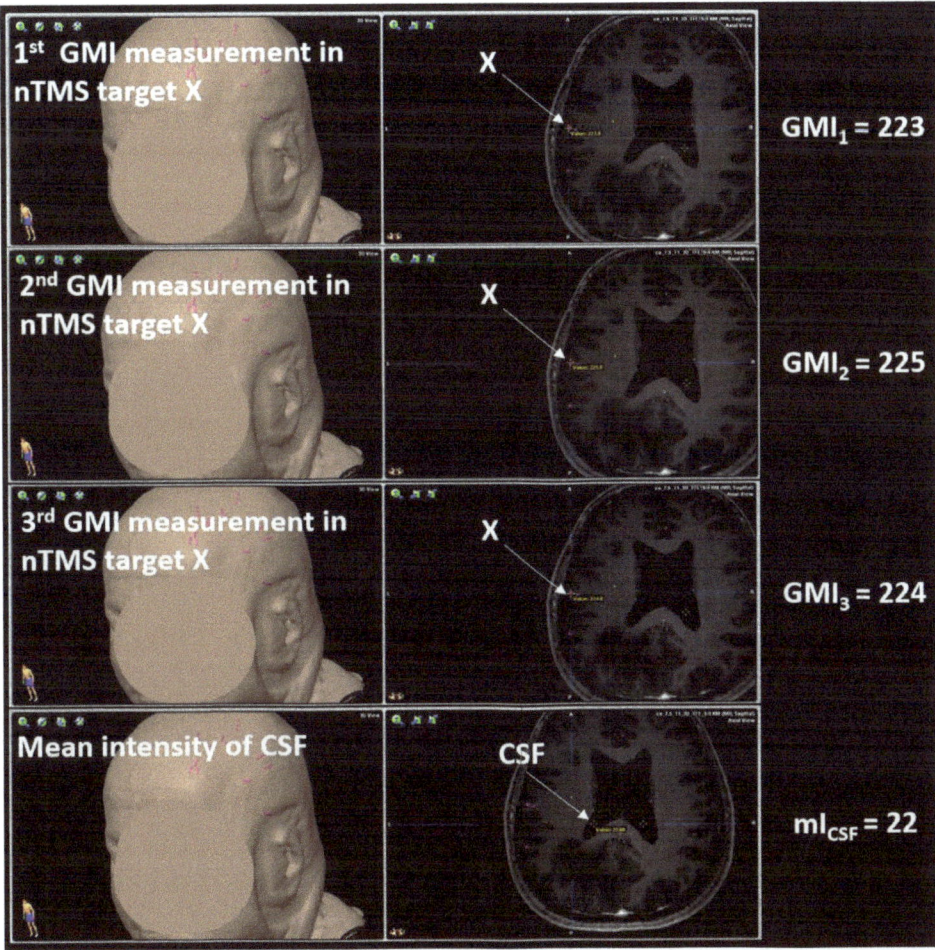

Figure 3. Analysis of gray matter intensity (GMI). The GMI of language-positive or language-negative spots, determined by language mapping using navigated transcranial magnetic stimulation (nTMS), was measured in T1-weighted sequences. Three pixels were randomly selected within a stimulation spot and used to calculate a mean GMI for language-positive nTMS spots (mGMI$_p$) and language-negative nTMS spots (mGMI$_n$). The signal intensity ratio (IR) was then calculated by dividing the mGMI$_p$ or mGMI$_n$ by the mean intensity of cerebrospinal fluid (mI$_{csf}$).

2.7.2. U-Fibers

Using nTMS-based DTI FT with the different FAT levels and a minimum FL of 30 and 3 mm, volumes of U-fibers were measured for fiber tractography considering language-positive or language-negative nTMS points as ROIs, respectively (Brainlab Elements, version 3.1.0; Brainlab AG, Munich, Germany; Figure 4). From assessing the difference in volumes between fibers with a minimum FL of 30 mm and fibers with a minimum FL of 3 mm, the volume of U-fibers was obtained (Figure 4). In this context, U-fibers are considered short association fibers, which connect cortical regions between adjacent gyri and typically have a length in the range of 3 to 30 mm [48].

A higher number of nTMS points considered during tractography should lead to more fibers, which was corrected for by dividing the fiber volume by N_p or N_n, resulting in the ratios $R_{Ufibers\_p}$ and

$R_{Ufibers\_n}$ for fibers with a length of 3 to 30 mm. The ratios in $R_{Ufibers\_p}$ and $R_{Ufibers\_n}$ were calculated as follows:

$$R_{Ufibers\_p} = \frac{V_{p(3)} - V_{p(30)}}{N_p} \quad (3)$$

$$R_{Ufibers\_n} = \frac{V_{n(3)} - V_{n(30)}}{N_n} \quad (4)$$

$V_{p(30)}$ indicates the volume of fibers with a minimum FL of 30 mm, as derived from nTMS-based DTI FT using the ROI of language-positive nTMS points, whereas $V_{p(3)}$ represents the respective volume for a minimum FL of 3 mm. Analogously, $V_{n(30)}$ indicates the volume of fibers with a minimum FL of 30 mm, as derived from nTMS-based DTI FT that considers the ROI of language-negative nTMS points, whereas $V_{n(3)}$ represents the respective volume for a minimum FL of 3 mm.

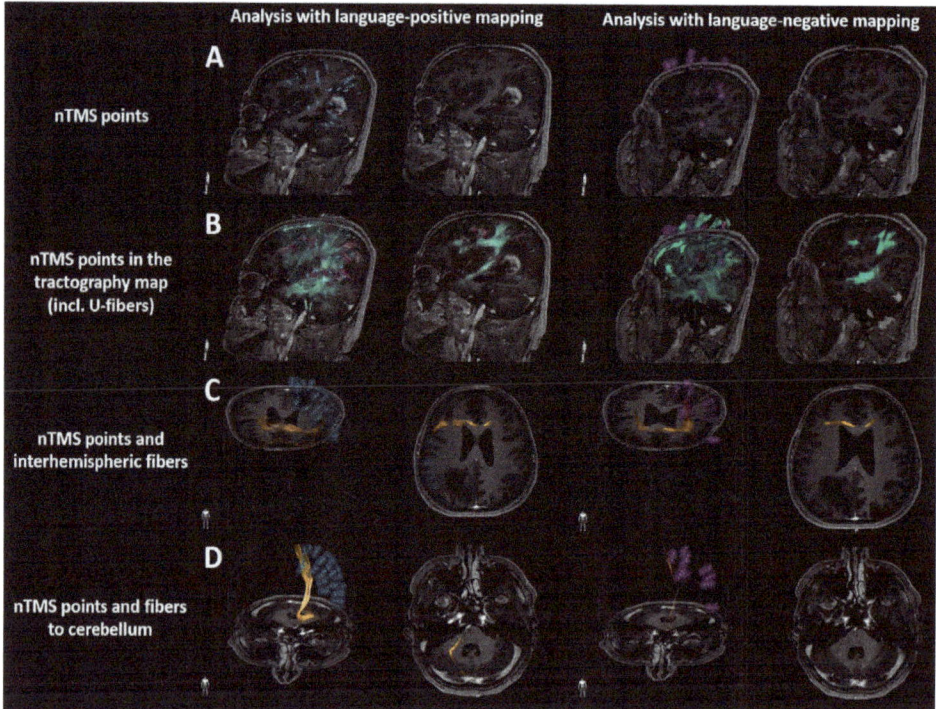

Figure 4. Analysis of subcortical fiber tracts. Tractography maps were generated based on language mapping data derived from navigated transcranial magnetic stimulation (nTMS), using the language-positive nTMS spots and language-negative nTMS spots as separate regions of interest (ROIs). Row (**A**) visualizes the language-positive nTMS points (blue) and language-negative nTMS points (purple) in an exemplary patient case with a left-hemispheric contrast-enhancing tumor with temporo-parietal location. Row (**B**) illustrates the complete picture of tractography (green fibers) with special emphasis on U-fibers (purple fibers). Row (**C**) depicts interhemispheric fiber courses (orange fibers) crossing the midline via the corpus callosum. Row (**D**) visualizes fibers projecting to the cerebellum (orange fibers).

2.7.3. Interhemispheric Fibers and Fibers Projecting to the Cerebellum

The highest percentage of patients presenting Cross-F or Cereb-F according to visual image inspection of tractography maps was achieved for 25% FAT, which was then taken as the adjustment used for evaluation of Cross-F and Cereb-F volumes, together with a minimum FL of 30 mm derived from language-positive or language-negative nTMS points as ROIs, respectively (Brainlab Elements, version 3.1.0; Brainlab AG, Munich, Germany; Figure 4).

The volumes of the Cross-F and Cereb-F for nTMS-based DTI FT with 25% FAT were recorded as V_{cross_p} and V_{cereb_p} when derived from nTMS-based DTI FT using the ROI of language-positive nTMS points, whereas V_{cross_n} and V_{cereb_n} represent the Cross-F and Cereb-F volumes for nTMS-based DTI FT conducted with language-negative nTMS points as the ROI. The ratios R_{cross_p} and R_{cereb_p} and R_{cross_n} and R_{cereb_n} were calculated as follows:

$$R_{cross\_p} = \frac{V_{cross\_p}}{N_p} \text{ and } R_{cereb\_p} = \frac{V_{cereb\_p}}{N_p} \quad (5)$$

$$R_{cross\_n} = \frac{V_{cross\_n}}{N_n} \text{ and } R_{cereb\_n} = \frac{V_{cereb\_n}}{N_n} \quad (6)$$

2.8. Statistical Analyses

GraphPad Prism (version 6.04; GraphPad Software Inc., La Jolla, CA, USA) was used for statistical data analyses and generation of graphs. Descriptive statistics using relative and absolute frequencies or mean, standard deviation (SD), and ranges were calculated for demographics and characteristics of language mapping and tractography. The Shapiro–Wilk normality test was used to assess the distribution of data, which indicated a non-Gaussian distribution for the majority of data.

To investigate differences between N_p and N_n, IR_p and IR_n, $R_{Ufibers\_p}$ and $R_{Ufibers\_n}$ (separately for nTMS-based DTI FT with 100% FAT, 75% FAT, 50% FAT, and 25% FAT), R_{cross_p} and R_{cross_n}, and R_{cereb_p} and R_{cereb_n}, Wilcoxon matched-pairs signed rank tests were performed. Furthermore, correlation analyses computing Spearman's rho were performed between N_p, IR_p, $R_{Ufibers\_p}$, R_{cross_p}, and R_{cereb_p} as well as, analogously, between N_n, IR_n, $R_{Ufibers\_n}$, R_{cross_n}, and R_{cereb_n} and the status of preoperative, postoperative, and follow-up aphasia, considering the four grades as derived from language function assessments at different time points. The correlation analyses were adjusted for multiple testing using the Benjamini–Hochberg procedure with a false discovery rate of 25%.

3. Results

3.1. Cohort Characteristics

Twenty patients (mean age: 63.2 ± 12.9 years, age range: 20.3–80.8 years, 4 females and 16 males, 16 right-handers according to EHI scores) were included, all diagnosed with a left-hemispheric GBM according to histopathological evaluation (Table 1). Ten patients (50%) showed preoperative aphasia, whereas eight patients (40%) showed aphasia during follow-up examinations three months after tumor resection. GTR according to postoperative MRI was achieved in 10 patients (50%).

3.2. Comparison between Language-Positive and Language-Negative Mapping and Tractography

Language mapping of the tumor-affected LH and nTMS-based DTI FT was possible in all enrolled patients. None of the patients showed adverse events in the course of stimulation.

There were statistically significant differences in almost all measures between mapping or tractography using language-positive or language-negative nTMS points as ROIs, respectively (Table 2). In detail, patients showed a higher mean number of language-negative nTMS points ($p = 0.0026$), whereas the IR of these points was elevated on average in comparison to language-positive nTMS points ($p = 0.0121$; Table 2). The ratios for U-fiber volumes as well as for long fibers projecting to the contralateral hemisphere and fibers projecting to the cerebellum were higher for tractography using language-positive nTMS points ($p \leq 0.0494$; Table 2).

Table 1. Cohort characteristics.

No.	Sex	Age	No. of Language-Positive CPS Sites	No. of Language-Negative CPS Sites	100% FAT for DTI FT with Language-Positive nTMS Points	100% FAT for DTI FT with Language-Negative nTMS Points	Awake Surgery	Aphasia Grading Preop	Aphasia Grading Postop	Aphasia Grading Follow-up
1	male	52.0	12	34	0.29	0.32	No	0	0	0
2	male	76.7	18	26	0.32	0.34	No	2	2	1
3	male	68.5	15	30	0.39	0.39	Yes	0	0	0
4	male	54.2	15	31	0.42	0.48	Yes	2	0	0
5	male	52.2	8	38	0.42	0.36	No	1	1	1
6	male	57.0	10	30	0.49	0.58	No	0	0	0
7	female	70.1	12	34	0.29	0.32	No	0	0	0
8	male	64.3	31	15	0.39	0.39	No	3	3	3
9	male	72.6	5	41	0.33	0.38	No	0	0	0
10	male	57.3	8	35	0.29	0.38	Yes	1	1	1
11	male	71.5	35	11	0.33	0.37	Yes	2	2	2
12	male	55.6	28	18	0.39	0.35	No	0	0	0
13	female	72.2	9	35	0.34	0.36	No	1	0	0
14	male	74.4	24	19	0.37	0.36	No	0	2	0
15	male	70.6	16	29	0.37	0.36	No	1	2	1
16	male	62.0	12	33	0.33	0.43	No	0	0	0
17	female	60.5	19	25	0.37	0.43	Yes	2	2	2
18	male	20.3	2	38	0.38	0.44	No	0	0	0
19	female	80.8	5	38	0.30	0.28	No	0	0	0
20	male	70.3	18	28	0.41	0.37	No	2	3	3

This table shows cohort details, including sex distribution and age (in years), information on the number of language-positive and language-negative sites according to the cortical parcellation system (CPS), the fractional anisotropy threshold (FAT) for diffusion tensor imaging fiber tracking (DTI FT) based on navigated transcranial stimulation (nTMS), and aphasia grades (at three different time points: preop = preoperative status, postop = postoperative status, follow-up = status during follow-up examinations three months after surgery). Language mapping aimed to cover 46 target points in total that were placed in relation to the CPS on the left hemisphere (LH), but a reduced number of targets was stimulated in patients with large tumor masses that precluded placement of all 46 target points. Thus, the numbers of language-positive and language-negative CPS sites do not necessarily add up to 46 in all enrolled patients.

Table 2. Comparison between language-positive and language-negative mapping and tractography.

Item			Language-Positive nTMS Points		Language-Negative nTMS Points		p
			Mean	SD	Mean	SD	
	N		15.1	8.9	29.4	8.3	**0.0026**
	IR		5.7	1.7	7.1	1.6	**0.0121**
$R_{Ufibers}$		100% FAT	0.3	0.2	0.1	0.1	**0.0012**
		75% FAT	0.5	0.2	0.3	0.3	**0.0020**
		50% FAT	0.6	0.3	0.4	0.2	**0.0056**
		25% FAT	0.7	0.5	0.4	0.3	0.1231
	R_{cross}		0.9	0.8	0.5	0.6	**0.0494**
	R_{cereb}		0.6	0.6	0.3	0.5	**0.0094**

This table shows the mean and standard deviation (SD) for the number (N) of language-positive and language-negative points as mapped by navigated transcranial magnetic stimulation (nTMS), intensity ratio (IR), ratio of volumes for U-fibers ($R_{Ufibers}$, as derived from tractography using 100%, 75%, 50%, and 25% of the individual fractional anisotropy threshold (FAT)), and ratio of volumes for interhemispheric fibers (R_{cross}, using tractography with 25% FAT) as well as fibers projecting to the cerebellum (R_{cereb}, using tractography with 25% FAT). Wilcoxon matched-pairs signed rank tests were conducted to assess differences in these characteristics between language-positive and language-negative mappings (level of statistical significance: $p < 0.05$). Statistically significant values are displayed in bold.

3.3. Associations with Aphasia Grading

For language-positive nTMS points, statistically significant positive correlations were revealed between their absolute frequency and aphasia for the preoperative (R = 0.4919, p = 0.0276), postoperative (R = 0.6183, p = 0.0037), and follow-up status (R = 0.4854, p = 0.0300; Table 3). The higher the number of language-positive nTMS points of the tumor-affected LH, the higher the aphasia grade. Furthermore, statistically significant negative correlations were observed between the ratio of U-fiber volumes (considering tractography with 100% FAT) and aphasia for the postoperative (R = −0.6102, p = 0.0043) as well as follow-up status (R = −0.4899, p = 0.0283; Table 3). Thus, the lower this ratio was, the higher the aphasia grade.

Regarding language-negative nTMS points, statistically significant negative correlations were revealed between their absolute frequency and aphasia for postoperative (R = −0.6097, p = 0.0043) and follow-up examinations (R = −0.4741, p = 0.0347; Table 3). Hence, the higher the number of language-negative nTMS points of the tumor-affected LH, the lower the aphasia grade.

Table 3. Associations with aphasia grading.

ROI	Item		Parameter	Aphasia Grading		
				Preoperative	Postoperative	Follow-Up
Language-Positive nTMS Points	N		rho	0.4919	0.6183	0.4854
			p	**0.0276**	**0.0037**	**0.0300**
	IR		rho	0.0138	−0.0806	0.0888
			p	0.9538	0.7354	0.7098
	$R_{Ufibers}$	100% FAT	rho	−0.3777	−0.6102	−0.4899
			p	0.1007	**0.0043**	**0.0283**
		75% FAT	rho	−0.2645	−0.4323	−0.4080
			p	0.2597	0.0570	0.0741
		50% FAT	rho	−0.0016	0.0590	−0.0854
			p	0.9946	0.8048	0.7205
		25% FAT	rho	−0.1823	0.0341	−0.0632
			p	0.4417	0.8866	0.7914
	R_{cross}		rho	−0.4590	−0.1629	−0.2288
			p	**0.0418**	0.4925	0.3320
	R_{cereb}		rho	−0.1717	−0.1347	−0.1110
			p	0.4691	0.5713	0.6414

Table 3. Cont.

ROI	Item		Parameter	Aphasia Grading		
				Preoperative	Postoperative	Follow-Up
Language-Negative nTMS Points	N		rho	−0.4521	−0.6097	−0.4741
			p	0.0454	**0.0043**	**0.0347**
	IR		rho	−0.1660	−0.1987	−0.2475
			p	0.4842	0.4011	0.2927
	R$_{Ufibers}$	100% FAT	rho	0.0733	−0.1297	−0.0956
			p	0.7589	0.5858	0.6885
		75% FAT	rho	−0.2784	−0.2752	−0.3141
			p	0.2347	0.2403	0.1774
		50% FAT	rho	−0.1457	0.1621	0.0768
			p	0.5400	0.4947	0.7475
		25% FAT	rho	0.1750	0.3267	0.2885
			p	0.4606	0.1598	0.2174
	R$_{cross}$		rho	0.0073	−0.0399	0.0598
			p	0.9755	0.8674	0.8024
	R$_{cereb}$		rho	0.0887	0.2959	0.1912
			p	0.7099	0.2052	0.4194

This table shows the correlation results between the number of language-positive and language-negative points as mapped by navigated transcranial magnetic stimulation (nTMS), intensity ratio (IR), ratio of volumes for U-fibers (R$_{Ufibers}$, as derived from tractography using 100%, 75%, 50%, and 25% of the individual fractional anisotropy threshold (FAT)), ratio of volumes for interhemispheric fibers (R$_{cross}$, using tractography with 25% FAT) as well as fibers projecting to the cerebellum (R$_{cereb}$, using tractography with 25% FAT) and the aphasia grades for the preoperative, postoperative, and follow-up status. Correlation coefficients are represented by Spearman's rho, and related p-values are given (level of statistical significance: $p < 0.05$). Statistically significant values that survived adjustments for multiple testing (Benjamini–Hochberg procedure with a false discovery rate of 25%) are depicted in bold.

4. Discussion

This study investigated the difference between language-positive and language-negative mappings as derived from presurgical nTMS and nTMS-based DTI FT in patients harboring supratentorial GBMs. There are three main results that can be taken from our analyses. First, regarding the cortical level, a significantly lower GMI was revealed for language-positive nTMS points compared to language-negative counterparts. Second, on the subcortical level, language-positive areas were characterized by an increased connectivity profile, i.e., such areas showed a significantly higher ratio in volumes for U-fibers, interhemispheric fibers, and fibers projecting to the cerebellum. Third, the number of language-positive nTMS points as well as the ratio in volumes for U-fibers were significantly associated with aphasia grading as derived from assessments at different time points.

4.1. Gray Matter Intensity

While there is evidence for alterations in GM distribution and volume related to language function, the role of the GMI to characterize language-involved areas has not been investigated to the authors' knowledge. Specifically, previous research has detected that language lateralization is predicted by the degree of GM lateralization [30]. Further, subjects diagnosed with dyslexia showed GM deficits, but the GM volume, however, can be subject to changes following training interventions in dyslexic children [31,50]. On the contrary, region-specific increases in GM volume have also been revealed for developmental language disorders, which might be interpreted as a result of compensatory mechanisms [31,32]. In the present study, lower GMI in T1-weighted sequences was revealed for language-positive nTMS points when compared to language-negative spots. This may probably reflect a sign of higher GM density, potentially suggesting increased functional involvement. Yet, this remains speculative until further studies using a similar setup can confirm our findings. For the present study, whether potentially increased functional involvement is due to higher intrinsic contribution of such areas to language function or related to compensatory mechanisms for language function at risk in our sample remains beyond the scope of investigation. However, the findings for GMI may provide a direct link between a functionally language-related area as mapped by nTMS and structural cortical characteristics.

4.2. White Matter Tractography

The ratios for fiber volumes of U-fibers, transcallosal fibers, and fibers coursing to the cerebellum were higher for language-positive areas compared to language-negative counterparts. Thus, language-related spots seem to be characterized by an enhanced connectivity profile, which exists focally for short fibers connecting cortical regions between adjacent gyri (increased $R_{Ufibers}$) as well as more remotely for long connecting fibers (increased R_{cross} and R_{cereb}). This may reflect either higher a-priori involvement in the cortico-subcortical language network, compensatory mechanisms in light of language function at risk in brain tumor patients, or a mixed picture of both. While the specificity of nTMS language mapping in comparison to intraoperative DES has shown to be comparatively low with 23.8%, the technique's negative predictive value was 83.9%, implicating that language-negative nTMS points are also mostly negative in intraoperative DES mapping [13]. The significantly lower connectivity profile of language-negative nTMS spots as revealed by the present study may resemble the "truly" absent or less involved character of these spots. Furthermore, a previous study on nTMS-based DTI FT revealed that there is a higher likelihood of subcortical connections with language-positive nTMS spots as compared to language-negative ones, with true-positive connections (connections for positive spots) being visualized up to four-fold more frequently than false-positive connections (connections for negative spots) [21]. The clear difference in this likelihood could reflect high reliability of nTMS-based DTI FT for the purpose of tracking parts of the human language network [21]. Likewise, the enhanced connectivity profile to adjacent gyri as well as the contralateral hemisphere and cerebellum may serve as a surrogate of good reliability.

4.3. Associations with Aphasia

The number of language-positive nTMS points was significantly associated with aphasia grading. The higher the aphasia grading is (i.e., the more severe language impairment is), the higher the number of language-positive nTMS spots. A higher frequency of language-positive nTMS spots among more impaired patients according to pre- and postoperative as well as follow-up examinations makes sense as a decline in language function should be associated with higher stimulation-induced errors, although this may bias the results of nTMS language mapping. A previous study is in good accordance with this finding and showed that aphasia as measured by the Berlin Aphasia Score correlated significantly with the incidence of errors during nTMS language mapping; yet, correlations were only evaluated for the preoperative status of language function [51]. Moreover, higher aphasia grading for the postoperative and follow-up status was associated with lower ratios in volume for U-fibers. Hence, this finding may underline the important role of short association fibers for keeping of language function as their impairment, e.g., in the direct perioperative course of tumor resection or related to even subtle perioperative ischemia, could result in language worsening.

4.4. Limitations and Perspectives

When interpreting the results of this study, the following limitations have to be acknowledged. First, the retrospective character and comparatively small sample size restrict the generalizability of the findings. Upcoming studies may include more patients and follow a prospective study design. Second, the finding of lower GMI for language-positive nTMS points when compared to language-negative nTMS points and related interpretation as a potential hint for higher GM density needs further validation. A potential explanation could also be linked to the increased connectivity profile of language-positive nTMS points. Future studies using imaging with higher resolution may provide evidence for our preliminary interpretations. Third, the technique of DTI has its inherent methodological shortcomings, which could lead to aberrant fiber reconstruction and visualization, particularly for crossing or kissing fibers and in the presence of edema, which is commonly observed in relation to brain tumors [52,53]. Other sequences and tracking algorithms are developed to become applicable in the clinical setting, but good alternatives to conventional DTI and its tractography-based analyses are still not available for clinical routine use [54,55]. Fourth, the tractography maps generated by nTMS-based DTI FT for depiction of the language network need further validation, ideally by intraoperative DES as the gold-standard method. In this context, previous studies have evaluated the agreement between preoperative nTMS language mapping and intraoperative DES [13,16,24]; however, on the subcortical level, such correlation analyses in representative samples are largely missing to date. First evidence of associations between preoperative nTMS-based tractography and intraoperative DES results or surgery-related aphasia has been obtained [20,37]. Upcoming studies may use more sophisticated approaches for diffusion-weighted MRI and fiber tractography, ideally combining it with functional data such as nTMS maps after further confirmatory studies. Furthermore, repeated investigations of the language network by serial nTMS language mappings and tractography after surgery may be of interest to track potential plastic effects as well as associations with aphasia grades on a longitudinal scale.

5. Conclusions

Language-positive and language-negative areas as determined by nTMS language mapping show differences in cortical and subcortical characteristics among patients diagnosed with GBMs. Specifically, language-positive areas demonstrate lower GMI and an enhanced connectivity profile with higher volume ratios for U-fibers, interhemispheric fibers, and fibers projecting to the cerebellum. While future studies may confirm these results in larger cohorts in the context of a prospective study design, these findings facilitate confidence in the technique of nTMS language mapping and nTMS-based tractography to detect language-involved structures of the human cortico-subcortical language network.

Author Contributions: Conceptualization, H.Z., S.S. and N.S.; methodology, H.Z., S.S., S.M.K. and N.S.; software, H.Z.; validation, S.M.K. and N.S.; formal analysis, H.Z., S.S., A.S. and N.S.; investigation, H.Z., S.S. and A.S.; resources, C.Z., B.M. and S.M.K.; data curation, H.Z. and N.S.; writing—original draft preparation, H.Z. and N.S.; writing—review and editing, S.M.K.; visualization, H.Z.; supervision, C.Z., B.M., S.M.K. and N.S.; project administration, S.M.K. and N.S.; funding acquisition, C.Z., B.M. and S.M.K. All authors have read and agreed to the published version of the manuscript.

Funding: This research received no external funding.

Conflicts of Interest: N.S. received honoraria from Nexstim Plc (Helsinki, Finland). S.M.K. is a consultant for Nexstim Plc (Helsinki, Finland) and Spineart Deutschland GmbH (Frankfurt, Germany) and received honoraria from Medtronic (Meerbusch, Germany) and Carl Zeiss Meditec (Oberkochen, Germany). S.M.K. and B.M. received research grants and are consultants for Brainlab AG (Munich, Germany). B.M. received honoraria, consulting fees, and research grants from Medtronic (Meerbusch, Germany), Icotec ag (Altstätten, Switzerland), and Relievant Medsystems Inc. (Sunnyvale, CA, USA); honoraria and research grants from Ulrich Medical (Ulm, Germany); honoraria and consulting fees from Spineart Deutschland GmbH (Frankfurt, Germany) and DePuy Synthes (West Chester, PA, USA); and royalties from Spineart Deutschland GmbH (Frankfurt, Germany). The other authors declare no conflicts of interest.

Abbreviations

CPS	Cortical parcellation system
CSF	Cerebrospinal fluid
DES	Direct electrical stimulation
DICOM	Digital Imaging and Communications in Medicine
DTI	Diffusion tensor imaging
DTI FT	Diffusion tensor imaging fiber tracking
EHI	Edinburgh Handedness Inventory
FA	Fractional anisotropy
FAT	Fractional anisotropy threshold
FL	Fiber length

FLAIR	Fluid attenuated inversion recovery
GBM	Glioblastoma multiforme
GM	Gray matter
GMI	Gray matter intensity
GTR	Gross total resection
IR	Intensity ratio
LH	Left hemisphere
MRI	Magnetic resonance imaging
nTMS	Navigated transcranial magnetic stimulation
ROI	Region of interest
SD	Standard deviation
TE	Echo time
TR	Repetition time

References

1. Duffau, H.; Mandonnet, E. The "onco-functional balance" in surgery for diffuse low-grade glioma: Integrating the extent of resection with quality of life. *Acta Neurochir.* **2013**, *155*, 951–957. [CrossRef]
2. Sanai, N.; Berger, M.S. Glioma extent of resection and its impact on patient outcome. *Neurosurgery* **2008**, *62*, 753–766. [CrossRef]
3. Eyupoglu, I.Y.; Buchfelder, M.; Savaskan, N.E. Surgical resection of malignant gliomas-role in optimizing patient outcome. *Nat. Rev. Neurol.* **2013**, *9*, 141–151. [CrossRef] [PubMed]
4. Hervey-Jumper, S.L.; Berger, M.S. Maximizing safe resection of low- and high-grade glioma. *J. Neuro-Oncol.* **2016**, *130*, 269–282. [CrossRef] [PubMed]
5. Schucht, P.; Beck, J.; Seidel, K.; Raabe, A. Extending resection and preserving function: Modern concepts of glioma surgery. *Swiss Med. Wkly.* **2015**, *145*, w14082. [CrossRef] [PubMed]
6. Ottenhausen, M.; Krieg, S.M.; Meyer, B.; Ringel, F. Functional preoperative and intraoperative mapping and monitoring: Increasing safety and efficacy in glioma surgery. *Neurosurg. Focus* **2015**, *38*, E3. [CrossRef] [PubMed]
7. Sanai, N.; Berger, M.S. Mapping the horizon: Techniques to optimize tumor resection before and during surgery. *Clin. Neurosurg.* **2008**, *55*, 14–19. [PubMed]
8. Duffau, H.; Lopes, M.; Arthuis, F.; Bitar, A.; Sichez, J.P.; Van Effenterre, R.; Capelle, L. Contribution of intraoperative electrical stimulations in surgery of low grade gliomas: A comparative study between two series without (1985-96) and with (1996-2003) functional mapping in the same institution. *J. Neurol. Neurosurg. Psychiatry* **2005**, *76*, 845–851. [CrossRef] [PubMed]
9. De Witt Hamer, P.C.; Robles, S.G.; Zwinderman, A.H.; Duffau, H.; Berger, M.S. Impact of intraoperative stimulation brain mapping on glioma surgery outcome: A meta-analysis. *J. Clin. Oncol. Off. J. Am. Soc. Clin. Oncol.* **2012**, *30*, 2559–2565. [CrossRef]
10. Mandonnet, E.; Winkler, P.A.; Duffau, H. Direct electrical stimulation as an input gate into brain functional networks: Principles, advantages and limitations. *Acta Neurochir.* **2010**, *152*, 185–193. [CrossRef]
11. Ruohonen, J.; Karhu, J. Navigated transcranial magnetic stimulation. *Neurophysiol. Clin. Clin. Neurophysiol.* **2010**, *40*, 7–17. [CrossRef]
12. Picht, T. Current and potential utility of transcranial magnetic stimulation in the diagnostics before brain tumor surgery. *CNS Oncol.* **2014**, *3*, 299–310. [CrossRef]
13. Picht, T.; Krieg, S.M.; Sollmann, N.; Rosler, J.; Niraula, B.; Neuvonen, T.; Savolainen, P.; Lioumis, P.; Makela, J.P.; Deletis, V.; et al. A comparison of language mapping by preoperative navigated transcranial magnetic stimulation and direct cortical stimulation during awake surgery. *Neurosurgery* **2013**, *72*, 808–819. [CrossRef]
14. Sollmann, N.; Kelm, A.; Ille, S.; Schroder, A.; Zimmer, C.; Ringel, F.; Meyer, B.; Krieg, S.M. Setup presentation and clinical outcome analysis of treating highly language-eloquent gliomas via preoperative navigated transcranial magnetic stimulation and tractography. *Neurosurg. Focus* **2018**, *44*, E2. [CrossRef] [PubMed]
15. Sollmann, N.; Meyer, B.; Krieg, S.M. Implementing Functional Preoperative Mapping in the Clinical Routine of a Neurosurgical Department: Technical Note. *World Neurosurg.* **2017**, *103*, 94–105. [CrossRef] [PubMed]

16. Tarapore, P.E.; Findlay, A.M.; Honma, S.M.; Mizuiri, D.; Houde, J.F.; Berger, M.S.; Nagarajan, S.S. Language mapping with navigated repetitive TMS: Proof of technique and validation. *NeuroImage* **2013**, *82*, 260–272. [CrossRef] [PubMed]
17. Sollmann, N.; Negwer, C.; Ille, S.; Maurer, S.; Hauck, T.; Kirschke, J.S.; Ringel, F.; Meyer, B.; Krieg, S.M. Feasibility of nTMS-based DTI fiber tracking of language pathways in neurosurgical patients using a fractional anisotropy threshold. *J. Neurosci. Methods* **2016**, *267*, 45–54. [CrossRef] [PubMed]
18. Negwer, C.; Ille, S.; Hauck, T.; Sollmann, N.; Maurer, S.; Kirschke, J.S.; Ringel, F.; Meyer, B.; Krieg, S.M. Visualization of subcortical language pathways by diffusion tensor imaging fiber tracking based on rTMS language mapping. *Brain Imaging Behav.* **2017**, *11*, 899–914. [CrossRef] [PubMed]
19. Sollmann, N.; Zhang, H.; Schramm, S.; Ille, S.; Negwer, C.; Kreiser, K.; Meyer, B.; Krieg, S.M. Function-specific Tractography of Language Pathways Based on nTMS Mapping in Patients with Supratentorial Lesions. *Clin Neuroradiol.* **2018**. [CrossRef] [PubMed]
20. Sollmann, N.; Giglhuber, K.; Tussis, L.; Meyer, B.; Ringel, F.; Krieg, S.M. nTMS-based DTI fiber tracking for language pathways correlates with language function and aphasia—A case report. *Clin. Neurol. Neurosurg.* **2015**, *136*, 25–28. [CrossRef] [PubMed]
21. Raffa, G.; Bahrend, I.; Schneider, H.; Faust, K.; Germano, A.; Vajkoczy, P.; Picht, T. A Novel Technique for Region and Linguistic Specific nTMS-based DTI Fiber Tracking of Language Pathways in Brain Tumor Patients. *Front. Neurosci.* **2016**, *10*, 552. [CrossRef] [PubMed]
22. Raffa, G.; Conti, A.; Scibilia, A.; Sindorio, C.; Quattropani, M.C.; Visocchi, M.; Germano, A.; Tomasello, F. Functional Reconstruction of Motor and Language Pathways Based on Navigated Transcranial Magnetic Stimulation and DTI Fiber Tracking for the Preoperative Planning of Low Grade Glioma Surgery: A New Tool for Preservation and Restoration of Eloquent Networks. *Acta Neurochir. Suppl.* **2017**, *124*, 251–261. [CrossRef] [PubMed]
23. Sollmann, N.; Zhang, H.; Fratini, A.; Wildschuetz, N.; Ille, S.; Schroder, A.; Zimmer, C.; Meyer, B.; Krieg, S.M. Risk Assessment by Presurgical Tractography Using Navigated TMS Maps in Patients with Highly Motor- or Language-Eloquent Brain Tumors. *Cancers* **2020**, *12*, 1264. [CrossRef] [PubMed]
24. Sollmann, N.; Kubitscheck, A.; Maurer, S.; Ille, S.; Hauck, T.; Kirschke, J.S.; Ringel, F.; Meyer, B.; Krieg, S.M. Preoperative language mapping by repetitive navigated transcranial magnetic stimulation and diffusion tensor imaging fiber tracking and their comparison to intraoperative stimulation. *Neuroradiology* **2016**, *58*, 807–818. [CrossRef]
25. Sollmann, N.; Negwer, C.; Tussis, L.; Hauck, T.; Ille, S.; Maurer, S.; Giglhuber, K.; Bauer, J.S.; Ringel, F.; Meyer, B.; et al. Interhemispheric connectivity revealed by diffusion tensor imaging fiber tracking derived from navigated transcranial magnetic stimulation maps as a sign of language function at risk in patients with brain tumors. *J. Neurosurg.* **2017**, *126*, 222–233. [CrossRef] [PubMed]
26. Chang, E.F.; Raygor, K.P.; Berger, M.S. Contemporary model of language organization: An overview for neurosurgeons. *J. Neurosurg.* **2015**, *122*, 250–261. [CrossRef] [PubMed]
27. Fedorenko, E.; Thompson-Schill, S.L. Reworking the language network. *Trends Cogn. Sci.* **2014**, *18*, 120–126. [CrossRef]
28. Dick, A.S.; Bernal, B.; Tremblay, P. The language connectome: New pathways, new concepts. *Neuroscientist* **2014**, *20*, 453–467. [CrossRef]
29. Yagmurlu, K.; Middlebrooks, E.H.; Tanriover, N.; Rhoton, A.L., Jr. Fiber tracts of the dorsal language stream in the human brain. *J. Neurosurg.* **2016**, *124*, 1396–1405. [CrossRef]
30. Josse, G.; Kherif, F.; Flandin, G.; Seghier, M.L.; Price, C.J. Predicting language lateralization from gray matter. *J. Neurosci. Off. J. Soc. Neurosci.* **2009**, *29*, 13516–13523. [CrossRef]
31. Vinckenbosch, E.; Robichon, F.; Eliez, S. Gray matter alteration in dyslexia: Converging evidence from volumetric and voxel-by-voxel MRI analyses. *Neuropsychologia* **2005**, *43*, 324–331. [CrossRef] [PubMed]
32. Pigdon, L.; Willmott, C.; Reilly, S.; Conti-Ramsden, G.; Gaser, C.; Connelly, A.; Morgan, A.T. Grey matter volume in developmental speech and language disorder. *Brain Struct. Funct.* **2019**, *224*, 3387–3398. [CrossRef]
33. Steinmann, S.; Mulert, C. Functional relevance of interhemispheric fiber tracts in speech processing. *J. Neurolinguistics* **2012**, *25*, 1–12. [CrossRef]
34. Fabbro, F. Introduction to language and cerebellum. *J. Neurolinguistics* **2000**, *13*, 83–94. [CrossRef]
35. Huber, W.; Poeck, K.; Willmes, K. The Aachen Aphasia Test. *Adv. Neurol.* **1984**, *42*, 291–303. [PubMed]

36. Sollmann, N.; Ille, S.; Hauck, T.; Maurer, S.; Negwer, C.; Zimmer, C.; Ringel, F.; Meyer, B.; Krieg, S.M. The impact of preoperative language mapping by repetitive navigated transcranial magnetic stimulation on the clinical course of brain tumor patients. *BMC Cancer* **2015**, *15*, 261. [CrossRef] [PubMed]

37. Negwer, C.; Beurskens, E.; Sollmann, N.; Maurer, S.; Ille, S.; Giglhuber, K.; Kirschke, J.S.; Ringel, F.; Meyer, B.; Krieg, S.M. Loss of Subcortical Language Pathways Correlates with Surgery-Related Aphasia in Patients with Brain Tumor: An Investigation via Repetitive Navigated Transcranial Magnetic Stimulation-Based Diffusion Tensor Imaging Fiber Tracking. *World Neurosurg.* **2018**, *111*, e806–e818. [CrossRef]

38. Oldfield, R.C. The assessment and analysis of handedness: The Edinburgh inventory. *Neuropsychologia* **1971**, *9*, 97–113. [CrossRef]

39. Ruohonen, J.; Ilmoniemi, R.J. Modeling of the stimulating field generation in TMS. *Electroencephalogr. Clin. Neurophysiol. Suppl.* **1999**, *51*, 30–40.

40. Krieg, S.M.; Lioumis, P.; Makela, J.P.; Wilenius, J.; Karhu, J.; Hannula, H.; Savolainen, P.; Lucas, C.W.; Seidel, K.; Laakso, A.; et al. Protocol for motor and language mapping by navigated TMS in patients and healthy volunteers; workshop report. *Acta Neurochir.* **2017**, *159*, 1187–1195. [CrossRef]

41. Lioumis, P.; Zhdanov, A.; Makela, N.; Lehtinen, H.; Wilenius, J.; Neuvonen, T.; Hannula, H.; Deletis, V.; Picht, T.; Makela, J.P. A novel approach for documenting naming errors induced by navigated transcranial magnetic stimulation. *J. Neurosci. Methods* **2012**, *204*, 349–354. [CrossRef] [PubMed]

42. Sollmann, N.; Tanigawa, N.; Ringel, F.; Zimmer, C.; Meyer, B.; Krieg, S.M. Language and its right-hemispheric distribution in healthy brains: An investigation by repetitive transcranial magnetic stimulation. *NeuroImage* **2014**, *102 Pt 2*, 776–788. [CrossRef] [PubMed]

43. Krieg, S.M.; Sollmann, N.; Tanigawa, N.; Foerschler, A.; Meyer, B.; Ringel, F. Cortical distribution of speech and language errors investigated by visual object naming and navigated transcranial magnetic stimulation. *Brain Struct. Funct.* **2016**, *221*, 2259–2286. [CrossRef] [PubMed]

44. Hernandez-Pavon, J.C.; Makela, N.; Lehtinen, H.; Lioumis, P.; Makela, J.P. Effects of navigated TMS on object and action naming. *Front. Hum. Neurosci.* **2014**, *8*, 660. [CrossRef]

45. Krieg, S.M.; Tarapore, P.E.; Picht, T.; Tanigawa, N.; Houde, J.; Sollmann, N.; Meyer, B.; Vajkoczy, P.; Berger, M.S.; Ringel, F.; et al. Optimal timing of pulse onset for language mapping with navigated repetitive transcranial magnetic stimulation. *NeuroImage* **2014**, *100*, 219–236. [CrossRef]

46. Sollmann, N.; Ille, S.; Negwer, C.; Boeckh-Behrens, T.; Ringel, F.; Meyer, B.; Krieg, S.M. Cortical time course of object naming investigated by repetitive navigated transcranial magnetic stimulation. *Brain Imaging Behav.* **2016**. [CrossRef]

47. Frey, D.; Strack, V.; Wiener, E.; Jussen, D.; Vajkoczy, P.; Picht, T. A new approach for corticospinal tract reconstruction based on navigated transcranial stimulation and standardized fractional anisotropy values. *NeuroImage* **2012**, *62*, 1600–1609. [CrossRef]

48. Song, A.W.; Chang, H.C.; Petty, C.; Guidon, A.; Chen, N.K. Improved delineation of short cortical association fibers and gray/white matter boundary using whole-brain three-dimensional diffusion tensor imaging at submillimeter spatial resolution. *Brain Connect.* **2014**, *4*, 636–640. [CrossRef]

49. Radbruch, A.; Weberling, L.D.; Kieslich, P.J.; Hepp, J.; Kickingereder, P.; Wick, W.; Schlemmer, H.P.; Bendszus, M. High-Signal Intensity in the Dentate Nucleus and Globus Pallidus on Unenhanced T1-Weighted Images: Evaluation of the Macrocyclic Gadolinium-Based Contrast Agent Gadobutrol. *Investig. Radiol.* **2015**, *50*, 805–810. [CrossRef]

50. Krafnick, A.J.; Flowers, D.L.; Napoliello, E.M.; Eden, G.F. Gray matter volume changes following reading intervention in dyslexic children. *NeuroImage* **2011**, *57*, 733–741. [CrossRef]

51. Schwarzer, V.; Bahrend, I.; Rosenstock, T.; Dreyer, F.R.; Vajkoczy, P.; Picht, T. Aphasia and cognitive impairment decrease the reliability of rnTMS language mapping. *Acta Neurochir.* **2018**, *160*, 343–356. [CrossRef] [PubMed]

52. Duffau, H. Diffusion tensor imaging is a research and educational tool, but not yet a clinical tool. *World Neurosurg.* **2014**, *82*, e43–e45. [CrossRef] [PubMed]

53. Conti Nibali, M.; Rossi, M.; Sciortino, T.; Riva, M.; Gay, L.G.; Pessina, F.; Bello, L. Preoperative surgical planning of glioma: Limitations and reliability of fMRI and DTI tractography. *J. Neurosurg. Sci.* **2019**, *63*, 127–134. [CrossRef] [PubMed]

54. Wende, T.; Hoffmann, K.T.; Meixensberger, J. Tractography in Neurosurgery: A Systematic Review of Current Applications. *J. Neurol. Surg. A Cent. Eur. Neurosurg.* **2020**. [CrossRef]
55. Nimsky, C.; Bauer, M.; Carl, B. Merits and Limits of Tractography Techniques for the Uninitiated. *Adv. Tech. Stand. Neurosurg.* **2016**, 37–60. [CrossRef]

© 2020 by the authors. Licensee MDPI, Basel, Switzerland. This article is an open access article distributed under the terms and conditions of the Creative Commons Attribution (CC BY) license (http://creativecommons.org/licenses/by/4.0/).

Article

Navigated TMS in the ICU: Introducing Motor Mapping to the Critical Care Setting

Severin Schramm [1,*], Alexander F. Haddad [2], Lawrence Chyall [2], Sandro M. Krieg [1,3], Nico Sollmann [3,4] and Phiroz E. Tarapore [2]

1. Department of Neurosurgery, Klinikum Rechts der Isar, Technische Universität München, Ismaninger Str. 22, 81675 Munich, Germany; Sandro.Krieg@tum.de
2. Department of Neurosurgery, University of California San Francisco, 1001 Potrero Ave, San Francisco, CA 94110, USA; Alexander.Haddad@ucsf.edu (A.F.H.); Lawrence.Chyall@sfdph.org (L.C.); Phiroz.Tarapore@ucsf.edu (P.E.T.)
3. TUM-Neuroimaging Center, Klinikum Rechts der Isar, Technische Universität München, 81675 Munich, Germany; Nico.Sollmann@tum.de
4. Department of Diagnostic and Interventional Neuroradiology, Klinikum Rechts der Isar, Technische Universität München, Ismaninger Str. 22, 81675 Munich, Germany
* Correspondence: Severin.Schramm@tum.de

Received: 23 November 2020; Accepted: 16 December 2020; Published: 18 December 2020

Abstract: Navigated transcranial magnetic stimulation (nTMS) is a modality for noninvasive cortical mapping. Specifically, nTMS motor mapping is an objective measure of motor function, offering quantitative diagnostic information regardless of subject cooperation or consciousness. Thus far, it has mostly been restricted to the outpatient setting. This study evaluates the feasibility of nTMS motor mapping in the intensive care unit (ICU) setting and solves the challenges encountered in this special environment. We compared neuronavigation based on computed tomography (CT) and magnetic resonance imaging (MRI). We performed motor mappings in neurocritical patients under varying conditions (e.g., sedation or hemicraniectomy). Furthermore, we identified ways of minimizing electromyography (EMG) noise in the interference-rich ICU environment. Motor mapping was performed in 21 patients (six females, median age: 69 years). In 18 patients, motor evoked potentials (MEPs) were obtained. In three patients, MEPs could not be evoked. No adverse reactions occurred. We found CT to offer a comparable neuronavigation to MRI (CT maximum e-field 52 ± 14 V/m vs. MRI maximum e-field 52 ± 11 V/m; $p = 0.6574$). We detailed EMG noise reduction methods and found that propofol sedation of up to 80 mcg/kg/h did not inhibit MEPs. Yet, nTMS equipment interfered with exposed pulse oximetry. nTMS motor mapping application and use was illustrated in three clinical cases. In conclusion, we present an approach for the safe and reliable use of nTMS motor mapping in the ICU setting and outline possible benefits. Our findings support further studies regarding the clinical value of nTMS in critical care settings.

Keywords: nTMS; brain stimulation; intensive care; motor mapping; ICU; neurocritical care; neuromonitoring; functional mapping; motor evoked potentials

1. Introduction

Navigated transcranial magnetic stimulation (nTMS) is an established technique for non-invasive brain stimulation [1,2]. At present, the technique is commonly applied in pre-operative mappings of patients undergoing neurosurgical procedures. In this study, we investigate whether nTMS may also play a role in the diagnosis and management of acute neurological injury in the inpatient setting. Because it is image-guided (and therefore highly accurate), it can provide detailed information about the level of intactness in specific neuromuscular pathways (useful in spinal cord injury). Unlike transcranial electrical motor-evoked potentials, it is non-invasive and painless, and can therefore be applied at bedside without anesthesia. Furthermore, it does not require patient participation, and may thus be used in patients who are sedated, or who are awake but unable to participate in detailed neurological examination. For these reasons, nTMS, if it can be successfully implemented in the neurocritical care setting, would offer a valuable new method for quantifiable neurological examination in patients with suspected or known neurological compromise.

TMS involves a strong magnetic field pulse that penetrates the skull and causes depolarization of underlying neuronal tissue via the principle of electromagnetic induction. When using a figure-of-eight coil it is highly focal and entirely noninvasive [2,3]. By combining the capacity for focal cortical stimulation with image-based, frameless stereotactic neuronavigation, nTMS allows for highly accurate, navigated interrogation of cortical function [1,4]. The broad applicability of this basic principle has created applications for nTMS in both the basic science and clinical realms. The technique has been utilized across neuroscience disciplines such as psychiatry, in physical medicine and rehabilitation, and neurosurgery [5–7].

As mentioned earlier, a common application of nTMS is the mapping of the primary motor cortical system (MCS). Here, stimulation of the cortical motor area leads to peripheral muscle activation (motor evoked potentials [MEPs]), which is recorded via electromyography (EMG) [1]. This process allows for attribution of a specific muscle response (i.e., abductor pollicis brevis or tibialis anterior muscles) to its respective cortical location, thereby generating an individualized functional map of the MCS [8,9]. Herein, the navigational aspect adds critical capabilities to the underlying TMS technology. Compared to non-navigated TMS of the motor cortex, where the stimulation site is usually determined in relation to anatomical landmarks [10], nTMS has been shown to be superior in evoking replicable and high-amplitude MEPs [11]. Importantly, nTMS allows optimal orientation of the electric field in relation to individual gyral and skull anatomy, a critical aspect in ensuring ideal MEP generation [12]. Furthermore, for any form of longitudinal monitoring, precise revisiting of previously identified sites is vital and, therefore, necessitates accurate navigation. The accuracy of nTMS motor mapping is superior to other noninvasive techniques such as functional magnetic resonance imaging (fMRI) and magnetoencephalography (MEG), and is highly comparable in accuracy to intraoperative direct cortical stimulation (DCS), the current gold-standard method for functional mapping [1,13].

In the clinical context, mapping of the MCS is often performed in preparation for brain tumor surgery, allowing for better presurgical planning by identifying eloquent motor regions and their relationship to the lesion [5,9,14,15]. While increasingly common in the outpatient setting, nTMS motor mapping has thus far remained underutilized in the acute inpatient setting. The use of other neurophysiological monitoring modalities in the ICU, most notably electroencephalography (EEG), has opened up a range of diagnostic innovations in recent years [16–20]. Similarly, nTMS could be a valuable tool for the inpatient setting. It can provide a highly accurate, low-cost, and relatively quick assessment of the functionality of the MCS without dedicated subject participation. It would thus be the ideal modality for evaluation of the corticospinal integrity and excitability in patients with altered mental status or coma. Alternatively, because it is painless, nTMS may also be used to assess the MCS in patients who are awake but unable or unwilling to participate in neurological examination.

However, to date, only few studies have addressed the application of nTMS motor diagnostics in the ICU context. While TMS/EEG combination setups have been used in exploring disorders of consciousness [20], nTMS motor mappings in the intensive care framework are at this time not reported in the literature. The reason for the current lack of exploration is likely the plethora of challenges faced in translating established protocols and workflows to the unique setting and patient population found in the ICU (Table 1). This study seeks to explore the feasibility of nTMS motor mappings in the intensive care framework, based on established protocol and safety guidelines [8,21]. By identifying the primary challenges, working through solutions, and sharing our results, we hope to encourage and enable further studies into the value of nTMS motor mapping in the neurocritical care setting.

Table 1. Translation of navigated transcranial magnetic stimulation from outpatient setting to critical care setting.

Factor	Outpatient Setting	Critical Care Setting	Challenge
Imaging	High-resolution MRI as standard imaging	CT as most common imaging modality	Does CT offer comparable navigation to MRI?
Safety	Very rare occurrences of serious adverse effects (i.e., seizures)	No published data	Does nTMS safety extend to critical neurological damage?
Setup	Patient usually without monitoring, uninterrupted workflow	Monitoring and adjustment of vital parameters required	Is nTMS compatible with multimodal monitoring and clinical workflow?
EMG	Little environmental background noise	High amount of background noise	Can EMG noise be adequately reduced for nTMS motor mapping?
Patient	Patient is usually awake, cooperative, with little impairment	Sedation, hemicraniectomy, ICP elevation	What is the influence of clinical factors on MEP evokability?

This table gives an overview over challenges in translating navigated transcranial magnetic stimulation (nTMS) motor mappings from the outpatient to the critical care setting. Other abbreviations: magnetic resonance imaging (MRI), computed tomography (CT), electromyography (EMG), motor evoked potential (MEP), intracranial pressure (ICP).

2. Materials and Methods

2.1. Study Setup and Patient Inclusion

This study was approved by the UCSF Institutional Review Board (16-20684, 1/9/2020). All research was conducted according to the Declaration of Helsinki. Informed consent was acquired from conscious patients or next of kin when known.

We conducted nTMS motor mappings on patients treated in the neurological ICU of Zuckerberg San Francisco General Hospital. Patients are referred to this unit when neurological damage necessitated continuous monitoring of vital parameters, close interval neurochecks, mechanical ventilation, vasoactive infusions, and other typical ICU indications. Available and commonly applied monitoring includes blood pressure (both invasive and noninvasive), pulse oximetry, electrocardiography (ECG), intracranial pressure (ICP), EEG, and interval neurochecks [22].

Inclusion criteria were defined as treatment in the ICU, confirmed presence of brain or spinal lesion and suspicion or presence of a motor deficit due to neurological damage. Exclusion criteria were denial of consent by patient or next of kin, uncontrolled epilepsy, presence of fixed nTMS-interacting devices close to or within the skull (e.g., cochlear implant, deep brain stimulation), acute isolation due to infectious disease, acute instability of vital parameters, and end-of-life care. The procedures were explained to the patient (if conscious) and/or next of kin (if known).

The nTMS system was deployed into the patient's room on the ICU, and motor mapping was performed. After the examination, clinical factors such as ICP and sedation were documented if applicable. Afterwards, the data were analyzed by experienced nTMS users (SS, PT) and a report on the procedure and results was written for internal use.

2.2. Image Acquisition and Processing

All CT imaging in this study was acquired with a Somatom 64-slice CT scanner (Siemens AG, Erlangen, Germany), all MRI imaging with a Magnetom Skyra 3T MRI scanner (Siemens AG, Erlangen, Germany). In order to compare the navigational accuracy of CT and MRI, we exported image stacks of both CT and MRI from 4 patients that had received both types of scans. For CT, we exported 4 scans of 2 mm slice thickness (3 sagittal, 1 axial). For MRI, we exported 4 T1-weighted scans with slice thickness ranging from 0.75 mm to 5 mm (3 sagittal, 1 axial).

To make CT scans viable for use in the nTMS system, we performed a series of preprocessing steps (removal of non-patient structures such as e.g., headrest, proper windowing, intensity rescaling) using the Aliza Medical Imaging and DICOM Viewer (Aliza 1.98.12, Copyright 2014–2020 Aliza Medical Imaging, Bonn, Germany), the details of which are available as Supplementary Material (Document S1). The exported DICOM files were loaded into the nTMS system (Nexstim eXimia NBS system, version 4.0; Nexstim Plc., Helsinki, Finland) and used to create head models (Figure 1A,B).

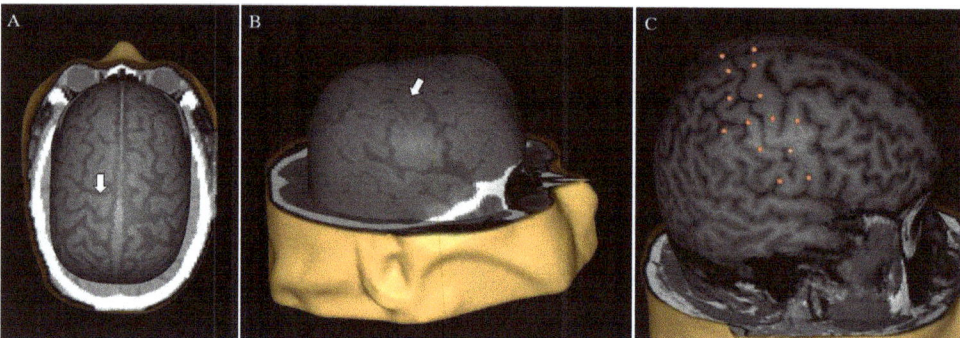

Figure 1. Head model based on computed tomography (CT). By using appropriate preprocessing, CT can serve as a viable basis for head model reconstruction in the nTMS system (**A,B**). Important anatomical structures such as the precentral gyrus are easy to identify (**A,B**, white arrows). For image testing, 12 targets were placed around the central sulcus on each hemisphere with two additional targets at the midline vertex (26 targets total per brain, **C**).

2.3. CT/MRI Comparison

In order to validate the use of CT imaging compared to the standard MRI imaging, we compared the e-field calculated by the system in both modalities in four sets of imaging data. For each head model, 26 stimulation targets were put on the cortical surface, distributed around the central sulcus (Figure 1C). The virtual model was co-registered to a dummy head in 3 cases and to the actual patient in 1 case. Each target was subsequently stimulated with an intensity of 30% of maximum stimulator output and the calculated local maximum e-field was recorded. This specific stimulation intensity was selected since 30% of maximum stimulator output is a common benchmark for motor thresholds in neurosurgical motor mappings.

2.4. Motor Mapping Protocol

For patient motor mappings, we followed the existing guidelines [8]. We employed an e-field-navigated TMS system (Nexstim eXimia NBS system, version 5.0; Nexstim Plc., Helsinki, Finland). The system provides an integration of all necessary subcomponents (navigation, figure-of-eight stimulation coil, EMG). For EMG recordings, surface electrodes were used (Neuroline 700; Ambu, Ballerup, Denmark). The nTMS system was positioned next to the patient's bed. The patient's skin at electrode placement spots was cleaned with alcohol swabs. EMG electrodes were applied to all muscles of interest, chosen according to the lesion location and predicted or known motor deficits.

EMG noise optimization was performed by finding optimal electrode placement spots, avoiding crossings of cables and eliminating all non-essential sources of electromagnetic noise [23]. Recorded muscles for the upper extremity were deltoid muscle, biceps brachii muscle, flexor carpi radialis muscle, and either abductor pollicis brevis muscle or adductor digiti minimi muscle, depending on accessibility. For lower extremities, recorded muscles were quadriceps femoris muscle, tibialis anterior muscle, and gastrocnemius muscle. MEPs were considered valid according to amplitude (above 50 µV), latency (10–35 ms for upper extremity musculature, 30–60 ms for lower limb musculature) [24], and plausible morphology.

Prepared imaging was loaded into the system and a head model was created. After successful co-registration, the head model was peeled to an individual depth allowing for visual identification of the precentral gyrus. During the entire mapping, the coil was positioned tangentially to local skull surface, aided by the neuronavigation system [3]. E-field orientation was kept at 90 degrees to local gyrus orientation [1]. A rough mapping was performed to give primary information on muscle representations of the upper extremity and possible motor hotspots. After identification of a given motor hotspot, said point was used for resting motor threshold (rMT) determination, defined as the stimulation intensity that results in a valid MEP in 50% of cases [8]. rMT determination was performed via a system-integrated procedure based on the maximum likelihood algorithm [25,26].

After determination of the rMT, the full extent of the MCS was mapped with 105% rMT. Then, lower extremity musculature was mapped with an intensity starting at 130% rMT. The distance between individual stimuli was roughly 5 mm. If no MEPs could be elicited, the stimulation intensity was raised stepwise until either MEPs were present, the patient reported discomfort, or no further raise was possible. For patients with prior hemicraniectomy (consisting of the temporary removal of a large fronto-temporo-parietal skull flap to allow for decompression of brain tissue) intensity was capped at 75% of maximum stimulator output due to safety considerations. In post-hoc analysis of motor mappings, the amplitude of EMG noise after signal optimization was measured and noted down for each mapped muscle for later descriptive statistical analysis.

The primary maxim during motor mapping was to avoid disrupting the nursing routine or clinical necessities. Extraordinary effort was made to explain the testing procedure to the bedside nurses and to ensure that testing would not disrupt their patient care responsibilities. Emphasis was made on the importance of open communication between the nursing team and the nTMS team. During testing, no active monitoring was removed from the patient and care was taken to minimize patient mobilization. If any other procedures had to be performed on the patient (e.g., neurological examination, suction, repositioning, and toileting), the examination was paused or aborted. After the examination, all nTMS equipment was removed from the patient, the room was returned to its original layout, and the system was cleaned and brought back to its designated storage place.

2.5. Statistics

Statistical analysis was performed in GraphPad Prism (version 8.0.0, GraphPad Software, San Diego, CA, USA). Due to the lack of prior investigations in the literature, we were hesitant to assume any given data distribution. Therefore, we tested for statistically significant differences between CT- and MRI-based e-field maxima using the non-parametric Wilcoxon matched-pairs signed rank test. The level of statistical significance was defined as $p < 0.05$.

3. Results

3.1. Patients

We performed motor mappings by nTMS in a group of 21 patients (6 females) with a range of diagnoses and etiologies (Table 2). Median age was 69 years (age range: 17–89 years). Seven patients were mapped under sedation via propofol (median dosage: 30 mcg/kg/min, dosage range: 15–80 mcg/kg/min). Nine patients underwent motor mapping of a hemisphere ipsilateral to hemicraniectomy. Five patients had ICP monitoring in place (two via drainage, three via cranial bolt). Average ICP in these cases was 15 ± 4 mmHg (median ICP: 17 mmHG, range ICP: 9–18 mmHg).

Table 2. Patient overview.

Patient	Age	Sex	Principal Diagnoses	Etiology	MEPs Elicited
01	69	m	Osteomyelitis of cervical spine	Chronic	Yes
02	45	m	SAH, SDH	Fall	Yes
03	76	m	SDH	Fall	Yes
04	82	m	Central Cord Syndrome	Fall	Yes
05	74	f	SDH	Fall	Yes
06	89	f	Central Cord Syndrome	Fall	Yes
07	64	m	Occlusion of ICA	Spontaneous	No
08	61	m	SAH, SDH	Fall	Yes
09	17	f	TBI	Vehicle accident	Yes
10	66	m	Central Cord Syndrome	Assault	Yes
11	49	m	SDH, IPH	Unknown	No
12	57	m	SDH	Fall	Yes
13	77	m	C2 fracture	Fall	Yes
14	75	f	Cervical spine injury	Vehicle accident	Yes
15	79	m	Spinal metastases	HCC	Yes
16	71	f	Spinal cord injury C1-C7	Fall	Yes
17	77	m	SDH	Fall	Yes
18	39	m	SDH	Assault	Yes
19	70	f	Hemorrhagic Stroke	Spontaneous	No
20	40	m	SDH	Unknown	Yes
21	67	m	C1 fracture	Assault	Yes

This table shows an overview over all patients included in this study. Abbreviations: subarachnoidal hemorrhage (SAH), subdural hemorrhage (SDH), internal carotid artery (ICA), traumatic brain injury (TBI), intraparenchymal hemorrhage (IPH).

3.2. Imaging Testing Results

The average difference of CT to MRI e-fields was −0.4 V/m ± 7.5 V/m (median difference: 0 V/m), and the maximum difference in e-fields between corresponding points was 18 V/m. In our statistical analysis of recorded e-field values, we found no statistically significant difference (mean CT e-field: 52 ± 14 V/m vs. mean MRI e-field: 52 ± 11 V/m; $p = 0.6574$).

3.3. Mapping Results

MEPs were successfully evoked in 18 patients. Average rMT (in percent of maximum stimulator output) was 63 ± 20% (median rMT: 68%, rMT range: 28–91%) for left hemisphere and 62 ± 22% (median rMT: 63%, rMT range: 31–100%) for right hemisphere. Within the sedated group of patients, MEPs were successfully evoked in 5 out of 7 patients, including the two cases with the highest observed sedation levels of 80 mcg/kg/min and 70 mcg/kg/min propofol, respectively. Excluding image preparation, time required per mapped hemisphere was approximately 90 min.

There were no adverse effects during or immediately after motor mapping. Conscious patients did not report the stimulation to be painful or uncomfortable. We observed no changes in continuously monitored autonomic parameters (O_2 saturation, blood pressure, heart frequency, ICP) attributable to stimulation. nTMS motor mapping did not prevent any other necessary procedure (e.g., neurochecks, application of medication, and suction) from being carried out. If necessary, the mapping was paused until the end of the procedure.

Intracranial access devices (e.g., subdural drainage and cranial bolt) interfered with motor mappings due to physical limitations imposed on coil placement. Cranial bolts specifically blocked access to sections of the respective hemisphere by taking up space. Motor mapping in these patients was, therefore, restricted to the hemisphere contralateral to the cranial bolt. ICP monitoring via drainages complicated but did not prohibit mappings of the ipsilateral cortex.

Additionally, we were able to demonstrate that nTMS navigation equipment interferes with pulse oximetry (Figure 2). This problem was solved by putting textiles over the probe, thereby blocking the interfering signal.

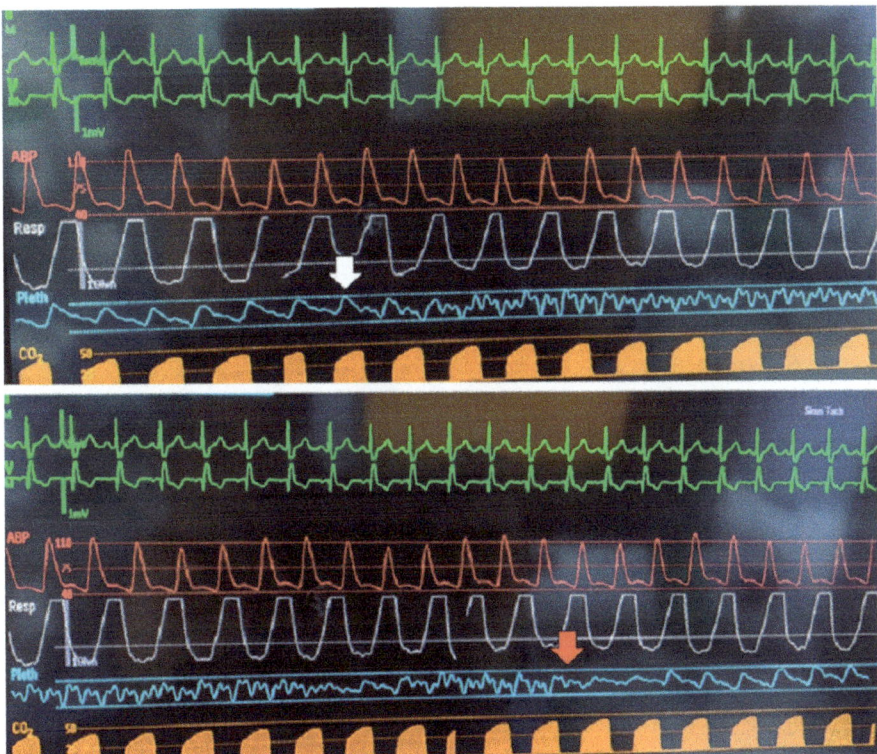

Figure 2. Interference of navigation equipment with pulse oximetry. This figure demonstrates the effect that the navigated transcranial magnetic stimulation (nTMS) navigational system imposes on pulse oximetry. The underlying mechanism is likely the infrared-spectrum based navigational camera creating interference with light used in pulse oximetry. The white arrow indicates the moment at which the camera was positioned for nTMS navigation. The red arrow indicates the moment at which a layer of textiles was put over the pulse oximetry probe.

3.4. EMG Noise Results

High amounts of noise were very common for patients in the ICU. With careful selection of electrode placement, however, it was usually possible to optimize noise sufficiently to allow for motor mapping (Figure 3). Average noise amplitude after optimization was 64 ± 58 µV (median noise amplitude: 43.5 µV). Trial-and-error approaches were often required for successful optimization, which entailed up to 10 different electrode placement trials per muscle. For each muscle group, particular anatomic locations for electrode placement (muscle reference electrode and grounding electrode) emerged. Despite identifying these sites as having a higher probability for noise mitigation, optimization via trial and error was nevertheless required in most cases (Figure 4). In many cases, noise optimization was the most time-consuming part of the procedure.

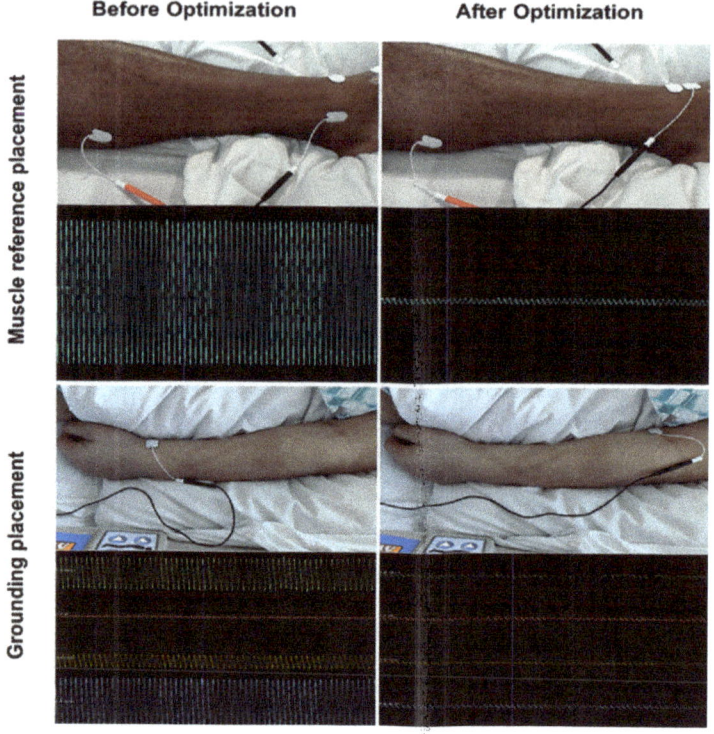

Figure 3. Electromyography (EMG) noise optimization. This figure illustrates the effect of electrode placement on noise level. All placement spots are common for neutral electrode placement, yet significant differences can be observed. At bedside, optimal positioning often requires testing of different spots until an adequate noise level is achieved. EMG scales are equal within the same row.

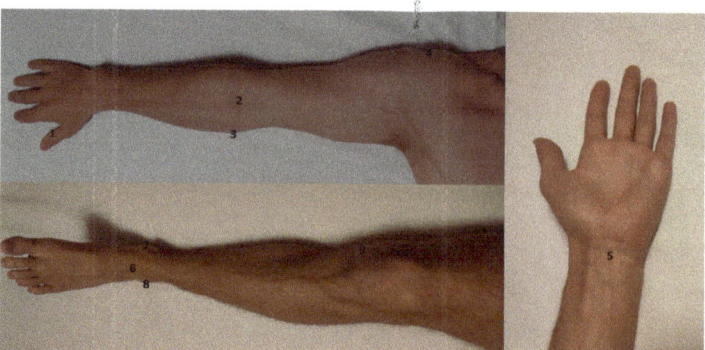

Figure 4. Placement spots for neutral electrodes. This figure illustrates anatomic landmarks that often emerged as viable for neutral electrode placement. 1: medial and lateral side of thumb, interphalangeal joint; 2: tendon of biceps brachii muscle; 3: medial epicondyle of humerus; 4: acromion; 5: tendons of hand flexors; 6: tendons of foot extensors; 7: medial malleolus of tibia; 8: lateral malleolus of fibula; 9: medial epicondyle of femur.

3.5. Illustrative Patient Cases

3.5.1. Sixty-One-Year-Old Male with Subarachnoid/Subdural Hemorrhage

A 61-year-old male presenting with left-sided subdural and subarachnoid hemorrhage after a fall was transferred to ICU. For clinical management, left-sided hemicraniectomy and sedation were required, the latter of which kept the patient in a comatose state. Due to asymmetric reaction to painful stimuli (left limbs moving against gravity, no movement on right side), motor mapping by nTMS was performed on the left hemisphere to test MCS function (right hemisphere was blocked due to cranial bolt). In contrast to the clinical examination, we were able to demonstrate intact MEPs in the adductor digiti minimi muscle (rMT: 36% of maximum stimulator output). Furthermore, by increasing mapping intensity to 125% of rMT, we observed additional MEPs in biceps brachii and flexor carpi radialis muscles, indicating that corticospinal tract innervation of those muscle groups was intact. As a result of this map, additional workup (CT angiography, EEG, MRI) of the patient's asymmetric examination was deferred. A follow-up motor mapping performed three days later showed a decrease in rMT to 28% of maximum stimulator output, indicating normalization of MCS threshold. Additionally, MEPs for tibialis anterior and gastrocnemius muscles were now detectable. Volitional movement started to return in the clinical examination six days after the initial mapping with slight withdrawals from painful stimuli. At discharge, 14 days after initial motor mapping, the patient displayed symptoms of hemineglect, but was able to move all extremities spontaneously.

3.5.2. Sixty-Seven-Year-Old Male with C1 Ring Fracture

A 67-year-old male patient presenting with comminuted/displaced C1 ring fracture after assault was treated on the ICU, presenting without any initial neurological symptoms. The patient underwent stabilization surgery, after which he developed aspiration pneumonia with acute respiratory distress syndrome (ARDS), necessitating prolonged sedation and ventilation under medically induced paralysis. Seventeen days after admission, the patient's ARDS improved and the paralysis was stopped. After the paralytics had worn off (as determined by intact train-of-four), the patient would open their eyes and track movement but remained unresponsive to central or peripheral stimulation. Motor mapping by nTMS was performed for both hemispheres. Despite the lack of movement in the clinical examination and a concurrent propofol infusion (10 mcg/kg/min), we were able to demonstrate MEPs in at least one muscle for every limb, albeit with very high rMT (81% and 99% of maximum stimulator output for left and right hemisphere, respectively). These findings effectively ruled out an occult spinal cord injury, and as a result, further workup with MRI and EEG was deferred. The next day, in a follow-up study with increased propofol dosage (40 mcg/kg/min), MEPs disappeared from both upper limbs and the left lower limb. The right lower limb still demonstrated MEPs but required a higher stimulation intensity of 100% versus 90% on the day before. These changes were thought to be a result of the increased propofol dose. In the subsequent days, the patient level of consciousness improved but he remained unable to follow commands or move volitionally, which was attributed to myoneuropathy of critical illness. Shortly thereafter, the patient developed a bowel perforation, and died of septic shock.

3.5.3. Seventy-Seven-Year-Old Male with Subdural Hematoma

A 77-year-old male was referred to the ICU with left holohemispheric subdural hematoma and midpontine hemorrhage after a fall. The patient remained non-verbal during his entire treatment. At admission, the patient did not follow commands and displayed hemiparesis on his right side but showed flexion to painful stimulation with his left side. The patient's family declined immediate surgical intervention, instead opting to wait for potential improvement over time. Four days after admission, the patient started following commands from his wife, but continued to only move his left side. Eight days after admission, the neurological state of the patient deteriorated, and he demonstrated left-sided decerebrate posturing. CT imaging demonstrated an increase in the size of the subdural hematoma and subfalcine as well as uncal herniation. After consulting with the patient's family, hemicraniectomy and evacuation of the hematoma were performed over the left hemisphere. Two days after the procedure, nTMS motor mapping was performed due to continued absence of movement on the right side. Contrary to the clinical presentation, MEPs were present bilaterally for both upper and lower extremities. Notably, the rMT on the left hemisphere was higher (42% of maximum stimulator output) than the rMT on the right hemisphere (34% of maximum stimulator output), despite the left sided hemicraniectomy. The results were therefore interpreted as pointing towards a damaged but still functional MCS, with hemineglect as a potential explanation for the present symptoms. Physical therapy, therefore, emphasized maintaining right-sided range of motion and targeted therapy to address right hemi-neglect. During this time, patient began to regain motor function on his right side. A follow-up mapping took place 8 days after the inital map. At this point, the patient had received cranioplasty and demonstrated an rMT of 37% of maximum stimulator output for the right hemisphere and 62% of maximum stimulator output on the left hemisphere. The patient's rehabilitation continued, his neglect improved, and he eventually demonstrated 4/5 muscle strength throughout his right side. Upon his discharge 26 days after admission, he was intermittently following commands, moving all his extremities purposefully, and was able to continue his physical and cognitive therapy in an acute rehabiliation institution.

4. Discussion

We performed nTMS motor mappings in 21 patients being treated in the neurological ICU for a variety of neurological injuries. All mappings were successfully and safely completed. In so doing, we identified common problems that occur in the ICU setting and, where possible, we arrived at solutions for these problems. This feasibility study will enable future studies to focus on establishing the clinical value nTMS-based motor mappings in the ICU and, more broadly, in the inpatient setting.

4.1. Safety

As with any new clinical test, particularly when dealing with the critically ill, patient safety is of utmost importance. nTMS in general is widely considered a safe modality, especially when single-pulse protocols are used [5,21]. The most serious adverse effect associated with TMS is the occurrence of epileptic seizures. These, however, are exceedingly rare, limited to isolated case reports often in individuals with prior history of epilepsy [27,28]. While we did exclude from our study any patient with uncontrolled epilepsy, none of our patients showed any adverse effects during or immediately after single-pulse nTMS motor mapping. No acute worsening of clinical state was attributable to motor mapping. This was also true for the nine patients who underwent mapping following prior hemicraniectomy, in whom the magnetic field incident upon the cortex is likely stronger than for patients with an intact skull. It should be noted again, however, that we capped the stimulation intensity for hemicraniectomy hemispheres to 75% of maximum stimulator output for precautionary reasons. Although these data are gathered from a limited number of cases, it is highly encouraging and indicates that nTMS is likely safe enough to be performed in patients with severe acute neurological damage and altered cerebral anatomy.

4.2. Neuronavigation Based on CT

A prerequisite for nTMS motor mapping is preexisting cranial imaging used for neuronavigation. In the conventional workflow, MRI offers unparalleled anatomical imaging of brain tissue with a high soft tissue contrast, and it is, therefore, the gold standard for navigational imaging [8]. MRI, however, is oftentimes not a viable option for ICU patients due to practical considerations or contraindications for scanning. Specifically, critically ill patients are often unable to tolerate lengthy MRI studies. Furthermore, intracranial monitoring devices such as cranial bolts and indwelling electrodes can be incompatible with the scanning environment [29]. Although it offers less detailed imaging of intracranial anatomy than MRI, CT is by far the more prevalent modality for patients in the ICU due to its rapid acquisition time and compatibility with clinical equipment [29,30]. Therefore, to enable widespread use of nTMS motor mapping in the ICU, we realized that CT must be adapted and validated as the basis for neuronavigation. Using CT for this purpose has previously only been described in a singular recent case which did not address possible navigation discrepancies between CT and MRI [31]. By preprocessing CT scans with a slice thickness of 2 mm, we were able to generate usable head models suited for patient co-registration (Figure 1). Comparing the calculated maximum e-field at corresponding points in CT head models to that in MRI head models, we found no significant difference in calculated e-field values. While the maximum difference between two points was 18 V/m, we believe that this discrepancy was likely an artifact due to the inherent difficulty of co-registering a given head model to a physical dummy with different measurements. Our results indicate that both CT and MRI can yield comparable neuronavigation in the employed nTMS system. Following our process enables targeted, replicable stimulation of given brain loci in patients lacking MRI data. This application of CT-based neuronavigation not only makes motor mapping in the ICU an option, but also extends nTMS usage to any patient unable to undergo MRI (because of retained metal fragments, implanted devices, etc.).

4.3. Compatibility with ICU Workflows

In our cases, translation of the outpatient workflow into the ICU setting did, for the most part, not pose significant problems. Nursing routines or monitoring were not impeded. In this regard, the flexibility of nTMS motor mappings is a relevant asset, as the examination takes place at the bedside and can be paused at any moment. Pulse oximetry, while being influenced by navigational equipment (Figure 2), can easily be maintained by physically covering the oximetry probe. The only major limitation was imposed by immovable devices connected to the patient's skull (i.e., cranial bolts placed for anchoring invasive ICP monitoring), as they tended to physically obstruct free movement of the coil. Overall, our experience demonstrated that nTMS motor mapping is compatible with the monitoring setup and clinical routine in ICU cases.

4.4. Optimization of EMG

For reliable and accurate motor maps, clear EMG readings are necessary to detect even MEPs with small amplitudes. Any active electrical device in proximity to the recording site (e.g., ventilators, perfusors, and monitoring) is a potential source of noise [23]. It is, therefore, unsurprising that noise levels in the ICU setting are far higher than in the normal outpatient setting. Usually, electronic noise presents as a uniform waveform with frequency of about 60 Hz (corresponding to the 60 Hz alternating current of standard US power outlets). Elimination of noise by deactivation of electrical devices is for the most part impossible since their continued functioning is essential for patient care. While this noise generally allows for MEP detection via disruption of the regular noise pattern (Figure 5), maximum signal-to-noise ratio is required to ensure validity and comparability of individual mappings. To minimize noise, general guiding principles of EMG should be followed, such as avoiding contacts between individual cables, cleaning the skin prior to electrode placement, and ensuring full surface contact of electrodes with skin.

The most important factor in our study, however, seemed to be the location of electrode placement, particularly regarding reference and grounding electrodes (Figure 3). Established anatomical landmarks that work in the outpatient setting may not work in the intensive care case. Anatomical positions often proved to be successful neutral electrode targets in this study (Figure 4). In our cases, we were able to achieve an average noise amplitude of 64 ± 58 µV. This distribution is slightly skewed by a few cases with persistent high-amplitude noise levels, which is reflected in the median level of 43.5 µV. Analyzing data at this level of noise is possible, since any valid MEP (threshold of 50 µV) will peak through the background noise. In higher noise levels, however, analysis gradually loses sensitivity and it becomes difficult to detect low-amplitude MEPs. Although this noise lowers the resolution of motor mappings, useful information regarding binary questions of motor system integrity (such as "is any activation present?") can still be obtained. One heuristic for detecting valid MEPs amongst regular noise is to subtract noise amplitude from the potential MEP. If the amplitude is still >50 µV, the MEP may be considered valid (Figure 5). Placement of multiple stimuli at a given location to verify replicability is also advisable in cases of questionable MEPs.

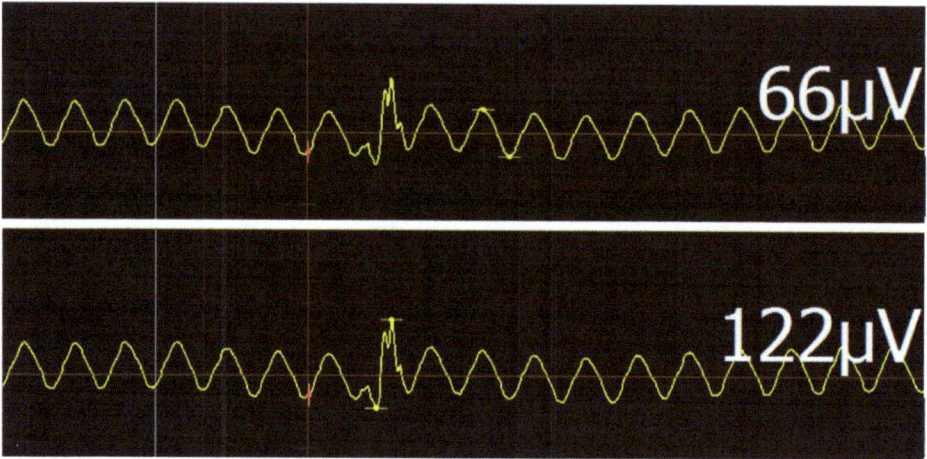

Figure 5. Motor evoked potential (MEP) disrupts 60 Hz noise. This figure shows regular 60 Hz noise disrupted by MEP occurrence. Measurements on the right correspond to amplitude between yellow measurement bars. One heuristic for detecting valid MEPs amongst regular noise is to subtract the noise amplitude from the potential MEP. If the amplitude is still >50 µV, the MEP may be considered valid (pictured). Additionally, placement of multiple stimuli at a given location can aid when MEPs are questionable. Note the two individual peaks within the MEP formed by the underlying noise reaching its minimum.

4.5. Pharmacological and Physiological Confounders

Patients in the ICU are subject to a variety of factors that most likely affect MEP elicitability, such as elevated ICP, hemicraniectomy, and propofol sedation. Propofol has previously been shown to decrease MEP amplitude, yet without impacting MEP latency [32–34]. One study compared the influence of different anesthetic agents on MEPs in animals and humans. The results indicated that even in high concentrations, propofol was unable to fully inhibit MEPs elicited by higher-than-threshold stimulation [34]. Additionally, it has been demonstrated that intravenous agents such as propofol impact MEPs less than volatile anesthetics [35]. The highest sedation levels in our study were propofol infusion of 80 mcg/kg/min and 70 mcg/kg/min. These doses did not prevent elicitation of MEPs in either case. While other patients may require higher doses, the current findings fit to the aforementioned literature, which indicates that MEP monitoring is still possible under propofol sedation. Based on the literature and our experience, we would recommend that patients under sedation showing no MEPs at 105% rMT should be tested with progressively higher stimulation intensities. While not tested in our approach, the literature indicates that TMS with repetitive pulse patterns might also be an option to elicit reliable MEPs under propofol sedation [34,36]. The dose-dependent input/output relationship between stimulus intensity and MEP amplitude might be exploited for individualized functional anesthesia monitoring. For nTMS applications in the ICU that are not specifically related to the MCS, it should be kept in mind that higher-order neuronal function such as network complexity are influenced by propofol as well [37]. Of note, opioids seem to only possess a small effect on MEPs, and we found no dampening of MEP due to opioids in our series [35].

Furthermore, ICP is another factor that could plausibly influence MEPs. Persistently elevated ICP is a known predictor of poor neurological outcome, as it lowers cerebral perfusion pressure resulting in cerebral hypoxia [38,39]. These effects are primarily observed at sustained ICPs of 25 mmHg and higher [38]. All of our cases were performed with an ICP lower than 20 mmHg, as patients are treated aggressively to maintain an ICP below this value. Therefore, while we can state that MEPs were still elicitable in the highest observed ICP of 18 mmHg, we can make no claims as to the influence of ICPs in higher ranges. Hemicraniectomy is a procedure often utilized to combat elevated ICP [40,41], and nine of our patients had undergone this procedure. While our main concern in mapping patients after hemicraniectomy was patient safety, it is worth reporting that MEPs could still be elicited in this abnormal anatomical state.

4.6. Implications and Applications

We believe that our reported findings serve as valid evidence for the general feasibility of nTMS motor mappings in the intensive care setting. In our series, we were able to map successfully 18 out of 21 patients, yielding reproducible MEPs, and resulting in no complications. These findings demonstrate that, for the critically ill patient with neurological injury, nTMS can deliver quantitative analysis of MCS functionality in a safe and reliable manner. nTMS motor mapping allows for objective investigations into the integrity of the motor system along its entire path, from upper motor neurons to lower motor neurons, and thereafter to the neuromuscular junction. Importantly, it does not rely on patient consciousness or cooperation. Furthermore, rMT specifically has been described to reflect neuronal excitability, which may be a metric of interest in a variety of acute neurological problems [42]. One example is traumatic brain injury, which has been linked to rises in rMT [43].

The potential applications and benefits of nTMS motor mapping on the ICU are numerous. Examinations can be performed on conscious and unconscious patients alike and yield detailed information about the function of individual corticospinal tracts. The procedure is painless, noninvasive and does not require (yet is possible during) sedation. Additionally, any trained member of the clinical team can perform a mapping without the need for additional personnel.

Compared to other, more established neurophysiological modalities such as EEG or somatosensory evoked potentials, nTMS offers the unique capability of linking effects to a specific, repeatedly targetable cortical site [16]. This enables detailed tracking of functional MCS changes over time, which is not present with non-navigated TMS (due to targeting inaccuracy) or EEG methods. Earlier diagnosis of specific motor deficits and prediction of motor recovery may be possible, since corticospinal integrity can be objectively demonstrated for any EMG-suited target muscle. For example, MEP recovery has recently been shown to predict neurological outcome in endovascular thrombectomy [44]. Muscular recruitment curves could serve as the basis of a standardized measure of cortical excitability. In a similar vein, individualized functional drug monitoring specifically for antiepileptic drugs or anesthesia seems plausible [33,45–47]. Protocols for specific types of monitoring could potentially be used for recognition and monitoring of transient complications, such as vasospasm in subarachnoid hemorrhage.

Of note, in all three clinical cases introduced above, nTMS motor mapping yielded information on the acute state of the MCS that was not visible in the clinical examination by demonstrating functional corticospinal motor connection despite lack of movement.

4.7. Limitations

While our findings confirm the general feasibility of nTMS motor mappings in the ICU setting and demonstrate some use cases, there are several limitations that need to be addressed. First, this study includes only 21 unique patients, which may be considered a small sample size. However, as a pilot feasibility study, we are confident that this sample is adequate to provide general guiding principles that may be built upon in future studies. Second, our imaging testing was limited both by number of cases and by the margin of error in registering a head model to a dummy head of different size. We believe, however, that with 104 total targets per modality, we would have been able to discover any meaningful systematic difference in e-field calculation if it had existed. A separate aspect of interest would be to investigate the comparability of CT and MRI not only based on calculated e-field, but also on physical coil position, which would require some form of system-separate coordinate measurement.

For a more extensive comparison, a large set of cases with CT and MRI datasets acquired within a small timeframe would be necessary. Since CT is inherently connected to radiation exposure, a study such as this could hardly be done with healthy volunteers. Even with patients, however, indications for both imaging modalities in short intervals remain rare, which unfortunately limits the pool of usable data. In conclusion, while the presented evidence may be lacking in robustness, it serves as a basis for more extensive systematic testing and, more importantly, demonstrates that repeatable targeting of specific cortical stimulation sites is possible with CT imaging. Third, this study took place in a single ICU at a single institution. Other centers may encounter other challenges based on respective local workflow or electrical noise environment. Fourth, in terms of clinical factors influencing MEPs, we were unable to collect data on high-level ICPs as well as sedation other than propofol. It is difficult to see a simple resolution to this limitation, since optimal treatment of the patient in terms of ICP control and best practice sedation is naturally of the highest priority. Data collection is, however, ongoing, and respective cases may appear to further our understanding of how clinical factors act upon MEPs. Specifically, a standardized I/O curve for varying levels of propofol could be very useful in interpreting mapping results achieved under varying levels of sedation. Fifth and lastly, while the navigational aspect of nTMS enables reproducible targeting of specific cortical sites, the resulting MEPs are also influenced by aspects aside from cortical stimulation site. This could complicate longitudinal patient-specific monitoring of nTMS based parameters such as rMT. Examples for these factors include underlying EEG activity, prior activity of the target muscle or limb posture [48–51]. While these factors are known confounders of MEP replicability, other studies have shown that by sampling 20 or more trials, good inter-session reliability can be achieved regarding measures of motor cortex excitability [48,49]. This should be kept in mind for any clinical application involving longitudinal measurement of MCS excitability.

5. Conclusions

This study demonstrated the feasibility of nTMS motor mappings in the ICU setting. In so doing, we have shown how a variety of confounding factors inherent to the ICU setting may be mitigated. We have demonstrated that MEPs can be obtained in spite of propofol and opioid sedation, that electrical interference due to ICU equipment can be reduced to the point where mapping is possible, and that CT may be used successfully for nTMS navigation. Moreover, we have demonstrated how nTMS can be successfully integrated into routine workflow and intensive patient care. Our data further suggest that single-pulse motor mapping is safe in a range of different diagnoses commonly encountered in the neurological ICU. We have also presented three case studies in which nTMS mapping provided valuable insight into the neurological state of the patient, and two cases changed the clinical management. These results demonstrate a need for future studies to evaluate systematically the clinical benefits provided by the use of nTMS in patients with acute neurological injury.

Supplementary Materials: The following are available online at http://www.mdpi.com/2076-3425/10/12/1005/s1, Document S1: Using cranial CT to construct a headmodel in Nexstim NBS.

Author Contributions: Conceptualization, S.M.K. and P.E.T.; Data curation, S.S.; Formal analysis, S.S.; Investigation, S.S. and A.F.H.; Methodology, S.S. and P.E.T.; Project administration, P.E.T.; Resources, L.C. and P.E.T.; Supervision, L.C., S.M.K. and P.E.T.; Validation, S.S.; Visualization, S.S.; Writing—original draft, S.S.; Writing—review & editing, S.S., S.M.K., N.S. and P.E.T.; All authors have read and agreed to the published version of the manuscript.

Funding: This research received no external funding.

Acknowledgments: We would like to thank the UCSF Biomagnetic Imaging Laboratory for their support of this study.

Conflicts of Interest: S.M.K. is a consultant for Brainlab AG (Munich, Germany) and for Nexstim Plc (Helsinki, Finland). N.S. received honoraria from Nexstim Plc (Helsinki, Finland).

Abbreviations

CT	Computed tomography
DCS	Direct cortical stimulation
ECG	Electrocardiography
EEG	Electroencephalography
EMG	Electromyography
fMRI	functional magnetic resonance imaging
ICA	Internal carotid artery
ICP	Intracranial pressure
ICU	Intensive care unit
IPH	Intraparenchymal hemorrhage
MCS	Motor cortical system
MEG	Magnetoencephalography
MEP	Motor evoked potential
MRI	Magnetic resonance imaging
nTMS	Navigated transcranial magnetic stimulation
rMT	Resting motor threshold
SAH	Subarachnoidal hemorrhage
SDH	Subdural hemorrhage
TBI	Traumatic brain injury
TMS	Transcranial magnetic stimulation

References

1. Krieg, S.M. *Navigated Transcranial Magnetic Stimulation in Neurosurgery*; Springer: Berlin/Heidelberg, Germany, 2017.
2. Rotenberg, A.; Horvath, J.C.; Pascual-Leone, A. The transcranial magnetic stimulation (TMS) device and foundational techniques. In *Transcranial Magnetic Stimulation*; Springer: Berlin/Heidelberg, Germany, 2014; pp. 3–13.
3. Sollmann, N.; Goblirsch-Kolb, M.F.; Ille, S.; Butenschoen, V.M.; Boeckh-Behrens, T.; Meyer, B.; Ringel, F.; Krieg, S.M. Comparison between electric-field-navigated and line-navigated TMS for cortical motor mapping in patients with brain tumors. *Acta Neurochir.* 2016, *158*, 2277–2289. [CrossRef] [PubMed]
4. Comeau, R. Neuronavigation for transcranial magnetic stimulation. In *Transcranial Magnetic Stimulation*; Springer: Berlin/Heidelberg, Germany, 2014; pp. 31–56.
5. Tarapore, P.E.; Picht, T.; Bulubas, L.; Shin, Y.; Kulchytska, N.; Meyer, B.; Berger, M.S.; Nagarajan, S.S.; Krieg, S.M. Safety and tolerability of navigated TMS for preoperative mapping in neurosurgical patients. *Clin. Neurophysiol.* 2016, *127*, 1895–1900. [CrossRef] [PubMed]
6. Lefaucheur, J.-P.; André-Obadia, N.; Antal, A.; Ayache, S.S.; Baeken, C.; Benninger, D.H.; Cantello, R.M.; Cincotta, M.; de Carvalho, M.; De Ridder, D. Evidence-based guidelines on the therapeutic use of repetitive transcranial magnetic stimulation (rTMS). *Clin. Neurophysiol.* 2014, *125*, 2150–2206. [CrossRef] [PubMed]
7. Huang, Y.-Z.; Edwards, M.J.; Rounis, E.; Bhatia, K.P.; Rothwell, J.C. Theta burst stimulation of the human motor cortex. *Neuron* 2005, *45*, 201–206. [CrossRef]
8. Krieg, S.M.; Lioumis, P.; Mäkelä, J.P.; Wilenius, J.; Karhu, J.; Hannula, H.; Savolainen, P.; Lucas, C.W.; Seidel, K.; Laakso, A. Protocol for motor and language mapping by navigated TMS in patients and healthy volunteers; workshop report. *Acta Neurochir.* 2017, *159*, 1187–1195. [CrossRef]
9. Tarapore, P.E.; Tate, M.C.; Findlay, A.M.; Honma, S.M.; Mizuiri, D.; Berger, M.S.; Nagarajan, S.S. Preoperative multimodal motor mapping: A comparison of magnetoencephalography imaging, navigated transcranial magnetic stimulation, and direct cortical stimulation. *J. Neurosurg.* 2012, *117*, 354–362. [CrossRef]
10. Herwig, U.; Satrapi, P.; Schönfeldt-Lecuona, C. Using the international 10–20 EEG system for positioning of transcranial magnetic stimulation. *Brain Topogr.* 2003, *16*, 95–99. [CrossRef]
11. Julkunen, P.; Säisänen, L.; Danner, N.; Niskanen, E.; Hukkanen, T.; Mervaala, E.; Könönen, M. Comparison of navigated and non-navigated transcranial magnetic stimulation for motor cortex mapping, motor threshold and motor evoked potentials. *Neuroimage* 2009, *44*, 790–795. [CrossRef]

12. Laakso, I.; Hirata, A.; Ugawa, Y. Effects of coil orientation on the electric field induced by TMS over the hand motor area. *Phys. Med. Biol.* **2013**, *59*, 203. [CrossRef]
13. Picht, T. Current and potential utility of transcranial magnetic stimulation in the diagnostics before brain tumor surgery. *CNS Oncol.* **2014**, *3*, 299–310. [CrossRef]
14. Lefaucheur, J.-P.; Picht, T. The value of preoperative functional cortical mapping using navigated TMS. *Neurophysiol. Clin./Clin. Neurophysiol.* **2016**, *46*, 125–133. [CrossRef] [PubMed]
15. Krieg, S.M.; Sabih, J.; Bulubasova, L.; Obermueller, T.; Negwer, C.; Janssen, I.; Shiban, E.; Meyer, B.; Ringel, F. Preoperative motor mapping by navigated transcranial magnetic brain stimulation improves outcome for motor eloquent lesions. *Neuro-Oncology* **2014**, *16*, 1274–1282. [CrossRef] [PubMed]
16. Guérit, J.-M.; Amantini, A.; Amodio, P.; Andersen, K.V.; Butler, S.; de Weerd, A.; Facco, E.; Fischer, C.; Hantson, P.; Jäntti, V. Consensus on the use of neurophysiological tests in the intensive care unit (ICU): Electroencephalogram (EEG), evoked potentials (EP), and electroneuromyography (ENMG). *Neurophysiol. Clin./Clin. Neurophysiol.* **2009**, *39*, 71–83. [CrossRef] [PubMed]
17. André-Obadia, N.; Zyss, J.; Gavaret, M.; Lefaucheur, J.-P.; Azabou, E.; Boulogne, S.; Guérit, J.-M.; Mcgonigal, A.; Merle, P.; Mutschler, V. Recommendations for the use of electroencephalography and evoked potentials in comatose patients. *Neurophysiol. Clin.* **2018**, *48*, 143–169. [CrossRef] [PubMed]
18. Theilen, H.J.; Ragaller, M.; Tschö, U.; May, S.A.; Schackert, G.; Albrecht, M.D. Electroencephalogram silence ratio for early outcome prognosis in severe head trauma. *Crit. Care Med.* **2000**, *28*, 3522–3529. [CrossRef] [PubMed]
19. Azabou, E.; Fischer, C.; Guerit, J.M.; Annane, D.; Mauguiere, F.; Lofaso, F.; Sharshar, T. Neurophysiological assessment of brain dysfunction in critically ill patients: An update. *Neurol. Sci.* **2017**, *38*, 715–726. [CrossRef] [PubMed]
20. Casarotto, S.; Comanducci, A.; Rosanova, M.; Sarasso, S.; Fecchio, M.; Napolitani, M.; Pigorini, A.G.; Casali, A.; Trimarchi, P.D.; Boly, M.; et al. Stratification of unresponsive patients by an independently validated index of brain complexity. *Ann. Neurol.* **2016**, *80*, 718–729. [CrossRef]
21. Rossi, S.; Hallett, M.; Rossini, P.M.; Pascual-Leone, A.; Safety of TMS Consensus Group. Safety, ethical considerations, and application guidelines for the use of transcranial magnetic stimulation in clinical practice and research. *Clin. Neurophysiol.* **2009**, *120*, 2008–2039. [CrossRef]
22. Le Roux, P.; Menon, D.K.; Citerio, G.; Vespa, P.; Bader, M.K.; Brophy, G.M.; Diringer, M.N.; Stocchetti, N.; Videtta, W.; Armonda, R. Consensus summary statement of the international multidisciplinary consensus conference on multimodality monitoring in neurocritical care. *Neurocritical Care* **2014**, *21*, 1–26. [CrossRef]
23. Reaz, M.B.I.; Hussain, M.S.; Mohd-Yasin, F. Techniques of EMG signal analysis: Detection, processing, classification and applications. *Biol. Proced. Online* **2006**, *8*, 11–35. [CrossRef]
24. Sollmann, N.; Bulubas, L.; Tanigawa, N.; Zimmer, C.; Meyer, B.; Krieg, S.M. The variability of motor evoked potential latencies in neurosurgical motor mapping by preoperative navigated transcranial magnetic stimulation. *BMC Neurosci.* **2017**, *18*, 5. [CrossRef] [PubMed]
25. Sollmann, N.; Hauck, T.; Obermuller, T.; Hapfelmeier, A.; Meyer, B.; Ringel, F.; Krieg, S.M. Inter- and intraobserver variability in motor mapping of the hotspot for the abductor policis brevis muscle. *BMC Neurosci.* **2013**, *14*, 94. [CrossRef] [PubMed]
26. Awiszus, F. TMS and threshold hunting. In *Supplements to Clinical Neurophysiology*; Elsevier: Amsterdam, The Netherlands, 2003; Volume 56, pp. 13–23.
27. Di Iorio, R.; Rossini, P.M. Safety considerations of the use of TMS. In *Navigated Transcranial Magnetic Stimulation in Neurosurgery*; Springer: Berlin/Heidelberg, Germany, 2017; pp. 67–83.
28. Schrader, L.M.; Stern, J.M.; Koski, L.; Nuwer, M.R.; Engel Jr, J. Seizure incidence during single-and paired-pulse transcranial magnetic stimulation (TMS) in individuals with epilepsy. *Clin. Neurophysiol.* **2004**, *115*, 2728–2737. [CrossRef] [PubMed]
29. Williamson, C.; Morgan, L.; Klein, J.P. Imaging in neurocritical care practice. *Semin. Respir. Crit. Care Med.* **2017**, *38*, 840–852. [CrossRef]
30. Algethamy, H.M.; Alzawahmah, M.; Young, G.B.; Mirsattari, S.M. Added value of MRI over CT of the brain in intensive care unit patients. *Can. J. Neurol. Sci.* **2015**, *42*, 324–332. [CrossRef]
31. Pinto, P.H.d.C.F.; Nigri, F.; Caparelli-Dáquer, E.M.; dos Santos Viana, J. Computed tomography-guided navigated transcranial magnetic stimulation for preoperative brain motor mapping in brain lesion resection: A case report. *Surg. Neurol. Int.* **2019**, *10*, 134. [CrossRef]

32. Jellinek, D.; Jewkes, D.; Symon, L. Noninvasive intraoperative monitoring of motor evoked potentials under propofol anesthesia: Effects of spinal surgery on the amplitude and latency of motor evoked potentials. *Neurosurgery* **1991**, *29*, 551–557. [CrossRef]
33. Nathan, N.; Tabaraud, F.; Lacroix, F.; Mouliès, D.; Viviand, X.; Lansade, A.; Terrier, G.; Feiss, P. Influence of propofol concentrations on multipulse transcranial motor evoked potentials. *Br. J. Anaesth.* **2003**, *91*, 493–497. [CrossRef]
34. Scheufler, K.-M.; Zentner, J. Total intravenous anesthesia for intraoperative monitoring of the motor pathways: An integral view combining clinical and experimental data. *J. Neurosurg.* **2002**, *96*, 571–579. [CrossRef]
35. Wang, A.C.; Than, K.D.; Etame, A.B.; La Marca, F.; Park, P. Impact of anesthesia on transcranial electric motor evoked potential monitoring during spine surgery: A review of the literature. *Neurosurg. Focus* **2009**, *27*, E7. [CrossRef]
36. Sollmann, N.; Zhang, H.; Kelm, A.; Schröder, A.; Meyer, B.; Pitkänen, M.; Julkunen, P.; Krieg, S.M. Paired-pulse navigated TMS is more effective than single-pulse navigated TMS for mapping upper extremity muscles in brain tumor patients. *Clin. Neurophysiol.* **2020**, *131*, 2887–2898. [CrossRef] [PubMed]
37. Varley, T.F.; Luppi, A.I.; Pappas, I.; Naci, L.; Adapa, R.; Owen, A.M.; Menon, D.K.; Stamatakis, E.A. consciousness & Brain functional complexity in propofol Anaesthesia. *Sci. Rep.* **2020**, *10*, 1–13.
38. Treggiari, M.M.; Schutz, N.; Yanez, N.D.; Romand, J.-A. Role of intracranial pressure values and patterns in predicting outcome in traumatic brain injury: A systematic review. *Neurocritical Care* **2007**, *6*, 104–112. [CrossRef]
39. Raboel, P.; Bartek, J.; Andresen, M.; Bellander, B.; Romner, B. Intracranial pressure monitoring: Invasive versus non-invasive methods—A review. *Crit. Care Res. Pract.* **2012**, *2012*, 950393. [CrossRef] [PubMed]
40. Lilja-Cyron, A.; Andresen, M.; Kelsen, J.; Andreasen, T.H.; Fugleholm, K.; Juhler, M. Long-term effect of decompressive craniectomy on intracranial pressure and possible implications for intracranial fluid movements. *Neurosurgery* **2020**, *86*, 231–240.
41. Lilja-Cyron, A.; Andresen, M.; Kelsen, J.; Andreasen, T.H.; Petersen, L.G.; Fugleholm, K.; Juhler, M. Intracranial pressure before and after cranioplasty: Insights into intracranial physiology. *J. Neurosurg.* **2019**, *1*, 1–11. [CrossRef]
42. Ziemann, U.; Lönnecker, S.; Steinhoff, B.; Paulus, W. Effects of antiepileptic drugs on motor cortex excitability in humans: A transcranial magnetic stimulation study. *Ann. Neurol. Off. J. Am. Neurol. Assoc. Child Neurol. Soc.* **1996**, *40*, 367–378. [CrossRef]
43. Tallus, J.; Lioumis, P.; Hämäläinen, H.; Kähkönen, S.; Tenovuo, O. Long-lasting TMS motor threshold elevation in mild traumatic brain injury. *Acta Neurol. Scand.* **2012**, *126*, 178–182. [CrossRef]
44. Greve, T.; Wagner, A.; Ille, S.; Wunderlich, S.; Ikenberg, B.; Meyer, B.; Zimmer, C.; Shiban, E.; Kreiser, K. Motor evoked potentials during revascularization in ischemic stroke predict motor pathway ischemia and clinical outcome. *Clin. Neurophysiol.* **2020**, *131*, 2307–2314. [CrossRef]
45. Ziemann, U.; Reis, J.; Schwenkreis, P.; Rosanova, M.; Strafella, A.; Badawy, R.; Müller-Dahlhaus, F. TMS and drugs revisited 2014. *Clin. Neurophysiol.* **2015**, *126*, 1847–1868. [CrossRef]
46. Lotto, M.L.; Banoub, M.; Schubert, A. Effects of anesthetic agents and physiologic changes on intraoperative motor evoked potentials. *J. Neurosurg. Anesthesiol.* **2004**, *16*, 32–42. [CrossRef] [PubMed]
47. Andreasson, A.C.; Sigurdsson, G.V.; Pegenius, G.; Thordstein, M.; Hallböök, T. Cortical excitability measured with transcranial magnetic stimulation in children with epilepsy before and after antiepileptic drugs. *Dev. Med. Child Neurol.* **2020**, *62*, 793–798. [CrossRef] [PubMed]
48. Goldsworthy, M.; Hordacre, B.; Ridding, M. Minimum number of trials required for within-and between-session reliability of TMS measures of corticospinal excitability. *Neuroscience* **2016**, *320*, 205–209. [CrossRef] [PubMed]
49. Biabani, M.; Farrell, M.; Zoghi, M.; Egan, G.; Jaberzadeh, S. The minimal number of TMS trials required for the reliable assessment of corticospinal excitability, short interval intracortical inhibition, and intracortical facilitation. *Neurosci. Lett.* **2018**, *674*, 94–100. [CrossRef]
50. Darling, W.G.; Wolf, S.L.; Butler, A.J. Variability of motor potentials evoked by transcranial magnetic stimulation depends on muscle activation. *Exp. Brain Res.* **2006**, *174*, 376–385. [CrossRef] [PubMed]

51. Vargas, C.; Olivier, E.; Craighero, L.; Fadiga, L.; Duhamel, J.; Sirigu, A. The influence of hand posture on corticospinal excitability during motor imagery: A transcranial magnetic stimulation study. *Cereb. Cortex* **2004**, *14*, 1200–1206. [CrossRef] [PubMed]

Publisher's Note: MDPI stays neutral with regard to jurisdictional claims in published maps and institutional affiliations.

 © 2020 by the authors. Licensee MDPI, Basel, Switzerland. This article is an open access article distributed under the terms and conditions of the Creative Commons Attribution (CC BY) license (http://creativecommons.org/licenses/by/4.0/).

Case Report

Preoperative Repetitive Navigated TMS and Functional White Matter Tractography in a Bilingual Patient with a Brain Tumor in Wernike Area

Valentina Baro [1,*], Samuel Caliri [1], Luca Sartori [1], Silvia Facchini [2], Brando Guarrera [1], Pietro Zangrossi [1], Mariagiulia Anglani [3], Luca Denaro [1], Domenico d'Avella [1], Florinda Ferreri [4] and Andrea Landi [1]

1. Academic Neurosurgery, Department of Neuroscience, University of Padova, 35128 Padova, Italy; samucaliri91@gmail.com (S.C.); sartori.luca.92@gmail.com (L.S.); brandoguarrera@gmail.com (B.G.); pietro.zangrossi@gmail.com (P.Z.); luca.denaro@unipd.it (L.D.); domenico.davella@unipd.it (D.d.); andrea.landi@unipd.it (A.L.)
2. Department of Neuroscience DNS, University of Padova, 35128 Padova, Italy; facchini.silvia@gmail.com
3. Unit of Neuroradiology, Padova University Hospital, 35128 Padova, Italy; mariagiulia.anglani@aopd.veneto.it
4. Unit of Neurology and Neurophysiology, Department of Neuroscience, University of Padova, 35128 Padova, Italy; florinda.ferreri@unipd.it
* Correspondence: valentina.baro@unipd.it

Abstract: Awake surgery and intraoperative neuromonitoring represent the gold standard for surgery of lesion located in language-eloquent areas of the dominant hemisphere, enabling the maximal safe resection while preserving language function. Nevertheless, this functional mapping is invasive; it can be executed only during surgery and in selected patients. Moreover, the number of neuro-oncological bilingual patients is constantly growing, and performing awake surgery in this group of patients can be difficult. In this scenario, the application of accurate, repeatable and non-invasive preoperative mapping procedures is needed, in order to define the anatomical distribution of both languages. Repetitive navigated transcranial magnetic stimulation (rnTMS) associated with functional subcortical fiber tracking (nTMS-based DTI-FT) represents a promising and comprehensive mapping tool to display language pathway and function reorganization in neurosurgical patients. Herein we report a case of a bilingual patient affected by brain tumor in the left temporal lobe, who underwent rnTMS mapping for both languages (Romanian and Italian), disclosing the true eloquence of the anterior part of the lesion in both tests. After surgery, language abilities were intact at follow-up in both languages. This case represents a preliminary application of nTMS-based DTI-FT in neurosurgery for brain tumor in eloquent areas in a bilingual patient.

Keywords: transcranial magnetic stimulation; brain tumor; bilingual; language; preoperative mapping; case report

1. Introduction

Surgical resection of lesions involving the language pathway remains a major challenge for the neurosurgeon, harboring a risk of new functional deficits. Repetitive navigated transcranial magnetic stimulation (rnTMS) has proven to provide a reliable non-invasive preoperatory cortical mapping for language function, showing a good overall correlation with intraoperative direct cortical stimulation (DCS) [1–5]. Nevertheless, its sensitivity, specificity, negative and positive predicting values varies widely among studies. Therefore, rnTMS speech mapping is the only method that can replace DCS when the latter cannot be performed [6–9]. Subcortical tracts can be identified by diffusion tensor imaging-fiber tracking (DTI-FT) based on rnTMS mapping, obtaining an accurate and functionally oriented white matter preoperative study. In fact, it allows planning of the best surgical strategy for resection, improving postoperative outcome, especially in patients who are not eligible for awake surgery [8,10–15]. A detailed preoperative mapping of the language pathway is mandatory, especially in case of bilingual patients, a peculiar subgroup that can present different patterns of cortical representation of the languages. In fact, the first language (L1) and the second language (L2) are processed both by shared brain areas as well as language-specific areas [16]. Moreover, even in L1 and L2 shared areas distinct language-specific neural population for the different languages have been identified by rnTMS [17]. Furthermore, Tussis et al. studied the cortical distribution of L1 and L2 in the non-dominant hemisphere with rnTMS, disclosing the involvement of dorsal precentral and middle precentral gyrus especially for L1, and triangular inferior frontal gyrus for L2 [18]. Whereby, a comprehensive preoperative understanding of the language pathway may be useful also in patients eligible for awake surgery, enabling a custom tailored craniotomy size and a faster and safer cortical mapping [2]. Herein we present the case of a 54-year-old Romanian woman affected by a primary brain tumor in the left angular gyrus who underwent preoperative rnTMS mapping to explore both Romanian and Italian languages. In the following, neurosurgical planning, surgical intervention and outcome are described and discussed.

2. Case Presentation

2.1. Patient Information, Clinical and Radiological Findings

A right-handed, bilingual 54-year-old woman was admitted at the emergency department for a generalized tonic clonic seizure sustained by a primitive brain tumor located between the posterior part of the superior and middle temporal gyri and the anterior part of the angular gyrus in the left hemisphere. The lesion did not enhance after contrast medium administration and it was hypometabolic at 18F-fluorodeoxyglucose PET/MRI. The functional MRI (fMRI) confirmed that the lesion was located in the dominant hemisphere (Figure 1). Due to the anxiety of the patient, mostly related to the diagnosis of brain tumor, the fMRI was performed testing only her mother tongue, i.e., Romanian. Interictal EEG showed an irritative activity in left centro–parietal derivations.

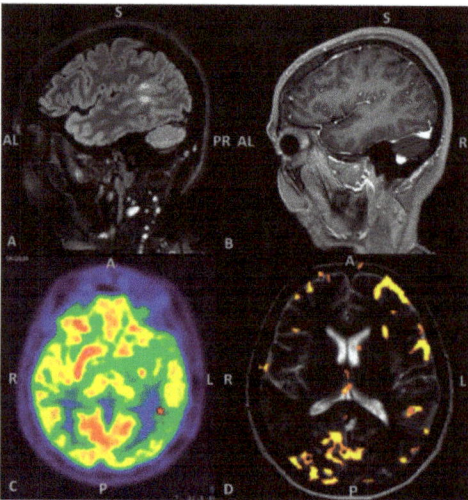

Figure 1. (**A**) 3D FLAIR (fluid attenuated inversion recovery) image discloses a primitive brain tumor located between the posterior part of the superior and middle temporal gyri and the anterior part of the angular gyrus in the left hemisphere; (**B**) the lesion does not enhance after contrast medium administration; (**C**) the 18F-fluorodeoxyglucose PET/MRI reveals the hypometabolism of the tumor (*). (**D**) Axial T2w image fused with the BOLD (blood oxygenation level dependent) signal activation map obtained during word generation task shows a focal cortical activation in the superior–anterior part of the lesion (*).

2.2. Neuropsychological Evaluation

Concerning the social and work surrounding the patient had been living in Italy for 17 years with her family, perfectly integrated in the social context, working as a housekeeper. Previously, she had 13 years of education, graduating in a vocational school in her home country.

The patient underwent a comprehensive battery of standardized neuropsychological tests performed in Italian, in order to evaluate the impact of the tumor on cognitive functions. A standardized evaluation of Romanian language was not executable because native language versions of the tests were not available and because none of the team spoke Romanian. The assessment was composed of tests covering different cognitive domains. The Oxford Cognitive Screen [19,20], a brief screening instrument composed of tasks on language, visual attention, spatial neglect, praxis abilities, visual and verbal memory, calculation, number reading and executive functions. Specific tests were also administered to better evaluate different cognitive functions. The Prose Memory Test (immediate and delayed recall) and Interference Memory test [21] were used as a measure of verbal memory. Forward and backward digit span and the Corsi block-tapping test were administered to measure short-term memory and working memory both for the verbal and visuospatial components [22]. Selective attention and switching abilities were measured using the Trail-Making-Test, forms A and B [21]. Different components of language abilities were assessed through specific tests: Phonemic Fluency test [21], the Boston Naming Test for visual naming ability [23], verbal comprehension of words and sentences and repetition of words and non-words [24]. Concerning language domain, the baseline preoperative assessment showed an impaired performance in naming and verbal fluency, whereas the other language abilities were normal (Table 1). Furthermore, the patient refused the proposition of an awake surgery. Therefore, we decided to test the patient for both languages by means of rnTMS integrated with DTI-FT. Due to her anxious state only the dominant hemisphere was evaluated, focused on the surgical planning.

Table 1. Neuropsychological assessment.

Assessment Test	Pre-Operative CS	Pre-Operative Performance	Post-Operative CS	Post-Operative Performance	Follow-Up 1 Month CS	Follow-Up 1 Month Performance	Follow-Up 4 Months CS	Follow-Up 4 Months Performance
GLOBAL COGNITIVE FUNCTIONS								
Oxford Cognitive Screen (OCS)								
Denomination	3	Impaired	3	Impaired	3	Impaired	4	Normal
Semantics	3	Normal	3	Normal	3	Normal	3	Normal
Orientation	4	Normal	4	Normal	4	Normal	4	Normal
Visual field	4	Normal	4	Normal	4	Normal	4	Normal
Reading	15	Normal	13	Impaired	15	Normal	15	Normal
Number writing	3	Normal	2	Impaired	3	Normal	3	Normal
Calculation	3	Borderline	3	Borderline	4	Normal	3	Normal
Visual search	47	Normal	47	Normal	46	Normal	49	Normal
egocentric neglect	−1	Normal	1	Normal	2	Normal	1	Normal
allocentric neglect	0	Normal	0	Normal	0	Normal	0	Normal
Imitation								
Right hand	11	Normal	8	Impaired	12	Normal	12	Normal
Left hand	12	Normal	12	Normal	12	Normal	12	Normal
Memory								
Verbal	3	Normal	3	Normal	2	Impaired	3	Normal
Episodic	4	Normal	4	Normal	4	Normal	4	Normal
Executive functions	−1	Normal	−2	Normal	0	Normal	0	Normal

Table 1. Cont.

Assessment	Pre-Operative		Post-Operative		Follow-Up 1 Month		Follow-Up 4 Months	
Test	CS	Performance	CS	Performance	CS	Performance	CS	Performance
LANGUAGE								
Boston Naming Test (15 items) E.N.P.A.	5	Impaired	3	Impaired	6	Impaired	6	Impaired
Verbal comprehension (words)	18.4	Normal	18.4	Normal	18.4	Normal	20	Normal
Verbal comprehension (sentences)	14	Normal	14	Normal	14	Normal	14	Normal
Repetition (words)	10	Normal	10	Normal	10	Normal	10	Normal
Repetition (nonwords)	5	Normal	5	Normal	5	Normal	5	Normal
Phonemic Fluency (Mondini, 2011) [20]	7.7	Impaired	1.7	Impaired	1.7	Impaired	4.3	Impaired
ATTENTION								
Trail Making Test								
A	26″	Normal	37″	Normal	51″	Normal	46″	Normal
B	167″	Impaired	167″	Impaired	156″	Impaired	133″	Normal
MEMORY								
Digit span								
Forward	4.75	Normal	2.75	Impaired	4.75	Normal	4.75	Normal
Backward	3.71	Normal	0	Impaired	3.71	Normal	3.79	Normal
Corsi Test								
Forward	6.74	Normal	5.74	Normal	4.74	Normal	3.81	Normal
Backward	5.67	Normal	5.67	Normal	5.67	Normal	3.79	Normal
Prose Memory								
Immediate	9	Normal	5	Impaired	12	Normal	10	Normal
Delayed	12	Normal	NE	Impaired	15	Normal	17	Normal
Memory Interference								
10 s	8	Normal	5	Normal	8	Normal	8	Normal
30 s	7	Normal	6	Normal	7	Normal	8	Normal

TCS: correct score (the raw score is adjusted for age and education basing on Italian-normative data from the literature, when appropriate). E.N.P.A.: Esame neuropsicologico per l'afasia (i.e., neuropsychological examination for aphasia). NE: not executable. The impairment of the performance is defined basing on cut-off, from normative data from the literature.

2.3. Patient's Informed Consent

The patient signed specific informed consent for MRI acquisition, rnTMS tests, neuropsychological evaluation and surgical intervention.

2.4. MRI Acquisition

The patient underwent brain MRI according to a specific protocol designed for the nTMS and DTI-FT using a 3T scanner (Ingenia 3T, Philips Healthcare) to obtain 3D T1-weighted images (TR/repetition time = 8, TE/echo time = 3.7); 3D FLAIR/fluid attenuated inversion recovery (TR = 4800, TE = 299, TI/inversion time = 1650, flip angle = 40, matrix = 240 × 240 mm^2, voxel = 1 × 1 × 1 mm^3, 196 slices, 4.05 min of acquisition time); diffusion weighted sequences (DWI with 32 directions, TR = 8736, TE = 91; single shell, b = 800 s/mm^2) for DTI-FT.

2.5. nTMS Language Cortical Mapping and Off-Line Analysis

The 3D T1-weighted sequence was imported into the nTMS system (NBS system 4.3—Nexstim Oy, Elimäenkatu 9 B, Helsinki, Finland) for language mapping, performed thorough a repetitive stimulation (rnTMS) according to the most update indications [25,26]. The patient's resting motor threshold (RMT) was determined by applying nTMS to the left motor cortex representing the hand, detecting the motor response of the m. abductor pollicis brevis. The patient performed the language assessment (base-line test, rnTMS mapping) first in Romanian (in the presence of an interpreter) and then in Italian. The base-line test was performed twice without stimulation, in order to cross out from the list the unfamiliar words, possible confounding variables in error analysis. A total of 80 black-and-white drawings of high and low frequency objects were presented on a 17-inch monitor placed 1 m in front of the patient for the picture naming task. Display and inter-picture time were set at 700 ms and 2500 ms, further adjusted to 2 s and 4 s for both languages. The patient was asked to say aloud the initial phrase "this is a ... " to distinguish between a speech arrest and anomia [27]. At the end of the base-line test, 70 and 67 figures were considered for Romanian and Italian mapping, respectively. The rnTMS stimulation frequency was set at the beginning at 5 pulses at 5 Hz at 110% RMT and then increased to 10 pulses at 10 Hz at 100% RMT because with the previous parameters of stimulation we did not obtain any error. The stimulation coil was randomly moved between the presentation of the images in about 1-cm steps over the perisylvian and peritumoral cortex. The rnTMS pulse train automatically triggered with picture presentation (0 ms) [26]. The entire mapping session was recorded on video for off-line data analysis, performed by an expert neuropsychologist (S.F.), helped by an interpreter for the review of the test performed in Romanian. The errors were classified according to Corina et al.: semantic paraphasias, circumlocutions, phonological paraphasias, neologisms, performance errors and no response errors [28]. We considered a site as language-eloquent if at least two of three stimulations caused an error response [25]. The stimulation sessions were well tolerated with a minimal discomfort reported (Visual Analogue Scale 2/10).

The off-line analysis highlighted 39 performance errors in Romanian (320 spots tested) of which a group of 5 was located in the superior–anterior and posterior–inferior border of the lesion. In Italian, 2 semantic and 15 performance errors were detected (271 sites tested), 3 of them located in the anterior part of the tumor. The language maps showed a convergence of the errors in the anterior middle temporal gyrus, middle middle temporal gyrus, posterior middle temporal gyrus, ventral precentral gyrus and anterior supramarginal gyrus according to the cortical parcellation system as described in Corina et al. [29] (Figure 2). The latest convergence corresponds to the anterior part of the tumor.

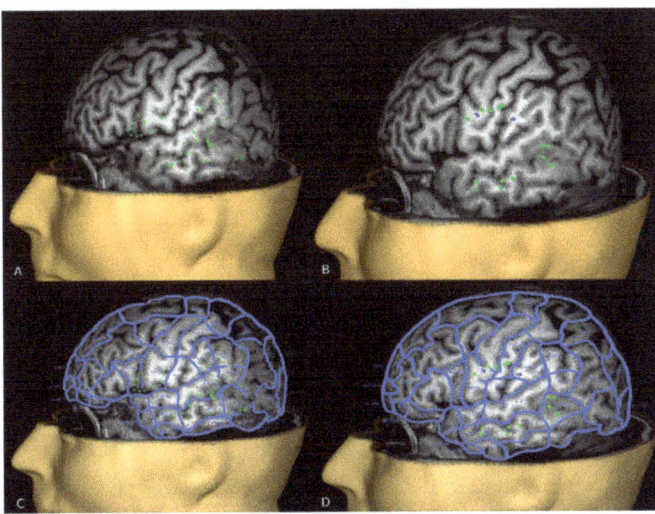

Figure 2. (**A**) Romanian rnTMS mapping (green spots: performance errors) and (**B**) Italian rnTMS mapping (green spots: performance errors, blue spots: semantic errors). (**C,D**) show the anatomical distribution of all errors according to the parcellization system area described by Corina et al., in Romanian and Italian, respectively.

2.6. nTMS Based DTI-FT of Language Pathway

The rnTMS cortical mapping was used to obtain the nTMS-based DTI-FT of the principal subcortical pathways of language function: arcuate fascicle (AF), frontal aslant tract (FAT), inferior fronto–occipital fascicle (IFOF), inferior longitudinal fascicle (ILF), superior longitudinal fascicle (SLF), uncinate fascicle (UF) [13,30,31]. The workflow for DTI-FT was performed on the StealthStation S7 navigation system by using StealthViz software (Medtronic Navigation, Coal Creek Circle Louisville, CO, USA). A deterministic approach based on the fiber assignment by continuous tracking (FACT) algorithm was used, with these parameters: FA cut off value = 0.15; vector step length = 0.5 mm; minimum fiber length = 30 mm; seeding density = 1.0; max directional change 90°. All language positive spots were imported into the planning station and used to create an overall object with an additional 5-mm border for each cortical spot. Subsequently, the object was exploited like a single ROI for tracking and the StealthViz software created a directionally encoded color map and then a 3D volume of white matter fibers originating from the cortical positive spots previously selected [11]. nTMS-based DTI-FT was able to identify the subcortical network for both languages, consisting of 533 and 293 fibers for Romanian and Italian, respectively. The 3D volumes were then manually elaborated to better visualize the single language-related tracts included in the reconstruction (i.e AF, SLF and ILF) under constant supervision of an expert neuroradiolgist [32–34] (Figure 3). White matter reconstruction displayed an overlap of AF in both languages with the anterior part of the lesion.

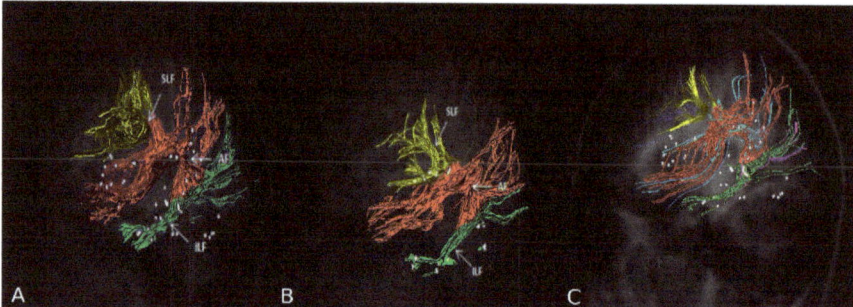

Figure 3. nTMS-based DTI-FT reconstructions of the language subcortical network in Romanian (**A**) and Italian (**B**) identifies the arcuate fascicle (AF, red), the superior longitudinal fascicle (SLF, yellow) and the inferior longitudinal fascicle (ILF, green) for both languages. In (**C**) the overlap of subcortical tracts is depicted. Italian color code—AF: red; SLF: yellow, ILF: green; Romanian color code—AF: light blue, SLF: blue, ILF violet.

2.7. Presurgical Planning

According to the rnTMS results, the tumor was divided into an eloquent and non-eloquent part, the latter identified as our surgical target. Using the planning station, the anterior part was highlighted in red and the target in violet. Then, the final reconstruction of language network was imported into the neuronavigation system to assist surgery (Figure 4).

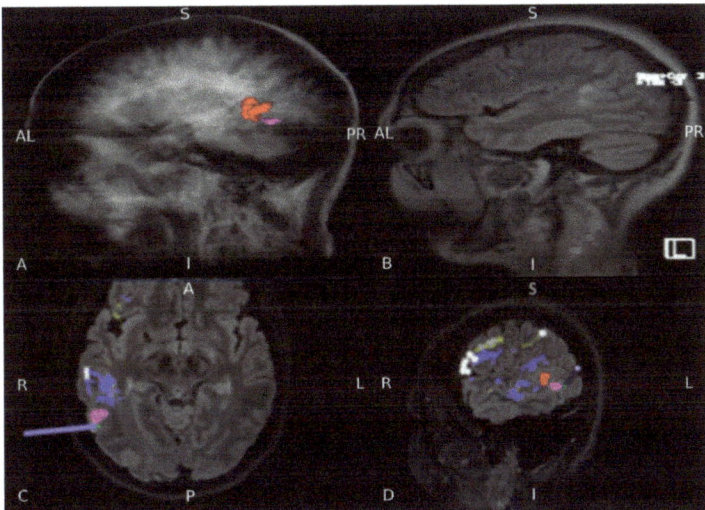

Figure 4. rnTMS-based planning using StealthViz software. (**A**) The anterior and eloquent part of the lesion identified by nTMS is colored in red and the posterior non-eloquent part in violet, (**B**) 3D FLAIR anatomical images for comparison. (**C,D**) A subtotal resection was planned and guided by neuronavigation (in violet the non-eloquent part).

2.8. Surgical Intervention and Neuropsychological Follow Up

Surgical resection of the posterior non-eloquent part was achieved by neuronavigation because the lesion was not clearly distinguished from normal brain parenchyma. Integrated histological and molecular diagnosis disclosed a WHO-grade IV gliomas [35]. She received perioperative antiepileptic drugs prophylaxis. Moreover, the patient received dexamethasone 4 mg four times daily for one week followed by gradual tapering. Postoperative neuropsychological assessment, performed after one week, showed a global worsening of the performance in language tasks (reading, number writing) and in other cognitive functions (praxical function in right hand, short- and long-term verbal memory, verbal working memory). However, the follow-up evaluations performed at 1 and 4 months after surgery, revealed a restoration of functions through the time. The performance at four months after the surgery was comparable with the baseline (Table 1). Relatives reported intact native language performance as well. The patient underwent whole brain radiotherapy (60 Gy/30 fractions) and medicated with Temozolomide (two cycles). The patient did not present seizures at last follow-up (10 months).

3. Discussion

Despite awake surgery associated with DCS still represents the gold standard for language mapping, an accurate preoperative assessment of language pathway is required to establish the best surgical strategy, for the risk–benefit balance and for the patient's counselling [36–38]. This is mandatory especially in case of patients with lesions located in eloquent areas who are not eligible for awake surgery [39]. Commonly, fMRI is the most accessible and applied preoperative mapping technique, providing the identification of eloquent cortical areas for different types of functions. Nevertheless, the indirect signal of area activation provided near a brain lesion could be undermined by a metabolic uncoupling induced by the lesion itself, determining a reduced fMRI signal in perilesional eloquent cortex [40–42]. This phenomenon, associated to a normal or increased activity in homologous brain regions, can simulate a reorganization of the function [43]. Moreover, previous studies have not clarified the reliability of fMRI for preoperative language mapping in tumors located in language-eloquent areas [42,44,45] thus, the use of fMRI in adjunct to other mapping methods is suggested [3,6,46].

nTMS is a recent and promising preoperative mapping technique for cortical functions localization and the development of nTMS-based DTI-FT allows a functionally oriented white matter reconstruction. In fact, the white matter reconstruction based on the rnTMS mapping showed a more accurate and reliable reconstruction of the subcortical language pathway compared to the standard anatomical technique [10,13]. Nonetheless, few centers have a broad experience with this technique and the language mapping has been less investigated compared to the motor nTMS mapping [6]. This may reflect the fact that language function is the result of a complex cortical and subcortical network which is more difficult to localize and challenge to map [28,47–49]. Currently, rnTMS combined with nTMS-based DTI-FT could be remarkably useful for patients who are not eligible for awake surgery, providing information concerning the true eloquence of the lesion with a high specificity of rnTMS in localizing language-negative areas. Furthermore, this technique can identify the presence of intra-hemispheric tumor-induced plasticity [14] or inter-hemispheric function reorganization and/or migration involving the non-dominant hemisphere [50–52].

For bilingual and multilingual neurosurgical patients affected by lesion located in language-eloquent areas, the preoperative mapping and languages preservation represent an additional major goal. At present, bilingual (and multilingual) neurosurgical patients have been investigated mostly with DCS and fMRI, as highlighted in a very recent review by Polczynska and Bookheimer [16]. This review suggests several principles concerning languages organization in bilingual patients, which may be useful in predicting the likelihood of separate versus converging representation of languages (i.e., age of L2 acquisition, proficiency level of L2 and linguistic distance between L1 and L2). Nonetheless, fMRI may falsely identify certain brain regions as potentially eloquent as above mentioned. Moreover, DCS evaluates a restricted coverage within language areas, mostly focused on sites in the frontal or posterior languages eloquent pathway. Unexpectedly, rnTMS has not been applied to study neurosurgical bilingual patients so far and late bilingual population has not been investigated.

We described the case of a bilingual patient affected by brain tumor located in language-eloquent region. The patient refused to undergo an awake craniotomy. Therefore, in order to deal with the aim of a safe resection, we applied an alternative method that could offer an accurate mapping of both languages. Preoperative languages mapping was obtained by rnTMS and nTMS-based DTI-FT language assessment according to the protocol for rnTMS language mapping used at our institution and established in the literature [11,25,53]. Language mapping for Romanian (L1) and Italian (L2) showed a convergence in the posterior areas of language pathways [48]. This overlapping may be explained by the high proficiency of L2 identified by extensive neuropsychological assessment and by the common derivation from Romance language as previously described. Furthermore, the language mapping disclosed the true eloquence of the anterior part of the lesion for both L1 and L2, limiting the surgical target to the non-eloquent region. Despite the worsening of language tasks and in some cognitive functions, the short-term follow-up highlighted a restoring of functions, comparable with the baseline. Probably this transient worsening was imputable to the surgical manipulation of subcortical fibers producing a functional, rather than anatomical damage. When performing language assessment in a bilingual patient, the setting should consider the presence of an interpreter both during rnTMS and off-line analysis, if the L1 is not properly known by one of the clinical staff members. Ideally, the native language version of the neuropsychological tests should be available and administered by a properly educated interpreter, to achieve a greater accuracy. This can be considered as an intrinsic limitation and possible bias when analyzing different languages, but it can be overcome by the advantages of an accurate mapping of the currently speaking languages, which requires to be preserved. Moreover, rnTMS still presents other pitfalls that need to be assessed. In fact, the likelihood of detecting language-positive spots is still low, drawing attention to the necessity of a revision of current stimulation protocols [49]. Furthermore, the interpretation of the hesitation errors varies among authors, constituting a matter of debate [49,54,55] and, in addition, the pre-existing moderate aphasia or severe cognitive impairment could undermine the reliability of the examination, entailing an accurate patients selection [56]. Nonetheless, the use of the initial sentence during picture naming, helping to distinguish between speech arrest and anomia is not routinely applied [25,27]. Another point is the influence of the antiepileptic drugs on the cortical excitability, which may influence the stimulation threshold as reported for the motor cortex but not investigated for extra-motor cortex [57]. Regarding the functional tractography obtained from the language mapping, a meaningful and debatable protocol should be assessed [10,11,13].

4. Conclusions

Our experience showed the reliability of rnTMS mapping in a bilingual patient who required surgery for a language-eloquent lesion for both languages. The potentials of this technique are different. First of all, the clinical application in safe neurosurgical practice is clear, because it represents a good tool for pre-surgical mapping, when awake surgery is not applicable for different reasons and rnTMS may allow filling of this gap. Furthermore, in the specific case of brain tumor, the preoperative mapping with nTMS-based DTI allows a better comprehension of language pathway reorganization and plasticity. A second important application concerns the neural basis of language, and bilingualism in particular, which remain still unclear in the literature. In this context, further studies with rnTMS on bilingual patients and healthy subjects are advocated to a comprehensive study of languages organization and plasticity.

Author Contributions: V.B.: methodology, data curation, formal analysis, writing—original draft, writing—review and editing, S.C. and L.S.: methodology, data curation, formal analysis, software, investigation, S.F.: data curation, formal analysis, investigation, writing—review and editing, B.G.: resources, software, writing—review and editing, P.Z.: resources, software, M.A.: supervision, validation, L.D., D.d.: supervision, F.F., A.L.: supervision, validation, writing—review and editing. All authors have read and agreed to the published version of the manuscript.

Funding: The authors declare that the research was conducted in the absence of any commercial or financial relationships that could be construed as a potential conflict of interest.

Informed Consent Statement: The patient signed her specific informed consent for MRI acquisition, nTMS tests, neuropsychological evaluation and surgical intervention. Moreover, she authorized the publication of this case report after anonymization of data and images.

Data Availability Statement: The data presented in this study are available on request from the corresponding author.

Acknowledgments: The authors thank Francesca Baro for the linguistic revision of the manuscript and Gianluigi De Nardo, neurophysiology technician, for his assistance during the tests.

Conflicts of Interest: The authors declare no conflict of interest.

Ethical Approval: For this type of paper, a case report, no approval of the local ethical committee was deemed necessary. The work is in accordance with the declaration of Helsinki and its later amendments, for as far applicable.

Abbreviations

AF: arcuate fascicle; DCS: direct cortical stimulation; DTI-FT: diffusion tensor imaging-fiber tracking; FAT: frontal aslant tract; IFOF: inferior fronto-occipital fascicle; ILF: inferior longitudinal fascicle; fMRI: functional MRI; L1: first language; L2: second language; nTMS: navigated navigate transcranial magnetic stimulation; RMT: resting motor threshold; rTMS: repetitive transcranial magnetic stimulation; rnTMS: repetitive navigate transcranial magnetic stimulation; SLF: superior longitudinal fascicle; UF: uncinate fascicle.

References

1. Tarapore, P.E.; Findlay, A.M.; Honma, S.M.; Mizuiri, D.; Houde, J.F.; Berger, M.S.; Nagarajan, S.S. Language mapping with navigated repetitive TMS: Proof of technique and validation. *Neuroimage* **2013**, *82*, 260–272. [CrossRef]
2. Picht, T.; Krieg, S.M.; Sollmann, N.; Rösler, J.; Niraula, B.; Neuvonen, T.; Savolainen, P.; Lioumis, P.; Mäkelä, J.P.; Deletis, V.; et al. A comparison of language mapping by preoperative navigated transcranial magnetic stimulation and direct cortical stimulation during awake surgery. *Neurosurgery* **2013**, *72*, 808–819. [CrossRef]
3. Ille, S.; Sollmann, N.; Hauck, T.; Maurer, S.; Tanigawa, N.; Obermueller, T.; Negwer, C.; Droese, D.; Boeckh-Behrens, T.; Meyer, B.; et al. Impairment of preoperative language mapping by lesion location: A functional magnetic resonance imaging, navigated transcranial magnetic stimulation, and direct cortical stimulation study. *J. Neurosurg.* **2015**, *123*, 314–324. [CrossRef] [PubMed]
4. Babajani-Feremi, A.; Narayana, S.; Rezaie, R.; Choudhri, A.F.; Fulton, S.P.; Boop, F.A.; Wheless, G.W.; Papanicolaou, A.C. Language mapping using high gamma electrocorticography, fMRI, and TMS versus electrocortical stimulation. *Clin. Neurophysiol.* **2016**, *127*, 1822–1836. [CrossRef] [PubMed]
5. Lehtinen, H.; Mäkelä, J.P.; Mäkelä, T.; Lioumis, P.; Metsähonkala, L.; Hokkanen, L.; Wilenius, J.; Gaily, E. Language mapping with navigated transcranial magnetic stimulation in pediatric and adult patients undergoing epilepsy surgery: Comparison with extraoperative direct cortical stimulation. *Epilepsia Open* **2018**, *3*, 224–235. [CrossRef]
6. Jeltema, H.R.; Ohlerth, A.K.; de Wit, A.; Wagemakers, M.; Rofes, A.; Bastiaanse, R.; Drost, G. Comparing navigated transcranial magnetic stimulation mapping and "gold standard" direct cortical stimulation mapping in neurosurgery: A systematic review. *Neurosurg Rev.* **2020**. Online ahead of print. [CrossRef]
7. Senova, S.; Lefaucheur, J.P.; Brugières, P.; Ayache, S.S.; Tazi, S.; Bapst, B.; Abhay, K.; Langeron, O.; Edakawa, K.; Palfi, S.; et al. Case Report: Multimodal Functional and Structural Evaluation Combining Pre-operative nTMS Mapping and Neuroimaging With Intraoperative CT-Scan and Brain Shift Correction for Brain Tumor Surgical Resection. *Front. Hum. Neurosci.* **2021**, *15*, 646268. [CrossRef] [PubMed]
8. Haddad, A.F.; Young, J.S.; Berger, M.S.; Tarapore, P.E. Preoperative Applications of Navigated Transcranial Magnetic Stimulation. *Front. Neurol.* **2021**, *11*, 628903. [CrossRef] [PubMed]
9. Hazem, S.R.; Awan, M.; Lavrador, J.P.; Patel, S.; Wren, H.M.; Lucena, O.; Semedo, C.; Irzan, H.; Melbourne, A.; Ourselin, S.; et al. Middle Frontal Gyrus and Area 55b: Perioperative Mapping and Language Outcomes. *Front. Neurol.* **2021**, *12*, 646075. [CrossRef]
10. Sollmann, N.; Zhang, H.; Schramm, S.; Ille, S.; Negwer, C.; Kreiser, K.; Meyer, B.; Krieg, S.M. Function-specific Tractography of Language Pathways Based on nTMS Mapping in Patients with Supratentorial Lesions. *Clin. Neuroradiol.* **2020**, *30*, 123–135. [CrossRef] [PubMed]
11. Sollmann, N.; Negwer, C.; Ille, S.; Maurer, S.; Hauck, T.; Kirschke, J.S.; Ringel, F.; Meyer, B.; Krieg, S.M. Feasibility of nTMS-based DTI fiber tracking of language pathways in neurosurgical patients using a fractional anisotropy threshold. *J. Neurosci. Methods* **2016**, *267*, 45–54. [CrossRef]
12. Sollmann, N.; Ille, S.; Hauck, T.; Maurer, S.; Negwer, C.; Zimmer, C.; Ringel, F.; Meyer, B.; Krieg, S.M. The impact of preoperative language mapping by repetitive navigated transcranial magnetic stimulation on the clinical course of brain tumor patients. *BMC Cancer* **2015**, *15*, 261. [CrossRef] [PubMed]
13. Raffa, G.; Bährend, I.; Schneider, H.; Faust, K.; Germanò, A.; Vajkoczy, P.; Picht, T. A novel technique for region and linguistic specific nTMS-based DTI fiber tracking of language pathways in brain tumor patients. *Front. Neurosci.* **2016**, *11*, 552. [CrossRef] [PubMed]
14. Raffa, G.; Quattropani, M.C.; Scibilia, A.; Conti, A.; Angileri, F.F.; Esposito, F.; Sindorio, C.; Cardali, S.M.; Germanò, A.; Tomasello, F. Surgery of language-eloquent tumors in patients not eligible for awake surgery: The impact of a protocol based on navigated transcranial magnetic stimulation on presurgical planning and language outcome, with evidence of tumor-induced intra-hemispheric plasticity. *Clin. Neurol. Neurosurg.* **2018**, *168*, 127–139. [CrossRef] [PubMed]
15. Sollmann, N.; Zhang, H.; Fratini, A.; Wildschuetz, N.; Ille, S.; Schröder, A.; Zimmer, C.; Meyer, B.; Krieg, S.M. Risk assessment by presurgical tractography using navigated tms maps in patients with highly motor-or language-eloquent brain tumors. *BMC Cancer* **2020**, *12*, 1124. [CrossRef] [PubMed]
16. Połczyńska, M.M.; Bookheimer, S.Y. Factors modifying the amount of neuroanatomical overlap between languages in Bilinguals—a systematic review of neurosurgical language mapping studies. *Brain Sci.* **2020**, *10*, 983. [CrossRef] [PubMed]
17. Hämäläinen, S.; Mäkelä, N.; Sairanen, V.; Lehtonen, M.; Kujala, T.; Leminen, A. TMS uncovers details about sub-regional language-specific processing networks in early bilinguals. *Neuroimage* **2018**, *171*, 209–221. [CrossRef]
18. Tussis, L.; Sollmann, N.; Boeckh-Behrens, T.; Meyer, B.; Krieg, S.M. The cortical distribution of first and second language in the right hemisphere of bilinguals—An exploratory study by repetitive navigated transcranial magnetic stimulation. *Brain Imaging Behav.* **2020**, *14*, 1034–1049. [CrossRef] [PubMed]
19. Demeyere, N.; Riddoch, M.J.; Slavkova, E.D.; Bickerton, W.-L.; Humphreys, G.W. The Oxford Cognitive Screen (OCS): Validation of a stroke-specific short cognitive screening tool. *Psychol. Assess.* **2015**, *27*, 883–894. [CrossRef]
20. Mancuso, M.; Varalta, V.; Sardella, L.; Capitani, D.; Zoccolotti, P.; Antonucci, G. Italian normative data for a stroke specific cognitive screening tool: The Oxford Cognitive Screen (OCS). *Neurol. Sci.* **2016**, *37*, 1713–1721. [CrossRef]
21. Mondini, S.; Mapelli, D.; Vestri, A.; Arcara, G.; Bisacchi, P.S. *Esame Neuropsicologico Breve—Una Batteria di Test per lo Screening Neuropsicologico*; Raffaello Cortina Editore: Milano, Italy, 2011.

22. Monaco, M.; Costa, A.; Caltagirone, C.; Carlesimo, G.A. Forward and backward span for verbal and visuo-spatial data: Standardization and normative data from an Italian adult population. *Neurol. Sci.* **2013**, *34*, 749–754. [CrossRef]
23. Kaplan, E.F.; Goodglass, H.; Weintraub, S. *The Boston Naming Test: The Experimental Edition*, 2nd ed.; Lea & Fabiger: Philadelphia, PA, USA, 1983.
24. Capasso, R.; Miceli, M. *Esame Neuropsicologico per l'Afasia*, 1st ed.; Springer: Mailand, Italy, 2001.
25. Krieg, S.M.; Lioumis, P.; Mäkelä, J.P.; Wilenius, J.; Karhu, J.; Hannula, H.; Savolainen, P.; Lucas, C.W.; Seidel, K.; Laakso, A.; et al. Protocol for motor and language mapping by navigated TMS in patients and healthy volunteers; workshop report. *Acta Neurochir.* **2017**, *159*, 1187–1195. [CrossRef] [PubMed]
26. Krieg, S.M.; Tarapore, P.E.; Picht, T.; Tanigawa, N.; Houde, J.; Sollmann, N.; Meyer, B.; Vajkoczy, P.; Berger, M.S.; Ringel, F.; et al. Optimal timing of pulse onset for language mapping with navigated repetitive transcranial magnetic stimulation. *Neuroimage* **2014**, *15*, 219–236. [CrossRef] [PubMed]
27. Mandonnet, E.; Sarubbo, S.; Duffau, H. Proposal of an optimized strategy for intraoperative testing of speech and language during awake mapping. *Neurosurg. Rev.* **2017**, *40*, 29–35. [CrossRef] [PubMed]
28. Corina, D.P.; Loudermilk, B.C.; Detwiler, L.; Martin, R.F.; Brinkley, J.F.; Ojemann, G. Analysis of naming errors during cortical stimulation mapping: Implications for models of language representation. *Brain Lang.* **2010**, *115*, 101–112. [CrossRef] [PubMed]
29. Corina, D.P.; Gibson, E.K.; Martin, R.; Poliakov, A.; Brinkley, J.; Ojemann, G.A. Dissociation of action and object naming: Evidence from cortical stimulation mapping. *Hum. Brain. Mapp.* **2005**, *159*, 1187–1195. [CrossRef]
30. Fekonja, L.; Wang, Z.; Bährend, I.; Rosenstock, T.; Rösler, J.; Wallmeroth, L.; Vajkoczy, P.; Picht, T. Manual for clinical language tractography. *Acta Neurochir.* **2019**, *161*, 1125–1137. [CrossRef] [PubMed]
31. Machetanz, K.; Trakolis, L.; Leão, M.T.; Liebsch, M.; Mounts, K.; Bender, B.; Ernemann, U.; Gharabaghi, A.; Tatagiba, M.; Naros, G. Neurophysiology-Driven Parameter Selection in nTMS-Based DTI Tractography: A Multidimensional Mathematical Model. *Front. Neurosci.* **2019**, *13*, 1–10. [CrossRef]
32. Catani, M.; Howard, R.J.; Pajevic, S.; Jones, D.K. Virtual in Vivo interactive dissection of white matter fasciculi in the human brain. *Neuroimage* **2002**, *17*, 77–94. [CrossRef]
33. Catani, M.; Thiebaut de Schotten, M. A diffusion tensor imaging tractography atlas for virtual in vivo dissections. *Cortex* **2008**, *44*, 1105–1132. [CrossRef]
34. Kamali, A.; Flanders, A.E.; Brody, J.; Hunter, J.V.; Hasan, K.M. Tracing superior longitudinal fasciculus connectivity in the human brain using high resolution diffusion tensor tractography. *Brain. Struct. Funct.* **2014**, *219*, 269–281. [CrossRef]
35. Brat, D.J.; Aldape, K.; Colman, H.; Figrarella-Branger, D.; Fuller, G.N.; Giannini, C.; Holland, E.C.; Jenkins, R.B.; Kleinschmidt-DeMasters, B.; Komori, T.; et al. cIMPACT-NOW update 5: Recommended grading criteria and terminologies for IDH-mutant astrocytomas. *Acta Neuropathol.* **2020**, *139*, 603–608. [CrossRef]
36. Sanai, N.; Berger, M.S. Intraoperative stimulation techniques for functional pathway preservation and glioma resection. *Neurosurg. Focus.* **2010**, *28*. [CrossRef]
37. Sanai, N.; Mirzadeh, Z.; Berger, M.S. Functional Outcome after Language Mapping for Glioma Resection. *N. Engl. J. Med.* **2008**, *358*, 118–127. [CrossRef]
38. De Benedictis, A.; Moritz-Gasser, S.; Duffau, H. Awake mapping optimizes the extent of resection for low-grade gliomas in eloquent areas. *Neurosurgery* **2010**, *66*, 1074–1084. [CrossRef]
39. Kayama, T. The Guidelines for Awake CraniotomyGuidelines Committee of The Japan Awake Surgery Conference. *Neurol. Med. Chir.* **2012**, *52*, 119–141. [CrossRef]
40. Fujiwara, N.; Sakatani, K.; Katayama, Y.; Murata, Y.; Hoshino, T.; Fukaya, C.; Yamamoto, T. Evoked-cerebral blood oxygenation changes in false-negative activations in BOLD contrast functional MRI of patients with brain tumors. *Neuroimage* **2004**, *21*, 1464–1471. [CrossRef]
41. Aubert, A.; Costalat, R.; Duffau, H.; Benali, H. Modeling of Pathophysiological Coupling between Brain Electrical Activation, Energy Metabolism and Hemodynamics: Insights for the Interpretation of Intracerebral Tumor Imaging. *Acta Biotheor.* **2002**, *50*, 281–295. [CrossRef] [PubMed]
42. Giussani, C.; Roux, F.-E.; Ojemann, J.; Pietro, S.E.; Pirillo, D.; Papagno, C. Is Preoperative Functional Magnetic Resonance Imaging Reliable for Language Areas Mapping in Brain Tumor Surgery? Review of Language Functional Magnetic Resonance Imaging and Direct Cortical Stimulation Correlation Studies. *Neurosurgery* **2010**, *66*, 113–120. [CrossRef] [PubMed]
43. Ulmer, J.L.; Hacein-Bey, L.; Mathews, V.P.; Mueller, W.M.; DeYoe, E.A.; Prost, R.W.; Meyer, G.A.; Krouwer, H.G.; Schmainda, K.M. Lesion-induced Pseudo-dominance at Functional Magnetic Resonance Imaging: Implications for Preoperative Assessments. *Neurosurgery* **2004**, *55*, 569–579. [CrossRef] [PubMed]
44. Fitzgerald, D.B.; Cosgrove, G.R.; Ronner, S.; Jiang, H.; Buchbinder, B.R.; Belliveau, J.W.; Rosen, B.R.; Benson, R.R. Location of Language in the Cortex: A Comparison between Functional MR Imaging and Electrocortical Stimulation. *AJNR Am. J. Neuroradiol.* **1997**, *18*, 1529–1539. [PubMed]
45. Roux, F.E.; Boulanouar, K.; Lotterie, J.A.; Mejdoubi, M.; LeSage, J.P.; Berry, I. Language functional magnetic resonance imaging in preoperative assessment of language areas: Correlation with direct cortical stimulation. *Neurosurgery* **2003**, *52*, 1335–1347. [CrossRef]
46. Ottenhausen, M.; Krieg, S.M.; Meyer, B.; Ringel, F. Functional preoperative and intraoperative mapping and monitoring: Increasing safety and efficacy in glioma surgery. *Neurosurg. Focus.* **2015**, *38*, E3. [CrossRef] [PubMed]

47. Rofes, A.; Mandonnet, E.; de Aguiar, V.; Rapp, B.; Tsapkini, K.; Miceli, G. Language processing from the perspective of electrical stimulation mapping. *Cogn Neuropsychol.* **2019**, *36*, 117–139. [CrossRef] [PubMed]
48. Hickok, G.; Poeppel, D. Dorsal and ventral streams: A framework for understanding aspects of the functional anatomy of language. *Cognition* **2004**, *92*, 67–99. [CrossRef] [PubMed]
49. Bährend, I.; Muench, M.R.; Schneider, H.; Moshourab, R.; Dreyer, F.R.; Vajkoczy, P.; Picht, T.; Faust, K. Incidence and linguistic quality of speech errors: A comparison of preoperative transcranial magnetic stimulation and intraoperative direct cortex stimulation. *J. Neurosurg.* **2020**, *29*, 1–10. [CrossRef]
50. Rösler, J.; Niraula, B.; Strack, V.; Zdunczyk, A.; Schilt, S.; Savolainen, P.; Lioumis, P.; Mäkelä, P.; Vajkoczy, P.; Frey, D.; et al. Language mapping in healthy volunteers and brain tumor patients with a novel navigated TMS system: Evidence of tumor-induced plasticity. *Clin. Neurophysiol.* **2014**, *125*, 526–536. [CrossRef]
51. Krieg, S.M.; Sollmann, N.; Hauck, T.; Ille, S.; Foerschler, A.; Meyer, B.; Ringel, F. Functional Language Shift to the Right Hemisphere in Patients with Language-Eloquent Brain Tumors. *PLoS ONE* **2013**, *17*, e75403. [CrossRef] [PubMed]
52. Duffau, H. Brain plasticity and tumors. *Adv. Tech. Stand. Neurosurg.* **2008**, *33*, 3–33. [CrossRef]
53. Krieg, S.M.; Sollmann, N.; Hauck, T.; Ille, S.; Meyer, B.; Ringel, F. Repeated mapping of cortical language sites by preoperative navigated transcranial magnetic stimulation compared to repeated intraoperative DCS mapping in awake craniotomy. *BMC Neurosci.* **2014**, *159*, 1187–1195. [CrossRef]
54. Lioumis, P.; Zhdanov, A.; Mäkelä, N.; Lehtinen, H.; Wilenius, J.; Neuvonen, T.; Hannula, H.; Deletis, V.; Picht, T.; Mäkelä, J.P. A novel approach for documenting naming errors induced by navigated transcranial magnetic stimulation. *J. Neurosci. Methods* **2012**, *15*, 349–354. [CrossRef]
55. Schuhmann, T.; Schiller, N.O.; Goebel, R.; Sack, A.T. Speaking of which: Dissecting the neurocognitive network of language production in picture naming. *Cereb Cortex* **2012**, *22*, 701–709. [CrossRef] [PubMed]
56. Schwarzer, V.; Bährend, I.; Rosenstock, T.; Dreyer, F.R.; Vajkoczy, P.; Picht, T. Aphasia and cognitive impairment decrease the reliability of rnTMS language mapping. *Acta Neurochir.* **2018**, *160*, 343–356. [CrossRef] [PubMed]
57. Hamed, S.A.; Tohamy, A.M.; Mohamed, K.O.; el Mageed Abd el Zaher, M.A. The Effect of Epilepsy and Antiepileptic Drugs on Cortical Motor Excitability in Patients With Temporal Lobe Epilepsy. *Clin. Neuropharmacol.* **2020**, *43*, 175–184. [CrossRef] [PubMed]

Article

Bihemispheric Navigated Transcranial Magnetic Stimulation Mapping for Action Naming Compared to Object Naming in Sentence Context

Ann-Katrin Ohlerth [1,2,*], Roelien Bastiaanse [1,3], Chiara Negwer [4], Nico Sollmann [5,6,7], Severin Schramm [4], Axel Schröder [4] and Sandro M. Krieg [4,6]

1. Center for Language and Cognition Groningen, Oude Kijk in 't Jatstraat 26, 9712 EK Groningen, The Netherlands; y.r.m.bastiaanse@rug.nl
2. International Doctorate for Experimental Approaches to Language and Brain (IDEALAB), Universities of Groningen (NL), Newcastle (UK), Potsdam (GE), Macquarie University, Sydney (AU), Oude Kijk in 't Jatstraat 26, 9712 EK Groningen, The Netherlands
3. Center for Language and Brain, Higher School of Economics, National Research University, 20 Myasnitskaya Street, 101000 Moscow, Russia
4. Department of Neurosurgery, School of Medicine, Klinikum rechts der Isar, Technical University of Munich, Ismaninger Straße 22, 81675 Munich, Germany; Chiara.Negwer@tum.de (C.N.); severin.schramm@gmx.de (S.S.); Axel.Schroeder@mri.tum.de (A.S.); sandro.krieg@tum.de (S.M.K.)
5. Department of Diagnostic and Interventional Neuroradiology, School of Medicine, Klinikum rechts der Isar, Technical University of Munich, Ismaninger Straße 22, 81675 Munich, Germany; nico.sollmann@tum.de
6. TUM-Neuroimaging Center, Klinikum rechts der Isar, Technical University of Munich, Ismaninger Straße 22, 81675 Munich, Germany
7. Department of Diagnostic and Interventional Radiology, University Hospital Ulm, Albert-Einstein-Allee 23, 89081 Ulm, Germany
* Correspondence: a.ohlerth@rug.nl; Tel.: +31-50-36-35858

Abstract: Preoperative language mapping with navigated transcranial magnetic stimulation (nTMS) is currently based on the disruption of performance during object naming. The resulting cortical language maps, however, lack accuracy when compared to intraoperative mapping. The question arises whether nTMS results can be improved, when another language task is considered, involving verb retrieval in sentence context. Twenty healthy German speakers were tested with object naming and a novel action naming task during nTMS language mapping. Error rates and categories in both hemispheres were compared. Action naming showed a significantly higher error rate than object naming in both hemispheres. Error category comparison revealed that this discrepancy stems from more lexico-semantic errors during action naming, indicating lexico-semantic retrieval of the verb being more affected than noun retrieval. In an area-wise comparison, higher error rates surfaced in multiple right-hemisphere areas, but only trends in the left ventral postcentral gyrus and middle superior temporal gyrus. Hesitation errors contributed significantly to the error count, but did not dull the mapping results. Inclusion of action naming coupled with a detailed error analysis may be favorable for nTMS mapping and ultimately improve accuracy in preoperative planning. Moreover, the results stress the recruitment of both left- and right-hemispheric areas during naming.

Keywords: language mapping; navigated transcranial magnetic stimulation; picture naming; bihemispheric; action naming; object naming

1. Introduction

There is growing recognition for navigated transcranial magnetic stimulation (nTMS) language mapping in both neuroscientific research and in clinical application in neurosurgery. In this non-invasive mapping technique, a magnetic field is directed at the cortex, causing a temporary disruption of neural activity [1–3]. In combination with neuro-navigation, it is possible to precisely pinpoint a targeted area and test it for cortical functions. Areas in which stimulation results in a transient inhibition of a cognitive function are considered to support this function. If applied area-by-area, functional boundaries are delineated and functional maps covering almost the entire cortex can be acquired [4–7]. These maps have been of great benefit for understanding language organization in the brain. Clinically, preoperative nTMS maps are used for planning and executing brain tumor surgery in language-eloquent areas: the added information enables the clinicians to perform a more targeted craniotomy, maximize the extent of resection, preserve functional areas, and minimize postoperative deficits [5,8,9].

Intraoperative mapping with direct electrical stimulation (DES) still remains the gold standard for locating cortical function in relation to a tumor [10–13]. While maps generated by nTMS overlap with those from intraoperative mapping concerning sensitivity, low specificity between language-positive areas under nTMS compared to under DES has to be faced [5,7,14–17]. Improvement of the methodology is still necessary. A consensus has been met for most parameters such as stimulation intensity, frequency, duration, and directionality of the stimulation, resulting in more standardized nTMS language mapping protocols [18]. However, when considering functional tasks for language mapping, current protocols still lack linguistic depth by solely administering a noun task: The only common task used with nTMS is object naming. Naming a drawing of an object aligned with cortical stimulation has been shown efficient in detecting language areas that correlate with those under DES [5,7,14–17]. Nonetheless, it is disputable whether a task that solely triggers noun retrieval and production is sufficient at representing language. Especially when compared to the variety of language tests commonly seen in intraoperative situations [19–23] this difference in depth of protocols might factor into the lack of accuracy in the preoperative mapping.

The limiting testing parameters of nTMS, such as a time frame of only up to 2 s for most protocols, ensure safe application, but compared to DES mapping with up to 4 s of stimulation, require even shorter tasks to be targeted during stimulation. These parameters, hence, do not allow for extensive neuropsychological screening, including exhaustive linguistic protocols; however, they are compatible with another picture naming task, targeting verbs. A drawing of an action is presented, and the verb needs to be retrieved and produced while stimulation is applied. Literature from several domains point towards a theoretical benefit of including verb tasks, as this process seems to be recruiting an at least partially different neuronal network. Behavioral data from aphasia research suggest a dissociation between the two word classes: spared noun production but impaired verb production and, less frequently, vice-versa, could be found after brain damage [24–27]. During intraoperative mapping under DES, partially segregated regions for object and action naming have been reported, with action naming in some cases being the only task revealing positive areas and, thus, guiding resection [23,28]. Therefore, including this task in language mapping was shown beneficial.

The full potential of action naming is prompted in sentence context, including a short lead-in phrase in the stimulus. In this way, not only verb retrieval and production is required, but inflections for person, number, and tense are triggered, linguistic skills that are not involved in object naming. Moreover, it allows investigation of error types, which can relate to different production processes.

Several cognitive models were proposed to capture the production of words and whole sentences, most of them deriving from the Levelt and Indefrey–Levelt models [29–31]. While details of the different processing levels and their serial or parallel execution are still debated, most models and adaptations agree on the following broad levels of production processes (see also Figure 1) (see [32,33] for a review).

1. Conceptual retrieval (retrieving non-linguistic information about concepts).
2. Lexico-semantic word retrieval (retrieving words with respective meaning in an uttered phrase).
3. Grammatical encoding (assigning morpho-syntactic features to the words such as marking number and tense in phrase).

4. Phonological encoding (assigning the required sounds to the words).
5. Articulation (programming and executing the required motor muscle movements).

These processes have not been systematically studied under nTMS or intraoperative DES. However, the disruption of each level is reflected in various error types. Figure 1 summarizes the processing levels and the most common error types documented under nTMS [6,34–37].

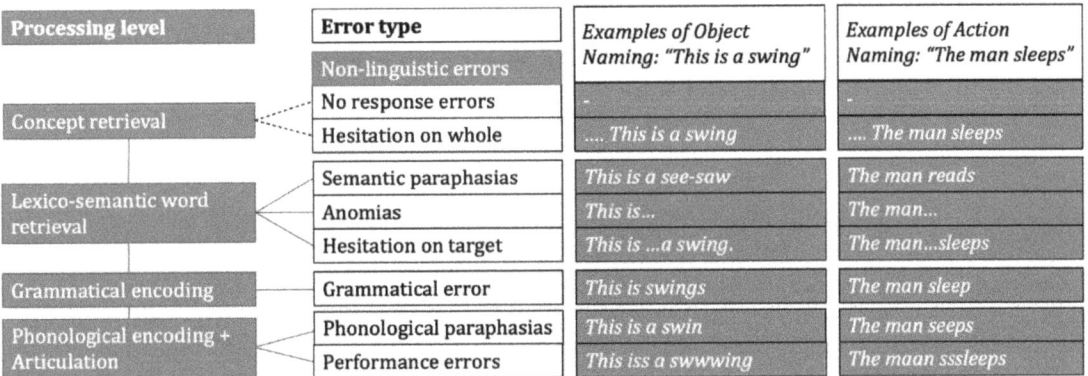

Figure 1. Underlying processing levels of sentence production on the left, leading to error types on the right, with examples for object naming and action naming.

No response errors and hesitations on the whole suggest an entire breakdown of speech and language production processes. While being the strongest effect of nTMS, it is not possible to pinpoint at which level of production the effect takes place, but it is likely that the conceptual level has already been affected. The error types with an intact lead-in phrase, but a missing target (anomia), delayed target (hesitation on target), and a semantically related, but incorrect target (semantic paraphasia), denote a disruption to lexico-semantic word retrieval or even concept retrieval; speech and automated production of the lead-in phrase may stay intact. Disrupted grammatical encoding may result in incorrect inflection (grammatical error). Disruption in phonological encoding and articulation may result in a target word that is still recognizable, but may omit or switch sounds (phonological paraphasia), or speech may be slurred or stuttered (performance errors) [29,38,39].

Object and action naming processes share these stages, but due to the more abstract nature of the concept and the more complex semantic structure of verbs as the head of the sentence, action naming in sentence context requires more complex conceptual and lexico-semantic processing (Figure 1: Steps 1 and 2). Moreover, due to grammatical processes, such as inflecting a verb for person, number, and tense, and integrating it into the sentence, grammatical encoding (Figure 1: Step 3) is cognitively more demanding for verbs than for nouns [40–42].

Based on these psycholinguistic assumptions, action naming in sentence context may ultimately be more difficult to execute and, hence, easier to disrupt by nTMS than object naming and, therefore, a more sensitive tool for language mapping. Although two previous experiments have looked into verb production under nTMS, both employed a simple action naming paradigm with single word answers, neither triggering grammatical processes nor differentiating between different errors [43,44].

The present study aims to evaluate the potential of object naming and action naming in sentence context by using both tasks with standardized stimuli and error type analysis in 20 healthy volunteers. To reveal differences in sensitivity of the tasks and in breakdown of noun and verb production, error rates and types as well as their neuroanatomical locations are compared in both hemispheres in all regions accessible by nTMS. Hesitations, while comprising a significant fraction of the error count with about 10–30% of errors in the literature and being the only elicitable error type in some cases [35–37], are still considered the most questionable category due to their subjective character. Current set-ups do not allow for objective classification of said errors, when evaluation is not tied to standardized measurements of voice latency through third-party data processing [37,45,46]. Therefore, in the paper at hand, these errors will be treated with caution and excluded for parts of the analysis to further investigate their role in mapping language. The questions we aim to answer are:

(1) Does the overall higher complexity of the verb task lead to a quantitative difference in error rates between object naming and action naming under stimulation?
(2) Is there a qualitative difference between the tasks as seen in error categories?
(3) Which anatomical regions are most prone to error elicitation under each of the tasks in the two hemispheres?
(4) How does excluding hesitation errors affect the error rates and maps?

2. Materials and Methods

2.1. Participants

Twenty healthy volunteers were tested. Inclusion criteria were German as a native language, and age of at least 18 years. Participants were excluded in case of contraindications for magnetic resonance imaging (MRI) at 3 Tesla or nTMS mapping (e.g., due to deep brain stimulation devices), presence of neurological, or psychiatric diseases, or pregnancy. Left-handedness was not an exclusion criterion, but was calculated through the Edinburgh Handedness Inventory (EHI) and noted [47]. Written informed consent was acquired from all participants, before testing commenced.

2.2. Magnetic Resonance Imaging

Anatomical MRI without intravenous contrast agent administration was acquired in a 3-Tesla MRI scanner (Achieva dStream; Philips Healthcare, Best, Amsterdam, The Netherlands). A three-dimensional (3D) T1-weighted gradient echo sequence (repetition time/echo time: 9/4 ms, 1 mm^3 isoVoxel covering the whole head with a flip angle of 8°) was acquired in all subjects. The images were used to construct 3D models of the individual's brain in the neuro-navigational system to guide placement of the coil during nTMS mapping.

2.3. Picture Naming Tasks

Previously standardized tasks from the German version of the Verb And Noun PeriOPerative test (VAN-POP [48]) entailing object and action naming in sentence context, were used for nTMS language mapping. The tasks consist of black-and-white drawings of 75 objects and 75 actions (Figure 2a,b). Lead-in phrases above the drawing prompt nouns and verbs in sentence context.

Das ist...

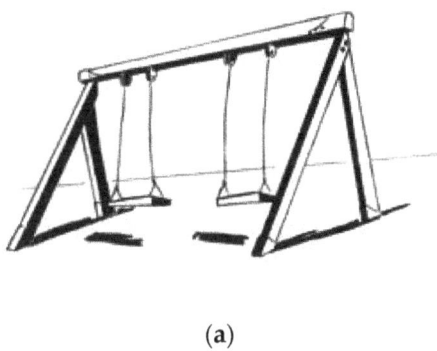

(a)

Der Mann...

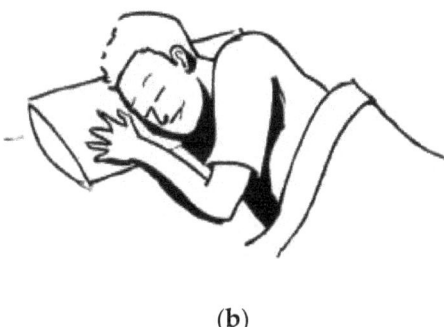

(b)

Figure 2. Example of object naming stimulus (**a**) and action naming stimulus (**b**). (**a**): Das ist . . . eine Schaukel, 'This is. . . a swing', (noun + article in German inflected for number, gender and case); (**b**): Der Mann. . . schläft, 'The man... sleeps', (verb in German inflected for person, number, tense).

All items showed naming agreement of at least 80% in previous phases of standardization by participants of various age groups and backgrounds under presentation parameters of nTMS [48]. Moreover, all items are balanced for linguistic factors known to influence naming, such as word frequency (see [48] for a full list).

2.4. Language Mappings

2.4.1. Setup

Before mapping started, the individual T1-weighted sequences were uploaded to the Nexstim eXimia NBS system version 4.3, (Nexstim Plc. Helsinki, Finland). Forty-six stimulation targets (placed in reference to the cortical parcellation system (CPS) [49]) were assigned to the 3D head models of each individual. They covered all cortical areas with the exception of the occipital lobe, frontal and temporal poles, and the inferior temporal regions due to inability to reach or high discomfort under stimulation. Figure 3 and Table 1 name all targeted areas. Using these predefined targets ensured the same cortical spot to be targeted in each task.

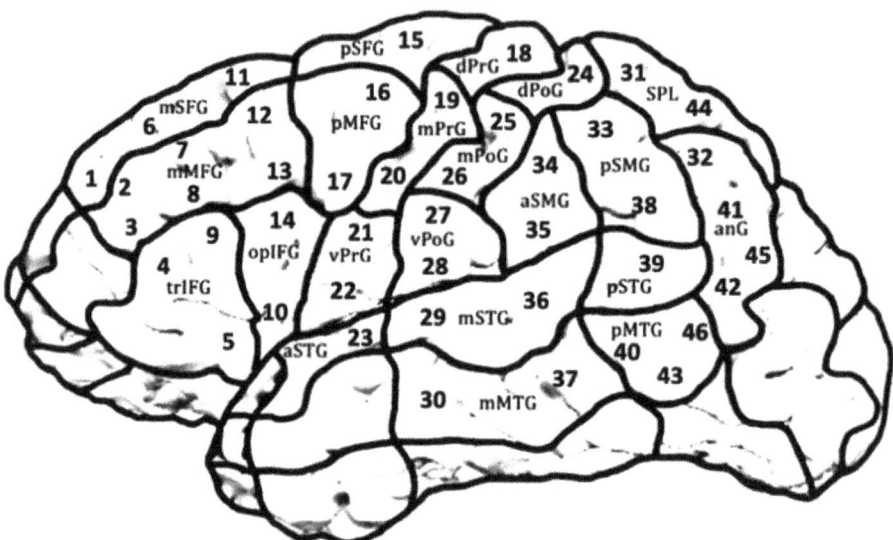

Figure 3. Template of the left hemisphere with 46 stimulation targets covering 21 cortical parcellation system (CPS) regions. The regions in the right hemisphere were mirror-inverted. Occipital areas, frontal and temporal poles, and inferior temporal regions were excluded due to inability of reach or high discomfort in combination with stimulation. See Table 1 for anatomical names corresponding to the abbreviations.

Table 1. Anatomical names and corresponding abbreviations of the 21 cortical parcellation system (CPS) regions as adapted from [49].

Abbreviation	Anatomy
anG	Angular gyrus
aSMG	Anterior supramarginal gyrus
aSTG	Anterior superior temporal gyrus
dPoG	Dorsal postcentral gyrus
dPrG	Dorsal precentral gyrus
mMFG	Middle middle frontal gyrus
mMTG	Middle middle temporal gyrus
mPoG	Middle postcentral gyrus
mPrG	Middle precentral gyrus
mSFG	Middle superior frontal gyrus
mSTG	Middle superior temporal gyrus
opIFG	Opercular inferior frontal gyrus
pMFG	Posterior middle frontal gyrus
pMTG	Posterior middle temporal gyrus
pSFG	Posterior superior frontal gyrus
pSMG	Posterior supramarginal gyrus
pSTG	Posterior superior temporal gyrus
SPL	Superior parietal lobe
trIFG	Triangular inferior frontal gyrus
vPoG	Ventral postcentral gyrus
vPrG	Ventral precentral gyrus

For nTMS, a focal figure-of-eight coil with upward handle position and automatic overheating protection was employed. It produced biphasic pulses (length: 230 µs) with a maximal electric field strength of 172 V/m ± 2% (at 25 mm depth beneath the coil in a spherical conductor model representing the head). The setup provided visualization of cortical areas to be targeted in relation to the coil's focal point as well as the e-field's orientation to the gyrus [50]. All pulse applications were tracked, controlled, and saved. Inaccurate pulse applications not on target were not saved by the software.

As a first step, the resting motor threshold (rMT) of each individual was established: surface electrodes for electromyography (EMG) were placed over the abductor pollicis brevis and abductor digiti minimi muscles. Single-pulse stimulation was applied over the anatomical hand knob to identify the most excitable spot with the electrical field perpendicular to the central sulcus. Using the built-in threshold-hunting algorithm, the lowest possible threshold to elicit at least five out of ten positive motor responses was defined as the rMT. For a more detailed description, see [18] as the most common approach used in the field as well as in the present set-up. Moreover, a rMT of 110% was employed as intensity for nTMS language mapping [18]. The rMT was defined separately for each hemisphere (for means see Table 2).

2.4.2. Baseline Naming

To determine the ideal set of picture stimuli per participant, baseline naming was performed. Participants sat about 60 cm from of a screen on which the pictures were presented with a picture presentation time (PPT) of 1000 ms and an inter-picture interval (IPI) of 3000 ms. The tasks appeared in blocks per task; the item order per task was randomized.

Two rounds of baseline naming were administered, in which participants had to name the entire set of 150 pictures without stimulation. The instruction was to name pictures using the entire sentence as quickly and precisely as possible. Those picture stimuli that were not named fluently and consistently with the same label within the given time window in two rounds of baseline testing were excluded from naming under stimulation. The number of incorrectly named stimuli was documented. The procedure was audio- and video-recorded for post-hoc analysis.

2.4.3. Mapping Procedure

The individualized set of stimuli resulting from the baseline naming was used per participant. Object and action naming tasks were administered in separate blocks. The task order and stimulation of hemispheres was balanced across participants, so that each task and hemisphere was targeted first in half of the participants and the respective other task and hemisphere in the other half of the participants. This was done to exclude fatigue as an influence of error rate. The same was done for stimuli order within each task: the stimuli order within each task was randomized. The individual and randomized list was used per participant and restarted, once it had reached its end during administration of each block to cover all areas.

The IPI and PPT were the same as during baseline testing. The onset of stimulation was synchronized with the onset of picture presentation; hence, the picture-to-trigger interval (PTI) was 0 ms. Stimulation consisted of 10 rTMS trains delivered at 5 Hz/5 pulses at 110% of the intensity established during motor threshold hunting.

During language mapping, the coil was placed over each of the 46 predefined stimulation target distributed over the majority of the cortical surface, while the participant named the items. Each target point was stimulated three times per round. There were two rounds per hemisphere and task, while alternating between left hemisphere (LH) and right hemisphere (RH). This means, that the overall mapping resulted in six data points per stimulation target.

The sequence of left- and right-hemispheric stimulation was randomized. The participants were instructed to report any pain and discomfort that may occur during stimulation to stop the mapping in case the procedure was intolerable.

2.4.4. Mapping Analysis

Through post-hoc analysis of the recorded and segmented videos, baseline-naming performances were compared to performance under stimulation in a side-by-side comparison. The stimulation videos were screened for any of the following speech and language errors compared to the baseline counterpart. The investigator carrying out the analysis was blinded to where stimulation had taken place. Errors due to pain, discomfort, or visible stimulation of peripheral facial nerves were excluded from the analysis.

Categories:

Conceptual (non-linguistic) errors:

1. No response: no intelligible answer or no speech output at all.
2. Hesitation (on whole sentence): noticeably delayed onset of correct answer compared to baseline recording or overall much slower sentence production.

Lexico-semantic errors:

3. Semantic paraphasia: intact lead-in phrase; incorrect, but often related target word, correctly pronounced.
4. Anomia: intact lead-in phrase, but target missing or uttered only after stimulation ended.
5. Hesitation on target: lead-in phrase intact and on time, but target word delayed compared to baseline.

Grammatical errors:

6. Grammatical error: for example, a missing or wrong inflection for verb or noun and article.

Phonological-articulatory errors:

7. Phonological paraphasia: target word recognizable, but missing or substituting speech sounds.
8. Performance errors: target word recognizable, but speech slurred or stuttered.

2.5. Statistical Analysis

All calculations were performed using R software (R Studio version 3.5.2; The R Foundation for Statistical Computing, Vienna, Austria). A *p*-value of <0.05 was considered statistically significant for all comparisons and correlations.

Baseline error rates were calculated by dividing the number of errors of the baseline by the number of items to name. To address research question 1, 3, and 4, error rates for object and action naming were calculated for the respective areas.

The error rate was defined as the number of errors divided by the total number of stimulations in a particular area. Two different overall error rates were calculated, one including all errors (categories 1–8), one without hesitation errors (excluding category 2 and 5). This was performed for errors per task, hemisphere, and CPS region in each hemisphere. Shapiro–Wilk normality tests suggested non-normal distribution of error rates. Hence, Mann–Whitney–Wilcoxon tests were conducted to assess differences in error rates between object and action naming in each of the regions. Moreover, Mann–Whitney–Wilcoxon tests were applied to compare error rates, when excluding or including hesitations, in the overall error rates.

To examine error categories regarding research question 2, error rates were calculated per category (non-linguistic speech errors, lexico-semantic errors, grammatical errors, phonological-articulatory errors) and, following the Shapiro–Wilk test, were compared between tasks per hemisphere using Mann–Whitney–Wilcoxon tests, and reported the included effect sizes. Moreover, error ratios per category were established as the errors per category divided by the overall number of errors (see Table 3 for example calculations).

Multivariate regression models were performed to evaluate the influence of baseline errors, rMT, handedness, and age on all errors, and on errors without hesitation in both hemispheres. Spearman's correlations were employed to reveal a relation between errors in the baseline and errors under stimulation for the individual tasks. Additionally, Mann–Whitney–Wilcoxon tests were used to compare error rates in the first and second round of mapping to evaluate the effect of fatigue as a potential confounding factor.

3. Results

3.1. Group Characteristics and Confounding Factors

Characteristics of the participants are summarized in Table 2. No participant had to be excluded due to pain or intolerance to nTMS or MRI acquisition; moreover, the mapping did not have to terminate early due to any such disturbance. All participants tolerated the stimulation well and reported no interference of the stimulation with the overall execution of the tasks.

Table 2. Demographics and error rates per round and hemisphere.

Number of Participants		20
Age (years in mean ± standard deviation (SD); range)		24.75 ± 6.980; 20–53
Gender (%)	Male/Female	40%/60%
Resting Motor Threshold (mean ± SD)	LH	35.15 ± 6.029
	RH	33.95 ± 6.074
Handedness by EHI (mean ± SD)	Right-handed (85%)	79.70 ± 10.82
	Left-handed (5%)	−100 ± 0
	Ambidextrous (10%)	32 ± 17.68
Error rate	LH	0.066 ± 0.039
	RH	0.073 ± 0.046
Error rate first round (mean ± SD)	LH	0.064 ± 0.043
	RH	0.076 ± 0.043
Error rate second round (mean ± SD)	LH	0.070 ± 0.046
	RH	0.070 ± 0.050

Baseline error rates for object naming amounted to 3.65 ± 2.11 and for action naming to 11.1 ± 4.424, meaning that during object naming under stimulation, 71.35 ± 2.11 remaining stimuli were used, and 63.9 ± 4.424 during action naming.

Multivariate regression models revealed that neither baseline errors (LH: $t = 0.505$, $p = 0.621$; RH: $t = 0.522$, $p = 0.609$), rMT (LH: $t = 0.422$, $p = 0.679$; RH: $t = -0.040$, $p = 0.968$), handedness (LH: $t = 0.815$, $p = 0.428$; RH: $t = -0.231$, $p = 0.820$), nor age (LH: $t = -0.601$, $p = 0.557$; RH: $t = -0.950$, $p = 0.357$) were significant predictors of the error rates in either of the hemispheres. The Mann–Whitney–Wilcoxon tests did not reveal a significant difference between the error rates of the first and second round of mapping (LH: $p = 0.349$, RH: $p = 0.422$; Table 2), nor a difference between error rates in the left and right hemisphere ($p = 0.227$).

3.2. Task Comparison of All Errors

Action naming demonstrated a significantly higher error rate than object naming in both hemispheres (LH: action naming mean error rate = 0.078, object naming mean error rate = 0.054 ($p = 0.015$, $r = -0.555$; RH: action naming mean error rate = 0.088, object naming mean error rate = 0.06 ($p = 0.040$, $r = -0.463$))). Significantly more pictures had to be excluded in the baseline naming of action naming (mean error rate = 0.124) compared to object naming (mean error rate = 0.045) ($p < 0.001$, $r = 0.877$), but no correlation between these error rates and the respective error rates under stimulation was found (LH: rho = 0.185, $p = 0.435$ for object naming; rho = 0.130, $p = 0.585$ for action naming; RH: rho = 0.052, $p = 0.830$ for object naming; rho = 0.339, $p = 0.143$ for action naming) for either of the tasks.

3.3. Comparison of Error Categories

3.3.1. Left Hemisphere

The most frequently induced errors in object naming occurred in the phonological-articulatory category (40%), with highest error occurrences in the mMTG, mSTG, pSMG, and vPrG, followed by lexico-semantic errors (39%) mainly in the anG, pSMG, pMFG, and mPrG. Fewer non-linguistic speech errors (19%) were found mainly in the aSMG and mMTG and with the lowest frequency, grammatical errors (2%) were found in the mMTG, mPrG, and SPL (Table 3).

Table 3. Error category rates ± standard deviation and ratios in the left hemisphere per cortical parcellation system (CPS) region. For instance, out of all 31 errors that occurred in the anG during object naming, four were of non-linguistic nature, resulting in a non-linguistic ratio of 12.9%, while the error category rate of 0.008 is based on those four errors out of the entire 480 trials in this region.

LH	Object Naming							Action Naming						
CPS Region	Non-Linguistic		Lexico-Semantic		Grammatical		Phonological-Articulatory		Non-Linguistic		Lexico-Semantic		Grammatical		Phonological-Articulatory
	Rate	Ratio	Rate	Ratio	Rate	Ratio	Rate	Ratio	Rate	Ratio	Rate	Ratio	Rate	Ratio	Rate	Ratio
AnG	0.008 ± 0.022	4/31 12.9%	0.038 ± 0.079	18/31 58%	0.000	0/31 0%	0.019 ± 0.032	9/31 29%	0.010 ± 0.027	5/33 15.1%	0.033 ± 0.046	16/33 48.5%	0.000	0/33 0%	0.021 ± 0.029	10/33 30.3%
ASMG	0.029 ± 0.062	7/14 50%	0.012 ± 0.031	3/14 21.4%	0.000	0/14 0%	0.017 ± 0.034	4/14 28.6%	0.010 ± 0.027	1/6 6.3%	0.033 ± 0.046	9/16 56.3%	0.000	0/16 0%	0.021 ± 0.029	6/16 37.5%
ASTG	0.000	0/4 0%	0.025 ± 0.082	3/4 75%	0.000	0/4 0%	0.008 ± 0.037	1/4 25%	0.025 ± 0.061	3/8 37.5%	0.008 ± 0.037	1/8 12.5%	0.000	0/8 0%	0.033 ± 0.087	4/8 50%
DPoG	0.000	0/3 0%	0.033 ± 0.103	3/3 100%	0.000	0/3 0%	0.000	0/3 0%	0.017	2/9 22.2%	0.042 ± 0.074	5/9 55.6%	0.000	0/9 0%	0.008	1/9 11.1%
DPrG	0.000	0/5 0%	0.025 ± 0.061	3/5 60%	0.000	0/5 0%	0.017 ± 0.051	2/5 40%	0.000	0/10 0%	0.067 ± 0.126	8/10 80%	0.000	0/10 0%	0.017 ± 0.051	2/10 20%
MMFG	0.015 ± 0.029	11/43 25.6%	0.025 ± 0.030	18/43 41.9%	0.000	0/43 0%	0.019 ± 0.033	14/43 32.6%	0.011 ± 0.017	8/61 13.1%	0.035 ± 0.034	25/61 41%	0.003 ± 0.009	2/61 3.3%	0.031 ± 0.034	22/61 36.1%
MMTG	0.029 ± 0.082	7/19 36.8%	0.012 ± 0.031	2/19 10.5%	0.004 ± 0.019	1/19 5.8%	0.038 ± 0.069	9/19 47.4%	0.021	5/20 25%	0.038 ± 0.083	9/20 45%	0.000	0/20 0%	0.025 ± 0.055	6/20 30%
MPoG	0.008 ± 0.026	2/9 22.2%	0.021	5/9 55.6%	0.000	0/9 0%	0.008 ± 0.026	2/9 22.2%	0.017 ± 0.044	4/23 17.3%	0.050 ± 0.091	12/23 52.2%	0.004 ± 0.019	1/23 4.3%	0.025 ± 0.077	6/23 26%
MPrG	0.008 ± 0.037	2/17 11.8%	0.029 ± 0.061	8/17 47%	0.004 ± 0.019	1/17 5.8%	0.025 ± 0.039	6/17 35.3%	0.004 ± 0.019	1/19 5.3%	0.046 ± 0.092	11/19 57.9%	0.000	0/19 0%	0.025 ± 0.039	6/19 31.6%
MSFG	0.008 ± 0.027	3/14 21.4%	0.017 ± 0.041	6/14 42.9%	0.000	0/14 0%	0.014 ± 0.031	5/14 35.1%	0.017 ± 0.045	6/24 25%	0.019 ± 0.033	7/24 0.292	0.003 ± 0.012	1/24 4.2%	0.022 ± 0.038	8/24 33.3%
MSTG	0.004 ± 0.019	1/15 6.7%	0.021 ± 0.037	5/15 33.3%	0.000	0/15 0%	0.038 ± 0.069	9/15 60%	0.029 ± 0.073	7/26 26.9%	0.029 ± 0.056	7/26 26.9%	0.000	0/26 0%	0.038 ± 0.050	9/26 34.6%
OpIFG	0.017 ± 0.032	6/18 33.4%	0.008 ± 0.020	3/18 16.7%	0.003 ± 0.012	1/18 5.6%	0.022 ± 0.033	8/18 44.4%	0.017 ± 0.051	6/18 33.3%	0.017 ± 0.032	6/18 33.3%	0.000	0/18 0%	0.008 ± 0.020	3/18 16.7%
PMFG	0.004 ± 0.010	1/10 10%	0.033 ± 0.078	8/10 80%	0.000	0/10 0%	0.004 ± 0.019	1/10 10%	0.004 ± 0.019	1/20 5%	0.038 ± 0.083	9/20 45%	0.004 ± 0.019	1/20 5%	0.038 ± 0.050	9/20 45%
PMTG	0.008 ± 0.020	3/23 13%	0.025 ± 0.052	9/23 39.1%	0.003 ± 0.012	1/23 4.3%	0.028 ± 0.049	10/23 43.5%	0.011 ± 0.023	4/23 17.3%	0.031 ± 0.046	11/23 47.8%	0.000	0/23 0%	0.022 ± 0.042	8/23 34.8%

Table 3. Cont.

LH		Object Naming								Action Naming							
		Non-Linguistic		Lexico-Semantic		Grammatical		Phonological-Articulatory		Non-Linguistic		Lexico-Semantic		Grammatical		Phonological-Articulatory	
CPS Region		Rate	Ratio	Rate	Ratio	Rate	Ratio	Rate	Ratio	Rate	Ratio	Rate	Ratio	Rate	Ratio	Rate	Ratio
PSFG		0.000	0/1 0%	0.000	0/1 0%	0.000	0/1 0%	0.008 ± 0.037	1/1 100%	0.008 ± 0.037	1/5 20%	0.017 ± 0.051	2/5 40%	0.008 ± 0.037	1/5 20%	0.008 ± 0.037	1/5 20%
PSMG		0.004 ± 0.019	1/18 5%	0.038 ± 0.095	9/18 50%	0.000	0/18 0%	0.038 ± 0.050	8/18 44.4%	0.029 ± 0.068	7/22 31.8%	0.042 ± 0.074	10/22 45.5%	0.000	0/22 0%	0.021 ± 0.037	5/22 22.7%
PSTG		0.008 ± 0.037	1/6 16.7%	0.008 ± 0.037	1/6 16.7%	0.000	0/6 0%	0.033 ± 0.087	4/6 66.7%	0.008 ± 0.037	1/11 9%	0.067 ± 0.100	8/11 72.7%	0.000	0/11 0%	0.008 ± 0.037	1/11 9.1%
SPL		0.008 ± 0.026	2/10 20%	0.012 ± 0.041	3/13 23.1%	0.004 ± 0.019	1/10 10%	0.017 ± 0.034	4/10 40%	0.004 ± 0.019	1/13 7.7%	0.029 ± 0.049	7/13 53.8%	0.000	0/13 0%	0.017 ± 0.034	4/13 30.8%
TrIFG		0.017 ± 0.034	4/13 28.6%	0.017 ± 0.051	3/13 28.6%	0.000	0/13 0%	0.025 ± 0.039	6/13 46.2%	0.004 ± 0.019	1/11 9%	0.029 ± 0.049	7/11 63.6%	0.000	0/11 0%	0.008 ± 0.026	2/11 18.2%
VPoG		0.004 ± 0.019	1/10 10%	0.012 ± 0.031	3/10 30%	0.000	0/10 0%	0.025 ± 0.061	6/10 60%	0.017 ± 0.044	4/25 16%	0.046 ± 0.063	11/25 44%	0.000	0/25 0%	0.033 ± 0.078	8/25 32%
VPrG		0.008 ± 0.026	2/17 19.2%	0.012 ± 0.041	4/17 23.5%	0.000	0/17 0%	0.046 ± 0.083	10/17 58.8%	0.021 ± 0.046	5/32 15.6%	0.029 ± 0.095	7/32 21.9%	0.004 ± 0.019	1/32 3.1%	0.071 ± 0.095	17/32 53.1%

In action naming, a similar pattern of error category frequencies appeared with lexico-semantic errors being most frequent (46%) and found mainly in the dPrG and pSTG, followed by phonological-articulatory errors (34%) being elicited for the vPrG, pMFG, mSTG, and aSTG. Again, fewer non-linguistic speech errors (18%) were found and appeared mostly in the aSTG, mSTG and pSMG, and grammatical errors with the lowest frequency (2%), found in the pSFG, mPoG and vPrG. After applying Mann–Whitney–Wilcoxon tests, action naming overall elicited significantly more errors at the lexico-semantic level than object naming ($p = 0.013$, $r = -0.560$; Table 4). Due to the small numbers of errors per CPS region, no meaningful comparison in error categories per region was achieved.

Table 4. Comparisons of error category rates ± standard deviation in left hemisphere (LH) and right hemisphere (RH). Statistical significance is marked in bold and with an asterisk (*).

Category		Object Naming LH	Action Naming LH	p-Value	Object Naming RH	Action Naming RH	p-Value
Non-linguistic errors		0.006 ± 0.006	0.007 ± 0.011	0.925	0.005 ± 0.008	0.009 ± 0.013	0.083
Lexico-semantic errors	No response	0.001 ± 0.004	0.001 ± 0.002	0.999	0.001 ± 0.003	0.003 ± 0.004	**0.047 ***, $r = -0.48$
	Hesitation whole	0.010 ± 0.012	0.012 ± 0.021	0.900	0.009 ± 0.013	0.015 ± 0.023	0.191
		0.007 ± 0.009	0.011 ± 0.010	**0.013 ***, $r = -0.560$	0.008 ± 0.012	0.013 ± 0.009	**0.022 ***, $r = -0.518$
	Hesitation target	0.016 ± 0.025	0.025 ± 0.026	**0.008 ***, $r = -0.615$	0.020 ± 0.032	0.030 ± 0.023	**0.027 ***, $r = -0.51$
	Anomia	0.002 ± 0.004	0.004 ± 0.006	0.360	0.002 ± 0.004	0.005 ± 0.008	**0.031 ***, $r = -0.462$
	Semantic paraphasia	0.003 ± 0.006	0.005 ± 0.007	0.214	0.003 ± 0.004	0.005 ± 0.009	0.436
Grammatical		0.001 ± 0.002	0.001 ± 0.002	0.482	0.001 ± 0.002	0.001 ± 0.002	0.233
Phonological-Articulatory errors		0.022 ± 0.023	0.025 ± 0.024	0.276	0.023 ± 0.031	0.025 ± 0.024	0.296

3.3.2. Right Hemisphere

In the RH, for object naming, the most frequent errors were in the lexico-semantic category (41%) mostly in the vPrG, vPoG, and aSMG, followed by phonological-articulatory errors (39%) in the aSTG, mMTG, and mSFG. Fewer errors were elicited in the non-linguistic speech category (18%) in the pSTG, TrIFG, and mMFG and hardly any incidences in the grammatical category (2%) in the mMTG, mPoG, and pMFG (Table 5).

Table 5. Error category rates ± standard deviation and ratios in the right hemisphere per cortical parcellation system (CPS) region. For instance, out of all 26 errors that occurred in the anG during object naming, six were of non-linguistic nature, resulting in a non-linguistic ratio of 23.1%, while the error category rate of 0.012 is based on those six errors out of the entire 480 trials in this region.

RH CPS Region	Object Naming								Action Naming							
	Non-Linguistic		Lexico-Semantic		Grammatical		Phonological-Articulatory		Non-Linguistic		Lexico-Semantic		Grammatical		Phonological-Articulatory	
	Rate	Ratio	Rate	Ratio	Rate	Ratio	Rate	Ratio	Rate	Ratio	Rate	Ratio	Rate	Ratio	Rate	Ratio
AnG	0.012 ± 0.027	6/26 23.1%	0.021 ± 0.032	10/26 38.5%	0.002 ± 0.009	1/26 3.8%	0.019 ± 0.025	9/26 34.6%	0.015 ± 0.034	7/39 17.9%	0.048 ± 0.045	23/29 59%	0.000	0/39 0%	0.015 ± 0.024	7/39 17.9%
ASMG	0.004 ± 0.019	1/16 6.3%	0.042 ± 0.079	10/16 62.5%	0.000	0/16 0%	0.021 ± 0.060	5/16 31.3%	0.012 ± 0.041	3/17 17.6%	0.029 ± 0.049	7/17 41.2%	0.000	0/17 0%	0.029 ± 0.049	7/17 41.2%
ASTG	0.008 ± 0.037	1/8 12.5%	0.008 ± 0.037	1/8 12.5%	0.000	0/8 0%	0.050 ± 0.188	6/8 75%	0.092 ± 0.166	11/19 57.9%	0.033 ± 0.068	4/19 21.1%	0.000	0/19 0%	0.033 ± 0.087	4/19 21.1%
DPoG	0.000	0/5 0%	0.025 ± 0.082	3/5 60%	0.000	0/5 0%	0.017 ± 0.051	2/5 40%	0.017 ± 0.051	2/8 25%	0.025 ± 0.061	3/8 37.5%	0.000	0/8 0%	0.017 ± 0.051	2/8 25%
DPrG	0.000	0/4 0%	0.008 ± 0.037	1/4 25%	0.000	0/4 0%	0.025 ± 0.061	3/4 75%	0.008 ± 0.037	1/13 7.7%	0.067 ± 0.100	8/13 61.5%	0.000	0/13 0%	0.017 ± 0.051	2/13 15.4%
MMFG	0.021 ± 0.043	15/55 27.3%	0.028 ± 0.052	20/55 36.4%	0.000	0/55 0%	0.028 ± 0.042	20/55 36.4%	0.010 ± 0.021	7/53 13.2%	0.032 ± 0.034	23/52 43.3%	0.001 ± 0.006	1/53 1.8%	0.022 ± 0.032	16/53 30.2%
MMTG	0.008 ± 0.037	2/21 9.5%	0.029 ± 0.062	7/21 33.3%	0.008 ± 0.037	2/21 9.5%	0.042 ± 0.099	10/21 47.6%	0.033 ± 0.068	8/29 27.6%	0.054 ± 0.078	13/29 44.8%	0.000	0/29 0%	0.029 ± 0.068	7/29 24.1%
MPoG	0.004 ± 0.019	1/10 10%	0.021 ± 0.060	5/10 50%	0.004 ± 0.019	1/10 10%	0.012 ± 0.056	3/10 30%	0.025 ± 0.048	6/24 25%	0.050 ± 0.068	12/24 50%	0.000	0/24 0%	0.025 ± 0.048	6/24 25%
MPrG	0.004 ± 0.019	1/12 8.3%	0.029 ± 0.049	7/12 58.3%	0.000	0/12 0%	0.017 ± 0.058	4/12 33.3%	0.029 ± 0.062	7/28 25%	0.062 ± 0.097	15/28 53.6%	0.004 ± 0.019	1/28 3.6%	0.021 ± 0.046	5/28 17.9%
MSFG	0.008 ± 0.027	3/26 11.5%	0.028 ± 0.069	10/26 38.5%	0.000	0/26 0%	0.036 ± 0.045	13/26 50%	0.011 ± 0.023	4/19 21.1%	0.028 ± 0.038	10/19 52.6%	0.000	0/19 0%	0.014 ± 0.031	5/19 26.3%
MSTG	0.008 ± 0.026	2/12 16.7%	0.017 ± 0.051	4/12 33.3%	0.000	0/12 0%	0.025 ± 0.067	6/12 50%	0.012 ± 0.041	3/25 12%	0.054 ± 0.078	13/25 52%	0.000	0/25 0%	0.033 ± 0.057	8/25 32%
OpIFG	0.014 ± 0.035	5/17 29.4%	0.011 ± 0.023	4/17 23.5%	0.003 ± 0.012	1/17 5.9%	0.019 ± 0.033	7/17 41.2%	0.025 ± 0.087	9/41 22%	0.036 ± 0.037	13/41 31.7%	0.003 ± 0.012	1/41 2.4%	0.050 ± 0.065	18/41 43.9%
PMFG	0.008 ± 0.026	2/12 16.7%	0.021 ± 0.046	5/12 41.7%	0.004 ± 0.019	1/12 8.3%	0.017 ± 0.044	4/12 33.3%	0.004 ± 0.019	1/15 6.7%	0.029 ± 0.049	7/15 46.7%	0.000	0/15 0%	0.025 ± 0.039	6/15 40%
PMTG	0.003 ± 0.012	1/13 7.7%	0.019 ± 0.045	7/13 53.8%	0.000	0/13 0%	0.014 ± 0.031	5/13 38.5%	0.014 ± 0.031	5/20 25%	0.019 ± 0.033	7/20 35%	0.000	0/20 0%	0.019 ± 0.033	7/20 35%

Table 5. Cont.

RH	Object Naming								Action Naming							
CPS Region	Non-Linguistic		Lexico-Semantic		Grammatical		Phonological-Articulatory		Non-Linguistic		Lexico-Semantic		Grammatical		Phonological-Articulatory	
	Rate	Ratio	Rate	Ratio	Rate	Ratio	Rate	Ratio	Rate	Ratio	Rate	Ratio	Rate	Ratio	Rate	Ratio
PSFG	0.025 ± 0.061	3/4 75%	0.008 ± 0.037	1/4 25%	0.000	0/4 0%	0.000	0/4 0%	0.008 ± 0.037	1/6 16.7%	0.017 ± 0.051	2/6 33.3%	0.000	0/6 0%	0.017 ± 0.051	2/6 33.3%
PSMG	0.008 ± 0.037	2/16 12.5%	0.033 ± 0.099	8/16 50%	0.004 ± 0.019	1/16 6.3%	0.021 ± 0.046	5/16 31.3%	0.017 ± 0.044	4/19 21.1%	0.033 ± 0.057	8/19 42.1%	0.000	0/19 0%	0.025 ± 0.048	6/19 31.6%
PSTG	0.017 ± 0.051	2/7 28.6%	0.008 ± 0.037	1/7 14.3%	0.000	0/7 0%	0.033 ± 0.068	4/7 57.1%	0.017 ± 0.051	2/14 14.3%	0.042 ± 0.074	5/14 35.7%	0.008 ± 0.037	1/14 7.1%	0.042 ± 0.119	5/14 35.7%
SPL	0.004 ± 0.019	1/12 8.3%	0.021 ± 0.037	5/12 41.7%	0.000	0/12 0%	0.025 ± 0.048	6/12 50%	0.012 ± 0.041	3/13 23.1%	0.021 ± 0.053	5/13 38.5%	0.000	0/13 0%	0.021 ± 0.037	5/13 38.5%
TrIFG	0.025 ± 0.067	6/19 31.6%	0.029 ± 0.062	7/19 36.8%	0.000	0/19 0%	0.025 ± 0.048	6/19 31.6%	0.025 ± 0.077	6/28 21.4%	0.058 ± 0.090	14/28 50%	0.000	0/28 0%	0.029 ± 0.056	7/28 25%
VPoG	0.008 ± 0.026	2/18 11.1%	0.042 ± 0.092	10/18 55.6%	0.000	0/18 0%	0.025 ± 0.048	6/18 33.3%	0.025 ± 0.077	6/30 20%	0.075 ± 0.085	18/30 60%	0.000	0/30 0%	0.025 ± 0.048	6/30 20%
VPrG	0.008 ± 0.026	2/18 11.1%	0.046 ± 0.063	11/18 61.1%	0.000	0/18 0%	0.021 ± 0.037	5/18 27.8%	0.017 ± 0.034	4/24 16.7%	0.033 ± 0.057	10/24 41.7%	0.000	0/24 0%	0.038 ± 0.048	9/24 37.5%

Action naming also displayed most errors in the lexico-semantic category, mainly in the vPoG, dPrG, mPrG, and trIFG (47%), and as the second most frequent, errors in the phonological-articulatory category (30%), mainly in the opIFG, pSTG, and vPrG. Fewer errors were found in the non-linguistic speech category (22%), mostly in the aSTG, mMTG, and mPrG, and errors in the grammatical category (1%) in the pSTG, mPrG, and opIFG. Again, action naming elicited more errors in the lexico-semantic category in the RH ($p = 0.022$, r = -0.518) (see Table 4).

3.4. Area-Wise Comparison of Tasks for All Errors

Action naming had a significantly higher error rate than object naming in both hemispheres. Table 6 shows the error rates according to the hemisphere and CPS region. In the LH, none of the CPS regions showed a significant difference between error rates for object and action naming, but a trend for a higher error rate for action naming was found in the mSTG ($p = 0.065$) and in the vPoG ($p = 0.051$). In the RH, action naming elicited significantly more errors in the aSTG ($p = 0.036$, r = -0.455), dPrG ($p = 0.026$, r = -0.547), mPoG ($p = 0.020$, r = -0.617), mPrG ($p = 0.012$, r = -0.557), mSTG ($p = 0.034$, r = -0.410) and a trend in the opIFG ($p = 0.050$). Figure 4 depicts the cortical distribution of error rates per CPS for object naming (a) and action naming (b) in heat maps for both hemispheres.

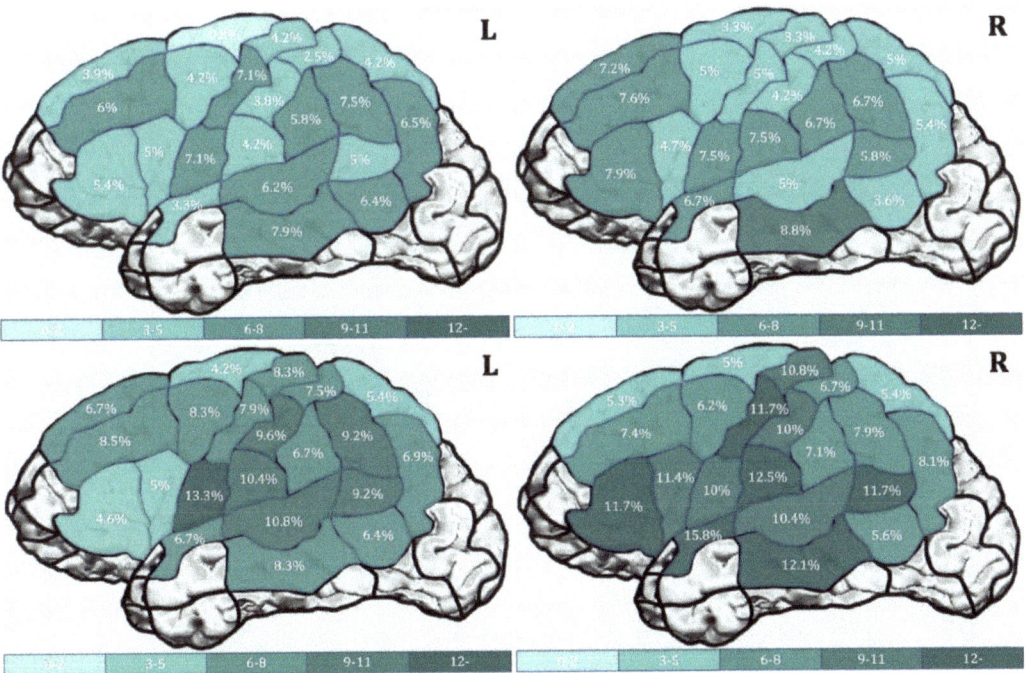

Figure 4. Error rates in percentage according to Table 6 comprised of all errors per cortical parcellation system (CPS) regions in the left hemisphere and right hemisphere during object naming (**upper row**) and action naming (**lower row**).

Table 6. Error rates ± standard deviation per region during object naming (ON) and action naming (AN) in the left hemisphere (LH) and right hemisphere (RH) and their differences. For instance, 31 errors in the anG out of 480 trials (20 participants per four stimulation targets per six stimulations) resulted in an error rate of 0.065 ± 0.087. Statistical significance is marked in bold and with an asterisk (*).

Region	Error Rate Object Naming in LH	Error Rate Action Naming in LH	p-Value	Error Rate Object Naming in RH	Error Rate Action Naming in RH	p-Value
overall	0.054 ± 0.043	0.078 ± 0.045	**0.015 *, r = −0.555**	0.060 ± 0.057	0.088 ± 0.052	**0.040 *, r = −0.463**
AnG	0.065 ± 0.087	0.069 ± 0.045	0.604	0.054 ± 0.045	0.081 ± 0.064	0.144
ASMG	0.058 ± 0.072	0.067 ± 0.075	0.813	0.067 ± 0.096	0.071 ± 0.078	0.745
ASTG	0.033 ± 0.103	0.067 ± 0.113	0.518	0.067 ± 0.190	0.158 ± 0.206	**0.036 *, r = −0.455**
DPoG	0.025 ± 0.082	0.075 ± 0.127	0.152	0.042 ± 0.092	0.067 ± 0.100	0.437
DPrG	0.042 ± 0.092	0.083 ± 0.148	0.359	0.033 ± 0.068	0.108 ± 0.156	**0.026 *, r = −0.547**
MMFG	0.060 ± 0.057	0.085 ± 0.057	0.079	0.076 ± 0.089	0.074 ± 0.054	0.825
MMTG	0.079 ± 0.119	0.083 ± 0.094	0.937	0.088 ± 0.17	0.121 ± 0.122	0.305
MPoG	0.038 ± 0.079	0.096 ± 0.109	0.091	0.042 ± 0.092	0.100 ± 0.075	**0.020 *, r = −0.617**
MPrG	0.071 ± 0.087	0.079 ± 0.116	0.827	0.050 ± 0.074	0.117 ± 0.106	**0.012 *, r = −0.557**
MSFG	0.039 ± 0.054	0.067 ± 0.064	0.131	0.072 ± 0.094	0.053 ± 0.058	0.594
MSTG	0.062 ± 0.097	0.108 ± 0.112	0.065	0.050 ± 0.087	0.104 ± 0.108	**0.034 *, r = −0.410**
OpIFG	0.050 ± 0.047	0.050 ± 0.057	0.923	0.047 ± 0.06	0.114 ± 0.124	0.050
PMFG	0.042 ± 0.079	0.083 ± 0.101	0.105	0.050 ± 0.074	0.062 ± 0.081	0.683
PMTG	0.064 ± 0.101	0.064 ± 0.075	0.393	0.036 ± 0.055	0.056 ± 0.062	0.223
PSFG	0.008 ± 0.037	0.042 ± 0.074	0.129	0.033 ± 0.087	0.050 ± 0.095	0.660
PSMG	0.075 ± 0.104	0.092 ± 0.014	0.649	0.067 ± 0.126	0.079 ± 0.083	0.570
PSTG	0.050 ± 0.095	0.092 ± 0.114	0.110	0.058 ± 0.135	0.117 ± 0.203	0.348
SPL	0.042 ± 0.063	0.054 ± 0.056	0.351	0.050 ± 0.074	0.054 ± 0.091	0.958
TrIFG	0.054 ± 0.062	0.046 ± 0.057	0.666	0.079 ± 0.113	0.117 ± 0.154	0.435
VPoG	0.042 ± 0.063	0.104 ± 0.124	0.051	0.075 ± 0.127	0.125 ± 0.128	0.198
VPrG	0.071 ± 0.099	0.133 ± 0.165	0.104	0.075 ± 0.071	0.100 ± 0.131	0.605

3.5. Hesitation Error Exclusion

The number of errors and resulting error rates in each task differed significantly, when hesitation errors were excluded. Table 7 summarizes these comparisons for each task and hemisphere.

Table 7. Comparisons of error rates ± standard deviation including all errors vs. excluding hesitations. Statistical significance is marked in bold and with an asterisk (*).

		Including	Excluding	*p*-Value
All tasks	All	0.070 ± 0.042	0.036 ± 0.027	**1.91 × 10⁶ *, r = 0.877**
Object Naming		0.057 ± 0.050	0.030 ± 0.031	**9.55 × 10⁵ *, r = 0.877**
	in LH	0.054 ± 0.043	0.029 ± 0.027	**0.0002 *, r = 0.865**
	In RH	0.060 ± 0.057	0.031 ± 0.036	**0.0001 *, r = 0.873**
Action Naming		0.083 ± 0.044	0.041 ± 0.030	**9.56 × 10⁵ *, r = 0.877**
	in LH	0.078 ± 0.045	0.040 ± 0.031	**9.29 × 10⁵ *, r = 0.878**
	In RH	0.088 ± 0.052	0.043 ± 0.033	**9.50 × 10⁵ *, r = 0.877**

For a second analysis, hesitation errors (both hesitation on the whole phrase and hesitation on the target) were excluded for a separate comparison. Error rates without hesitations differed significantly between tasks: action naming demonstrated a higher error rate than object naming in both hemispheres (LH: mean error rate action naming = 0.040, mean error rate object naming = 0.029 ($p = 0.042$, $r = -0.472$); RH: mean error rate action naming = 0.043, mean error rate object naming = 0.031 ($p = 0.035$, $r = -0.472$)). Again, error rates on baseline naming did not correlate significantly with the error rates without hesitation in either of the tasks or hemispheres (LH: rho = 0.080, $p = 0.737$ for object naming, rho = 0.180, $p = 0.447$ for action naming; RH: rho = 0.148, $p = 0.551$ for object naming, rho = 0.224, $p = 0.342$ for action naming).

Multivariate regression models revealed that neither baseline errors (LH: $t = 1.231$, $p = 0.237$; RH: $t = 0.880$, $p = 0.393$), rMT (LH: $t = -0.680$, $p = 0.507$; RH: $t = -0.281$, $p = 0.782$), handedness (LH: $t = 1.065.23$, $p = 0.304$; RH: $t = 0.783$, $p = 0.446$) nor age (LH: $t = -0.811$, $p = 0.430$; RH: $t = -0.986$, $p = 0.340$) were significant predictors for the error rates in either of the hemispheres.

3.6. Area-Wise Comparison of Error Rates without Hesitations

When analyzed separately per hemisphere, a significantly higher error rate was observed for action naming over object naming in both hemispheres (LH: $p = 0.042$, $r = -0.472$; RH: $p = 0.035$, $r = -0.472$). Table 8 depicts the error rates per hemisphere and for the two tasks and their comparisons. In the LH, a significantly higher error rate was found for action naming in the pMFG ($p = 0.037$, $r = -0.561$); in the RH, in the opIFG ($p = 0.020$, $r = -0.536$) and a trend in mSTG ($p = 0.067$). Figure 5 depicts the cortical distribution of error rates per CPS region for object naming (a) and action naming (b) in heat maps for both hemispheres.

Table 8. Error rates ± standard deviation excluding hesitations per region during object naming (ON) and action naming (AN) in the left hemisphere (LH) and right hemisphere (RH) and their differences. For instance, 15 errors in the anG out of 480 trials (20 participants per four stimulation targets per six stimulations) resulted in an error rate of 0.031 ± 0.052. Statistical significance is marked in bold and with an asterisk (*).

Region	Error Rate Object Naming in LH	Error Rate Action Naming in LH	p-Value	Error Rate Object Naming in RH	Error Rate Action Naming in RH	p-Value
overall	0.029 ± 0.027	0.040 ± 0.031	**0.042 *,** **r = −0.472**	0.031 ± 0.036	0.043 ± 0.033	**0.035 *,** **r = −0.472**
AnG	0.031 ± 0.052	0.038 ± 0.030	0.421	0.027 ± 0.031	0.029 ± 0.053	0.706
ASMG	0.017 ± 0.034	0.033 ± 0.050	0.129	0.033 ± 0.063	0.038 ± 0.050	0.824
ASTG	0.017 ± 0.051	0.033 ± 0.087	0.572	0.050 ± 0.188	0.067 ± 0.166	0.572
DPoG	0.000 ± 0.000	0.017 ± 0.075	0.999	0.017 ± 0.051	0.033 ± 0.068	0.484
DPrG	0.025 ± 0.082	0.033 ± 0.068	0.850	0.025 ± 0.061	0.058 ± 0.098	0.203
MMFG	0.029 ± 0.037	0.049 ± 0.043	0.182	0.035 ± 0.057	0.036 ± 0.046	0.656
MMTG	0.042 ± 0.074	0.042 ± 0.057	0.999	0.054 ± 0.133	0.062 ± 0.097	0.751
MPoG	0.021 ± 0.037	0.042 ± 0.083	0.430	0.025 ± 0.061	0.042 ± 0.063	0.340
MPrG	0.042 ± 0.051	0.054 ± 0.095	0.642	0.025 ± 0.061	0.046 ± 0.057	0.303
MSFG	0.019 ± 0.033	0.039 ± 0.045	0.168	0.044 ± 0.061	0.022 ± 0.033	0.272
MSTG	0.042 ± 0.074	0.054 ± 0.056	0.240	0.029 ± 0.068	0.071 ± 0.078	0.067
OpIFG	0.033 ± 0.046	0.025 ± 0.034	0.507	0.022 ± 0.033	0.069 ± 0.078	**0.020 *,** **r = −0.536**
PMFG	0.017 ± 0.058	0.054 ± 0.068	**0.037 *,** **r = −0.561**	0.025 ± 0.048	0.029 ± 0.049	0.851
PMTG	0.031 ± 0.055	0.025 ± 0.042	0.999	0.017 ± 0.032	0.031 ± 0.042	0.073
PSFG	0.008 ± 0.037	0.017 ± 0.051	0.773	0.000 ± 0.000	0.025 ± 0.061	0.149
PSMG	0.038 ± 0.057	0.029 ± 0.049	0.565	0.029 ± 0.049	0.033 ± 0.057	0.824
PSTG	0.033 ± 0.087	0.033 ± 0.087	0.999	0.033 ± 0.068	0.083 ± 0.167	0.281
SPL	0.021 ± 0.046	0.025 ± 0.048	0.766	0.029 ± 0.049	0.021 ± 0.037	0.530
TrIFG	0.033 ± 0.050	0.012 ± 0.041	0.236	0.042 ± 0.069	0.038 ± 0.069	0.821
VPoG	0.029 ± 0.062	0.050 ± 0.083	0.314	0.046 ± 0.074	0.054 ± 0.078	0.778
VPrG	0.058 ± 0.086	0.108 ± 0.156	0.189	0.025 ± 0.039	0.067 ± 0.133	0.131

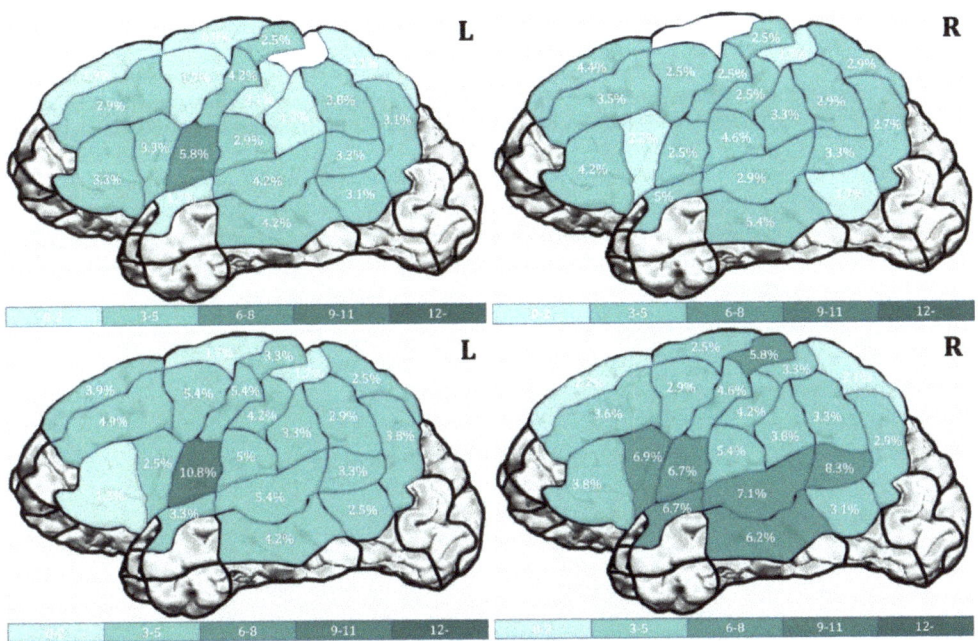

Figure 5. Error rates in percentage according to Table 8 comprised of errors excluding hesitations per cortical Parcellation System (CPS) region in left and right hemisphere during object naming (**upper row**) and action naming (**lower row**).

4. Discussion

The objective of this study was to evaluate the potential of tasks in sentence context, specifically a verb-targeting language task, for error elicitation in different cortical surfaces under nTMS. By extensively mapping 20 healthy participants in both hemispheres using the novel task action naming in sentence context, together with object naming in sentence context, task sensitivities were compared. Quantitative differences in error elicitation of the two tasks were investigated overall and per small cortical area. Moreover, we aimed to understand the breakdown in language production caused by nTMS through a detailed qualitative error analysis. Lastly, the effect of hesitation errors on mapping results was examined through a separate analysis to define its significance further.

4.1. Overall Task Comparison

As for the primary comparison, action naming delivered a higher error rate than object naming in both hemispheres. This leads to the conclusion that retrieving a verb in sentence context is more easily disrupted under nTMS than retrieving a noun in sentence context. No significant correlation between the number of baseline errors and errors under stimulation was found. Moreover, all error-prone verb stimuli are removed during the baseline process. As a result, errors under stimulation should be considered true positives. The higher error rate of action naming, therefore, cannot be explained by this task being more error-prone, per se, but points towards a higher vulnerability of the more complex process of verb versus noun retrieval in sentence context under stimulation. This finding seems contrary to previous studies, which have argued that verb tasks under nTMS did not reach the same sensitivity as object naming and were, hence, not worth including for nTMS language mapping [43,44].

The discrepancy with the present findings may be attributed to the designs of the tasks across studies. Firstly, the picture-naming paradigm differed. Former studies employed well-established databases for the object naming variant [51], while using homemade drawings or photos for the verb task. This hampers the direct comparison. Our design entails stimuli for action naming of similar complexity and style as object naming. Moreover, our stimuli had been previously tested for a high naming agreement. Secondly, whereas the former studies targeted single word retrieval for both tasks ('ball' for object naming, 'throwing' for action naming), the current study made use of object naming and action naming in sentence context. Next to target word retrieval, this required inflection and embedding of the target in a short lead-in phrase. The higher cognitive effort needed for our action naming task is likely to be differently affected by nTMS and may have resulted in the higher error rate under stimulation in our sample. This finding is entirely in line with data from DES mapping, arguing that verb tasks are more sensitive under stimulation [23,28,52]. The following sections will provide a closer look at the root of this sensitivity by looking at error types and specific cortical locations.

4.2. Error Category Comparison

In most protocols for nTMS language mapping, different error types are used for a more detailed mapping depiction [6,18,44]. However, these classifications are hardly ever used to further unravel the origin of the errors. In the present study, we employed the common error types found in word production under stimulation, projected them on errors in sentence context, and assigned them to the level of production disruption they indicate (see Figure 1). This classification was used to better understand the difference in error rates on the two tasks under investigation. Additionally, testing in sentence context allows for screening of more subtle errors than testing a single word and allows categorization of the errors into different stages of breakdown in the sentence production process.

When taking together all tasks and hemispheres, a similar pattern of error category frequency appeared: errors at the phonological-articulatory and lexico-semantic level were most prevalent, whereas fewer non-linguistic errors and even fewer grammatical errors were induced (Tables 3–5). This pattern is consistent with previous reports [6,34–37]. While intraoperative DES, possibly due to its higher frequency and direct application on the exposed cortex, easily elicits full disruptions of speech [38], nTMS is known to hinder mainly the phonological-articulatory processes and lexico-semantic retrieval [6,34–37]. The effect of nTMS versus intraoperative DES is not yet well understood. It has been established, however, that the timing of the nTMS pulses in relation to the stimulus to name can alter the error pattern [53]. The common protocol of a delay of 0 ms was found to produce the best mapping results when compared to the intraoperative gold standard [16,18]. This protocol evokes about 40% of errors on the sound level, as confirmed by our study with 257/729 errors in the LH and 269/815 error in the RH on the sound level.

Regardless of the task, nTMS seems to disrupt the later levels of word and sentence production with varying location in both hemispheres. Analysis of errors at this level, thus, did not reveal differences between object and action naming. Due to the additional inflectional effort needed to embed a target verb in a sentence compared to embedding a target noun, one expects errors at the grammatical level to be more pronounced for action naming than for object naming. However, no significant difference was detected. Instead, the error rates on the two tasks differed at the lexico-semantic level (Table 4). These errors occurred more frequently for the verbs than the nouns. Therefore, nTMS seems to affect verb retrieval more than noun retrieval. A possible reason for this is the verb's more complex conceptual and lexico-semantic information. As head of the sentence, lexical entries of verbs carry information about argument structure of the sentence. Moreover, higher abstractness of actions compared to nouns and objects adds to the semantic complexity and may result in a higher vulnerability for verbs. This known distinction has been reported after brain damage [54,55] and seems to hold for nTMS mapping as well.

We cannot rule out that lexico-semantic errors indicate the difficulty to retrieve even the concept of the target or to inflect the target. Inflection in sentence context may require more cognitive effort, even though the error is not grammatical in nature, but rather leads to hesitations of the inflected target or to anomia. While the present setup cannot disentangle this further, elicitation of the targets in sentence context allows distinguishing between a full speech breakdown and disturbance in target retrieval. Narrowing down the origin of the errors revealed a higher vulnerability for action naming at this level under nTMS. The inclusion of a small lead-in phrase is, hence, not only informative, but also crucial for sophisticated error classification and creating a more effortful task, that is evidently easier to disturb with nTMS.

4.3. Area-Wise Comparison in the Left Hemisphere

Smaller scaled comparisons per predefined CPS region were performed to reveal whether one of the tasks elicited a higher error rate in a more localized cortical surface area. In the LH, none of the comparisons between object and action naming reached significance for any CPS region (Table 6). Whereas on the entire hemisphere, action naming seems to be more easily disrupted; this could not be localized to a specific area.

Taken together with the fact that, during both tasks, at least a few errors appeared in every CPS region, this is in line with the body of navigated stimulation mapping studies that do not support a classical double dissociation of noun and verb production in the LH [25,54,56] (for reviews see [57,58]). Literature on stroke-induced aphasia suggests a left-sided temporal lobe hub for comprehension and production of object names and a left-sided frontal lobe hub for tasks related to action/verb production [26,27,59] (for a recent review see [60]). This claim was questioned by data from many methodologies [57,58], including mapping studies under nTMS and intraoperative DES.

In nTMS mappings using single-word targets, no selective areas for either task were reported [43,44], but a widespread region in the perisylvian area, covering all three lobes for both object naming and action naming. The conclusion is similar in studies using intraoperative DES with single-word targets [49,52]. A double dissociation could be delineated in single cases; however, this distinction did not surface at the group level. The same conclusion holds for intraoperative mappings with object and action naming in sentence context. Single cases of an exclusive involvement of the opIFG in action naming have been described [23] and a more prefrontal/premotor network for action words [28], but no clear-cut group pattern of a double dissociation arose either.

The grand conclusion emerging from group analysis in intraoperative DES mapping in patients and nTMS mapping in healthy participants, that our data add to as well, is a mainly to entirely shared perisylvian network for verb and noun production. While this conclusion cannot help to resolve the decade-old debate about a neural, clear-cut segregation of nouns versus verbs in the brain, it stresses the usefulness of a two-task design: a few double dissociations were obtained in single cases [28,49,52]; moreover, cases have been described in which action naming was the only task to elicit errors [23]. These observations let the authors to conclude that action naming is a necessary addition to the standard object naming task when trying to avoid postoperative deficits. That this is not confirmed at the group level is likely due to high inter-individual variability [28,38,61–63], but becomes evident in the reported single cases as well as in the present data.

The group data in this experiment still gives insight into error patterns per task and hemisphere. In the LH, object naming errors were spread over all CPS regions (Table 6), but the highest error rates were found in the vPrG, mPrG, pSMG, and mMTG. The middle and ventral parts of the PrG are known as components in articulatory planning [29], which is reflected in our data in a high ratio of phonological-articulatory errors (Table 3) as well as in other nTMS [36,53] and intraoperative DES studies [64]). The function of the pSMG ranges from access to semantic representations as part of Gschwind's region [36,65] to phonological decision making [28,29,36,38], and showed no clear error association in our data either. In the mMTG as a presumed semantic hub [29], a high ratio for phonological-articulatory errors was elicited in our data and, therefore, cannot confirm this common function relation. With that said, it is important to point out that none of the areas were correlated with a specific error type, frequently occurring in that area. The areas rather seem to be frequent network hubs during language production. The above-named functional associations are, therefore, to be taken with caution regarding their presumed underlying processes.

Action naming highest error occurrences lay in the vPrG, vPoG, and mSTG, with trends for significantly higher error rates compared to object naming in the vPoG and mSTG (Table 6). Again, no CPS region appeared without errors. The ventral parts of the PrG and PoG play a role in the embodiment hypothesis [66] and, therefore, would be compatible with lexico-semantic errors during action naming. However, our data show several error types prevalent in these regions (Table 3), not predominantly lexico-semantic errors. The medial part of the STG, both found with high error rates in nTMS [43,44] and intraoperative DES studies [49] for action naming, is so far the only constant area throughout several studies that is essentially involved in action naming. Classically thought of as a semantic hub close to Wernicke's area, a mixed error ratio in our data cannot further specify its exact role in production (Table 3). However, regardless of the distinct error category, the persistent appearance of the mSTG throughout different methodologies stresses its role in action naming and may make the verb task a better candidate for mapping in this temporal area.

4.4. Area-Wise Comparison in the Right Hemisphere

Studies of mapping with a direct comparison of object naming and action naming are rare. The literature is even scarcer for a comparison of tasks in the RH. Since language functions are dominantly hosted by the LH, mapping of the contralateral hemisphere is usually deemed unnecessary [57] However, attention to the RH has been renewed after studies using multiple methodologies have suggested that the involvement of the RH in language has long been underestimated [67,68]. Recruitment of RH areas was found in functional MRI studies, ranging from domain-general processes, such as attention and working memory [69,70] to linguistic processes, such as sound to lexical meaning mapping in the IFG [71], bilateral conceptual knowledge in the anterior temporal lobe [72–74], bilateral phonological decision making in the IFG and SMG [75,76], and explicit impairment in comprehension after RH damage [67]. Studies using intraoperative DES have described a mirrored pattern of LH homologues in the RH with stimulation of frontal areas resulting in articulatory errors and speech arrest, and a temporal hub in the RH for conceptual and semantic knowledge, seen in semantic errors and anomias [61,64,68,77–80].

Only a handful of nTMS studies have mapped both the LH and RH according to segmentation of the brain by the CPS, all employing object naming. Overall, a comparably high error rate was reported for the RH [6,35,81,82], albeit lower than the LH counterpart. The detection of language areas in the RH by nTMS has therefore been described, but has not been compared between tasks. The present study is, hence, the first to systematically compare of object naming and action naming in the RH under nTMS.

In our sample, object naming errors were elicited in all CPS regions. The highest error rates were located rostrally to the Sylvian fissure, but also over all three lobes with the highest occurrences in the mMTG, vPoG, trIFG, mMFG, and vPrG (Table 6). Both the mMTG and vPrG mirror the pattern of higher error rates of the LH and align with findings from nTMS [6,81] and intraoperative DES [68,77,80]. However, error categories in our sample—phonological-articulatory errors in the mMTG and lexico-semantic errors in the vPrG (Table 5)—do not fit the described clear function allocation described by Duffau and colleagues [61,64,68,77,80] where frontal stimulation would result in speech motor errors and temporal stimulation in lexico-semantic errors. The vPoG as well as the trIFG and mMFG displaying mixed errors can be considered the RH homologues of known LH language regions, engaging in speech motor functions [81]. Bilateral activation for articulatory and speech motor functions is evident and falls in line with reports for a bilateral language recruitment from nTMS, intraoperative DES, and neuroimaging [6,64,68,70,72–76,79–81].

Action naming was most frequently disturbed around the central sulcus and Sylvian fissure, with the aSTG, vPoG, and mMTG as the most receptive areas (Table 6). As part of the STG, the aSTG is to some degree mirroring the LH pattern. Being prone to non-linguistic speech errors (Table 5), the aSTG may be crucial for early conceptual processes and thereby in accordance with bilateral activation during conceptual knowledge recruitment found in fMRI and DES [68,72–74,78,80]. The function of the vPoG in bilateral sensorimotor activation during performance of actions may explain its involvement during action naming. The dominant error category in vPoG of lexico-semantic disruption underlines this further and points towards embodiment in the RH. A high ratio of lexico-semantic errors in the mMTG in our sample aligns with reports of semantic errors in this region under intraoperative DES [61,68,77,80] and is a clear indicator of this area's role in bilateral recruitment for naming through lexico-semantic involvement.

Differences between object naming and action naming were significant in the aSTG, dPrG, mPoG, mPrG, and mSTG for action naming being more easily disturbed in these areas (Table 6). Compared to mere trends for significance in the LH, even more areas were prone to errors in action than in object naming in the RH. The middle and dorsal parts of the primary motor and sensory cortex may once more be crucial for action-related word production, as part of the embodiment theory [66]. Accordingly, mostly lexico-semantic errors were elicited here (Table 5). Both the aSTG and mSTG are so far not known for their strong involvement in the RH during verb tasks, but the mSTG mirrors the consistent reports of involvement during verb tasks in the LH [43,44,49]. Overall, the RH's STG as a semantic hub under intraoperative DES [61,64,68,77,78,80] and nTMS [81] could be sensitive once more to action naming's higher lexico-semantic complexity. Mostly conceptual and lexico-semantic errors arising from this region in our data confirms this interpretation.

In conclusion, the even more pronounced recruitment of the RH in action naming could be rooted in a bilateral conceptual and lexico-semantic knowledge processing [61,68,72–74,77,78,80], manifesting itself in many lexico-semantic errors in these areas. The verb task's higher demand on conceptual and meaning retrieval could make it more sensitive for area detection in the RH. This finding may specifically be visible in the current setup, employing action naming and sentence context.

4.5. Overall Involvement of the Right Hemisphere

An overall mirrored pattern in the RH as compared to the LH of easily disrupted areas was discovered in our data, in other nTMS studies [6,81], and during intraoperative DES as well as fMRI [61,64,68,72–77,79,80]. However, a clear distinction to LH areas is usually made by the authors, claiming the RH regions may have to be considered language-involved in contrast to language-eloquent areas in the LH [5,81,83]. This translates to the distinction that resection of or damage to these language-involved areas would not result in the same drastic language impairments as damage to eloquent counterparts. Their involvement may be secondary.

The nature of nTMS to reveal involved areas and, hence, overcall positive areas is a known phenomenon and can, to some extent, explain the very spread effects in the RH in the present data. Moreover, it cannot be excluded that nTMS can activate long distance interhemispheric connectivity, as has been shown with dual-coil stimulation [84]. Stimulation of involved right hemispheric areas to eloquent left hemispheric parts may have contributed to the current findings.

Two other uncertainties deliver possible explanations for the current RH data. Firstly, a training effect could be assumed. After administering two rounds of mapping with about 140 stimulations necessary to cover all CPS regions three times, the participant has to name each of the approximate 75 items about 10 times. It has been shown that for (novel) verb learning both the LH and to an even greater extent, the RH are recruited in the same area sensitive in our sample [85]. It can therefore not be excluded that repeated exposure to the same stimulus can result in similar effect as a training effect and thereby is related to the high involvement of the RH in those areas. Secondly, the timing of the nTMS onset in relation to the stimulus to name should be considered. The early onset of 0 ms used here as well as in common protocols is likely to disrupt the earliest stages of naming, namely conceptual retrieval [29,53], while delayed onsets have shown to elicit errors at the sound level [53]. While it is a necessary pre-linguistic step, disrupting this process may deliver conceptually involved areas on top of language-eloquent areas. As conceptual knowledge of a word is thought to be more holistically recruited from bihemispheric regions [68,72–74], the high error rate may be explained by this parameter choice of an early onset of stimulation.

None of these explanations for the RH can be ruled out at this point. A study using different paradigms with varying stimulation onset and potentially even more stimuli to avoid learning effects is needed to clear up the uncertainty between language-eloquent and language-involved areas in the RH.

4.6. Hesitation Errors

The majority of stimulation protocols refrain from including hesitation errors as positive error occurrences. As of now, no ready-made program is available to identify a hesitation at a subject-tailored level, but currently requires reprocessing of the video material and analysis through third-party programs [45]. This leaves it to the subjective opinion of the experimenter to draw a line between a normal response and a hesitation, when comparing baseline naming to naming under stimulation [37]. Due to the difficult quantification of these errors, separate analyses were conducted, excluding the categories in question: hesitation on the whole sentence as a weaker pronounced no response and hesitation of the target as a weaker pronounced anomia were excluded from the total error count in each task. Doing so decreased the number of errors significantly in both tasks and hemispheres (Table 7). However, even in the remaining errors, action naming demonstrated significantly more errors than object naming.

Two conclusions are to be drawn from this. Firstly, action naming's higher sensitivity is also apparent in a more conservative error count. This strengthens the claim to include the task in mapping. Secondly, it is still important to screen for subtler errors, such as hesitations, as they after all indicate a disruption in language processing [29], and constitute an essential part of the error rates. These errors may be the only elicitable error category in some individuals and, therefore, the only data on which to base a mapping.

4.7. Area-Wise Comparison Excluding Hesitations

As another approach, we performed an anatomical analysis in which hesitation errors were excluded. Task comparison per CPS regions revealed higher error rates for action naming in the pMFG in the LH and opIFG in the RH (Table 8). The pMFG as part of the prefrontal action related network and the opIFG as a bilateral production area are no surprising components. Since, however, no double dissociation was present and no specific error type pattern is evident in our data, the relations again remain tentative.

When describing the areas with the highest error rates, the following picture arose in both hemi-spheres: the areas with the highest error rate excluding hesitations are by large distributed in a similar pattern as the areas of maps including all errors (Figures 4 and 5, Tables 6 and 8: LH object naming in the vPrG, mMTG, and mPrG; LH action naming in the vPrG, mSTG, and vPoG; RH object naming in the mMTG and vPoG; RH action naming in the mSTG and opIFG). This study did not aim to quantify comparisons of all errors included versus excluding hesitations per CPS region. However, as seen from a descriptive analysis regarding this matter, similar CPS region patterns of high error rates result from both analyses. Hence, including all errors did not deliver any new error rate pattern in the CPS regions. It rather strengthens the error count per area, which would have been revealed in a more conservative map, excluding hesitations. This leads to the exploratory conclusion that including hesitations does, on the one hand, influence the error count significantly, on the other hand, it does not dull the mapping result by revealing an unexpected area pattern based on these more subjective errors.

4.8. Clinical Implications

Using a balanced and pretested set of tasks [48], the setup of the present study showed the value of action naming in sentence context as a more sensitive tool than object naming in sentence context under nTMS. This is relevant for clinical application: Cases of entire "zero maps"—no effect at all of stimulation while performing a language task like object naming—are a well-known phenomenon in clinical practice, at least when it comes to certain regions. In these instances, using a more sensitive task, such as action naming, may lead to a positive mapping result.

More specifically, action naming may be the more suitable task to use for these cases, where a tumor is infiltrating areas prone to action naming, such as the mSTG, as seen in the data here and in previous studies [28,43,44,49]; and in frontal regions, as seen in the analysis excluding hesitations (Table 8). Moreover, action naming may be a more accurate tool for RH nTMS mappings, as it was shown to be more sensitive specifically in this hemisphere.

On that premise, it is important to keep the cooperation of the two tasks in mind. We do not suggest that the action naming variant should replace the standard object naming, but it is considered an addition to the test battery. The aforementioned single cases of distinct functional allocation for object naming justify its importance in mapping as much as action naming [28,49,52]. Furthermore, marked aphasia in patients with brain tumors may not allow for testing a more complex task with verbs. In that case, object naming may be required for informative results. Conversely, some cases may demonstrate an inability to perform object naming and might do well during action naming. The ideal setup would, therefore, entail administration of both tasks under nTMS. If fatigue or other time constraining factors play in, one may choose to administer solely action naming.

Regarding error evaluation, clinical practice could opt to screen for more subtle errors such as hesitation on the target that become evident through integration of a lead-in phase. For those patients not demonstrating a strong effect of nTMS, as seen in full speech arrests, categories such as hesitations (on the target) may be the only error source to build on.

4.9. Limitations and Resulting Future Steps

The current setup only allowed comparing the tasks' relative sensitivity. To fully evaluate the sensitivity and reliability of the nTMS mappings, including tasks in sentence context, data from the gold standard of intraoperative mapping with DES are required. Only this direct comparison allows the conclusion that areas predicted to be positive in action naming and/or object naming proved to be critically involved under the gold standard. The highly invasive nature of DES precludes application in samples of healthy volunteers. Therefore, future studies are needed enrolling a group of patients who undergo both preoperative mapping with nTMS and intraoperative mapping with DES to fully validate this study's protocol.

As a second limitation, subjective measurements of a delayed response were used for defining hesitations. Future studies could profit from built-in response time measurements to establish participant-tailored cut-offs for objectively delayed responses. At present, this requires extensive reprocessing through external software, not compatible with the TMS set-up and hence not feasible in the workflow. However, to compensate for this shortcoming, the present study excluded hesitation errors for a separate analysis. Third, relations between error categories and function allocation per CPS region can only tentatively be drawn from the descriptive analyses used in this study. While our conclusions did not arise from a statistical comparison per CPS region, using error ratios per category gave nonetheless clear indication of the prevalent errors. Future investigations tailored for function allocation may tackle this in a bigger sample size.

Moreover, unbalanced handedness resulted in a more heterogeneous participant group than in other studies. However, no correlation between handedness and error rates overall and per hemisphere was obtained. This is in accordance with a low prevalence of RH dominance, in both right- and lefthanders [86–89]. Including left-handers should, therefore, not have altered the findings.

Lastly, we did not look into network effects of mutual areas involved in each task. While it would have shed more light on the interconnection of cortical areas and their dependence of each other, it would not directly help improve the mapping results using different tasks; hence, it exceeded the scope of the current study.

5. Conclusions

During language mapping of the healthy brain under nTMS, action naming in sentence context proved to be more easily disrupted than object naming in sentence context. Through inclusion of a lead-in phrase and error categorization, it was argued that this difference predominantly arises from the disruption at the lexico-semantic level, where verb retrieval seems to be more affected than noun retrieval. This was observed for both the left and the right hemisphere and the pattern persisted in a more conservative error count, excluding hesitation errors. The role of hesitation errors was more clearly defined as contributing significantly to the error rates, while, at the same time, not dulling the mapping results and, thus, worth including.

Author Contributions: A.-K.O., R.B., N.S. and S.M.K. conceived and designed the study. A.-K.O., C.N., S.S. and A.S. collected and analyzed the data. A.-K.O. prepared the manuscript, with contributions and editing by all authors. All authors have read and agreed to the published version of the manuscript.

Funding: The study was financed by institutional grants from the Department of Neurosurgery at the Klinikum rechts der Isar, Technical University of Munich, Munich, Germany, and received no external funding.

Institutional Review Board Statement: The study was approved by the local ethics committee (Institutional Review Board of the Technical University of Munich; registration number: 202/18 S) and was conducted in accordance with the Declaration of Helsinki. Written informed consent was obtained from all subjects prior to study inclusion.

Informed Consent Statement: Written informed consent was obtained from the participant(s) to publish this paper.

Data Availability Statement: The datasets generated and analyzed during the current study are available from the corresponding author upon reasonable request.

Acknowledgments: We gratefully acknowledge all participants for their time and efforts partaking in this study. Roelien Bastiaanse is partially supported by the Center for Language and Brain NRU Higher School of Economics, RF Government Grant, ag. No. 14.641.31.0004.

Conflicts of Interest: S.M.K. is a consultant for Brainlab AG (Munich, Germany). N.S. received honoraria from Nexstim Plc (Heldinki, Finland). None of the authors report conflicts of interest. The funders had no role in the design of the study; in the collection, analyses, or interpretation of data; in the writing of the manuscript, or in the decision to publish the results.

References

1. Barker, A.T.; Jalinous, R.; Freeston, I.L. Non-Invasive Magnetic Stimulation of Human Motor Cortex. *Lancet* **1985**, *1*, 1106–1107. [CrossRef]
2. Hallett, M. Transcranial Magnetic Stimulation and the Human Brain. *Nature* **2000**, *406*, 147–150. [CrossRef] [PubMed]
3. Ilmoniemi, R.J.; Virtanen, J.; Ruohonen, J.; Karhu, J.; Aronen, H.J.; Näätänen, R.; Katila, T. Neuronal Responses to Magnetic Stimulation Reveal Cortical Reactivity and Connectivity. *Neuroreport* **1997**, *8*, 3537–3540. [CrossRef]
4. Lioumis, P.; Zhdanov, A.; Mäkelä, N.; Lehtinen, H.; Wilenius, J.; Neuvonen, T.; Hannula, H.; Deletis, V.; Picht, T.; Mäkelä, J.P. A Novel Approach for Documenting Naming Errors Induced by Navigated Transcranial Magnetic Stimulation. *J. Neurosci. Methods* **2012**, *204*, 349–354. [CrossRef]

5. Picht, T.; Krieg, S.M.; Sollmann, N.; Rösler, J.; Niraula, B.; Neuvonen, T.; Savolainen, P.; Lioumis, P.; Mäkelä, J.P.; Deletis, V.; et al. A Comparison of Language Mapping by Preoperative Navigated Transcranial Magnetic Stimulation and Direct Cortical Stimulation during Awake Surgery. *Neurosurgery* **2013**, *72*, 808–819. [CrossRef] [PubMed]
6. Rösler, J.; Niraula, B.; Strack, V.; Zdunczyk, A.; Schilt, S.; Savolainen, P.; Lioumis, P.; Mäkelä, J.; Vajkoczy, P.; Frey, D.; et al. Language Mapping in Healthy Volunteers and Brain Tumor Patients with a Novel Navigated TMS System: Evidence of Tumor-Induced Plasticity. *Clin. Neurophysiol.* **2014**, *125*, 526–536. [CrossRef]
7. Tarapore, P.E.; Findlay, A.M.; Honma, S.M.; Mizuiri, D.; Houde, J.F.; Berger, M.S.; Nagarajan, S.S. Language Mapping with Navigated Repetitive TMS: Proof of Technique and Validation. *Neuroimage* **2013**, *82*, 260–272. [CrossRef]
8. Sollmann, N.; Ille, S.; Hauck, T.; Maurer, S.; Negwer, C.; Zimmer, C.; Ringel, F.; Meyer, B.; Krieg, S.M. The Impact of Preoperative Language Mapping by Repetitive Navigated Transcranial Magnetic Stimulation on the Clinical Course of Brain Tumor Patients. *BMC Cancer* **2015**, *15*. [CrossRef] [PubMed]
9. Sollmann, N.; Kelm, A.; Ille, S.; Schröder, A.; Zimmer, C.; Ringel, F.; Meyer, B.; Krieg, S.M. Setup Presentation and Clinical Outcome Analysis of Treating Highly Language-Eloquent Gliomas via Preoperative Navigated Transcranial Magnetic Stimulation and Tractography. *Neurosurg. Focus* **2018**, *44*. [CrossRef] [PubMed]
10. De Witt Hamer, P.C.; Robles, S.G.; Zwinderman, A.H.; Duffau, H.; Berger, M.S. Impact of Intraoperative Stimulation Brain Mapping on Glioma Surgery Outcome: A Meta-Analysis. *J. Clin. Oncol.* **2012**, *30*, 2559–2565. [CrossRef] [PubMed]
11. De Witte, E.; Mariën, P. The Neurolinguistic Approach to Awake Surgery Reviewed. *Clin. Neurol. Neurosurg.* **2013**, *115*, 127–145. [CrossRef]
12. Duffau, H. Contribution of Cortical and Subcortical Electrostimulation in Brain Glioma Surgery: Methodological and Functional Considerations. *Neurophysiol. Clin.* **2007**, *37*, 373–382. [CrossRef]
13. Duffau, H.; Lopes, M.; Arthuis, F.; Bitar, A.; Sichez, J.P.; Van Effenterre, R.; Capelle, L. Contribution of Intraoperative Electrical Stimulations in Surgery of Low Grade Gliomas: A Comparative Study between Two Series without (1985–1996) and with (1996–2003) Functional Mapping in the Same Institution. *J. Neurol. Neurosurg. Psychiatry* **2005**, *76*, 845–851. [CrossRef]
14. Bährend, I.; Muench, M.R.; Schneider, H.; Moshourab, R.; Dreyer, F.R.; Vajkoczy, P.; Picht, T.; Faust, K. Incidence and Linguistic Quality of Speech Errors: A Comparison of Preoperative Transcranial Magnetic Stimulation and Intraoperative Direct Cortex Stimulation. *J. Neurosurg.* **2020**, *134*, 1409–1418. [CrossRef]
15. Ille, S.; Sollmann, N.; Hauck, T.; Maurer, S.; Tanigawa, N.; Obermueller, T.; Negwer, C.; Droese, D.; Zimmer, C.; Meyer, B.; et al. Combined Noninvasive Language Mapping by Navigated Transcranial Magnetic Stimulation and Functional MRI and Its Comparison With Direct Cortical Stimulation. *J. Neurosurg.* **2015**, *123*, 1–14. [CrossRef]
16. Krieg, S.M.; Tarapore, P.E.; Picht, T.; Tanigawa, N.; Houde, J.; Sollmann, N.; Meyer, B.; Vajkoczy, P.; Berger, M.S.; Ringel, F.; et al. Optimal Timing of Pulse Onset for Language Mapping with Navigated Repetitive Transcranial Magnetic Stimulation. *Neuroimage* **2014**, *100*, 219–236. [CrossRef]
17. Sollmann, N.; Kubitscheck, A.; Maurer, S.; Ille, S.; Hauck, T.; Kirschke, J.S.; Ringel, F.; Meyer, B.; Krieg, S.M. Preoperative Language Mapping by Repetitive Navigated Transcranial Magnetic Stimulation and Diffusion Tensor Imaging Fiber Tracking and Their Comparison to Intraoperative Stimulation. *Neuroradiology* **2016**, *58*, 807–818. [CrossRef] [PubMed]
18. Krieg, S.M.; Lioumis, P.; Mäkelä, J.P.; Wilenius, J.; Karhu, J.; Hannula, H.; Savolainen, P.; Lucas, C.W.; Seidel, K.; Laakso, A.; et al. Protocol for Motor and Language Mapping by Navigated TMS in Patients and Healthy Volunteers; Workshop Report. *Acta Neurochir.* **2017**, *159*, 1187–1195. [CrossRef] [PubMed]
19. Bello, L.; Acerbi, F.; Giussani, C.; Baratta, P.; Taccone, P.; Songa, V.; Fava, M.; Stocchetti, N.; Papagno, C.; Gaini, S.M. Intraoperative Language Localization in Multilingual Patients with Gliomas. *Neurosurgery* **2006**, *59*, 115–123. [CrossRef] [PubMed]
20. Bello, L.; Gallucci, M.; Fava, M.; Carrabba, G.; Giussani, C.; Acerbi, F.; Baratta, P.; Songa, V.; Conte, V.; Branca, V.; et al. Intraoperative Subcortical Language Tract Mapping Guides Surgical Removal of Gliomas Involving Speech Areas. *Neurosurgery* **2007**, *60*, 67–80. [CrossRef] [PubMed]
21. De Witte, E.; Satoer, D.; Robert, E.; Colle, H.; Verheyen, S.; Visch-Brink, E.; Mariën, P. The Dutch Linguistic Intraoperative Protocol: A Valid Linguistic Approach to Awake Brain Surgery. *Brain Lang.* **2015**, *140*, 35–48. [CrossRef] [PubMed]
22. Rofes, A.; Miceli, G. Language Mapping with Verbs and Sentences in Awake Surgery: A Review. *Neuropsychol. Rev.* **2014**, *24*, 185–199. [CrossRef] [PubMed]
23. Rofes, A.; Spena, G.; Talacchi, A.; Santini, B.; Miozzo, A.; Miceli, G. Mapping Nouns and Finite Verbs in Left Hemisphere Tumors: A Direct Electrical Stimulation Study Mapping Nouns and Finite Verbs in Left Hemisphere Tumors: A Direct Electrical Stimulation Study International Doctorate in Experimental Approaches to Language. *Neurocase* **2017**, *23*, 105–113. [CrossRef]
24. Hillis, A.E.; Tuffiash, E.; Caramazza, A. Modality-Specific Deterioration in Naming Verbs in Nonfluent Primary Progressive Aphasia. *J. Cogn. Neurosci.* **2002**, *14*, 1099–1108. [CrossRef]
25. Mätzig, S.; Druks, J.; Masterson, J.; Vigliocco, G. Noun and Verb Differences in Picture Naming: Past Studies and New Evidence. *Cortex* **2009**, *45*, 738–758. [CrossRef]
26. Miceli, G.; Silveri, M.C.; Villa, G.; Caramazza, A. On the Basis for the Agrammatic's Difficulty in Producing Main Verbs. *Cortex* **1984**, *20*, 207–220. [CrossRef]
27. Zingeser, L.B.; Berndt, R.S. Grammatical Class and Context Effects in a Case of Pure Anomia: Implications for Models of Language Production. *Cogn. Neuropsychol.* **1988**, *5*, 473–516. [CrossRef]

28. Lubrano, V.; Filleron, T.; Démonet, J.-F.; Roux, F.-E. Anatomical Correlates for Category-Specific Naming of Objects and Actions: A Brain Stimulation Mapping Study. *Hum. Brain Mapp.* **2014**, *35*, 429–443. [CrossRef]
29. Indefrey, P. The Spatial and Temporal Signatures of Word Production Components: A Critical Update. *Front. Psychol.* **2011**, *2*, 1–16. [CrossRef]
30. Indefrey, P.; Levelt, W.J.M. The Neural Correlates of Language Production. In *The New Cognitive Neurosciences*, 2nd ed.; MIT Press: Cambridge, MA, USA, 2000; pp. 845–865.
31. Levelt, W.J.M. Producing Spoken Language: A Blueprint of the Speaker. In *The Neurocognition of Language*; Brown, C., Ed.; Oxford University Press: Oxford, UK, 1999; pp. 82–122. [CrossRef]
32. Bastiaanse, R.; Wieling, M.; Wolthuis, N. The Role of Frequency in the Retrieval of Nouns and Verbs in Aphasia. *Aphasiology* **2016**, *30*, 1221–1239. [CrossRef]
33. Thompson, C.K.; Faroqi-Shah, Y.; Lee, J. Models of Sentence Production. In *The Handbook of Adult Language Disorders*; Hillis, A.E., Ed.; Psychology Press: Hove, UK, 2015. [CrossRef]
34. Ille, S.; Sollmann, N.; Butenschoen, V.M.; Meyer, B.; Ringel, F.; Krieg, S.M. Resection of Highly Language-Eloquent Brain Lesions Based Purely on RTMS Language Mapping without Awake Surgery. *Acta Neurochir.* **2016**, *158*, 2265–2275. [CrossRef]
35. Krieg, S.M.; Sollmann, N.; Hauck, T.; Ille, S.; Foerschler, A.; Meyer, B.; Ringel, F. Functional Language Shift to the Right Hemisphere in Patients with Language-Eloquent Brain Tumors. *PLoS ONE* **2013**, *8*. [CrossRef]
36. Krieg, S.M.; Sollmann, N.; Tanigawa, N.; Foerschler, A.; Meyer, B.; Ringel, F. Cortical Distribution of Speech and Language Errors Investigated by Visual Object Naming and Navigated Transcranial Magnetic Stimulation. *Brain Struct. Funct.* **2016**, *221*, 2259–2286. [CrossRef] [PubMed]
37. Sollmann, N.; Hauck, T.; Hapfelmeier, A.; Meyer, B.; Ringel, F.; Krieg, S.M. Intra- and Interobserver Variability of Language Mapping by Navigated Transcranial Magnetic Brain Stimulation. *BMC Neurosci.* **2013**, *14*, 1–10. [CrossRef]
38. Corina, D.P.; Loudermilk, B.C.; Detwiler, L.; Martin, R.F.; Brinkley, J.F.; Ojemann, G. Analysis of Naming Errors during Cortical Stimulation Mapping: Implications for Models of Language Representation. *Brain Lang.* **2010**, *115*, 101–112. [CrossRef]
39. Ellis, A.W.; Young, A.W. *Human Cognitive Neuropsychology: A Textbook with Readings*; Psychology Press: Hove, UK, 1988.
40. Bastiaanse, R.; Van Zonneveld, R. Broca's Aphasia, Verbs and the Mental Lexicon. *Brain Lang.* **2004**, *90*, 198–202. [CrossRef]
41. Caramazza, A. How Many Levels of Processing Are There in Lexical Access? *Cogn. Neuropsychol.* **1997**, *14*, 177–208. [CrossRef]
42. Kemmerer, D.; Tranel, D. Verb Retrieval in Brain-Damaged Subjects: 1. Analysis of Stimulus, Lexical, and Conceptual Factors. *Brain Lang.* **2000**, *73*, 347–392. [CrossRef] [PubMed]
43. Hauck, T.; Tanigawa, N.; Probst, M.; Wohlschlaeger, A.; Ille, S.; Sollmann, N.; Maurer, S.; Zimmer, C.; Ringel, F.; Meyer, B.; et al. Task Type Affects Location of Language—Positive Cortical Regions by Repetitive Navigated Transcranial Magnetic Stimulation Mapping. *PLoS ONE* **2015**, *10*, e0125298. [CrossRef] [PubMed]
44. Hernandez-Pavon, J.C.; Mäkelä, N.; Lehtinen, H.; Lioumis, P.; Mäkelä, J.P. Effects of Navigated TMS on Object and Action Naming. *Front. Hum. Neurosci.* **2014**, *8*. [CrossRef]
45. Schramm, S.; Tanigawa, N.; Tussis, L.; Meyer, B.; Sollmann, N.; Krieg, S.M. Capturing Multiple Interaction Effects in L1 and L2 Object-Naming Reaction Times in Healthy Bilinguals: A Mixed-Effects Multiple Regression Analysis. *BMC Neurosci.* **2020**, *21*, 1–26. [CrossRef]
46. Sollmann, N.; Zhang, H.; Schramm, S.; Ille, S.; Negwer, C.; Kreiser, K.; Meyer, B.; Krieg, S.M. Function-Specific Tractography of Language Pathways Based on NTMS Mapping in Patients with Supratentorial Lesions. *Clin. Neuroradiol.* **2018**, 1–13. [CrossRef] [PubMed]
47. Oldfield, R.C. The Assessment and Analysis of Handedness: The Edinburgh Inventory. *Neuropsychologia* **1971**, *9*, 97–113. [CrossRef]
48. Ohlerth, A.-K.; Valentin, A.; Vergani, F.; Ashkan, K.; Bastiaanse, R. The Verb and Noun Test for Peri-Operative Testing (VAN-POP): Standardized Language Tests for Navigated Transcranial Magnetic Stimulation and Direct Electrical Stimulation. *Acta Neurochir.* **2020**, *162*, 397–406. [CrossRef] [PubMed]
49. Corina, D.P.; Gibson, E.K.; Martin, R.; Poliakov, A.; Brinkley, J.; Ojemann, G.A. Dissociation of Action and Object Naming: Evidence from Cortical Stimulation Mapping. *Hum. Brain Mapp.* **2005**, *24*, 1–10. [CrossRef] [PubMed]
50. Ruohonen, J.; Karhu, J. Navigated Transcranial Magnetic Stimulation. *Neurophysiol. Clin.* **2010**, *40*, 7–17. [CrossRef]
51. Snodgrass, J.G.; Vanderwart, M. A Standardized Set of 260 Pictures: Norms for Name Agreement, Image Agreement, Familiarity, and Visual Complexity. *J. Exp. Psychol. Hum. Learn. Mem.* **1980**, *6*, 174–215. [CrossRef]
52. Havas, V.; Gabarrós, A.; Juncadella, M.; Rifa-Ros, X.; Plans, G.; Acebes, J.J.; de Diego Balaguer, R.; Rodríguez-Fornells, A. Electrical Stimulation Mapping of Nouns and Verbs in Broca's Area. *Brain Lang.* **2015**, *145–146*, 53–63. [CrossRef]
53. Sollmann, N.; Ille, S.; Negwer, C.; Boeckh-Behrens, T.; Ringel, F.; Meyer, B.; Krieg, S.M.; Ringel, F.; Bernhard Meyer, T. Cortical Time Course of Object Naming Investigated by Repetitive Navigated Transcranial Magnetic Stimulation. *Brain Imaging Behav.* **2017**, *11*, 1192–1206. [CrossRef]
54. Crepaldi, D.; Berlingeri, M.; Paulesu, E.; Luzzatti, C. A Place for Nouns and a Place for Verbs? A Critical Review of Neurocognitive Data on Grammatical-Class Effects. *Brain Lang.* **2011**, *116*, 33–49. [CrossRef]
55. Luzzatti, C.; Raggi, R.; Zonca, G.; Pistarini, C.; Contardi, A.; Pinna, G.D. Verb-Noun Double Dissociation in Aphasic Lexical Impairments: The Role of Word Frequency and Imageability. *Brain Lang.* **2002**, *81*, 432–444. [CrossRef] [PubMed]

56. Pisoni, A.; Mattavelli, G.; Casarotti, A.; Comi, A.; Riva, M.; Bello, L.; Papagno, C. Object-Action Dissociation: A Voxel-Based Lesion-Symptom Mapping Study on 102 Patients after Glioma Removal. *NeuroImage Clin.* **2018**. [CrossRef] [PubMed]
57. Crepaldi, D.; Berlingeri, M.; Cattinelli, I.; Borghese, N.A.; Luzzatti, C.; Paulesu, E. Clustering the Lexicon in the Brain: A Meta-Analysis of the Neurofunctional Evidence on Noun and Verb Processing. *Front. Hum. Neurosci.* **2013**, *7*, 1–15. [CrossRef] [PubMed]
58. Vigliocco, G.; Vinson, D.P.; Druks, J.; Barber, H.; Cappa, S.F. Nouns and Verbs in the Brain: A Review of Behavioural, Electrophysiological, Neuropsychological and Imaging Studies. *Neurosci. Biobehav. Rev.* **2011**, *35*, 407–426. [CrossRef] [PubMed]
59. Baxter, D.M.; Warrington, E.K. Category Specific Phonological Dysgraphia. *Neuropsychologia* **1985**, *23*, 653–666. [CrossRef]
60. Pillon, A.; d'Honincthun, P. The Organization of the Conceptual System: The Case of the "Object versus Action" Dimension. *Cogn. Neuropsychol.* **2010**, *27*, 587–613. [CrossRef]
61. Chang, E.F.; Wang, D.D.; Perry, D.W.; Barbaro, N.M.; Berger, M.S. Homotopic Organization of Essential Language Sites in Right and Bilateral Cerebral Hemispheric Dominance: Clinical Article. *J. Neurosurg.* **2011**, *114*, 893–902. [CrossRef] [PubMed]
62. Maldonado, I.L.; Moritz-Gasser, S.; De Champfleur, N.M.; Bertram, L.; Moulinié, G.; Duffau, H. Surgery for Gliomas Involving the Left Inferior Parietal Lobule: New Insights into the Functional Anatomy Provided by Stimulation Mapping in Awake Patients: Clinical Article. *J. Neurosurg.* **2011**, *115*, 770–779. [CrossRef] [PubMed]
63. Ojemann, G.; Ojemann, J.; Lettich, E.; Berger, M. Cortical Language Localization in Left, Dominant Hemisphere. An Electrical Stimulation Mapping Investigation in 117 Patients. 1989. *J. Neurosurg.* **2008**, *108*, 411–421. [CrossRef]
64. Tate, M.C.; Herbet, G.; Moritz-Gasser, S.; Tate, J.E.; Duffau, H. Probabilistic Map of Critical Functional Regions of the Human Cerebral Cortex: Broca's Area Revisited. *Brain* **2014**, *137*, 2773–2782. [CrossRef]
65. Catani, M.; Jones, D.K.; Ffytche, D.H. Perisylvian Language Networks of the Human Brain. *Ann. Neurol.* **2005**, *57*, 8–16. [CrossRef]
66. Hauk, O.; Johnsrude, I.; Pulvermüller, F. Somatotopic Representation of Action Words in Human Motor and Premotor Cortex. *Neuron* **2004**, *41*, 301–307. [CrossRef]
67. Gajardo-Vidal, A.; Lorca-Puls, D.L.; Hope, T.M.H.; Parker Jones, O.; Seghier, M.L.; Prejawa, S.; Crinion, J.T.; Leff, A.P.; Green, D.W.; Price, C.J. How Right Hemisphere Damage after Stroke Can Impair Speech Comprehension. *Brain* **2018**, *141*, 3389–3404. [CrossRef]
68. Vilasboas, T.; Herbet, G.; Duffau, H. Challenging the Myth of Right Nondominant Hemisphere: Lessons from Corticosubcortical Stimulation Mapping in Awake Surgery and Surgical Implications. *World Neurosurg.* **2017**, *103*, 449–456. [CrossRef] [PubMed]
69. Baumgaertner, A.; Hartwigsen, G.; Roman Siebner, H. Right-Hemispheric Processing of Non-Linguistic Word Features: Implications for Mapping Language Recovery after Stroke. *Hum. Brain Mapp.* **2013**, *34*, 1293–1305. [CrossRef]
70. Vigneau, M.; Beaucousin, V.; Hervé, P.Y.; Jobard, G.; Petit, L.; Crivello, F.; Mellet, E.; Zago, L.; Mazoyer, B.; Tzourio-Mazoyer, N. What Is Right-Hemisphere Contribution to Phonological, Lexico-Semantic, and Sentence Processing? Insights from a Meta-Analysis. *Neuroimage* **2011**, *54*, 577–593. [CrossRef]
71. Bozic, M.; Tyler, L.K.; Ives, D.T.; Randall, B.; Marslen-Wilson, W.D. Bihemispheric Foundations for Human Speech Comprehension. *Proc. Natl. Acad. Sci. USA* **2010**, *107*, 17439–17444. [CrossRef]
72. Jung, J.Y.; Lambon Ralph, M.A. Mapping the Dynamic Network Interactions Underpinning Cognition: A CTBS-FMRI Study of the Flexible Adaptive Neural System for Semantics. *Cereb. Cortex* **2016**, *26*, 3580–3590. [CrossRef]
73. Ralph, M.A.L.; Jefferies, E.; Patterson, K.; Rogers, T.T. The Neural and Computational Bases of Semantic Cognition. *Nat. Rev. Neurosci.* **2016**, *18*, 42–55. [CrossRef] [PubMed]
74. Rice, G.E.; Ralph, M.A.L.; Hoffman, P. The Roles of Left versus Right Anterior Temporal Lobes in Conceptual Knowledge: An ALE Meta-Analysis of 97 Functional Neuroimaging Studies. *Cereb. Cortex* **2015**, *25*, 4374–4391. [CrossRef] [PubMed]
75. Hartwigsen, G.; Baumgaertner, A.; Price, C.J.; Koehnke, M.; Ulmer, S.; Siebner, H.R. Phonological Decisions Require Both the Left and Right Supramarginal Gyri. *Proc. Natl. Acad. Sci. USA* **2010**, *107*, 16494–16499. [CrossRef]
76. Hartwigsen, G.; Saur, D.; Price, C.J.; Ulmer, S.; Baumgaertner, A.; Siebner, H.R. Perturbation of the Left Inferior Frontal Gyrus Triggers Adaptive Plasticity in the Right Homologous Area during Speech Production. *Proc. Natl. Acad. Sci. USA* **2013**, *110*, 16402–16407. [CrossRef] [PubMed]
77. Duffau, H.; Leroy, M.; Gatignol, P. Cortico-Subcortical Organization of Language Networks in the Right Hemisphere: An Electrostimulation Study in Left-Handers. *Neuropsychologia* **2008**, *46*, 3197–3209. [CrossRef]
78. Herbet, G.; Moritz-Gasser, S.; Duffau, H. Direct Evidence for the Contributive Role of the Right Inferior Fronto-Occipital Fasciculus in Non-Verbal Semantic Cognition. *Brain Struct. Funct.* **2017**, *222*, 1597–1610. [CrossRef] [PubMed]
79. Rolland, A.; Herbet, G.; Duffau, H. Awake Surgery for Gliomas within the Right Inferior Parietal Lobule: New Insights into the Functional Connectivity Gained from Stimulation Mapping and Surgical Implications. *World Neurosurg.* **2018**, *112*, e393–e406. [CrossRef] [PubMed]
80. Sarubbo, S.; Tate, M.; De Benedictis, A.; Merler, S.; Moritz-Gasser, S.; Herbet, G.; Duffau, H. Mapping Critical Cortical Hubs and White Matter Pathways by Direct Electrical Stimulation: An Original Functional Atlas of the Human Brain. *Neuroimage* **2020**, *205*, 116237. [CrossRef]
81. Sollmann, N.; Tanigawa, N.; Ringel, F.; Zimmer, C.; Meyer, B.; Krieg, S.M. Language and Its Right-Hemispheric Distribution in Healthy Brains: An Investigation by Repetitive Transcranial Magnetic Stimulation. *Neuroimage* **2014**, *102*, 776–788. [CrossRef] [PubMed]

82. Sollmann, N.; Tanigawa, N.; Tussis, L.; Hauck, T.; Ille, S.; Maurer, S.; Negwer, C.; Zimmer, C.; Ringel, F.; Meyer, B.; et al. Cortical Regions Involved in Semantic Processing Investigated by Repetitive Navigated Transcranial Magnetic Stimulation and Object Naming. *Neuropsychologia* **2015**, *70*, 185–195. [CrossRef]
83. Negwer, C.; Ille, S.; Hauck, T.; Sollmann, N.; Maurer, S.; Kirschke, J.S.; Ringel, F.; Meyer, B.; Krieg, S.M. Visualization of Subcortical Language Pathways by Diffusion Tensor Imaging Fiber Tracking Based on RTMS Language Mapping. *Brain Imaging Behav.* **2017**, *11*, 899–914. [CrossRef]
84. Lafleur, L.P.; Tremblay, S.; Whittingstall, K.; Lepage, J.F. Assessment of Effective Connectivity and Plasticity with Dual-Coil Transcranial Magnetic Stimulation. *Brain Stimul.* **2016**, *9*, 347–355. [CrossRef]
85. Weber, K.; Christiansen, M.H.; Petersson, K.M.; Indefrey, P.; Hagoort, P. FMRI Syntactic and Lexical Repetition Effects Reveal the Initial Stages of Learning a New Language. *J. Neurosci.* **2016**, *36*, 6872–6880. [CrossRef]
86. Knecht, S.; Dräger, B.; Deppe, M.; Bobe, L.; Lohmann, H.; Flöel, A.; Ringelstein, E.-B.; Hennigsen, H. Handedness and Hemispheric Language Dominance in Healthy Humans. *Brain* **2000**, *123*, 2512–2518. [CrossRef] [PubMed]
87. Mazoyer, B.; Zago, L.; Jobard, G.; Crivello, F.; Joliot, M.; Perchey, G.; Mellet, E.; Petit, L.; Tzourio-Mazoyer, N. Gaussian Mixture Modeling of Hemispheric Lateralization for Language in a Large Sample of Healthy Individuals Balanced for Handedness. *PLoS ONE* **2014**, *9*, 9–14. [CrossRef] [PubMed]
88. Pujol, J.; Deus, J.; Losilla, J.M.; Capdevila, A. Cerebral Lateralization of Language in Normal Left-Handed People Studied by Functional MRI. *Neurology* **1999**, *52*, 1038–1043. [CrossRef] [PubMed]
89. Van der Haegen, L.; Cai, Q.; Seurinck, R.; Brysbaert, M. Further FMRI Validation of the Visual Half Field Technique as an Indicator of Language Laterality: A Large-Group Analysis. *Neuropsychologia* **2011**, *49*, 2879–2888. [CrossRef]

Article

Brain Response Induced with Paired Associative Stimulation Is Related to Repetition Suppression of Motor Evoked Potential

Shohreh Kariminezhad [1,2,*], Jari Karhu [3], Laura Säisänen [2], Jusa Reijonen [1,2], Mervi Könönen [2,4] and Petro Julkunen [1,2]

1. Department of Applied Physics, University of Eastern Finland, 70211 Kuopio, Finland; jusa.reijonen@kuh.fi (J.R.); petro.julkunen@kuh.fi (P.J.)
2. Department of Clinical Neurophysiology, Kuopio University Hospital, 70029 Kuopio, Finland; laura.saisanen@kuh.fi (L.S.); mervi.kononen@kuh.fi (M.K.)
3. Nexstim Plc, 00510 Helsinki, Finland; jari.karhu@nexstim.com
4. Department of Clinical Radiology, Kuopio University Hospital, 70029 Kuopio, Finland
* Correspondence: Shohreh.kariminezhad@uef.fi

Received: 21 August 2020; Accepted: 25 September 2020; Published: 26 September 2020

Abstract: Repetition suppression (RS), i.e., the reduction of neuronal activity upon repetition of an external stimulus, can be demonstrated in the motor system using transcranial magnetic stimulation (TMS). We evaluated the RS in relation to the neuroplastic changes induced by paired associative stimulation (PAS). An RS paradigm, consisting of 20 trains of four identical suprathreshold TMS pulses 1 s apart, was assessed for motor-evoked potentials (MEPs) in 16 healthy subjects, before and following (at 0, 10, and 20 min) a common PAS protocol. For analysis, we divided RS into two components: (1) the ratio of the second MEP amplitude to the first one in RS trains, i.e., the "dynamic" component, and (2) the mean of the second to fourth MEP amplitudes, i.e., the "stable" component. Following PAS, five subjects showed change in the dynamic RS component. However, nearly all the individuals ($n = 14$) exhibited change in the stable component ($p < 0.05$). The stable component was similar between subjects showing increased MEPs and those showing decreased MEPs at this level ($p = 0.254$). The results suggest the tendency of the brain towards a stable state, probably free from the ongoing dynamics, following PAS.

Keywords: repetition suppression; neuroplasticity; transcranial magnetic stimulation; paired associative stimulation

1. Introduction

Owing to its dynamicity, the brain responds to an intense, novel stimulus with enhanced, transient neural activity. This rapid response, referred to as a startle, is considered to play a critical function in promoting survival [1]. However, exposure to a higher number of identical sensory stimuli yields attenuation of neural activity in the responding network, a phenomenon known as repetition suppression (RS) [2]. RS has been well-characterized across several brain regions, employing various stimulus categories and modalities [3–6]. In the motor system, RS has been demonstrated as a decrement in the amplitude of motor-evoked potentials (MEPs) when transcranial magnetic stimulation (TMS) is applied to an optimal motor cortex location [7,8]. Although it has been suggested that the attenuation observed in RS may serve to provide an energy-efficient neuronal information processing [9], the exact mechanisms underlying RS have remained elusive. RS was initially portrayed merely as an expression of bottom-up mechanisms [2,3,10,11]. However, more recent theories have emphasized the role of top–down mechanisms within a predictive coding scheme, relying on iterative comparison between prior expectations and sensory inputs [12]. Interestingly, RS of MEPs have been demonstrated to be closely associated with neuroplasticity [13].

Neuroplasticity is considered one of the key mechanisms that grants living organisms the ability to adapt and respond flexibly in the face of changing environmental demands [14]. Depending on the speed of these changes, neuroplasticity can take different forms and occur at different timescales. Neuroplasticity is considered the keystone of learning, memory, and recovery from (mild) brain injuries [15–17]. Aberrant neuroplasticity has been put forth as the pathophysiological basis of several neuropsychiatric disorders, such as schizophrenia, depression, and chronic pain [18–20]. Long-term potentiation (LTP) consists of persistent synaptic activity, which is often considered as the cellular basis in the mediation of these functions [21].

Currently, TMS provides the opportunity to study neuroplasticity at the system level, ranging from synaptic plasticity to network-level plasticity [22]. The shifts towards either elevated excitation or diminished inhibition have been proposed as potential underlying mechanisms of neuroplasticity, with the short-term plasticity most likely mediated by the reduction of GABAergic inputs onto excitatory synapses [23].

A well-established and widely used TMS paradigm to induce short-term, topographically specific plasticity in the motor cortex is paired associative stimulation (PAS), in which electrical peripheral nerve stimulation is paired with cortical stimulation [24,25]. If the peripheral input precedes the cortical stimulation, PAS can lead to elevated cortical excitability that manifests itself via an increase in the MEP amplitude (LTP-like plasticity) [26,27]. By contrast, if the order of the arrival of inputs is reversed, depression of cortical excitability is likely to occur (long-term depression (LTD)-like plasticity) [24]. Due to its dependency on timing, PAS has been suggested to induce spike-timing dependent plasticity [28].

In the present paper, to investigate neuroplastic effects induced with PAS, RS is hypothesized to represent the interplay of two states: (1) one reflecting the efficient processing of a novel input, "dynamic RS", indexed by the initial decrement from the first amplitude to the second one, and (2) one reflecting the overall cortical excitability free from the ongoing dynamics, "stable RS". Stable RS, described here as the suppressed amplitude level of the second to the fourth MEPs within the RS trials, might potentially display the capacity of the brain to maintain the processed input as an initial "memory trace". We investigated the dynamic and stable RS prior to and following a common PAS-LTP protocol [24]. We hypothesized that the brain would demonstrate a trend towards a state with low variation in MEP amplitude, which we consider the target level of neuronal network excitability as it is independent from reactive dynamics within the network. As an implication, for long-term neuroplastic effects, the modulation of this stable level could potentially be targeted by neuromodulation, and to create optimal conditions for adaptive neural changes.

2. Materials and Methods

2.1. Subjects

Sixteen healthy, right-handed volunteers with no history of neuropsychiatric disorders participated in this study (seven male, age range: 22–42 years, mean ± SD: 30 ± six years). All subjects provided a written informed consent prior to the experiment. This study was approved by the research ethics committee of the Kuopio University Hospital (256/2017).

2.2. Transcranial Magnetic Stimulation (TMS)

To enable neuronavigation for TMS, structural T1-weighted magnetic resonance images (MRIs) were obtained with a 3T MRI scanner (Philips Achieva 3.0T TX, Philips, Eindhoven, The Netherlands) with the following parameters: repetition time (TR) = 8.2 ms, echo time (TE) = 3.7 ms, flip angle = 8°, voxel size = $1 \times 1 \times 1$ mm^3. TMS was conducted using NBS System 4.3 (Nexstim Plc, Helsinki, Finland) with an air-cooled figure-of-eight coil and biphasic pulses.

The stimulation procedure was initiated by locating the optimal motor representation of the right abductor pollicis brevis (APB) muscle, i.e., APB "hotspot", with the corresponding optimized coil orientation. The hotspot was defined as the cortical site repeatedly eliciting the greatest peak-to-peak MEP responses compared to adjacent stimulation sites. Once the hotspot was determined, the resting motor threshold (rMT) was identified at this cortical site using a system-integrated iterative threshold assessment tool [29]. In the RS paradigm, trials of four TMS stimuli were applied over the APB hotspot, with an inter-stimulus interval (ISI) of 1 s, at an intensity of 120% rMT. The RS paradigm, comprising 20 trials of four single biphasic TMS pulses, was employed with an inter-train interval (ITI) of 17 s [30], before (RS-baseline) and immediately (0 min), 10 min, and 20 min after the PAS intervention (Figure 1).

We recorded MEPs via an integrated electromyography (EMG) system (Nexstim Plc) at a sampling frequency of 3 kHz. A pair of disposable Ag–Cl electrodes was utilized, with the active electrode over the belly of the APB muscle while the reference electrode was placed over the joint distal to the active electrode (Figure 1). The MEP data were processed offline in MATLAB (version R2017b, MathWorks Inc., Natick, MA, USA), and only the MEPs with no preceding muscle activation and peak-to-peak amplitude greater than 50 µV were included as responses.

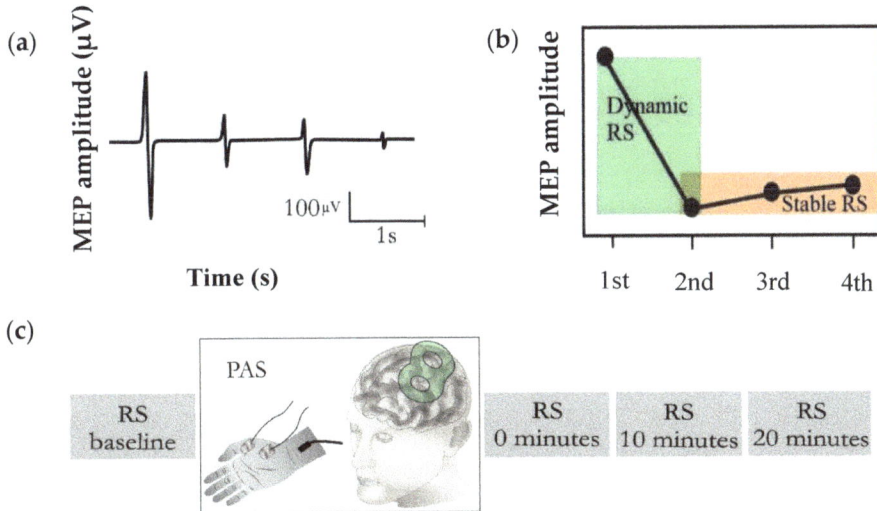

Figure 1. Repetition suppression (RS) paradigm. (**a**) Typical RS. Motor evoked potentials (MEPs) were recorded during four identical TMS pulses. (**b**) The RS paradigm was divided into two components for the analysis: "dynamic RS", i.e., the ratio of the second MEP to the first one, and "stable RS", i.e., the mean of the second, third, and fourth MEPs. (**c**) RS applied before (baseline) and after PAS intervention (at 0 min, 10 min, and 20 min). In the PAS intervention, electrical stimulation of the median nerve-innervated APB muscle was delivered prior to TMS at an ISI of 25 ms to generate plasticity.

2.3. Paired Associative Stimulation (PAS)

PAS consisting of 180 single stimuli was applied over the right median nerve at an intensity of 300% of the sensory threshold (ST) [31]. A bipolar stimulation electrode was placed over the median nerve, and the ST was measured by adjusting the stimulation current until the subject indicated sensation of the stimulus. The pairing with TMS at the APB hotspot was implemented with a self-built triggering and delayer device. To generate a plasticity effect, median nerve stimulation at a frequency of 0.2 Hz was delivered 25 ms prior to TMS [24], with the TMS pulses delivered at 120% of rMT. The median nerve stimulation was conducted using a constant-current electrical stimulator (Digitimer model DS7A, Digitimer, Welwyn Garden City, Herts, UK), using a rectangular pulse form (0.2 ms, maximum voltage of 300 V).

2.4. Statistical Analysis

The MEP amplitudes of each subject were first averaged based on their ordinal position in a trial, i.e., the first, second, third, and fourth. To evaluate the dynamic component of RS, the average of the second stimulus MEP amplitudes was divided by the average of the first stimulus MEP amplitudes. Further, to assess the stable component of RS, the mean of the averaged responses was computed over the second, third, and fourth stimuli per subject.

Considering the inherent heterogeneity of the neurophysiological characteristics, the analysis for identifying significant PAS-effects was initially performed at the individual level using the non-parametric Wilcoxon signed rank test. Individuals with a statistically significant increase in MEP amplitudes at a stable level at 0 min were identified as those showing LTP-like plasticity as an immediate response to PAS (as a higher MEP amplitude is considered as an index of elevated cortical excitability), and clustered as the "LTP-like group". In addition, individuals with decreased MEP amplitudes at a stable level were considered as those exhibiting LTD-like plasticity as an immediate effect to PAS and clustered as the "LTD-like group".

To test the change of the dynamic RS and stable RS over a time course of 20 min, the Friedman test was employed. Post hoc comparisons were performed using the Wilcoxon signed rank test.

A comparison of the two clusters prior to and following PAS was made using the Mann–Whitney U test. Statistical analysis was conducted using SPSS (v. 25.0, SPSS Inc., IBM Company, Armonk, NY, USA) and MATLAB (version R2017b, MathWorks Inc., Natick, MA, USA), and $p < 0.05$ indicated statistical significance.

3. Results

Eleven subjects showed no significant change of dynamic RS following PAS ($p > 0.1$). Only one subject exhibited significantly milder dynamic RS (a lower drop from the second MEP to the first one) ($p < 0.05$), and four subjects showed significantly stronger dynamic RS (a higher drop from the second to the first MEP) ($p < 0.05$) (Figure 2a).

Fourteen subjects exhibited a significant change at stable RS (Figure 2b). The stable RS levels were significantly higher in six subjects immediately following PAS compared to those before PAS ($p < 0.05$). This heightened post-intervention MEP amplitude was assumed to be linked to LTP-like plasticity. However, eight subjects demonstrated significantly diminished MEP amplitudes at stable RS (i.e., the LTD-like group) ($p < 0.05$), and two subjects showed no significant change ($p > 0.1$). A non-parametric Friedman test revealed no change in the trend over the time of measurement following PAS (0, 10, and 20 min). One subject demonstrated delayed LTP-like plasticity at 20 min after exhibiting no effect at earlier time points.

Furthermore, a Mann–Whitney U test revealed that the dynamic and stable RS at the baseline was significantly higher in the LTD-like group compared to the LTP-like group ($p < 0.05$). Following PAS, no significant difference in these two components was observed between the two groups ($p > 0.1$) (Figure A1).

The STs were 2.1 ± 0.5 mA, rMTs were 35 ± 8%-maximum stimulator output (MSO) and MEP latencies were 22.8 ± 1.7 ms. No difference in rMT ($p = 0.845$), ST ($p = 0.244$), and MEP latency ($p = 0.825$) was observed between the two groups.

The low between-group and high within-group homogeneities were observed at stable RS prior and following the PAS, respectively (Figure 3).

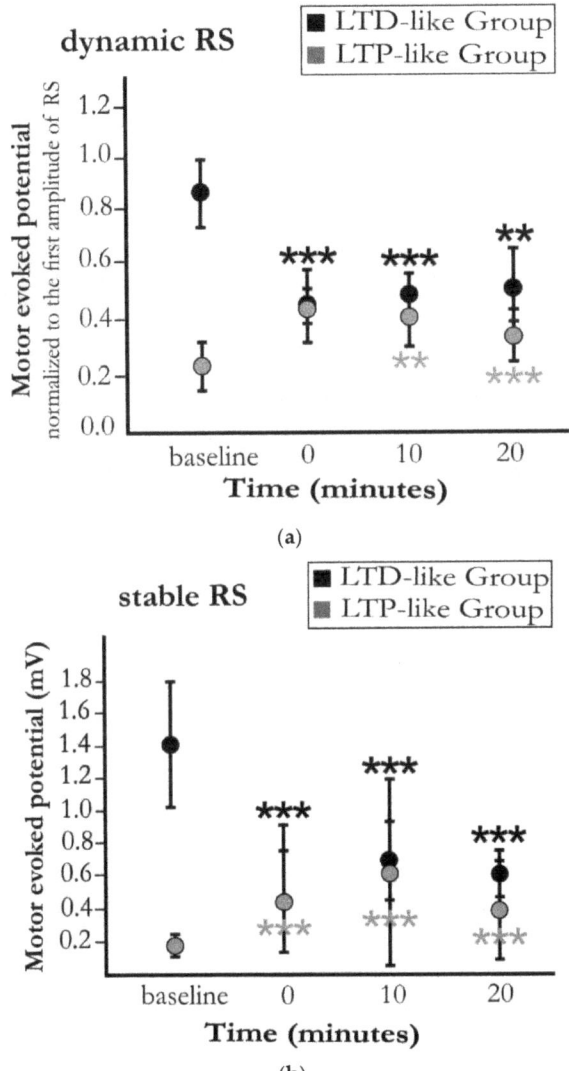

Figure 2. Changes in (**a**) dynamic component (mean ± standard error) and (**b**) stable component of RS (mean ± standard error), across the subjects from two clusters, within the trials before (baseline) and at 0, 10, and 20 min after the induction of short-term plasticity with PAS. Although dynamic RS was mild in the LTD-like group at baseline, a trend towards a stronger suppression and low variability at this level was observed in the LTD-like group following PAS. However, this trend in dynamic RS was towards recovery in the LTP-like group, as this component became milder with time. Similar trends, i.e., sustaining of the suppression and recovery from it, were also observed in stable RS after PAS in the LTD-like group and the LTP-like group, respectively. An asterisk indicates significant differences for pairwise comparisons between time points ($p < 0.01$ for *** and $p < 0.05$ for **).

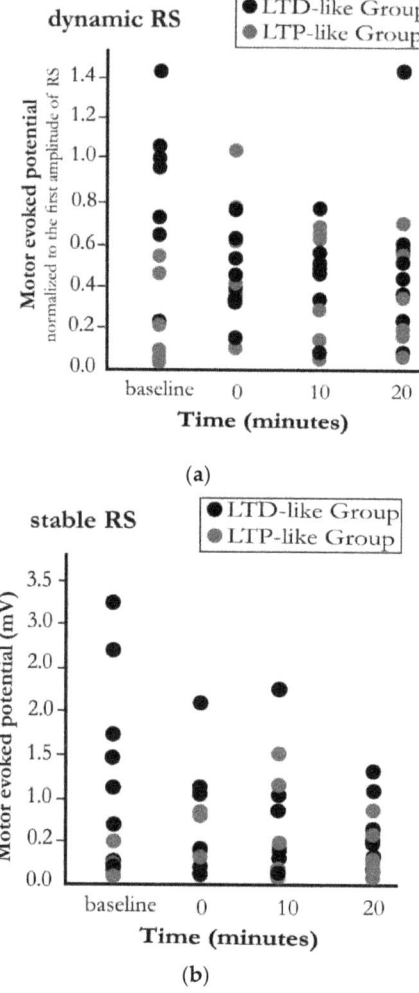

Figure 3. Scatter plot indicating (**a**) the dynamic RS component and (**b**) the stable RS component in all subjects. The low between-group and high within-group homogeneity can be observed in stable RS following applying the PAS intervention, whereas this is not evident for the dynamic RS.

4. Discussion

Our study investigated two distinct components of RS as a measure of neuroplasticity: (1) immediate changes in motor response upon the first repetition ("dynamic RS") and (2) the suppressed level of RS ("stable RS"). Surprisingly, induction of plasticity with PAS with a 25 ms ISI resulted in different trends whereof one was rather LTD-like. Irrespective of such a discrepancy, the brain demonstrated an overall tendency towards a common level in stable RS following PAS intervention (Figure A2).

Minimizing the surprise encountered in the face of a novel stimulus is the principle behind the free energy principle [12]. According to this principle, to maintain its integrity, any adaptive biological system, like the brain, seeks to minimize its free energy [32]. It has been proposed that minimizing the free energy rests on either changing the top–down predictions, which are the conceptual internal models, or the bottom–up predicted sensory inputs [32]. In this regard, the dynamic RS depicts an update of the predictions in response to a twice repeated stimulus, ensuring an efficient sensory processing in a known environment. In other terms, the attenuation of the stimulus-evoked motor response upon the first repetition of the stimulus reduces the prediction errors originating from the incoming sensory information. In this respect, the suppressed level of the MEPs during the stable, suppressed part of RS may reflect a level of cortical excitability that is relatively free from ongoing dynamics in cortical excitability, which exhibits as a characteristically high intra-individual variance in the MEPs and may affect the sensitivity of MEPs to reveal longitudinal changes in excitability due to long-term neuromodulation and -plasticity.

RS has been demonstrated to last over short timescales in the visual and auditory systems, indicating a memory trace of the recently viewed or heard stimulus [33]. This short-term storage of information is reflected in our findings in the stable RS. A potential explanation for the observed stable RS might go back to the existence of a short-term internal representation of the perceived involuntary movement ("automatic memory"). Evidence consistent with this postulate is the lack of RS while an ITI of less than 3 s was employed in a TMS study, with the RS being more pronounced with longer ITIs [30]. Apart from the initial motor response, the subsequent responses elicited by TMS are modulated by sensory feedback, i.e., their magnitudes are controlled by the sensory inputs onto the motor neurons. The brain embodies a dynamic interconnected hierarchal processing organization that enables the reciprocal influence of current and past information. A plethora of positive and negative-feedback connections at both the cellular and network levels is central to sustaining the encoded sensory information on a timescale of seconds [34,35]. Hence, to maintain the automatic memory over a short timescale in RS, a negative feedback probably needs to be provided via recruiting inhibitory pathways to sustain the underlying neural activity. These pathways include the intracortical sensory areas and subcortical areas, among which the basal ganglia and thalamus play a key role. It has been demonstrated via RS that this stable state cannot be achieved in patients with progressive myoclonus type 1, who have impaired neuroplasticity in the thalamo-cortical connections [13,36].

Both dynamic and stable RS might reflect alterations in synaptic efficacy. The persistent changes in synaptic efficacy serve as a window into the formation of synaptic plastic changes, a candidate mechanism through which PAS works [26]. If the neuronal network is provided with only a positive feedback loop, that is, the spiking activity of the presynaptic neuron is correlated with the spiking activity of the postsynaptic neuron, its stability gets disrupted. In fact, this unidirectional process reduces the threshold for the presynaptic neuron to stimulate the postsynaptic neuron, thus precluding the stability and reversibility of the system. To counteract this instability and to tune the neuronal activity within a functional dynamic boundary, the brain employs an array of homeostatic mechanisms [37]. Homeostatic plasticity provides the necessary negative feedback loop to prevent the neural circuits from hyper- or hypo-activity.

A well-established proposed mechanism for homeostatic plasticity is the Bienenstock, Cooper, and Munro (BCM) model [38]. This model assumes a bidirectional synaptic plasticity, where the threshold for LTP/LTD induction varies as a function of the dynamic state of the brain. Considering this model, the more excitable the corticospinal pathway is, the more capacity for inhibition may be required. This can in part explain the reversal of the LTP-like plasticity effect to LTD-like plasticity in individuals showing higher pre-intervention MEP amplitudes (baseline). The degree of the modifications of neuronal plasticity depends on updating the synaptic efficacy. Thus, assessing the RS in the mentioned terms, i.e., are dynamic RS and stable RS, can provide us with information on how the alteration in synaptic efficacy following PAS can be reflected in RS.

A few limitations need to be acknowledged in this study. First, we applied the PAS paradigm using a fixed ISI of 25 ms. Inter-individual variability in responses has been reported for PAS due to non-optimized timing of the peripheral stimulus [39]. The potential decrease in this variability might have been achieved by employing an individualized ISI [40]. However, we did not measure individual sensory evoked potential to optimize PAS for LTP-like effects. This was by design to enable more inter-individual latency variance in the induced PAS effect, and to make the sessions shorter for the subjects. Second, in spite of having a sample size within the range of other studies in this field, the number of subjects was still small to account for generalization in large populations or in patients. We consider this a successful proof-of-concept study, but for application in patient groups, a larger-scale trial is required considering more inter-individual heterogeneities. Thirdly, we identified the PAS effect from the suppressed responses of RS (stable RS) to avoid the dynamicity of causing variance in the identification of the plastic effects, as we observed in the case of the first responses in the RS trials. This is not common practice with PAS. However, since no previous studies have been conducted with PAS in relation to RS, we had no point of reference.

Author Contributions: P.J. and J.K. designed the experiments; S.K., P.J. and L.S. conducted the experiments; J.R. performed MRI imaging; S.K. conducted the formal analysis of the data; S.K. wrote the original manuscript; P.J., J.K., L.S., J.R., and M.K. participated in editing the manuscript. All authors have read and agreed to the published version of the manuscript.

Funding: This work was supported by State Research Funding (project 5041763, Kuopio, Finland) and Orion Research Foundation (Espoo, Finland). S.K. was funded by the Alfred Kordelin Foundation and Kuopio University Foundation. J.R. was funded by the Vilho, Yrjö and Kalle Väisälä Foundation of the Finnish Academy of Science and Letters. P.J. and L.S. are supported by the Academy of Finland (grant no: 322423).

Conflicts of Interest: P.J. has received consulting fees and travel support from Nexstim Plc and has an unrelated patent with Nexstim Plc. J.K. is employed part-time by Nexstim Plc, manufacturer of navigated TMS systems. S.K., L.S., J.R., and M.K. declare no conflict of interest.

Appendix A

Figure A1, Comparison of LTD-like group and LTP-like group prior (baseline) and following PAS (at 0, 10, and 20 min).

Figure A2, RS at 0, 10, and 20 min after PAS in LTD-like and LTP-like groups.

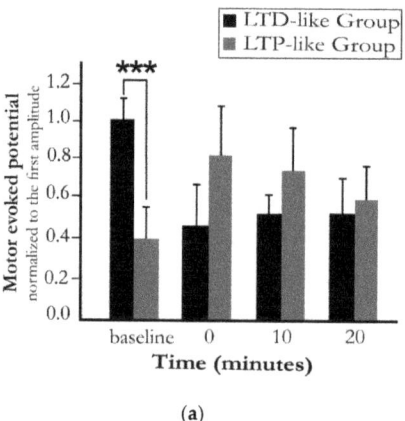

(a)

Figure A1. *Cont.*

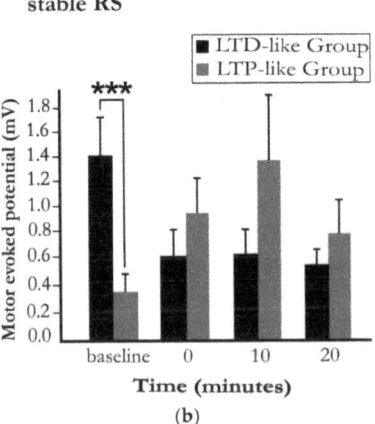

Figure A1. (**a**) Normalized MEP amplitude (mean ± standard error) at dynamic level of RS and (**b**) MEP amplitude (mean ± standard error) at stable level of RS, across the subjects from the LTP/LTD-like groups within the trials before (baseline) and at 0, 10, and 20 min after the induction of short-term plasticity with PAS. Subjects in the LTD-like group exhibited significantly higher pre-intervention MEP amplitudes (baseline) in dynamic RS ($p = 0.005$), and stable RS ($p = 0.007$). No significant difference was observed between the two clusters following PAS in either component. *** indicates significant differences for pairwise comparisons ($p < 0.01$).

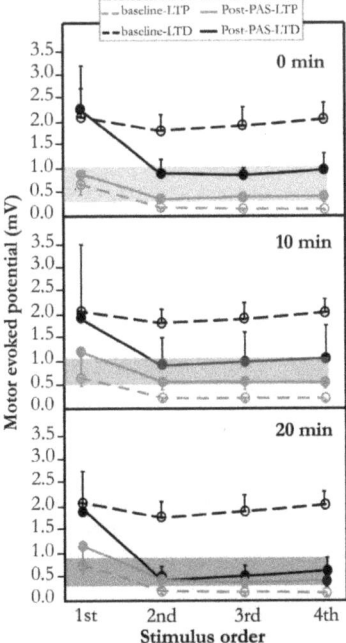

Figure A2. RS at 0, 10, and 20 min after the induction of short-term plasticity with PAS (post-PAS). Effects of LTP/LTD-like plasticity were measured with respect to "baseline" pre-intervention neural activity. Irrespective of the dynamic state of RS, the brain shows a tendency towards a suppressed amplitude level of the second to the fourth MEPs within trials upon repetition of the stimulus. This static status was maintained within a narrow range in two groups after PAS.

References

1. Moruzzi, G.; Magoun, H.W. Brain stem reticular formation and activation of the EEG. *Electroencephalogr. Clin. Neurophysiol.* **1949**, *1*, 455–473. [CrossRef]
2. Grill-Spector, K.; Henson, R.; Martin, A. Repetition and the brain: Neural models of stimulus-specific effects. *Trends Cogn. Sci.* **2006**, *10*, 14–23. [CrossRef] [PubMed]
3. Desimone, R. Neural mechanisms for visual memory and their role in attention. *Proc. Natl. Acad. Sci. USA* **1996**, *93*, 13494–13499. [CrossRef] [PubMed]
4. Miller, E.K.; Li, L.; Desimone, R. Activity of neurons in anterior inferior temporal cortex during a short-term memory task. *J. Neurosci.* **1993**, *13*, 1460–1478. [CrossRef]
5. Krekelberg, B.; Boynton, G.M.; van Wezel, R.J.A. Adaptation: From single cells to BOLD signals. *Trends Neurosci.* **2006**, *29*, 250–256. [CrossRef] [PubMed]
6. Näätänen, R.; Picton, T. The N1 Wave of the Human Electric and Magnetic Response to Sound: A Review and an Analysis of the Component Structure. *Psychophysiology* **1987**, *24*, 375–425. [CrossRef]
7. Löfberg, O.; Julkunen, P.; Tiihonen, P.; Pääkkönen, A.; Karhu, J. Repetition suppression in the cortical motor and auditory systems resemble each other—A combined TMS and evoked potential study. *Neuroscience* **2013**, *243*, 40–45. [CrossRef]
8. Löfberg, O.; Julkunen, P.; Pääkkönen, A.; Karhu, J. The auditory-evoked arousal modulates motor cortex excitability. *Neuroscience* **2014**, *274*, 403–408. [CrossRef]
9. Friston, K.; Kilner, J.; Harrison, L. A free energy principle for the brain. *J. Physiol. Paris* **2006**, *100*, 70–87. [CrossRef]
10. Li, L.; Miller, E.K.; Desimone, R. The representation of stimulus familiarity in anterior inferior temporal cortex. *J. Neurophysiol.* **1993**, *69*, 1918–1929. [CrossRef]
11. Sobotka, S.; Ringo, J.L. Mnemonic responses of single units recorded from monkey inferotemporal cortex, accessed via transcommissural versus direct pathways: A dissociation between unit activity and behavior. *J. Neurosci.* **1996**, *16*, 4222–4230. [CrossRef] [PubMed]
12. Friston, K. A theory of cortical responses. *Philos. Trans. R. Soc. Lond. B. Biol. Sci.* **2005**, *360*, 815–836. [CrossRef] [PubMed]
13. Julkunen, P.; Löfberg, O.; Kallioniemi, E.; Hyppönen, J.; Kälviäinen, R.; Mervaala, E. Abnormal motor cortical adaptation to external stimulus in Unverricht-Lundborg disease (progressive myoclonus type 1, EPM1). *J. Neurophysiol.* **2018**, *120*, 617–623. [CrossRef] [PubMed]
14. Pascual-Leone, A.; Amedi, A.; Fregni, F.; Merabet, L.B. The plastic human brain cortex. *Annu. Rev. Neurosci.* **2005**, *28*, 377–401. [CrossRef]
15. Rioult-Pedotti, M.-S.; Friedman, D.; Hess, G.; Donoghue, J.P. Strengthening of horizontal cortical connections following skill learning. *Nat. Neurosci.* **1998**, *1*, 230–234. [CrossRef]
16. Stefan, K.; Wycislo, M.; Gentner, R.; Schramm, A.; Naumann, M.; Reiners, K.; Classen, J. Temporary occlusion of associative motor cortical plasticity by prior dynamic motor training. *Cereb. Cortex* **2006**, *16*, 376–385. [CrossRef]
17. Rossini, P.M.; Dal Forno, G. Neuronal post-stroke plasticity in the adult. *Restor. Neurol. Neurosci.* **2004**, *22*, 193–206.
18. Daskalakis, Z.J.; Christensen, B.K.; Fitzgerald, P.B.; Chen, R. Dysfunctional neural plasticity in patients with Schizophrenia. *Arch. Gen. Psychiatry* **2008**, *65*, 378–385. [CrossRef]
19. Player, M.J.; Taylor, J.L.; Weickert, C.S.; Alonzo, A.; Sachdev, P.; Martin, D.; Mitchell, P.B.; Loo, C.K. Neuroplasticity in depressed individuals compared with healthy controls. *Neuropsychopharmacology* **2013**, *38*, 2101–2108. [CrossRef]
20. Flor, H. Cortical reorganisation and chronic pain: Implications for rehabilitation. *J. Rehabil. Med.* **2003**, 66–72. [CrossRef]
21. Voronin, L.L. Long-term potentiation in the hippocampus. *Neuroscience* **1983**, *10*, 1051–1069. [CrossRef]
22. Pascual-Leone, A.; Tarazona, F.; Keenan, J.; Tormos, J.M.; Hamilton, R.; Catala, M.D. Transcranial magnetic stimulation and neuroplasticity. *Neuropsychologia* **1999**, *37*, 207–217. [CrossRef]
23. Chen, R.; Cohen, L.G.; Hallett, M. Nervous system reorganization following injury. *Neuroscience* **2002**, *111*, 761–773. [CrossRef]

24. Stefan, K.; Kunesch, E.; Cohen, L.G.; Benecke, R.; Classen, J. Induction of plasticity in the human motor cortex by paired associative stimulation. *Brain* **2000**, *123 Pt 3*, 572–584. [CrossRef]
25. Tolmacheva, A.; Savolainen, S.; Kirveskari, E.; Brandstack, N.; Mäkelä, J.P.; Shulga, A. Paired associative stimulation improves hand function after non-traumatic spinal cord injury: A case series. *Clin. Neurophysiol. Pract.* **2019**, *4*, 178–183. [CrossRef]
26. Wolters, A.; Sandbrink, F.; Schlottmann, A.; Kunesch, E.; Stefan, K.; Cohen, L.G.; Benecke, R.; Classen, J. A temporally asymmetric Hebbian rule governing plasticity in the human motor cortex. *J. Neurophysiol.* **2003**, *89*, 2339–2345. [CrossRef]
27. Stefan, K.; Kunesch, E.; Benecke, R.; Cohen, L.G.; Classen, J. Mechanisms of enhancement of human motor cortex excitability induced by interventional paired associative stimulation. *J. Physiol.* **2002**, *543*, 699–708. [CrossRef]
28. Müller-Dahlhaus, F.; Ziemann, U.; Classen, J. Plasticity resembling spike-timing dependent synaptic plasticity: The evidence in human cortex. *Front. Synaptic Neurosci.* **2010**, *2*, 34. [CrossRef]
29. Awiszus, F.; Borckardt, J. TMS Motor Threshold Assessment Tool 2.0. 2012. Available online: http://clinicalresearcher.org/software.htm (accessed on 19 October 2012).
30. Pitkänen, M.; Kallioniemi, E.; Julkunen, P. Effect of inter-train interval on the induction of repetition suppression of motor-evoked potentials using transcranial magnetic stimulation. *PLoS ONE* **2017**, *12*, e0181663. [CrossRef]
31. Hamada, M.; Strigaro, G.; Murase, N.; Sadnicka, A.; Galea, J.M.; Edwards, M.J.; Rothwell, J.C. Cerebellar modulation of human associative plasticity. *J. Physiol.* **2012**, *590*, 2365–2374. [CrossRef]
32. Friston, K. The free-energy principle: A unified brain theory? *Nat. Rev. Neurosci.* **2010**, *11*, 127–138. [CrossRef]
33. Ranganath, C.; Rainer, G. Neural mechanisms for detecting and remembering novel events. *Nat. Rev. Neurosci.* **2003**, *4*, 193–202. [CrossRef] [PubMed]
34. Lim, S.; Goldman, M.S. Balanced cortical microcircuitry for maintaining information in working memory. *Nat. Neurosci.* **2013**, *16*, 1306–1314. [CrossRef] [PubMed]
35. Frank, M.J.; Loughry, B.; O'reilly, R.C. Interactions between frontal cortex and basal ganglia in working memory: A computational model. *Cogn. Affect. Behav. Neurosci.* **2001**, *1*, 137–160. [CrossRef]
36. Koskenkorva, P.; Khyuppenen, J.; Niskanen, E.; Könönen, M.; Bendel, P.; Mervaala, E.; Lehesjoki, A.E.; Kälviäinen, R.; Vanninen, R. Motor cortex and thalamic atrophy in Unverricht-Lundborg disease: Voxel-based morphometric study. *Neurology* **2009**, *73*, 606–611. [CrossRef] [PubMed]
37. Turrigiano, G.G.; Leslie, K.R.; Desai, N.S.; Rutherford, L.C.; Nelson, S.B. Activity-dependent scaling of quantal amplitude in neocortical neurons. *Nature* **1998**, *391*, 892–896. [CrossRef] [PubMed]
38. Bienenstock, E.L.; Cooper, L.N.; Munro, P.W. Theory for the development of neuron selectivity: Orientation specificity and binocular interaction in visual cortex. *J. Neurosci.* **1982**, *2*, 32–48. [CrossRef] [PubMed]
39. López-Alonso, V.; Cheeran, B.; Río-Rodríguez, D.; Fernández-del-Olmo, M. Inter-individual variability in response to non-invasive brain stimulation paradigms. *Brain Stimul.* **2014**, *7*, 372–380. [CrossRef]
40. Campana, M.; Papazova, I.; Pross, B.; Hasan, A.; Strube, W. Motor-cortex excitability and response variability following paired-associative stimulation: A proof-of-concept study comparing individualized and fixed inter-stimulus intervals. *Exp. Brain Res.* **2019**, *237*, 1727–1734. [CrossRef]

© 2020 by the authors. Licensee MDPI, Basel, Switzerland. This article is an open access article distributed under the terms and conditions of the Creative Commons Attribution (CC BY) license (http://creativecommons.org/licenses/by/4.0/).

Article

Comparing the Impact of Multi-Session Left Dorsolateral Prefrontal and Primary Motor Cortex Neuronavigated Repetitive Transcranial Magnetic Stimulation (nrTMS) on Chronic Pain Patients

Sascha Freigang [1,*], Christian Lehner [1], Shane M. Fresnoza [2,3], Kariem Mahdy Ali [1], Elisabeth Hlavka [1], Annika Eitler [1], Istvan Szilagyi [4], Helmar Bornemann-Cimenti [5], Hannes Deutschmann [6], Gernot Reishofer [6], Anže Berlec [1], Senta Kurschel-Lackner [1], Antonio Valentin [7], Bernhard Sutter [1], Karla Zaar [1] and Michael Mokry [1]

[1] Department of Neurosurgery, Medical University Graz, 8036 Graz, Austria; christian.lehner@medunigraz.at (C.L.); kariem.mahdy-ali@medunigraz.at (K.M.A.); elisabeth.hlavka@stud.medunigraz.at (E.H.); annika.eitler@stud.medunigraz.at (A.E.); Anze.Berlec@uniklinikum.kages.at (A.B.); senta.kurschel@medunigraz.at (S.K.-L.); bernhard.sutter@medunigraz.at (B.S.); Karla.Zaar@klinikum-graz.at (K.Z.); michael.mokry@medunigraz.at (M.M.)
[2] Institute of Psychology, University of Graz, 8010 Graz, Austria; shane.fresnoza@uni-graz.at
[3] BioTechMed, 8010 Graz, Austria
[4] Department of Paediatric Surgery, Medical University Graz, 8036 Graz, Austria; istvan.szilagyi@medunigraz.at
[5] Department of Anaesthesiology, Critical Care and Pain Medicine, Medical University Graz, 8036 Graz, Austria; helmar.bornemann@medunigraz.at
[6] Department of Radiology, Clinical Division of Neuroradiology, Vascular and Interventionial Radiology, Medical University of Graz, 8036 Graz, Austria; hannes.deutschmann@medunigraz.at (H.D.); gernot.reishofer@medunigraz.at (G.R.)
[7] Department of Basic & Clinical Neuroscience, Institute of Psychiatry, Psychology and Neuroscience, King's College London, London SE5 9RT, UK; antonio.valentin@kcl.ac.uk
* Correspondence: sascha.freigang@medunigraz.at; Tel.: +43-316-385-81935

Citation: Freigang, S.; Lehner, C.; Fresnoza, S.M.; Mahdy Ali, K.; Hlavka, E.; Eitler, A.; Szilagyi, I.; Bornemann-Cimenti, H.; Deutschmann, H.; Reishofer, G.; et al. Comparing the Impact of Multi-Session Left Dorsolateral Prefrontal and Primary Motor Cortex Neuronavigated Repetitive Transcranial Magnetic Stimulation (nrTMS) on Chronic Pain Patients. Brain Sci. 2021, 11, 961. https://doi.org/10.3390/brainsci 11080961

Academic Editors: Nico Sollmann and Petro Julkunen

Received: 14 June 2021
Accepted: 19 July 2021
Published: 22 July 2021

Publisher's Note: MDPI stays neutral with regard to jurisdictional claims in published maps and institutional affiliations.

Copyright: © 2021 by the authors. Licensee MDPI, Basel, Switzerland. This article is an open access article distributed under the terms and conditions of the Creative Commons Attribution (CC BY) license (https://creativecommons.org/licenses/by/4.0/).

Abstract: Repetitive transcranial stimulation (rTMS) has been shown to produce an analgesic effect and therefore has a potential for treating chronic refractory pain. However, previous studies used various stimulation parameters (including cortical targets), and the best stimulation protocol is not yet identified. The present study investigated the effects of multi-session 20 Hz (2000 pulses) and 5 Hz (1800 pulses) rTMS stimulation of left motor cortex (M1-group) and left dorsolateral prefrontal cortex (DLPFC-group), respectively. The M1-group ($n = 9$) and DLPFC-group ($n = 7$) completed 13 sessions of neuronavigated stimulation, while a Sham-group ($n = 8$) completed seven sessions of placebo stimulation. The outcome was measured using the German Pain Questionnaire (GPQ), Depression, Anxiety and Stress Scale (DASS), and SF-12 questionnaire. Pain perception significantly decreased in the DLPFC-group (38.17%) compared to the M1-group (56.11%) ($p \leq 0.001$) on the later sessions. Health-related quality of life also improved in the DLPFC-group (40.47) compared to the Sham-group (35.06) ($p = 0.016$), and mental composite summary ($p = 0.001$) in the DLPFC-group (49.12) compared to M1-group (39.46). Stimulation of the left DLPFC resulted in pain relief, while M1 stimulation was not effective. Nonetheless, further studies are needed to identify optimal cortical target sites and stimulation parameters.

Keywords: chronic pain; low back pain; repetitive transcranial magnetic stimulation; neuromodulation; dorsolateral prefrontal cortex; primary motor cortex

1. Introduction

Pain is recently redefined by the International Association for the Study of Pain (IASP) as "an unpleasant sensory and emotional experience associated with, or resembling that associated with, actual or potential tissue damage" [1]. Pain is considered chronic if it persists or recurs for more than 3 months regardless of whether it is the sole complaint (chronic primary pain) or secondary to an underlying disease (chronic secondary pain) [2]. Worldwide, chronic pain is one of the leading causes of years lived with disability (YLDs) and reduced quality of life (QoL). In Europe, high prevalence rates were reported for back/neck (40%), hand/arm (22%), and foot/leg (21%) pain [3]. Over the past 30 years, although the prevalence of most diseases showed a pattern of steady decline as measured by age-standardised disability-adjusted life-years (DALYs) rates, chronic low back pain (LBP) remained in the top ten (fourth) causes of DALYs for children and younger adults. Low back pain in childhood predicts low back pain in adult life and is more common in female than male individuals at all ages [4]. As modern medicine extends the population age, it is most likely that the global prevalence of LBP will further increase in the following decades. Therefore, research to develop safe and effective interventions is needed to improve health and alleviate the socioeconomic burden of chronic pain patients.

LBP with unidentifiable pathoanatomical and pathophysiological causes is the most common form of chronic pain condition and is termed non-specific low back pain (NSLBP) [5]. With no specific treatment, management of NSLBP focuses on limiting risk exposure (e.g., lifting heavy objects), patient education, and interventions such as exercise and physical therapy to reduce pain. Current literature also suggests that pharmacological treatments with non-steroidal anti-inflammatory drugs (NSAIDs) and acetaminophen, as well as antidepressants, muscle relaxants, and opioid analgesics are effective for chronic LBP [6–8]. Nonetheless, non-pharmacological therapies may fail in some patients, and few trials have investigated their effectiveness [4]. With medications, pain relief is achievable but insufficient because any benefit is likely to be temporary, and symptoms will recur when medication is stopped [9]. Treatment-emergent adverse events (e.g., skeletal muscle relaxants-induced sedation) and long-term use related side effects (e.g., increased risk of vascular events for NSAIDs) are also serious setbacks of pharmacotherapy [10,11]. Moreover, with regard to surgical management, there are still no well-defined clinical practice guidelines related to surgical intervention for chronic LBP in the absence of serious anatomical problems [12].

The most challenging issue in managing chronic pain, including NSLBP, is that the underlying mechanisms are still poorly understood. In NSLBP, pain sensations do not necessarily reflect the presence of a peripheral noxious stimulus because neurons in the pain pathway can be activated by a low threshold, innocuous or non-noxious inputs [13]. The neurobiological cause is thought to be maladaptive plasticity such as central sensitization, which manifests as distort or amplify (hyperalgesia and allodynia), increase degree or duration (after sensations and temporal summation), and spatial extent (expansion of the receptive field), as well as a reduced conditioned pain modulation [13,14]. In chronic pain patients, the prevalent expectation for brain activity is a sustained or enhanced activation of areas already identified for acute pain [15]. For instance, increased functional connectivity between sensorimotor and frontoparietal networks could reflect sustained attention to bodily sensations and hypervigilance to somatic sensations [16,17]. Furthermore, compared with healthy controls, patients also exhibit greater resting-state electroencephalography (EEG) alpha oscillations (8.5–12.5 Hz) at the parietal region, which could be relevant with attenuated sensory information gating and excessive integration of pain-related information [17]. The early evoked magnetic field elicited by stimulation of the painful back is also elevated in very chronic patients [18]. Chronicity-dependent cortical reorganization, regardless of aetiology, is also reported in the primary somatosensory (SI) cortex of chronic pain patients [18–20].

In contrast to SI, the evidence of altered structural, organizational, and functional alternation in the primary motor cortex (M1) for neuropathic and non-neuropathic pain conditions is conflicting [21]. However, several studies suggest the association of cortical reorganization of muscle representation in M1 with deficits in postural control, such as impaired anticipatory activation of trunk muscles [22,23]. In addition, similar to SI and other pain-relevant brain regions, enhanced neuronal activity/excitability as measured with transcranial magnetic stimulation (TMS); evoked peripheral muscle potentials (motor evoked potentials or MEPs) are reported in M1 [24]. Enhanced cortical excitability is thought to be secondary to M1 disinhibition as indicated by reduced GABA-mediated short-interval intracortical inhibition (SICI) and cortical silent period (CSP), as well as an increase in the glutamatergic-mediated short-interval intracortical facilitation (SICF) [21,24]. A decrease in the level of thalamic and M1 N-acetylaspartate (NAA) is also considered an index of neuronal depression and altered neuronal-glial interactions in chronic pain patients [25,26]. Higher levels of glutamate/glutamine compounds in the amygdala are also observed in fibromyalgia patients compared to healthy controls [27]. Overall, these studies suggest the potential of the cerebral cortex as a promising target in treating chronic pain.

In the past decades, the development of non-invasive brain stimulation methods such as TMS and several variants of transcranial electrical stimulation allows the identification of the causal role of different cortical regions and neuromodulation of these structures to treat pathological conditions such as chronic pain. Application of repetitive magnetic pulses at a specific frequency (repetitive TMS or rTMS) or in a burst of 3–5 pulses delivered at theta frequencies (theta burst stimulation or TBS) can modulate cortical excitability during and beyond the period of stimulation [28,29]. Induction of neuroplasticity, such as long-term potentiation (LTP) and long-term depression (LTD) at the synaptic level is thought to be the neurophysiological mechanism behind the after-effects of rTMS and TBS paradigms [30]. Several studies applied these paradigms to disrupt or reverse maladaptive and enhanced adaptive neuroplasticity associated with chronic pain. Systematic reviews of rTMS studies for chronic pain with known etiological factors (e.g., fibromyalgia) suggest a beneficial effect of a single or repetitive dose of high-frequency stimulation of M1 [31–33]. A meta-analytical study suggests that five-sessions of high-frequency (5, 10, and 20 Hz) rTMS on M1 has a maximal analgesic effect lasting up to 1 month in chronic neuropathic pain patients [34]. However, another meta-analysis reported that low-frequency rTMS is ineffective in treating chronic pain, while single doses of high-frequency rTMS of M1 are considered to have no clinical significance for chronic pain due to its negligible effect [35]. On the other hand, the evidence is still insufficient for chronic pain of unknown origin, such as NSLBP. So far, only one study has shown that one week of 20 Hz rTMS applied to the left M1/S1 hand area can decrease pain perception. Significant reduction in visual analogue scale (VAS) and Short Form McGill pain questionnaire (SF-MPQ) scores were observed in the rTMS-treated group but not in the sham group, as well as lower mean pain score compared to patients treated with physical therapy [36].

This study was undertaken to explore the efficacy of rTMS for NSLBP and improve the available protocol in managing chronic pain. We aimed to replicate the beneficial effect of multi-session left M1 rTMS in the study of Ambriz-Tututi and colleagues (2016). In addition to M1, we also applied neuronavigated rTMS (nrTMS) on the left dorsolateral prefrontal cortex (DLPFC) because stimulation of this brain area is reported to change pain perception in healthy subjects and has analgesic effects in acute postoperative pain, fibromyalgia, and traumatic spinal cord injury patients [37–39]. Therefore, we hypothesized that left M1 and DLPFC stimulation would reduce pain perception in NSLBP patients. To our knowledge, this is so far the first report exploring the effect of left DLPFC and M1 nrTMS in NSLBP patients in a single study.

2. Materials and Methods

2.1. Patients

Thirty-four chronic pain patients participated in the study (19 females and 15 males, mean age ± SD: 54 ± 11 years). They were either previous neurosurgical patients or regular pain clinic patients at the University Hospital Graz-Austria. All have no prior knowledge about TMS and had no planned pain-related interventions during the study. The sample size was a priori calculated using G *Power 3.1.9 and is based on a planned repeated measure ANOVA (with within-between interaction) on the numerical pain scale data. We expected an effect size of d = 0.20, power = 0.95, and a = 0.05. Inclusion criteria were age between 18 and 80 years, clinical diagnosis of chronic LBP and or neck pain, average resting pain-level greater than 3 in the Numeric Rating Scale (0–10), no changes in pain medication 4 weeks before baseline measurements, and no single or multiple surgical procedures in the head and lower back in the last two years. Patients with the following characteristics were excluded from the study: metallic and electronic implants in the head, neck and chest; intake of opioid analgesics (>100 mg orally per day), tetracyclic antidepressants, antiviral, and antipsychotic drugs; history of frequent headache or tinnitus and alcohol or drug abuse; confirmed or suspected pregnancy and breastfeeding. All participants provided written informed consent before the experimental procedures. The Ethics Committee of the Medical University of Graz approved the study (registration number: 30-459-ex 17/18), and all procedures conform to the Declaration of Helsinki regarding human experimentation.

2.2. Study Design and Procedure

The study was conducted in a single-blinded, randomized, partial placebo-controlled design. It was retrospectively registered at clinicaltrails.gov (registration number: NCT04934150). The experiments took place in the outpatient clinic of the Department of Neurosurgery (Medical University Graz) between February 2019 and March 2020. The patients were allocated into an "M1-group" (5 males, 6 females; mean age ± SD: 53.8 ± 12.7 years), "DLPFC-group" (6 males, 6 females; mean age ± SD: 56.8 + 9.6 years), and "sham-group" (4 males, 7 females; mean age ± SD: 52.5 ± 12.5 years) using permuted block randomization on the online software random.org. Patients were blinded to their assigned group. Each patient in the M1-group and DLPFC-group underwent 13 nrTMS experimental sessions. The first 5 sessions were conducted every day for 5 consecutive days without a break (1 session per day). One week later, the remaining 7 sessions were conducted in a span of 9 months (week 3, 4, 6, 8, 12, 20, 28, and 36). The sham-group followed the same schedule; however, the experiment was stopped after the seventh session (4th week) because of ethical considerations. Each session started with head/brain and TMS coil co-registration. Subsequently, stimulation intensity was determined, and target areas underwent stimulation. Pain assessments before and after stimulation using numerical pain rating scales (NPRS) were conducted on the first (baseline), 7th (4th week), and 13th (36th week) experimental sessions (Figure 1). NPRS scores were documented through an interview before and after stimulation. An experimental session including the preparations lasted for approximately 30 min.

2.3. Neuronavigated Repetitive Transcranial Magnetic Stimulation (nrTMS)

TMS was administered using a figure-of-eight coil (MCF-B65) connected to a MagPro X100 stimulator (MagVenture A/S, Farum, Denmark). Patients were seated on a reclining chair with head and neck support and were asked to relax. For precise coil placement and stimulation, neuronavigation (line-navigated) was performed using Localite TMS Navigator software (LOCALITE Biomedical Visualization Systems GmbH, Sankt Augustin, Germany) that tracts the coil movement with an infrared stereo-optical tracking camera (Polaris Spectra, Northern Digital Inc., Waterloo, Ontario, Canada). The tracking system monitors the location of passive marker spheres attached to the TMS coil and head in real-time. Each patient's T1-weighted MRI scan (MPRAGE, TR = 1650, TE = 1.82 ms, matrix = 256 × 256, FOV = 256 mm, 192 sagittal slices, in-plane resolution: 1 mm × 1 mm, slice thickness: 1 mm, 0.5 mm gap) was used for head and coil registration, target planning, and neuronavigation during stimulation (Siemens Medical Systems, Erlangen, Germany). TMS parameters were consistent with Ambriz-Tututi et al. (2016): 2000 biphasic pulses at an intensity of 95% resting motor threshold (RMT) applied (10 trains with 28 s inter-train interval (ITI)) for 10 s at 20 Hz. RMT was determined by electromyographic recording over the abductor pollicis brevis muscle and defined as the minimum stimulator output that elicits a 50 uV motor-evoked potential (MEPs) in 5 out of 10 single-pulse TMS stimulation of M1 at rest. Anatomically defined targets over the left M1 were marked by a 5 × 2 grid overlay with 10 mm between-target spacing (Figure 2A), while for the left DLPFC, targets were marked by a 3 × 4 grid overlay (Figure 2B). For the DLPFC-group, 12 trains of 1800 TMS pulses (150 pulses per train, ITI 10 s) were delivered at 5 Hz and 90% RMT [39]. During the stimulations, the coil was held tangentially to the scalp at an angle of 45° to the midsagittal plane generating a current with posterior-anterior direction. TMS was administered over the left M1 in the sham group but with the coil tilted approximately 45 degrees away from the scalp. Therefore, patients could still hear and feel the typical TMS sound and vibration, respectively, without active stimulation. Sham stimulation was limited to seven sessions and used the same frequency, quantity of stimuli, and ITI as the M1-group.

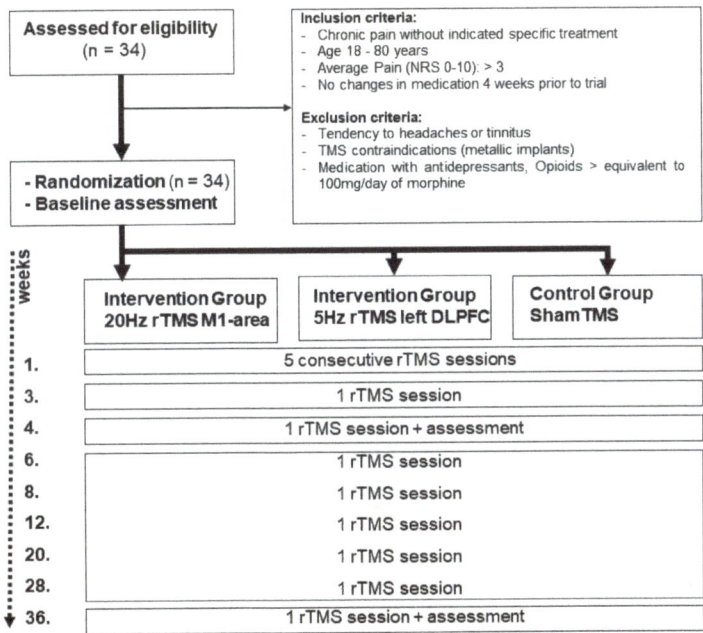

Figure 1. Flowchart depicting the course of the study (CONSORT 2010).

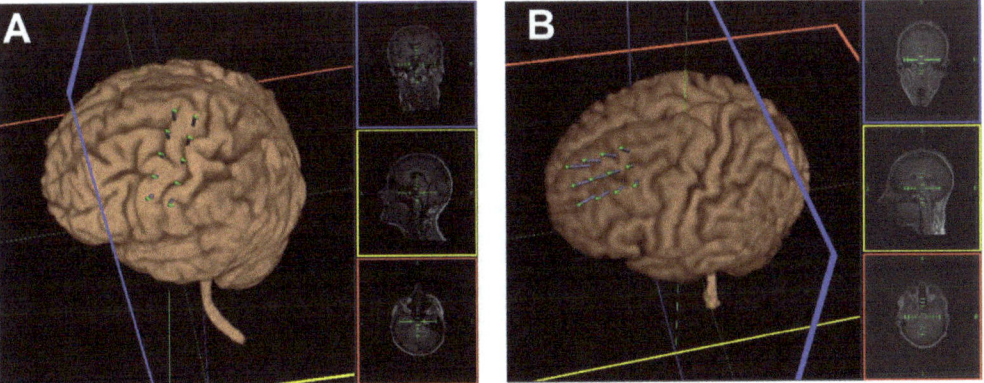

Figure 2. Target arrays for left M1 (**A**) and DLPFC (**B**) rTMS stimulation. The target grid for M1 was a 2 × 5 points array with a 10 mm interpoint distance. For the DLPFC, the target grid was a 3 × 4 points array with a 10 mm interpoint distance. For each participant, the target array was positioned based on anatomical landmarks.

2.4. Outcome Assessments

The primary outcome variable was derived from the German Pain Questionnaire (GPQ). At baseline and for each experimental session, patients were asked to verbally rate their perceived greatest pain intensity before stimulation and bearable pain intensity after stimulation using the GPQ-NPRS ranging from 0 ("no pain") to 10 ("worst pain imaginable") [40]. As secondary outcome measures, depression, anxiety, and stress scores were obtained using the self-report "Depression, Anxiety, and Stress Scale" (DASS) questionnaire also at baseline, after the 4th and after the 36th week of stimulation [41]. DASS contains 21 items (7 per category: depression, anxiety and stress), and patients were asked to score every item on a scale from 0 to 3 (0 = never, 1 = sometimes, 2 = often, 3 = almost always). The item scores were added, and each category can have a total DASS score of 21. There are separate severity ratings for depression (0 to 4 = normal, 5 to 6 = mild, 7 to 10 = moderate, 11 to 13 = severe, >14 = extremely severe), anxiety (0 to 3 = normal, 4 to 5 = mild, 6 to 7 = moderate, 8 to 9 = severe, >10 = extremely severe), and stress (0 to 7 = normal, 8 to 9 = mild, 10 to 12 = moderate, 13 to 16 = severe, >17 = extremely severe) [42]. To assess the impact of the stimulation on health-related quality of life (HRQoL) measures, the patients answered the 12-item short-form questionnaire (SF-12 version 1) [43]. The SF-12 questionnaire uses eight domains on a 100-point scale which include physical function (PF), role limitations caused by physical problems (RP), pain (BP), general health (GH), vitality/energy (VT), social function (SF), mental health/emotional well-being (MH), and role limitations caused by emotional problems/mental health (RE) [44]. The items refer to perceived health status during the last four weeks; a higher score indicates a better perceived health state. The PF, RP, BP, and GH dimensions were summarized into a physical composite summary (PCS), and the VT, SF, MH, and RE dimensions were summarized into a mental composite summary (MCS) [44]. Unlike the NPRS, DASS and SF-12 scores were only recorded at baseline, after the 4th and 36th week of stimulation.

2.5. Statistical Analysis

Statistical analyses were performed using SPSS version 26 software (IBM Corp., Armonk, NY, USA), and figures were generated using Graphpad Prism (GraphPad Prism version 9.0.0 for Windows, GraphPad Software, San Diego, CA, USA). All data were analysed using linear mixed-effects modelling (LMM) with random intercept. We decided to use LMMs for various statistical reasons. First, LMMs are well suited for unbalanced datasets such as ours (sham-group only has a baseline until 4th week measurements, while the M1-group and DLPFC-group have a baseline until 36th week measurements) [45,46]. Second, unlike the regular ANOVA, LMMs does not perform listwise deletion since it has an automatically in-built implicit imputation that assumes the "Missing completely at random (MCAR)" function. For this reason, incomplete data from patients who drop out can still be incorporated into the models. Third, LMMs is a suitable and robust statistical approach because multilevel models tolerate the heterogeneity of variances (due to unequal sizes) between groups. Finally, compared to the traditional ANOVA for a study with repeated measures that only compare the variances between the group means and therefore does not consider the interindividual differences, LMM considers the inter-individual differences by incorporating the participants as a "random factor" in the model. This feature is useful for pain studies because it accounts for the large inter-individual patient's subjective rating variability.

For the analysis, the GPQ-NPRS, DASS, and SF-12 scores served as the dependent variables. We first analysed GPQ-NPRS data from time points similar to when the DASS and SF-12 scores were obtained (baseline, after 4th and 36th week of stimulation). Before the analysis, the ordinal GPQ-NPRS scores were converted into percentages of the 11-point scale (e.g., "10" = 100% and "0" = 9.09%). The model contained the between-subject factor "group" (sham, M1, and DLPFC) and the within-subject factor "session" (baseline, after 4th and after 36th week of stimulation) and "time" ("before stimulation" = percentage of greatest perceived pain and "after stimulation" = percentage of bearable pain) as fixed factors. This analysis does not give us a complete picture of how multi-session stimulation affected pain intensity; therefore, we plotted baseline and post-stimulation GPQ-NPRS scores from each experimental session and calculated the area under the curve (AUC) using a trapezoidal method in Microsoft Excel (2019). The AUC indicates the summation of pain relief over time. Smaller AUC values indicate a greater decrease in pain intensity. Each patient's AUC was determined, and the group average was obtained. Using an independent paired t-test, we first compared the three groups (sham, M1, and DLPFC) average AUC for the first five days of stimulation. Next, we compared the average AUC from the 4th to the 36th week of stimulation of the M1 and DLPFC-group.

The DASS and SF-12 scores were treated as continuous variables and therefore not converted. For the DASS and SF-12 scores, the model does not contain the within-subject factor "time" since patients only answered these questionnaires after stimulation. Each patient was specified as a random factor (random intercept model) in all models. Normality (data distribution) and homogeneity of variance test were conducted using Shapiro–Wilk and Levene's test, respectively. In cases of normality violation, all modelled data underwent logarithmic transformation (log10). We also performed a Pearson Chi-Square test to assess the model's goodness of fit to the data. We calculated Cohen's d as a measure of effect size (<0.2—trivial, >0.2—small, >0.5—medium and >0.8—large). Significant findings from the models were explored with post hoc comparisons (paired t-test, two-tailed, Bonferroni adjusted for multiple comparisons). Lastly, we tested collinearity in the final models by determining the tolerance and variance inflation factors. A p-value of <0.05 was considered significant for all statistical analyses. All values are expressed as the mean ± standard error of the mean (SEM).

3. Results

In total, 34 patients were enrolled in the study (sham- and M1-group: 11, DLPFC-group: 12). Mean age, gender distribution, and chronic pain duration were comparable between the groups (Table 1). Unexpectedly, there were more patients with clinically diagnosed depression in the sham-group than in the treatment groups. All patients tolerated the experimental procedure well. There were no reports of dizziness, headaches, or nausea. However, for reasons unrelated to the stimulation, three patients in the sham-group (all after the first session), two patients in the M1-group (one after the 7th session and one after the 11th session), and five patients in the DLPFC-group (one after the 1st session, one after the 2nd session, one after the 6th session, and two after the 11th session) dropped out from the study. Reasons included demographic challenges such as a longer commute to the treatment facility and the anticipated difficulty following the specific intervals between experimental sessions. Therefore, all modelled data comprised 90.34% of the expected dataset. Shapiro–Wilk test indicates that the NPRS score ($p \leq 0.001$), DASS scores (depression: $p \leq 0.001$, anxiety: $p \leq 0.001$, stress: $p = 0.032$) and SF-12 scores (PCS: $p = 0.040$, MCS: $p = 0.025$) are not normally distributed, hence all were log-transformed. In all models, Levene's test showed that variances were equal for each group (all $p > 0.05$). Tolerance range and variance inflation factors were equal to 1.000 in the final models indicating that multicollinearity did not affect the findings.

Table 1. Demographic information of patients who completed the experimental protocols.

	Sham	M1	DLPFC
Number of patients	8	9	7
Mean age (in years)	52.9	51.2	62.6
Gender	F: 2, M: 6	F: 5, M: 4	F: 4, M: 3
Duration of chronic pain (in years)	6.4	5.2	5.3
Patients diagnosed w/depression	6	1	0

3.1. GPQ-NPRS

For the GPQ-NPRS, we interpreted a full model because the Pearson Chi-Square test indicated the goodness of fit of this model to the data [X^2 (147) = 4.034, p = 0.058]. The results showed that overall pain intensity decreases after rTMS stimulation as indicated by the significant main effect of factor time (F (1, 122.52) = 194.14, $p \leq 0.001$, d = 0.961) and interaction of time and session (F (2, 122.52) = 3.73, p = 0.027, d = 0.440) (Figure 3A–C). The overall reduction in pain intensity was significant toward the end of the study as indicated by the factor session's significant main effect (F (2, 134.25) = 4.77, p = 0.010, d = 0.527). Overall pain intensity on the 4th week (50.19%, p = 0.086) of stimulation was comparable to baseline (57.72%), while pain intensity on the 36th week of stimulation was significantly lower (47.19%, p = 0.033) than baseline. The main effect of the factor group was only nearly significant (Table 2); however, its interaction with the factor session was significant (F (3, 133.89) = 3.99, p = 0.009, d = 0.457). Bonferroni corrected post hoc t-test showed that at baseline and 4th week after stimulation, there were no significant differences in overall pain intensity between the groups (Figure 4A). However, on the 36th week, pain intensity in the DLPFC-group (38.17%) was significantly lower ($p \leq 0.001$) than in the M1-group (56.11%) (Figure 4B). Additional exploratory post hoc t-test showed that the significant differences between the DLPFC- and M1-group's pain intensity on the 36th week was driven by the significantly lower ($p \leq 0.001$) pain intensity in the former (47.83%) than the latter (75.81%) before stimulation (Figure 4B). After stimulation, although pain intensity was still lower in the DLPFC-group (28.51%) than the M1-group (36.41%), the differences did not reach significance (p = 0.085).

Table 2. Results of the linear mixed models (LMM) performed for the GPQ-NPRS, DASS, and SF-12 scores.

	Numerator df	Denominator df	F-Value	p-Value	Cohen's d
GPQ-NPRS scores					
Group	2	32.53	3.08	0.059	0.343
Session	2	134.25	4.77	0.010 *	0.527
Time	1	122.52	194.14	<0.001 *	0.961
Group × session	3	133.89	3.99	0.009 *	0.457
Group × time	2	122.52	0.50	0.612	0.251
Session × time	2	122.52	3.73	0.027 *	0.440
Group × session × time	3	122.52	0.55	0.645	0.243
DASS (depression) score					
Group	2	30.17	1.43	0.254	0.292
Session	2	35.08	1.76	0.188	0.338
Group × session	3	35.16	1.35	0.274	0.237
DASS (anxiety) score					
Group	2	37.03	2.18	0.127	0.295
Session	2	43.19	6.90	0.003 *	0.433
Group × session	3	42.910	0.51	0.678	0.272
DASS (stress) score					
Group	2	34.81	0.48	0.625	0.108
Session	2	40.91	12.13	<0.001 *	0.315
Group × session	3	41.04	0.94	0.431	0.141
SF-12					
Group	2	154	3.75	0.026 *	0.213
Session	2	154	6.36	0.002 *	0.456
Composite summary	1	154	61.65	<0.001 *	0.642
Group × session	3	154	1.25	0.293	0.222
Group × composite summary	2	154	2.46	0.089	0.241
Session × composite summary	2	154	0.41	0.668	0.359
Group × session × composite summary	3	154	0.221	0.882	0.227

* = indicate significant results ($p < 0.05$), df = Degrees of freedom.

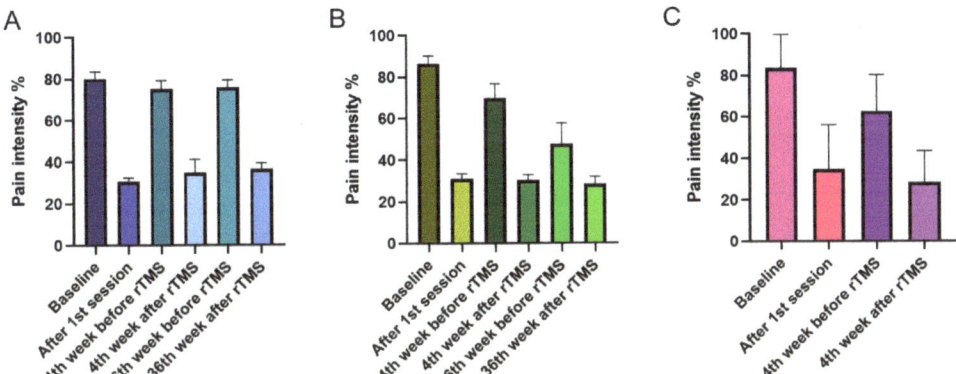

Figure 3. rTMS effects on pain intensity in the M1 (**A**), DLPFC (**B**), and sham group (**C**). The y-axis indicated the mean pain intensity expressed as a percentage (%). The x-axis indicates the time points (in weeks) of conducted measurements. Error bars depict the standard error of mean (SEM).

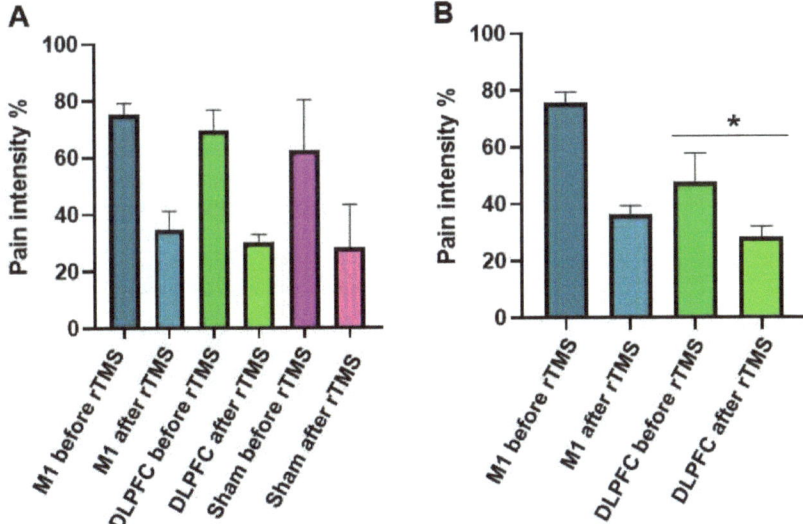

Figure 4. rTMS effects on pain intensity between the groups on the 4th week (**A**) and 36th week (**B**). The y-axis indicates the mean pain intensity expressed as a percentage (%). The x-axis depicts the time points (in weeks) of conducted measurements. The * symbols indicate significant group differences (Bonferroni corrected, two-tailed, paired t-test, $p \leq 0.05$). Error bars depict the standard error of mean (SEM).

For the AUC, there were no significant group differences (all $p \geq 0.05$) from baseline until the end of the daily (5th day) experimental session (Figure 5B). In contrast, the AUC from the 6th to the 13th experimental session was significantly smaller in the DLPFC-group compared to the M1-group ($p = 0.007$) (Figure 5B).

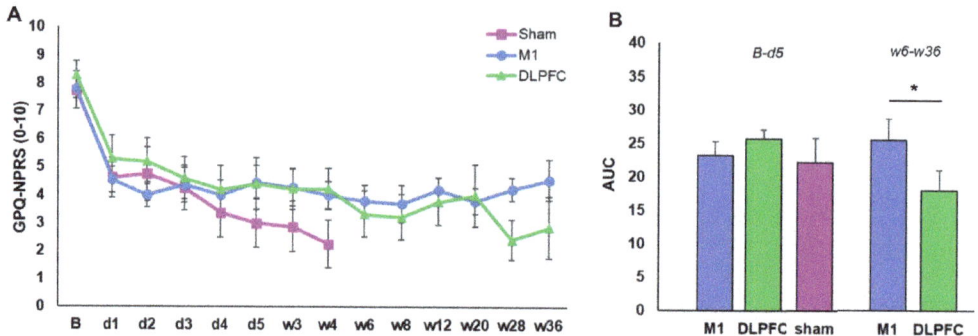

Figure 5. rTMS effects on pain intensity. (**A**) Mean GPQ-NPRS across sessions and stimulation conditions. (**B**) The area under the curve (AUC) across stimulation conditions from baseline up to the 5th day of stimulation (left) and from the 6th week until the 36th of stimulation (right). B = baseline; d = day; w = week; GPQ-NPRS = German Pain Questionnaire-Numerical Pain Rating scale; M1 = primary motor cortex; DLPFC = dorsolateral prefrontal cortex; Error bars denote standard error of mean (SEM). The * symbol indicates significant group differences ($p \leq 0.05$).

3.2. DASS

Pearson Chi-Square test results indicated goodness of fit of the models for the depression score [$X^2 (61) = 7.043$, $p = 0.115$], anxiety score [$X^2 (62) = 6.851$, $p = 0.111$], and stress score [$X^2 (65) = 6.809$, $p = 0.105$]. For the depression score, the model revealed no significant results (Table 2). In contrast, the main effect of the factor session on the anxiety score was significant ($F (2, 43.19) = 6.90$, $p = 0.003$, $d = 0.433$). Bonferroni corrected post hoc t-test showed a significant decrease ($p = 0.001$) in anxiety score on the 4th week (4.09) of stimulation compared to baseline (6.47). This indicates a change of overall anxiety level from moderate to mild (Figure 6B). The main effect of the factor session on the stress score was also significant ($F (2, 40.91) = 12.19$, $p \leq 0.001$, $d = 0.315$). Bonferroni corrected post hoc t-test showed a significant decrease in stress score on the 4th week (6.84, $p = 0.001$) and 36th week (7.51, $p = 0.018$) compared to baseline (10.17). These results indicate a change in overall stress level from moderate at baseline to normal on the 4th and 36th week of stimulation (Figure 6C).

3.3. SF-12

For the SF-12 questionnaire, we initially modelled PCS and MCS, separately. However, Pearson Chi-Square test indicated that the goodness of fit of this model to the PCS [$X^2 (69) = 0.820$, $p = 0.012$] and MCS [$X^2 (69) = 0.955$, $p = 0.014$] data were violated. Thus, we modelled the two data sets together. In this full model, the between-subject factor "group" (sham, M1, and DLPFC) and the within-subject factor "session" (baseline, after 4th and after 36th week), and "composite summary" (physical and mental) are the fixed factors. The goodness-of-fit of the full model is better than individual models [$X^2 (140) = 16435.639$, $p \geq 0.999$]. The results of the full model showed a significant main effect of group ($F (2, 154) = 3.75$, $p = 0.026$, $d = 0.213$). Bonferroni corrected post hoc t-test indicated a significantly higher ($p = 0.016$) overall score in the DLPFC-group (40.47) than the sham group (35.06). The main effect of session was also significant ($F (2, 154) = 3.36$, $p = 0.002$, $d = 0.456$) as indicated by the significantly higher overall score in the 4th (39.34, $p = 0.012$) and 36th week (40.69, $p = 0.031$) compared to baseline (35.87). Similarly, the main effect of composite summary was significant ($F (1, 154) = 61.65$, $p \leq 0.001$, $d = 0.642$) as indicated by the significantly higher ($p \leq 0.001$) MCS (44.89) than PCS (31.86) (Figure 7A,B). Additional exploratory post hoc t-test showed that the significantly higher ($p = 0.001$) MCS (Figure 7B) in the DLPFC-group (49.12) than the M1-group (39.46) is the one driving the marginally significant interaction effect of group × composite summary ($F (2, 154) = 2.46$, $p = 0.089$, $d = 0.241$).

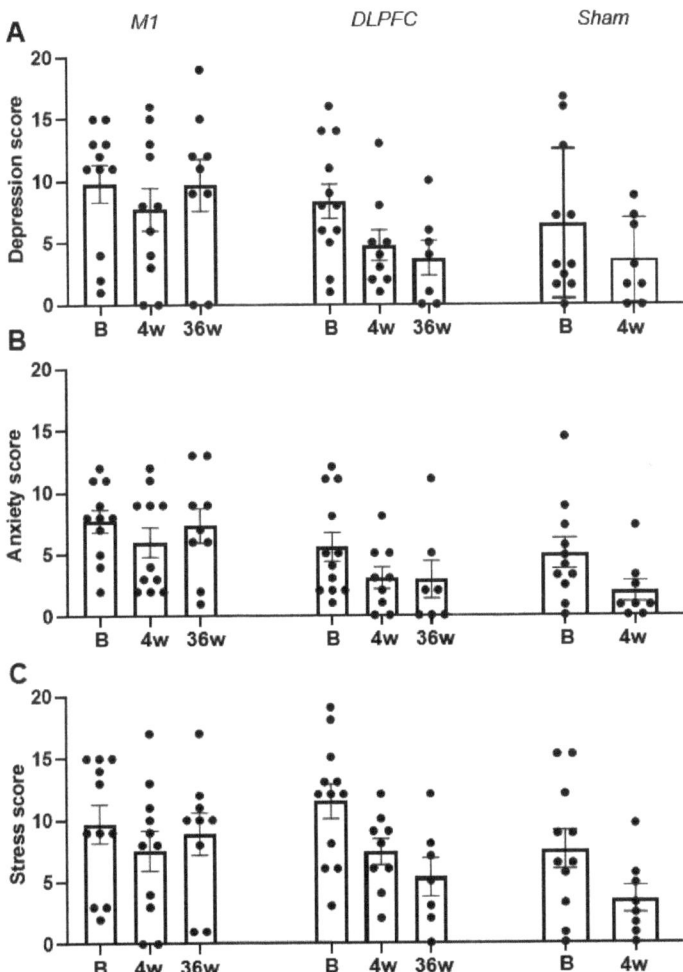

Figure 6. rTMS effects on DASS score for depression (**A**), anxiety (**B**), and stress (**C**). The y-axis indicates the mean score, and the x-axis depicts the time points (in weeks) of conducted measurements. Each dot shows the individual patients' score. B = baseline; w = week; M1 = primary motor cortex; DLPFC = dorsolateral prefrontal cortex; Error bars depict the standard error of mean (SEM).

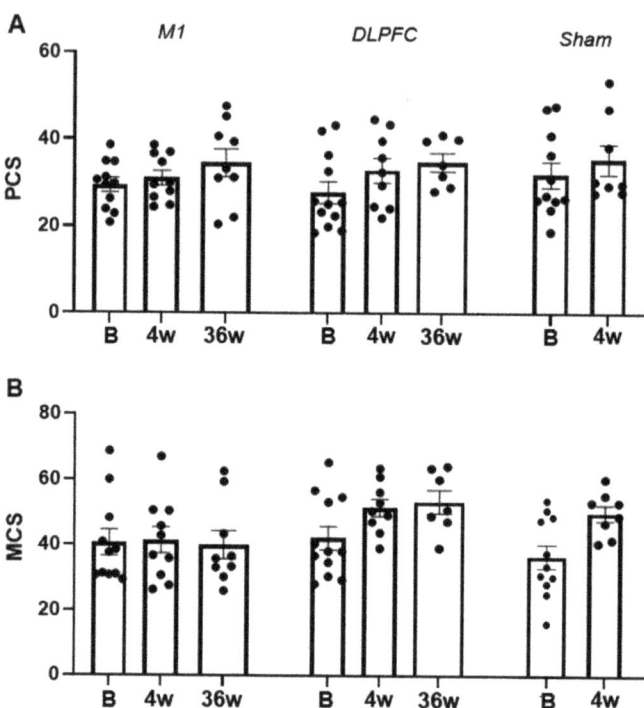

Figure 7. rTMS effects on SF-12 physical composite summary (**A**) and mental composite summary (**B**). The y-axis indicates the mean score, and the x-axis indicates the time points (in weeks) of conducted measurements. Each dot indicates the individual patients' summary. PCS = physical composite summary; MCS = mental composite summary; B = baseline; w = week; M1 = primary motor cortex; DLPFC = dorsolateral prefrontal cortex; Error bars denote mean ± SEM.

4. Discussion

This study investigated the potential of multi-session high-frequency rTMS over M1 and DLPFC to treat chronic pain patients. To our knowledge, this is the first investigation of analgesic effects of rTMS applied to these brain regions using an MRI-guided neuronavigation method in a single study. In the active treatment groups (M1 and DLPFC-group), we monitored changes in the sensory-discriminative aspect of pain utilising the GPQ and affective/emotional aspect using the DASS and SF-12 questionnaires for 36 weeks. Assessment in the Sham-group ended after 4 weeks. Compared to left M1 stimulation, the results showed that 36 weeks of left DLPFC stimulation could reduce pain perception and improve health-related quality of life.

In the earlier periods of stimulation, pain perception is reduced as indicated by the reduction in pain intensity scale at baseline and on the 4th week post-stimulation, as well as by the decrease in the AUC covering the first five days of stimulation. However, the reduction in pain perception was comparable between the three groups, which can be due to a robust placebo effect in the sham group or minimal effect of active stimulation in the M1 and DLPFC-group. Robust placebo effect cannot be rule out since a non-significant trend in pain reduction is evident in the sham group in the early period of the study. Placebo effect is common in pain therapies and is thought to arise from expectancy-induced analgesia [47,48]. For instance, the initial decline in GPQ scores in all groups from baseline to day 1 can be attributed to expectancy-induced analgesia because single-dose high-frequency rTMS stimulation only has minimal effects for chronic pain [35]. In brain stimulation studies, the placebo effect may also arise from the modulation of pain perception due to attentional bias caused by the clicking sound of the TMS coil, which may distract or pull the patients' attention away from the pain. Salient stimuli can disengage the patients from pain signals resulting in altered pain ratings and variations in pain responses [49]. This scenario is possible in our sham group since the tilted coil (active sham) produces a clicking sound even at reduced stimulation intensity.

The possibility of M1 and DLPFC stimulation having minimal effects on pain perception during the early sessions, on the other hand, find support from a meta-analysis of pain studies showing that high-frequency multiple-dose rTMS (e.g., five consecutive days of stimulation) only had minor short-term effects on chronic pain [35]. The authors concluded that the effects do not clearly exceed the predetermined threshold of minimal clinical significance. In contrast, four other review papers reported pain improvement after rTMS treatment, especially in M1 [31–34]. However, the overall findings of these reviews must be taken with caution because of highly variable rTMS parameters and types of targeted pain across trials. Nevertheless, studies that used stimulation parameters similar to our study for M1 stimulation (20 Hz rTMS at 80–90% RMT applied consecutively for five days) reported pain relief in patients with phantom pain [50], irritable bowel syndrome but limited to those who are hypersensitive [51], diabetic neuropathy [52], central pain after stroke [53], orofacial pain [54], and bladder pain syndrome [55]. For LBP, although there were studies included in the reviews that showed pain relief with 1 Hz and 10 Hz rTMS over M1, the evidence for the efficacy of 20 Hz rTMS for treating chronic LBP is only reported by Ambriz-Tututi and colleagues (2016) and therefore still insufficient. Nonetheless, the non-significant trend in pain reduction we observed in the M1-group (relative to sham) could be reminiscent of the effect shown in their study and requires further exploration.

Concerning the effect of stimulation in later sessions, we observed that pain perception reduction is lower in magnitude in the M1-group than the DLPFC-group. For the M1-group, the weaker effect was persistent since pre- and post-stimulation pain perception at baseline, 4th and 36th week has a waxing and waning pattern (Figure 3A), indicating that pain relief was temporary or only within-session. The absence of an observable reduction in the AUC at later sessions in this group was also suggestive of the transient analgesic effect of M1 stimulation. The impact of repeated M1 stimulation (13 sessions) on pain, particularly in later sessions, is difficult to reconcile with the findings of previous reviews since no study aside from Ambriz-Tututi et al. (2016) have the same experimental design. Nonetheless, the results in the M1-group resemble the short-term reduction in pain intensity reported by studies that used single-dose high-frequency stimulation [31,35]. This could suggest that a cumulative effect of repeated M1 stimulation was not achieved in our study, which is not in accordance with the significant build-up of analgesic effect shown by Ambriz-Tututi et al. (2016). Our methodological approach and stimulation parameters were comparable to their study; therefore, nonconforming results may have been influenced by sample size differences. Participants in their M1-group ($n = 28$) constituted a relatively larger sample size than in our group ($n = 11$). Alternatively, the effectiveness of rTMS for pain relief can be influenced by pain chronicity [56]. Patients with various pain duration histories may respond differently to the stimulation due to the differences in the degree of motor-cortex reorganization or excitability changes (increased excitability and decreased intracortical inhibition). In principle, patients with an extensive reorganization of trunk-muscle representation in M1 may not be amenable to plastic changes induced by rTMS. This scenario is remote since the mean pain duration of patients in our M1-group (5.2 years) is lower than in their group (7.1 years). An alternative theoretical explanation would be that pain-induced functional remodelling of M1 is already finished in patients with longer chronicity giving rTMS a more stable neuronal network to induce plasticity. In contrast, in patients with shorter pain chronicity, M1 is still undergoing functional remodelling, making it an unstable neuronal network to induce plasticity. Moreover, the duration of the lack of somatosensory input, disuse of the limb, and loss of muscle targets may lead to differential changes in M1 excitability between patients [56].

For the DLPFC-group, pain perception reduction was robust compared to that of the M1-group at later sessions. This was indicated by the significantly lower pain intensity scale compared to baseline and compared to those of the M1 group at the same time point. In the DLPFC-group, although post-stimulation pain perception at baseline, 4th and 36th weeks were comparable, there was an evident decline in the pre-stimulation pain perception suggesting that pain no longer reverts to baseline level after each session (Figure 3B). There was also a steady reduction in AUC, further indicating pain relief over time. In summary, these results suggest that multi-session stimulation of the left DLPFC has a cumulative analgesic effect. The DLPFC plays a role in "keeping pain out of mind" by modulation of the cortico-subcortical and cortico-cortical pathways, employing both somatosensory (non-emotional) areas and areas that process emotionally salient stimuli [49,57]. For example, stimulation of the DLPFC may transynaptically modulate the medial prefrontal cortex (mPFC), the brain region best reflecting high magnitude of back pain and the anterior cingulate (ACC), which is dubbed as the main brain region signalling pain, or emotional pain [58]. Moreover, high left DLPFC activity has been shown to reduce the inter-regional correlation of midbrain and medial thalamic activity through a "top-down" mode of inhibition. Therefore, high-frequency rTMS stimulation of the left DLPFC may dampen the effective connectivity of the midbrain-medial thalamic pathway that convey greater affective reactions [57]. High-frequency rTMS of the left DLPFC is also reported to induce dopamine release in several pain-relevant brain areas, including the ipsilateral ACC, medial orbitofrontal cortex, and caudate nucleus [59,60]. Dopamine can have two possible sites of action: peripheral and central. Basal ganglia dopaminergic activity is involved in pain processing and variations in the emotional aspects of pain stimuli, the nigrostriatal dopamine D2 receptor activation to the sensory aspect of pain, while mesolimbic dopamine D2/D3 receptor activity is related to negative affect and fear [49,61,62]. Peripherally, dopaminergic activity may alter pain response due to its potential effect on blood flow and nociception [61,63,64].

Concerning the modulation of pain's affective/emotional aspect by rTMS, the DASS survey only showed overall improvement in anxiety (moderate to mild) and stress (moderate to normal). A significant change in anxiety level is observed between baseline and the 4th week without group-specificity. For stress, the significant decrease on the 36th week compared to baseline is only driven by the M1 and DLPFC-group since there were no measurements for the sham-group at this period. Although the differences between the groups did not differ statistically, it was evident that the mean stress level on the 36th week was lower in the DLPFC group (6.50) than the M1-group (8.50), indicating normal and mild stress levels, respectively. Improvement in stress level was only present in the DLPFC group, from moderate at baseline to normal on the 36th week. In contrast, in the M1-group, stress levels did not change from baseline to 36th week (both mild). Modulation of brain structures linked to the affective/emotional aspect of pain, such as the cingulate cortex through cortico-subcortical pathways, can directly account for stress level improvement in the DLPFC-group. Imaging studies provided evidence that left DLPFC rTMS also affects blood flow and metabolism in the ACC [65]. The ACC is suggested to be involved in anticipation of pain and higher activity in its anterior and middle segments (including those in the insula) at rest is considered a sign of distorted resting-state network in chronic pain patients [58,66]. Pain anticipation in chronic pain patients is stressful because it is cognitively demanding and may lead to sustained emotional suffering [58]. Reduction in stress level and pain perception in the DLPFC-group may explain why patients in this group (relative to sham) reported a significantly better overall health state in the SF-12 questionnaires. In the 36th week, the DLPFC-group has a superior mental composite summary than the M1-group, which is somehow expected because the emotional and social functioning aspect of pain (vitality/energy, social function, mental health/emotional well-being, and role limitations) is more accessible through DLPFC than M1 stimulation.

5. Conclusions

The results of the present study indicate that multi-session rTMS of the left DLPFC leads to significant improvement in pain perception and stress level reduction. These effects are better than those obtained from left M1 stimulation, where no effective pain relief was elicited. This indicates an advantage of the DLPFC as a target area for pain rehabilitation by multi-session rTMS. However, the following limitations of our study must be taken into account. First, there were no measurements from the sham group at later sessions. We considered this a significant drawback of the study because comparisons in those time points are only limited between the M1- and DLPFC-group. In our opinion, data comparisons are not entirely non-trivial because the sham and M1 stimulation (both stimulated left M1) have comparable effects at early time points, while the comparison of data from two separate brain areas (M1 vs. DLPFC) finally revealed significant differences. Second, our sample size was relatively small; hence, further studies with a larger population are warranted. Finally, the patients' maintenance medications (e.g., selective serotonin and norepinephrine reuptake inhibitor antidepressants (SSNRI) and analgesics) were not discontinued during the study. There are reports that analgesics (e.g., Tramadol) affects cortical excitability [67]. At the same time, serotogenic and adrenergic drugs were shown to modulate plasticity induced by other brain stimulation techniques such as transcranial direct current stimulation (tDCS) and paired associative stimulation (PAS) [68,69]. The impact of these medications on the after-effect of rTMS is unexplored. Still, we cannot entirely rule out their influence on our findings since brain stimulation paradigms share physiological underpinnings. Future studies must replicate the present results in patients who are off-medication at least 24 h before plasticity induction. In conclusion, the present study emphasizes the potential of other pain-related brain regions as treatment targets in chronic pain patients. The study also highlights the importance of brain stimulation methods to investigate the relationship between pain-related brain regions.

Author Contributions: Conceptualization, S.F., C.L., E.H., A.E., H.B.-C., H.D., G.R., K.Z., I.S. and M.M.; Data curation, S.F.; Formal analysis, S.F., S.M.F. and K.Z.; Investigation, C.L., K.M.A., E.H., A.E., G.R., B.S. and K.Z.; Methodology, S.F., S.M.F., H.B.-C., H.D., I.S. and K.Z.; Project administration, M.M.; Resources, S.F., H.B.-C., H.D., G.R., I.S. and M.M.; Supervision, K.Z. and M.M.; Validation, S.F., K.Z. and M.M.; Visualization, S.F.; Writing—original draft, S.F.; Writing—review and editing, C.L., S.F., K.M.A., A.B., S.K.-L. and A.V. All authors have read and agreed to the published version of the manuscript.

Funding: This research received no external funding.

Institutional Review Board Statement: The study was conducted according to the guidelines of the Declaration of Helsinki and approved by the Institutional Ethics Committee of the Medical University Graz, Austria (30-459-ex 17/18).

Informed Consent Statement: Informed consent was obtained from all subjects involved in the study.

Data Availability Statement: The data presented in this study are available on request from the corresponding author. The data are not publicly available due to privacy and data protection declaration of the trial.

Conflicts of Interest: The authors declare no conflict of interest.

References

1. Raja, S.N.; Carr, D.B.; Cohen, M.; Finnerup, N.B.; Flor, H.; Gibson, S.; Keefe, F.J.; Mogil, J.S.; Ringkamp, M.; Sluka, K.A.; et al. The revised International Association for the Study of Pain definition of pain: Concepts, challenges, and compromises. *Pain* **2020**, *161*, 1976–1982. [CrossRef] [PubMed]
2. Treede, R.D.; Rief, W.; Barke, A.; Aziz, Q.; Bennett, M.I.; Benoliel, R.; Cohen, M.; Evers, S.; Finnerup, N.B.; First, M.B.; et al. Chronic pain as a symptom or a disease: The IASP Classification of Chronic Pain for the International Classification of Diseases (ICD-11). *Pain* **2019**, *160*, 19–27. [CrossRef] [PubMed]
3. Todd, A.; McNamara, C.L.; Balaj, M.; Huijts, T.; Akhter, N.; Thomson, K.; Kasim, A.; Eikemo, T.A.; Bambra, C. The European epidemic: Pain prevalence and socioeconomic inequalities in pain across 19 European countries. *Eur. J. Pain* **2019**, *23*, 1425–1436. [CrossRef] [PubMed]

4. Maher, C.; Underwood, M.; Buchbinder, R. Non-specific low back pain. *Lancet* **2017**, *389*, 736–747. [CrossRef]
5. Manek, N.J.; MacGregor, A.J. Epidemiology of back disorders: Prevalence, risk factors, and prognosis. *Curr. Opin. Rheumatol.* **2005**, *17*, 134–140. [CrossRef]
6. Peck, J.; Urits, I.; Peoples, S.; Foster, L.; Malla, A.; Berger, A.A.; Cornett, E.M.; Kassem, H.; Herman, J.; Kaye, A.D.; et al. A Comprehensive Review of Over the Counter Treatment for Chronic Low Back Pain. *Pain Ther.* **2020**. [CrossRef]
7. Skelly, A.C.; Chou, R.; Dettori, J.R.; Turner, J.A.; Friedly, J.L.; Rundell, S.D.; Fu, R.; Brodt, E.D.; Wasson, N.; Kantner, S.; et al. Noninvasive Nonpharmacological Treatment for Chronic Pain: A Systematic Review. *AHRQ Comp. Eff. Rev.* **2018**. [CrossRef]
8. Chou, R. Nonpharmacologic Therapies for Low Back Pain. *Ann. Intern. Med.* **2017**, *167*, 604–605. [CrossRef]
9. Mens, J.M.A. The use of medication in low back pain. *Best Pract. Res. Clin. Rheumatol.* **2005**, *19*, 609–621. [CrossRef] [PubMed]
10. Kearney, P.M.; Baigent, C.; Godwin, J.; Halls, H.; Emberson, J.R.; Patrono, C. Do selective cyclo-oxygenase-2 inhibitors and traditional non-steroidal anti-inflammatory drugs increase the risk of atherothrombosis? Meta-analysis of randomised trials. *BMJ* **2006**, *332*, 1302–1308. [CrossRef] [PubMed]
11. Chou, R.; Huffman, L.H. Medications for acute and chronic low back pain: A review of the evidence for an American Pain Society/American College of Physicians clinical practice guideline. *Ann. Intern. Med.* **2007**, *147*, 505–514. [CrossRef]
12. Chopko, B.; Liu, J.C.; Khan, M.K. Anatomic surgical management of chronic low back pain. *Neuromodulation* **2014**, *17*, 46–51. [CrossRef] [PubMed]
13. Younger, J.; Mccue, R.; Mackey, S. Pain Outcomes: A Brief Review of Instruments and Techniques. *Curr. Pain Headache Rep.* **2009**, *13*, 39–43. [CrossRef] [PubMed]
14. Echeita, J.A.; Preuper, H.R.S.; Dekker, R.; Stuive, I.; Timmerman, H.; Wolff, A.P.; Reneman, M.F. Central Sensitisation and functioning in patients with chronic low back pain: Protocol for a cross-sectional and cohort study. *BMJ Open* **2020**, *10*, 1–13. [CrossRef]
15. Apkarian, A.V. Chronic Pain and Neuroplasticity. *NIH Public Access* **2012**, *152*, 1–35. [CrossRef]
16. Zhao, Z.; Huang, T.; Tang, C.; Ni, K.; Pan, X.; Yan, C.; Fan, X.; Xu, D.; Luo, Y. Altered resting-state intra- and inter- network functional connectivity in patients with persistent somatoform pain disorder. *PLoS ONE* **2017**, *12*, e0176494. [CrossRef] [PubMed]
17. Ye, Q.; Yan, D.; Yao, M.; Lou, W.; Peng, W. Hyperexcitability of Cortical Oscillations in Patients with Somatoform Pain Disorder: A Resting-State EEG Study. *Neural Plast.* **2019**, *2019*. [CrossRef] [PubMed]
18. Flor, H.; Braun, C.; Elbert, T.; Birbaumer, N. Extensive reorganization of primary somatosensory cortex in chronic back pain patients. *Neurosci. Lett.* **1997**, *224*, 5–8. [CrossRef]
19. Flor, H. The modification of cortical reorganization and chronic pain by sensory feedback. *Appl. Psychophysiol. Biofeedback* **2002**, *27*, 215–227. [CrossRef] [PubMed]
20. Vartiainen, N.; Kirveskari, E.; Kallio-Laine, K.; Kalso, E.; Forss, N. Cortical Reorganization in Primary Somatosensory Cortex in Patients With Unilateral Chronic Pain. *J. Pain* **2009**, *10*, 854–859. [CrossRef]
21. Chang, W.J.; O'Connell, N.E.; Beckenkamp, P.R.; Alhassani, G.; Liston, M.B.; Schabrun, S.M. Altered Primary Motor Cortex Structure, Organization, and Function in Chronic Pain: A Systematic Review and Meta-Analysis. *J. Pain* **2018**, *19*, 341–359. [CrossRef]
22. Tsao, H.; Galea, M.P.; Hodges, P.W. Reorganization of the motor cortex is associated with postural control deficits in recurrent low back pain. *Brain* **2008**, *131*, 2161–2171. [CrossRef]
23. Tsao, H.; Hodges, P.W. Immediate changes in feedforward postural adjustments following voluntary motor training. *Exp. Brain Res.* **2007**, *181*, 537–546. [CrossRef]
24. Parker, R.S.; Lewis, G.N.; Rice, D.A.; McNair, P.J. Is Motor Cortical Excitability Altered in People with Chronic Pain? A Systematic Review and Meta-Analysis. *Brain Stimul. Basic Transl. Clin. Res. Neuromodulation* **2016**, *9*, 488–500. [CrossRef] [PubMed]
25. Fukui, S.; Matsuno, M.; Inubushi, T.; Nosaka, S. N-Acetylaspartate concentrations in the thalami of neuropathic pain patients and healthy comparison subjects measured with 1H-MRS. *Magn. Reson. Imaging* **2006**, *24*, 75–79. [CrossRef]
26. Sharma, N.K.; Brooks, W.M.; Popescu, A.E.; VanDillen, L.; George, S.Z.; McCarson, K.E.; Gajewski, B.J.; Gorman, P.; Cirstea, C.M. Neurochemical analysis of primary motor cortex in chronic low back pain. *Brain Sci.* **2012**, *2*, 319–331. [CrossRef] [PubMed]
27. Valdés, M.; Collado, A.; Bargalló, N.; Vázquez, M.; Rami, L.; Gómez, E.; Salamero, M. Increased glutamate/glutamine compounds in the brains of patients with fibromyalgia: A magnetic resonance spectroscopy study. *Arthritis Rheum.* **2010**, *62*, 1829–1836. [CrossRef]
28. Huang, Y.; Chen, R.; Rothwell, J.; Wen, H. The after-effect of human theta burst stimulation is NMDA receptor dependent. *Clin. Neurophysiol.* **2007**, *118*, 1028–1032. [CrossRef] [PubMed]
29. Chen, R.; Classen, J.; Gerloff, C.; Celnik, P.; Wassermann, E.M.; Hallett, M.; Cohen, L.G. Depression of motor cortex excitability by low-frequency transcranial magnetic stimulation. *Neurology* **1997**, *48*, 1398–1403. [CrossRef]
30. Pascual-Leone, A.; Amedi, A.; Fregni, F.; Merabet, L.B. The Plastic Human Brain Cortex. *Annu. Rev. Neurosci.* **2005**, *28*, 377–401. [CrossRef]
31. Goudra, B.; Shah, D.; Balu, G.; Gouda, G.; Balu, A.; Borle, A.; Singh, P. Repetitive transcranial magnetic stimulation in chronic pain: A meta-analysis. *Anesth. Essays Res.* **2017**, *11*, 751. [CrossRef] [PubMed]
32. Yang, S.; Chang, M.C. Effect of Repetitive Transcranial Magnetic Stimulation on Pain Management: A Systematic Narrative Review. *Front. Neurol.* **2020**, *11*, 114. [CrossRef]

33. Galhardoni, R.; Correia, G.S.; Araujo, H.; Yeng, L.T.; Fernandes, D.T.; Kaziyama, H.H.; Marcolin, M.A.; Bouhassira, D.; Teixeira, M.J.; De Andrade, D.C. Repetitive transcranial magnetic stimulation in chronic pain: A review of the literature. *Arch. Phys. Med. Rehabil.* **2015**, *96*, S156–S172. [CrossRef] [PubMed]
34. Jin, Y.; Xing, G.; Li, G.; Wang, A.; Feng, S.; Tang, Q.; Liao, X.; Guo, Z.; McClure, M.A.; Mu, Q. High frequency repetitive transcranial magnetic stimulation therapy for chronic neuropathic pain: A meta-analysis. *Pain Physician* **2015**, *18*, E1029–E1046. [PubMed]
35. O'Connell, N.E.; Wand, B.M.; Marston, L.; Spencer, S.; Desouza, L.H. Non-invasive brain stimulation techniques for chronic pain. *Cochrane Database Syst. Rev.* **2010**, CD008208. [CrossRef]
36. Ambriz-Tututi, M.; Alvarado-Reynoso, B.; Drucker-Colín, R. Analgesic effect of repetitive transcranial magnetic stimulation (rTMS) in patients with chronic low back pain. *Bioelectromagnetics* **2016**, *37*, 527–535. [CrossRef]
37. Nardone, R.; Höller, Y.; Langthaler, P.B.; Lochner, P.; Golaszewski, S.; Schwenker, K.; Brigo, F.; Trinka, E. RTMS of the prefrontal cortex has analgesic effects on neuropathic pain in subjects with spinal cord injury. *Spinal Cord* **2017**, *55*, 20–25. [CrossRef]
38. Borckardt, J.J.; Smith, A.R.; Reeves, S.; Madan, A.; Shelley, N.; Branham, R.; Nahas, Z.; George, M.S. A pilot study investigating the effects of fast left prefrontal rTMS on chronic neuropathic pain. *Pain Med.* **2009**, *10*, 840–849. [CrossRef] [PubMed]
39. Brighina, F.; De Tommaso, M.; Giglia, F.; Scalia, S.; Cosentino, G.; Puma, A.; Panetta, M.; Giglia, G.; Fierro, B. Modulation of pain perception by transcranial magnetic stimulation of left prefrontal cortex. *J. Headache Pain* **2011**, *12*, 185–191. [CrossRef]
40. Casser, H.R.; Hüppe, M.; Kohlmann, T.; Korb, J.; Lindena, G.; Maier, C.; Nagel, B.; Pfingsten, M.; Thoma, R. German pain questionnaire and standardised documentation with the KEDOQ-Schmerz. A way for quality management in pain therapy. *Schmerz* **2012**, *26*, 168–175. [CrossRef]
41. Nilges, P.; Essau, C. Die Depressions-Angst-Stress-Skalen: Der DASS—Ein Screeningverfahren nicht nur für Schmerzpatienten. *Schmerz* **2015**, *10*, 649–657. [CrossRef] [PubMed]
42. Lovibond, P.F.; Lovibond, S.H. The structure of negative emotional states: Comparison of the Depression Anxiety Stress Scales (DASS) with the Beck Depression and Anxiety Inventories. *Behav. Res. Ther.* **1995**, *33*, 335–343. [CrossRef]
43. Turner-Bowker, D.; Hogue, S.J. Short Form 12 Health Survey (SF-12). In *Encyclopedia of Quality of Life and Well-Being Research*; Springer: Dordrecht, The Netherlands, 2014; pp. 5954–5957.
44. Pagels, A.A.; Söderkvist, B.; Medin, C.; Hylander, B.; Heiwe, S. Health-related quality of life in different stages of chronic kidney disease and at initiation of dialysis treatment. *Health Qual. Life Outcomes* **2012**, *10*, 71. [CrossRef] [PubMed]
45. Warton, D.I.; Lyons, M.; Stoklosa, J.; Ives, A.R. Three points to consider when choosing a LM or GLM test for count data. *Methods Ecol. Evol.* **2016**, *7*, 882–890. [CrossRef]
46. Searle, S.R. Mixed models and unbalanced data: Wherefrom, whereat and whereto? *Commun. Stat. Theory Methods* **1988**, *17*, 935–968. [CrossRef]
47. Colloca, L. The placebo effect in pain therapies. *Annu. Rev. Pharmacol. Toxicol.* **2019**, *59*, 191–211. [CrossRef] [PubMed]
48. Colloca, L.; Lopiano, L.; Lanotte, M.; Benedetti, F. Overt versus covert treatment for pain, anxiety, and Parkinson's disease. *Lancet Neurol.* **2004**, *3*, 679–684. [CrossRef]
49. Ahmad, A.H.; Abdul Aziz, C.B. The Brain In Pain. *Back Lett.* **2008**, *23*, 1. [CrossRef]
50. Ahmed, M.A.; Mohamed, S.A.; Sayed, D. Long-term antalgic effects of repetitive transcranial magnetic stimulation of motor cortex and serum beta-endorphin in patients with phantom pain. *Neurol. Res.* **2011**, *33*, 953–958. [CrossRef]
51. Melchior, C.; Gourcerol, G.; Chastan, N.; Verin, E.; Menard, J.F.; Ducrotte, P.; Leroi, A.M. Effect of transcranial magnetic stimulation on rectal sensitivity in irritable bowel syndrome: A randomized, placebo-controlled pilot study. *Color. Dis.* **2014**, *16*, O104–O111. [CrossRef]
52. Onesti, E.; Gabriele, M.; Cambieri, C.; Ceccanti, M.; Raccah, R.; Di Stefano, G.; Biasiotta, A.; Truini, A.; Zangen, A.; Inghilleri, M. H-coil repetitive transcranial magnetic stimulation for pain relief in patients with diabetic neuropathy. *Eur. J. Pain* **2013**, *17*, 1347–1356. [CrossRef] [PubMed]
53. Khedr, E.M.; Kotb, H.; Kamel, N.F.; Ahmed, M.A.; Sadek, R.; Rothwell, J.C. Longlasting antalgic effects of daily sessions of repetitive transcranial magnetic stimulation in central and peripheral neuropathic pain. *J. Neurol. Neurosurg. Psychiatry* **2005**, *76*, 833–838. [CrossRef] [PubMed]
54. Fricová, J.; Klírová, M.; Masopust, V.; Novák, T.; Vérebová, K.; Rokyta, R. Repetitive Transcranial Magnetic Stimulation in the Treatment of Chronic Orofacial Pain. *Physiol. Res.* **2013**, S125–S134. [CrossRef] [PubMed]
55. Cervigni, M.; Onesti, E.; Ceccanti, M.; Gori, M.C.; Tartaglia, G.; Campagna, G.; Panico, G.; Vacca, L.; Cambieri, C.; Libonati, L.; et al. Repetitive transcranial magnetic stimulation for chronic neuropathic pain in patients with bladder pain syndrome/interstitial cystitis. *Neurourol. Urodyn.* **2018**, *37*, 2678–2687. [CrossRef] [PubMed]
56. Mercier, C.; Léonard, G. Interactions between Pain and the Motor Cortex: Insights from Research on Phantom Limb Pain and Complex Regional Pain Syndrome. *Physiother. Can.* **2011**, *63*, 305–314. [CrossRef]
57. Lorenz, J.; Minoshima, S.; Casey, K.L. Keeping pain out of mind: The role of the dorsolateral prefrontal cortex in pain modulation. *Brain* **2003**, *126*, 1079–1091. [CrossRef]
58. Apkarian, V.A.; Hashmi, J.A.; Baliki, M.N. Pain and the brain: Specificity and plasticity of the brain in clinical chronic pain. *Pain* **2011**, *152*, S49–S64. [CrossRef]
59. Cho, S.S.; Strafella, A.P. rTMS of the Left Dorsolateral Prefrontal Cortex Modulates Dopamine Release in the Ipsilateral Anterior Cingulate Cortex and Orbitofrontal Cortex. *PLoS ONE* **2009**, *4*, e6725. [CrossRef]

60. Strafella, A.P.; Paus, T.; Barrett, J.; Dagher, A. Repetitive Transcranial Magnetic Stimulation of the Human Prefrontal Cortex Induces Dopamine Release in the Caudate Nucleus. *J. Neurosci.* **2001**, *21*, RC157. [CrossRef]
61. Haddad, M.; Pud, D.; Treister, R.; Suzan, E.; Eisenberg, E. The effects of a dopamine agonist (apomorphine) on experimental and spontaneous pain in patients with chronic radicular pain: A randomized, double-blind, placebo-controlled, cross-over study. *PLoS ONE* **2018**, *13*, e0195287. [CrossRef]
62. Scott, D.J.; Heitzeg, M.M.; Koeppe, R.A.; Stohler, C.S.; Zubieta, J.-K. Variations in the Human Pain Stress Experience Mediated by Ventral and Dorsal Basal Ganglia Dopamine Activity. *J. Neurosci.* **2006**, *26*, 10789–10795. [CrossRef]
63. Charbit, A.R.; Akerman, S.; Goadsby, P.J. Comparison of the Effects of Central and Peripheral Dopamine Receptor Activation on Evoked Firing in the Trigeminocervical Complex. *J. Pharmacol. Exp. Ther.* **2009**, *331*, 752–763. [CrossRef] [PubMed]
64. Main, D.C.J.; Waterman, A.E.; Kilpatrick, I.C.; Jones, A. An assessment of the peripheral antinociceptive potential of remoxipride, clonidine and fentanyl in sheep using the forelimb tourniquet. *J. Vet. Pharmacol. Ther.* **1997**, *20*, 220–228. [CrossRef] [PubMed]
65. Kimbrell, T.A.; Dunn, R.T.; George, M.S.; Danielson, A.L.; Willis, M.W.; Repella, J.D.; Benson, B.E.; Herscovitch, P.; Post, R.M.; Wassermann, E.M. Left prefrontal-repetitive transcranial magnetic stimulation (rTMS) and regional cerebral glucose metabolism in normal volunteers. *Psychiatry Res.* **2002**, *115*, 101–113. [CrossRef]
66. Malinen, S.; Vartiainen, N.; Hlushchuk, Y.; Koskinen, M.; Ramkumar, P.; Forss, N.; Kalso, E.; Hari, R. Aberrant temporal and spatial brain activity during rest in patients with chronic pain. *Proc. Natl. Acad. Sci. USA* **2010**, *107*, 6493–6497. [CrossRef] [PubMed]
67. Khedr, E.M.; Gabra, R.H.; Noaman, M.; Abo Elfetoh, N.; Farghaly, H.S.M. Cortical excitability in tramadol dependent patients: A transcranial magnetic stimulation study. *Drug Alcohol Depend.* **2016**, *169*, 110–116. [CrossRef] [PubMed]
68. Awiszus, F. Chapter 2 TMS and threshold hunting. In *Transcranial Magnetic Stimulation and Transcranial Direct Current Stimulation*; Paulus, W., Tergau, F., Nitsche, M.A., Rothwell, J.G., Ziemann, U., Hallett, M., Eds.; Elsevier: Amsterdam, The Netherlands, 2003; Volume 56, pp. 13–23.
69. Ziemann, U. TMS and drugs. *Clin. Neurophysiol.* **2004**, *115*, 1717–1729. [CrossRef] [PubMed]

Article

Capturing Neuroplastic Changes after iTBS in Patients with Post-Stroke Aphasia: A Pilot fMRI Study

Shuo Xu [1,†], Qing Yang [1,†], Mengye Chen [1], Panmo Deng [2], Ren Zhuang [3], Zengchun Sun [4], Chong Li [5], Zhijie Yan [6], Yongli Zhang [7] and Jie Jia [1,*]

1. Department of Rehabilitation Medicine, Huashan Hospital, Fudan University, Shanghai 200040, China; xus20@fudan.edu.cn (S.X.); 07301010208@fudan.edu.cn (Q.Y.); babybreathks@126.com (M.C.)
2. Department of Rehabilitation Medicine, Jingan District Central Hospital Affiliated to Fudan University, Shanghai 200040, China; xyz3325127@163.com
3. Department of Rehabilitation Medicine, Changzhou Dean Hospital, Changzhou 213000, China; zr2003@163.com
4. Sichuan Bayi Rehabilitation Center, Affiliated Sichuan Provincial Rehabilitation Hospital of Chengdu University of TCM, Chengdu 610075, China; mikezsun@outlook.com
5. Faculty of Sport and Science, Shanghai University of Sport, Shanghai 200040, China; lichongsus@163.com
6. The Third Affiliated Hospital, Xinxiang Medical University, Xinxiang 453003, China; yzj2020mail@126.com
7. Institute of Rehabilitation, Fujian University of Traditional Chinese Medicine, Fuzhou 350122, China; islilyong@163.com
* Correspondence: shannonjj@126.com; Tel.: +86-136-1172-2357
† These authors contributed equally to this work.

Abstract: Intermittent theta-burst stimulation (iTBS) is a high-efficiency transcranial magnetic stimulation (TMS) paradigm that has been applied to post-stroke aphasia (PSA). However, its efficacy mechanisms have not been clarified. This study aimed to explore the immediate effects of iTBS of the primary motor cortex (M1) of the affected hemisphere, on the functional activities and connectivity of the brains of PSA patients. A total of 16 patients with aphasia after stroke received iTBS with 800 pulses for 300 s. All patients underwent motor, language, and cognitive assessments and resting-state functional MRI scans immediately before and after the iTBS intervention. Regional, seed-based connectivity, and graph-based measures were used to test the immediate functional effects of the iTBS intervention, including the fractional amplitude of low-frequency fluctuation (fALFF), degree centrality (DC), and functional connectivity (FC) of the left M1 area throughout the whole brain. The results showed that after one session of iTBS intervention, the fALFF, DC, and FC values changed significantly in the patients' brains. Specifically, the DC values were significantly higher in the right middle frontal gyrus and parts of the left parietal lobe ($p < 0.05$), while fALFF values were significantly lower in the right medial frontal lobe and parts of the left intracalcarine cortex ($p < 0.05$), and the strength of the functional connectivity between the left M1 area and the left superior frontal gyrus was reduced ($p < 0.05$). Our findings provided preliminary evidences that the iTBS on the ipsilesional M1 could induce neural activity and functional connectivity changes in the motor, language, and other brain regions in patients with PSA, which may promote neuroplasticity and functional recovery.

Keywords: stroke; aphasia; iTBS; fMRI; neuroplasticity; rehabilitation

1. Introduction

Patients with post-stroke aphasia (PSA) often have impaired upper extremity (UE) motor and cognitive function [1–3]. Non-invasive brain stimulation (NIBS), specifically transcranial magnetic stimulation (TMS) and transcranial direct current stimulation (tDCS), are emerging as promising NIBS modalities in treating the language and motor dysfunction of stroke patients [4–8]. Intriguingly, not only have the NIBS targeting the "motor" or "language" areas been proven beneficial for improving the patients' motor or language performances, respectively [2], but recent studies have also reported that NIBS targeting the motor area (i.e., M1) could also improve the language function of the PSA patients [9]. As the M1 area is far easier to be located than classical language areas such as Broca's area, this finding highlighted a potentially practical way to "kill two birds with one stone", in treating the language and motor impairments of patients with stroke. However, although the structural and functional connectivity have been identified between the brain's language and motor areas in healthy volunteers and the patients with various neurological disorders [10–13], it is still not clear if and how the NIBS targeting the M1 would affect the language function via these connections. Therefore, we conducted this preliminary functional MRI study to investigate the immediate brain functional effects of the left M1 stimulation with intermittent theta-burst stimulation (iTBS), which is a popular TMS method, in treating PSA using resting-state functional magnetic resonance (fMRI) images.

Intermittent theta-burst stimulation (iTBS) is a popular new TMS intervention paradigm, with a high within burst frequency usually at 50 Hz and a between burst frequency usually at 5 Hz. With seconds of intervals between trains of intervention, it has been proposed to be able to induce excitatory effects in the area of stimulation [14–17]. A standard iTBS session usually takes about three minutes to apply, and this time duration is easier for the patients to cooperate. This quick application is a major advantage of iTBS that makes it increasingly applied and reported in treating patients with various neurological disorders including stroke [17–19]. iTBS have been reported to be able to alter cortical excitability at the site of stimulation, as well as in its surrounding and even connected remote regions, [20]. Such effects have been also reported in other TMS paradigms. For example, Wang et al. [21] revealed the enhanced cortico-hippocampal functional connectivity (FC) by multi-session excitatory TMS over the lateral parietal cortex. Hawco et al. [22] reported spread TMS-induced cortical changes that were related to the FC between the stimulated site and salience network.

Interestingly, evidence from behavioral brain imaging and brain stimulation studies suggests that primary motor cortex (M1) stimulation by tDCS may represent a promising and clinically feasible approach to enhance language therapy outcomes in post-stroke aphasia [9]. A few studies have also found that repetitive TMS stimulating area M1 can improve speech functions such as verbal fluency and naming to varying degrees, as well as enhance cognitive function to some extent [23–25]. However, there is little research on whether iTBS stimulation of M1 in the affected cortex induces changes in the brains after stroke. Therefore, whether and how the iTBS targeting M1 would induce brain functional changes in PSA patients remains to be investigated.

Brain functional changes following iTBS can be detected using a series of non-invasive imaging and electrophysiological techniques [22,24,26]. In this study, we used resting-state fMRI, which is one of the most popular brain functional imaging methods, to measure the spontaneous neural activities of the brain, with a coverage of the whole cerebral cortex and subcortical structures at a millimeter-level resolution. Previous studies have proved that resting-state fMRI could reveal the spontaneous neural activities and functional connectivity patterns of the brain, which partially reflected the structural connectivity, and correlated with the brain activation pattern and behavioral performances under various task conditions [27,28]. In addition, the "resting-state" does not require the subject to perform specific tasks, which could be easier for patients to cooperate and facilitate comparisons across studies [29,30]. As iTBS and other TMS paradigms have been reported to be able to induce both regional neural activity and functional connectivity changes [31], we included both regional- and connectivity-based metrics to measure the functional changes induced by the M1 iTBS. Specifically, we used the fractional amplitude of low-frequency fluctuation (fALFF) [32] to reflect the regional spontaneous neural activity, while seed-based functional connectivity analyses were used to measure the functional dependence between the M1 and other regions in the brain; and we also adopted the degree centrality (DC), a graph theory-based metrics [33], to measure the relative importance of each region in the brain functional network [34].

Therefore, in our study, we used iTBS to stimulate the left M1 hand area, which is in the affected brain hemisphere of our patients with PSA, and observed its immediate effects on brain activity, using the resting-state fMRI data. This preliminary study aimed to explore the potential therapeutic mechanisms of motor area iTBS for treating patients with PSA.

2. Materials and Methods

2.1. Participants

Patients with post-stroke aphasia were recruited from Huashan Hospital, Fudan University. The inclusion criteria were as follows: (i) aged 18 or older; (ii) native Chinese speakers; (iii) clinical diagnosis as ischemic or hemorrhagic stroke at hospital discharge from the neurological department, confirmed with computed tomography (CT) or magnetic resonance images (MRI); (iv) able to complete the study; (v) aphasia quotient below 93.8 assessed by the Chinese version of western aphasia battery (WAB). The exclusion criteria were as follows: (i) previous stroke; (ii) severe psychiatric condition; (iii) epilepsy, (iv) other neurological disorders, such as Parkinson's disease and Alzheimer's diseases; (v) brain tumor(s) or brain injury; (vi) unstable vital sign(s) or severe heart or renal failure; (vii) metal implants, devices, or other conditions that would forbid the application of iTBS or MRI scanning. During the study, all patients received the same form and intensity of rehabilitation treatment, including speech and language therapies, upper extremity training, gait, and mobility-related functions and activities. All patients or their legal representative(s) provided their written informed consent.

Following recruitment, each participant accepted a TMS evaluation to determine the iTBS parameters before the first session of treatment. The three-step MRI scans were conducted as follows: in the first step, the structural MRI was scanned, while in the second step, fMRI (time point 1, TP1) was scanned, and during the third step, scanning continued with fMRI (time point 2, TP2). iTBS intervention was performed between the second and third steps. No other intervention or training was provided to the participant on the same day of the above MRI scans and iTBS treatment until the study procedure was completed. The participants accepted conventional medical and rehabilitative modes of treatment during hospitalization. This study was approved by the ethics board of HuaShan hospital Fudan University (2019-336), and all procedures conform to the Declaration of Helsinki regarding human experimentation. Our trail registration number is ChiCTR-TRC-2100041936.

2.2. Behavioral Assessments

The Chinese version of WAB, which is a widely used clinical evaluation tool of PSA, was used to assess assessing the presence, type, and severity of aphasia for each participant. The scores on the four subscales of the Western Aphasia Battery were combined to form the aphasia quotient (AQ), which quantified the severity of each patient's aphasia [35]. In addition, the fluency and content scores from the spontaneous speech, auditory comprehension, repetition, and naming subscales of WAB were also used to assess each participant's language abilities.

As the motor and cognitive abilities were often impaired in the PSA patients, and importantly, because we planned to investigate how motor-area target stimulation affects the patients' language performance, we also assessed the participants' motor and cognitive impairments. For motor function, the Fugl–Meyer motor assessment–upper extremity (FMA-UE) was used to assess the patients' upper limb motor function, including voluntary movement, velocity, coordination, and reflex activities. A total score was calculated combining scores of 33 items, resulting in a total maximum score of 66; higher scores indicated better motor function. For cognitive abilities, the non-language-based cognitive assessment (NLCA) was used; the scale is specifically designed and validated to assess the cognitive function of aphasic patients, which uses pictures and objects instead of language instructions and requires no speaking or writing outputs [36]. The NLCA evaluated each participants' cognitive abilities from five domains: memory, visual–spatial abilities, attention, logical reasoning, and executive function, with a total maximum score of 80; higher scores indicate better performances. In addition, the Barthel index (BI) was used to assess each patient's levels of activities of daily living (ADL), with a range from 0 (dependent) to 100 (independent) [36,37].

2.3. iTBS Procedure

iTBS was delivered using a MagPro X100 magnetic stimulator (Medtronic Co, Copenhagen, Denmark) with a figure-eight coil (MC-B70). Before the first session of iTBS, the individual resting motor threshold (RMT) was examined for each subject following the procedure as follows: first, the motor evoked potential (MEP) was measured from the hemiplegia abductor pollicis brevis muscle with surface electrodes (patients with unmeasured MEP on the hemiplegic side are measured with the contralateral side) [38], and then, the "hotspot" was determined using single-pulse TMS over the primary motor area (M1) where the largest MEP was evoked; then, the single-pulse TMS was given at the "hotspot" from low intensity, and with a stepwise increasing intensity until the amplitudes of 5 out of 10 trials exceeding 50 mV; this intensity was defined as the individual RMT. The hotspot location was marked on a positioning cap for each participant, which the participant wore while receiving iTBS over the M1 area [31,39]. After determining the RMT and stimulation point for each subject, the first session of left M1 iTBS was performed with the following parameters: intensity of 80% RMT, three pulses at 50 Hz in each burst at 5 Hz, and 2 s stimulation with 8 s interval; a total of 800 pulses were delivered for one session. The precise location of the iTBS-targeted stimulation site can be found in Figure 1. The TMS machine was placed outside the MRI room, and each participant accepted consecutive MRI scanning before, immediately after the iTBS session. We focused on the immediate brain functional effects of the iTBS in this study, so only the images from the first two time points (i.e., before and immediately after the iTBS session) were included in the analyses.

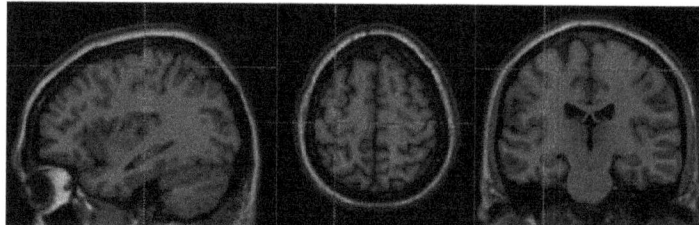

Figure 1. The core location of iTBS intervention ($-36, -21, 58$).

2.4. MRI Acquisition

Patients were scanned at the Jingan Branch of Huashan Hospital, Fudan University, with a 3.0T GE MR750 scanner. The T1-weighted images were acquired using 3D FSPGR with the following parameters: matrix size = 260×224, FOV = 200×200 mm^2, layer thickness = 1 mm, voxel size = $1 \times 1 \times 1$ mm^3, repetition time TR = 7800 ms, echo time TE = 5 ms, tilt angle = $12°$, and number of layers = 248 layers.

The resting-state functional images were acquired in the horizontal plane by EPI echo-planar imaging TR = 2000 ms, echo time, TE = 30 ms, and tilt angle = 248 layers. Imaging sequence was acquired in the horizontal plane with the following parameters: repetition time TR = 2000 ms, echo time, TE = 30 ms, inclination = $90°$, matrix size = 96×96, and voxel size = $3 \times 3 \times 3$ mm^3. The entire resting-state functional scanning procedure lasted 480 s, and a total of 240 functional images were obtained for each subject (one image every 2 s). Before scanning, each participant was instructed to keep still with eyes closed in the machine and to be relaxed with no systematic thinking.

2.5. Lesion Analysis

The lesions were manually segmented using ITK-SNAP tools [40] on individual structural images by inspection of the T1-weighted, T2-weighted, and FLAIR images, registered to Montreal Neurological Institution (MNI) brain using spm12 (https://www.fil.ion.ucl.ac.uk/spm/, accessed on 13 January 2021). A voxel-wise map of lesions was created by summing up all the individual lesions.

2.6. Preprocessing of Resting-State Functional MRI Data

The resting-state functional MR images were pre-processed following the conventional steps: (1) discarding the first 10 volumes of each image; (2) slice timing for systematic time shift; (3) head motion correction using Friston 24 parameters; (4) image reorientation for better normalization; (5) normalizing to MNI space using unified segmentation on T1-weighted images and being resliced into 3 mm × 3 mm × 3 mm resolution; (6) removing the linear signal trend; (7) regression of the nuisance variables including Friston 24 parameters, the white matter, and cerebrospinal fluid signal. Additionally, additional band-pass filtering of 0.01–0.10 Hz signal was performed for functional connectivity analyses. As spatial normalization might be influenced by the lesions, we used an enantiomorphic approach [41] to replace the lesioned brain tissue with the contralateral mirrored scans. The functional image pre-processing was all applied using the data processing assistant for resting-state fMRI, packed in the data processing and analysis for (resting-state) brain imaging (DPABI) procedures [41,42].

2.7. Resting-State Brain Functional MRI Data Analyses

We calculated the regional, seed-based connectivity, and graph-based measures to test the immediate functional effects of the left M1 iTBS in the aphasic brains more comprehensively. The analyses and comparisons were all restricted in a gray matter mask in accordance with the automated anatomical labeling (AAL) atlas, covering the cerebral cortex and nucleus and not including the brain stem or the cerebellum; we will refer to this mask as the "whole-brain mask" in the paragraphs below for the sake of brevity.

2.7.1. Regional Functional Activity Analyses

The fractional amplitude of low-frequency fluctuation (fALFF) was calculated to measure the regional spontaneous brain activity on voxel level, which reflects the strength of spontaneous neural activities in comparison with other non-neural biological signals or artifacts [43]. The fALFF value was computed as the ratio of the summed amplitudes of signals at the low-frequency range (0.01–0.08 Hz) to the amplitudes of the entire frequency range at each voxel, resulting in a spatial map of fALFF for each subject. Then, the individual spatial fALFF maps were z-normalized in the whole-brain mask at the voxel level.

2.7.2. Seed-Based Functional Connectivity Analyses

Seed-based connectivity (FC) analyses were performed to investigate the functional connectivity changes in the left M1 area, which was the site of the iTBS, with all the other voxels in the whole brain. We defined the left M1 seed as the peak activation point reported previously in a hand motion task [44] (MNI-coordinate: −36, −21, 58, Figure 1). The time courses of all voxels within a sphere of 4 mm radius around the center coordinate were averaged, and the connectivity maps were calculated by calculating the Fisher z-transformed correlation coefficient between this mean time course to the time course of each voxel in the whole-brain mask for each subject.

2.7.3. Degree Centrality (DC) Analyses

Degree centrality is a widely used graph-based nodal metric, which measures the importance of a node in a given brain network by calculating the number of other regions it connects within the network [45]. As previous studies have revealed that the functional connectivity undergoes distinguished changes within the same or between the two-brain hemisphere(s) following stroke, we calculated three voxel-wise DC metrics to measure the importance of each voxel in the resting-state brain functional network, including a whole-brain DC, DC within the same hemisphere, and DC with the contralateral hemisphere. First, the zero-lag Pearson's linear correlation coefficients were calculated between the time courses of each pair of voxels in the whole-brain mask. Next, the individual correlation coefficients (i.e., connections) were entered into an N × N adjacency matrix, where N is the number of voxels. The voxel network matrix was thresholded by $r > 0.25$, suppressing random correlations, and was subsequently z-transformed. The three DC metrics were then calculated for each voxel by summing up all the connections it had with (i) all the other voxels in the whole brain (the whole-brain DC), (ii) all the other voxels within the same hemisphere (within-hemispheric DC), and (iii) all the voxels in the contralateral hemisphere (interhemispheric DC).

2.8. Statistical Analysis

Statistical analyses of fMRI data were conducted using the statistical tools of DPABI based on SPM12 and running on MATLAB. Paired *t*-tests were used to test the changes in fALFF, DC, and FC metrics between pre- and post-iTBS treatment on the voxel level. AlphaSim corrections were adopted to adjust for the multiple comparisons [20,46,47], and the probability of false-positive detection was set to $p < 0.05$.

3. Results

3.1. Demographic and Clinical Information for Stroke Patients with PSA

A total of 20 individuals were recruited. Four patients did not complete the MRI scanning and were thus excluded from the assessments. Eventually, 16 patients (4 females and 12 males; age (mean ± SD): 55.6 ± 11.8 years, formal education (mean ± SD): 16.0 ± 2.3 years) were included in the presented study. Table 1 shows the demographic information, behavioral assessment scores, and RMTs of the 16 patients. The individual lesion sites can be found in Table 1, and the lesion overlap map for all patients is presented in Figure 2.

Table 1. Demographic and clinical information of the participants.

	Demographic Information							Functional Assessment				
Patient Code	Age (Years)	Sex	Education (Years)	TSI (Months)	Type of Stroke	Lesion Location in LH	Handedness	WAB-AQ	FMA-UE	BI	NLCA	RMT
P1	70	F	12	3	Ischemic	IFG, STG, BG	R	3	25	50	5	60
P2	40	M	16	2	Ischemic	IFG, PreCG	R	90.2	22	90	75	46
P3	58	F	12	2	Ischemic	IFG, BG	R	60	45	75	65	52
P4	68	M	12	8	Ischemic	BG, MTG, AG	R	61.3	37	35	61	43
P5	65	M	12	7	Ischemic	IFG, BG	R	62.4	9	40	45	40
P6	41	F	16	2	hemorrhage	AG, BG	R	89	25	90	75	34
P7	58	M	9	7	Ischemic	AG, PreCG	R	95.8	21	60	76	48
P8	69	M	12	6	Ischemic	IFG, BG	L	76.1	34	35	57	43
P9	43	M	15	8	Ischemic	IFG, PreCG	R	95	27	50	79	50
P10	60	M	9	9	Ischemic	STG, BG, IFG	R	7	7	10	15	60
P11	68	M	12	6	Ischemic	IFG, PreCG	R	85.6	24	60	70	34
P12	42	F	15	4	hemorrhage	STG, BG	L	18.1	24	55	53	76
P13	62	M	9	5	Ischemic	AG, BG	R	71.2	50	90	73	41
P14	41	M	12	3	hemorrhage	IFG, BG	R	68.5	40	60	65	46
P15	62	M	12	5	hemorrhage	STG, BG	R	45	33	50	60	65
P16	42	M	15	5	Ischemic	IFG, BG	R	69	37	65	72	60

Abbreviation: TSI, time since injury; WAB-AQ, Western Aphasia Battery aphasia quotient; FMA-UE, Fugl–Meyer assessment of upper extremity; BI, barthel index; NLCA, non-language-based cognitive assessment; RMT, resting motor threshold; LH, left hemisphere; IFG, inferior frontal gyrus; BG, basal ganglia; preCG, pre-central gyrus; MTG, middle temporal gyrus; AG, angular gyrus; STG, superior temporal gyrus.

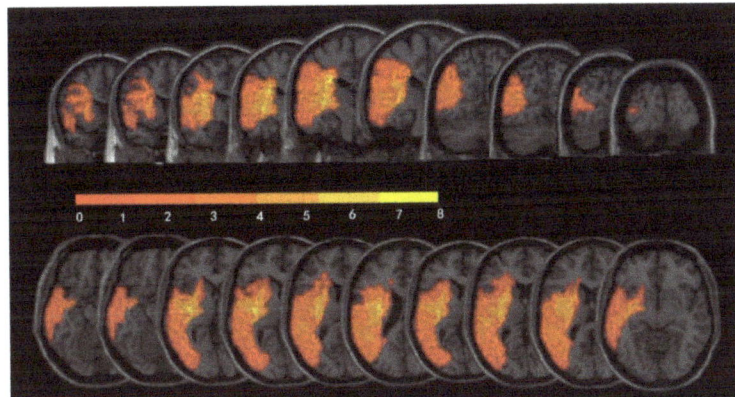

Figure 2. Lesion overlap map across 16 patients with PSA illustrating the distribution of lesions. The color scale in the spectrograms represents the injured brain locations' magnitude of a frequency.

3.2. The Immediate Effects of Left M1 iTBS on fALFF

After the left M1 area iTBS was performed, two clusters exhibited decreased fALFF values, as identified by the paired sample t-test in the brain regions including the right paracingulate gyrus (BA10) and the left intracalcarine cortex (BA17), respectively ($p < 0.05$, with AlphaSim correction; Table 2, Figure 3A).

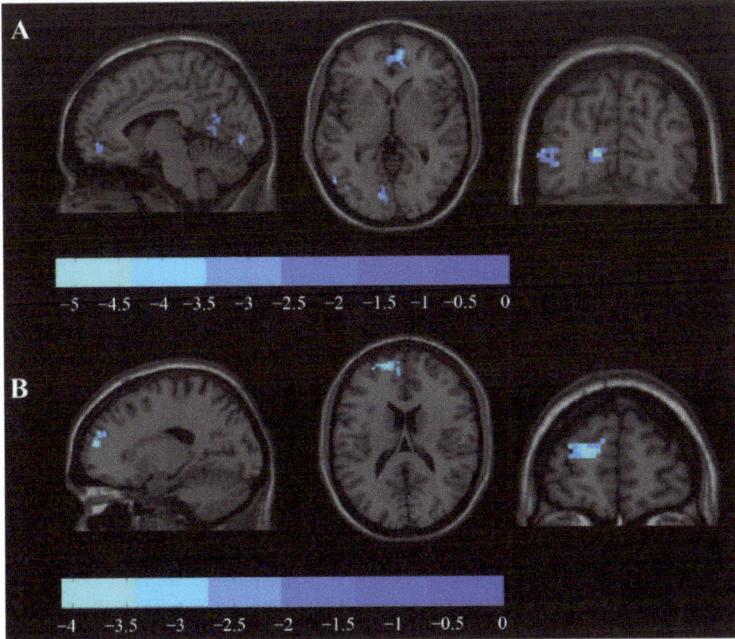

Figure 3. Changes in fALFF as well as FC before and after iTBS intervention: (**A**) differences in fALFF between TP1 and TP2 (paired t-test, $p < 0.05$, AlphaSim correction, cluster size ≥ 56 voxels); (**B**) differences in FC with left precentral gyrus (−36, 57,18) between TP1 and TP2 (paired t-test, $p < 0.05$, AlphaSim correction, cluster size ≥ 129 voxels). The blue concentration represents the degree of decreased regional fALFF or FC values. AAL: automated anatomical labeling atlas, BA: Brodmann area, R: right, L: left.

Table 2. Significant differences in regional fALFF between TP1 and TP2.

Brain Region (AAL)	Brain Region (BA)	Cluster Size (Voxels)	Peak MNI Coordinates (mm)			T Value
Paracingulate gyrus (R)	BA10_R	56	9	51	0	−5.32
Intracalcarine cortex (L)	BA17_L	81	−12	−72	9	−4.65

3.3. The Immediate Effects of Left M1 iTBS on Its Functional Connectivity

Immediately following the left M1 iTBS, the functional connectivity between the hand area of the left M1 and the left precentral gyrus, as well as the left frontal pole, was found to be significantly decreased, as tested by paired two-sample t-test ($p < 0.05$, with AlphaSim correction; Table 3, Figure 3B).

Table 3. Significant differences in the ROI-to-ROI functional connectivity between TP1 and TP2.

Seed Based Functional Connectivity	BA	Cluster SIZE (Voxels)	Peak MNI Coordinates (mm)			T Value
Frontal pole (L)	BA10_L	129	−18	57	18	−4.11

3.4. The Immediate Effects of Left M1 iTBS on DC

The three DC metrics showed different changes after the left M1 area iTBS in our PSA patients. For the whole-brain DC, one cluster in the right middle frontal gyrus and one cluster in the left middle frontal gyrus showed significant elevation from pre- to post-intervention ($p < 0.05$, AlphaSim corrected). Interhemispheric DC values were also revealed to be increased in the left central opercular cortex ($p < 0.05$, with AlphaSim correction), while no significant change was found for within-hemispheric DC (Table 4, Figure 4).

Table 4. Significant differences in regional DC between TP1 and TP2.

Brain Region	BA	Cluster Size (Voxels)	Peak MNI Coordinates (mm)			T Value
		Whole grey matter				
Frontal pole (R)	BA45_R	142	39	36	18	4.83
Superior parietal lobule (L)	BA40_L	158	−39	−45	51	8.07
		Interhemisphere				
Central opercular cortex (L)	BA48_L	169	−51	−3	3	4.19

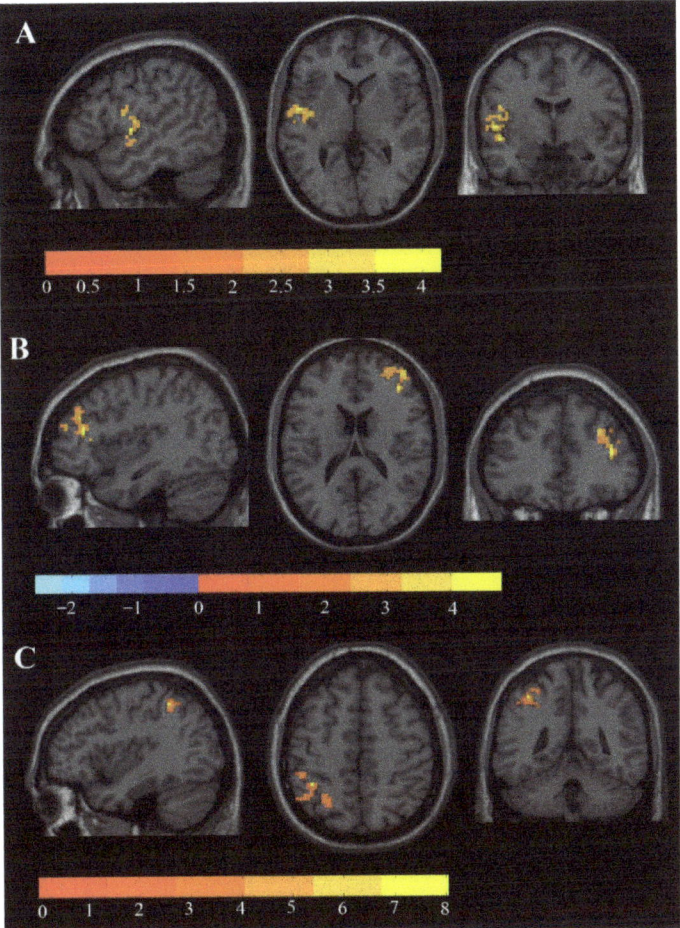

Figure 4. Changes in DC before and after iTBS intervention: (**A**) differences in interhemispheric DC between TP1 and TP2 (paired t-test, $p < 0.05$, AlphaSim correction, cluster size $\geq$ 169 voxels); (**B,C**) differences in DC of whole grey matter between TP1 and TP2 (paired t-test, $p < 0.05$, AlphaSim correction, cluster size $\geq$ 142 voxels). The blue areas represent the regions that have decreased DC values, while the yellow areas represent the regions that have increased DC values.

4. Discussion

To the best of our knowledge, iTBS can modulate cortical excitability and may serve as a potential tool for neuroplasticity of impaired motor, language, and cognition brain areas [48–51]. This study focused on immediate effects and changes of brain activity in patients with PSA, modulated by iTBS. After a combination of iTBS and fMRI, our results reveal that iTBS acting on the M1 area of the affected hemisphere can cause altered cortical excitability and functional connectivity of different regions, which may further lead to orientative neural plasticity in language- and cognition-related functional areas of the brain.

iTBS can increase ipsilesional cortical excitability and has been increasingly used in patients with stroke. Combining fALFF, FC, and DC of rs-fMRI datasets before and after a brain stimulation protocol allowed us to map the changes throughout the whole brain induced by iTBS, instead of only looking at the single change of excitability. We found that after 200 s iTBS intervention, fALFF, DC, and FC all changed to different degrees, which is an indication of the immediate effects of iTBS and is promising for neural plasticity after PSA. Such changes that occur in the neural network of the patient's brain for a long time will produce qualitative changes and cause behavioral progress. Regarding fALFF, we found decreased fALFF values in two clusters in the right frontal and left parietal lobes (Figure 3). The fALFF value stands for spontaneous neuronal activity in brain regions by directly observing the magnitude of baseline changes in the blood oxygenation level-dependent (BOLD) signal of functional brain activity [52]. After the iTBS intervention, the fALFF value in the right frontal part was reduced, suggesting that iTBS attenuated neuronal activity in the contralateral brain. According to the interhemispheric inhibition theory [53,54], the affected hemisphere is inhibited for a certain period after brain injury, while the healthy hemisphere is excited, a condition that is not conducive to functional recovery after stroke. iTBS acting on the left impaired M1 area can inhibit brain activity in the right medial frontal lobe, and some studies [55,56] have shown that right frontal lobe excitability is closely related to the recovery of language status after stroke. For example, A TMS study indicated that non-fluent aphasia after stroke is due to an imbalance in the functional network of the language brain in the bilateral hemispheres [57,58]. Further, 1 Hz of rTMS acting on the right Broca suppressed the excitatory state of this region and promoted language function recovery, especially naming and spontaneous speech ability in patients [59]. The altered right frontal lobe is close to the mirror area of the dominant hemisphere language area (Broca); therefore, this change might promote the recovery of language function and cognitive function in patients with PSA. It was confirmed that the fALFF value was positively correlated with cognitive evaluation scores [60], while the parietal fALFF value in the ipsilateral hemisphere decreased, suggesting that the immediate effect of iTBS might not improve the excitability of brain neurons in cognition-related parietal regions.

DC refers to the number of connections between a voxel and other voxels in the whole brain, and the centrality of a voxel in the whole brain is evaluated by the change in the number of connections. DC value can make full use of the whole brain signal and can avoid subjective selection of seed points, which is a more reliable fMRI analysis method. To find the immediate effect of iTBS in the whole brain and hemispheres accurately, we observed from three perspectives: whole brain, interhemispheric, and within hemispheric. In whole-brain gray matter analysis, the DC value in the right cortical area BA45 and the left BA40 area significantly increased after iTBS (Figure 4). Rao et al. [61] confirmed that after 7 days of rTMS in the affected Broca area rTMS intervention, the patients' DC values in both Broca areas increased significantly and gradually approached the normal level compared with the preintervention. Meanwhile, the patients' spontaneous speech and auditory comprehension functions were significantly improved compared with baseline. Our results indicated that iTBS, as a special mode of TMS, may have similar effects. In addition, consistent with our initial hypothesis, stimulation of the affected M1 area achieved such effects, which manifests that there may be some functional connection between the M1 area and the contralateral Broca's area. This phenomenon has been found in some studies [62]. Similarly, the role of the left BA40 area as part of Wernicke's area, which is traditionally considered to be the comprehension of language, is an innovative finding of our study. From the within-hemispheric perspective, we failed to observe changes in DC value in some brain regions. However, from the interhemispheric point of view, the significant growth of DC was found in the left BA48 area, which is similar to the results observed from whole-brain gray matter. The seed-to-seed FC intensity between the stimulation target and left BA10 was reduced (Figure 3). It is believed that the BA10 area has a close relationship with cognitive function [62–64], which may suggest that we should focus on cognitive functional changes after iTBS-stimulated area M1. Furthermore, the study by Hara et al. [26] found that the lateralization index of BA10 was closely related to the outcome of speech dysfunction treatment after speech training and TMS intervention.

There are some limitations to our study. Firstly, our study was designed as a single-arm study and without a proper control group; thus, results can only be compared between preintervention and postintervention within a group, which made it difficult to diminish the effects caused by confounding factors beyond the intervention of concern and may have caused uncritical results. Second, due to certain technical limitations, we did not perform further analysis of brain network connectivity, which will be completed in our next study. Third, the study adopted a traditional localization paradigm for the localization of the M1 region and did not use a precision navigation localization method. However, we performed an accurate individualized localization cap for each patient with a homogeneous localization pattern. We selected the MEP maximum amplitude location and obtained the same location after positioning by two specialized physicians before using it as the final stimulation site. In a future study, we will apply TMS navigation technology to further improve the accuracy of this study.

5. Conclusions

To the best of our knowledge, few studies have characterized the brain regional dynamics induced by iTBS manipulation on the left M1 area in patients with PSA. The fALFF and DC measurements consistently demonstrated a significant immediate iTBS effect around language- and cognition-related regions, while FC showed iTBS effect was prominent between the affected M1 and frontal pole region. From this perspective, iTBS on the left M1 region may be a promising rehabilitation tool to enhance language and cognition rehabilitation for patients with PSA. In the future, more clinical randomized controlled trials of iTBS are needed to verify the long-term therapeutic effect and to evaluate the progress of clinical behavior and the brain mechanism.

Author Contributions: Study design, S.X. and Q.Y.; patient recruitment and evaluation, S.X., R.Z. and Z.S.; evaluation and localization before iTBS treatment, P.D., Z.Y. and C.L.; iTBS intervention, S.X. and Q.Y.; data analyses, Q.Y., S.X., Y.Z. and M.C.; writing—original draft preparation, S.X. and Q.Y.; resources, J.J.; test management and quality control, J.J. All authors have read and agreed to the published version of the manuscript.

Funding: The study was funded by the National Key Research and Development Program of the Ministry of Science and Technology of the People's Republic of China (Grant Numbers 2018YFC2002300 and 2018YFC2002301); the National Natural Science Foundation of China (Grant Number 91948302); the Innovative Research Group Project of National Natural Science Foundation of China (Grant Number 82021002), and the Shanghai Sailing Program (No. 19YF1405200). The APC was funded by all the above funding.

Institutional Review Board Statement: The study protocol was approved by the medical ethics board of HuaShan hospital Fudan University (2019-336), and all procedures conform to the Declaration of Helsinki regarding human experimentation. Our trail registration number is ChiCTR-TRC-2100041936.

Informed Consent Statement: Informed consent was obtained from all subjects involved in the study.

Data Availability Statement: Data are available on reasonable request.

Acknowledgments: Thanks to Liang, Guan, Gao, and Wei of the Department of Medical Imaging for their equipment support, as well as to all patients and their families for their cooperation in completing the study.

Conflicts of Interest: The authors declare no conflict of interest.

References

1. Harnish, S.; Meinzer, M.; Trinastic, J.; Fitzgerald, D.; Page, S. Language changes coincide with motor and fMRI changes following upper extremity motor therapy for hemiparesis: A brief report. *Brain Imaging Behav.* **2014**, *8*, 370–377. [CrossRef] [PubMed]
2. Meister, I.G.; Sparing, R.; Foltys, H.; Gebert, D.; Huber, W.; Töpper, R.; Boroojerdi, B. Functional connectivity between cortical hand motor and language areas during recovery from aphasia. *J. Neurol. Sci.* **2006**, *247*, 165–168. [CrossRef] [PubMed]
3. Lazar, R.M.; Boehme, A.K. Aphasia As a Predictor of Stroke Outcome. *Curr. Neurol. Neurosci.* **2017**, *17*, 83. [CrossRef]
4. Cherney, L.R.; Babbitt, E.M.; Wang, X.; Pitts, L.L. Extended fMRI-Guided Anodal and Cathodal Transcranial Direct Current Stimulation Targeting Perilesional Areas in Post-Stroke Aphasia: A Pilot Randomized Clinical Trial. *Brain Sci.* **2021**, *11*, 306. [CrossRef] [PubMed]

5. Saxena, S.; Hillis, A.E. An update on medications and noninvasive brain stimulation to augment language rehabilitation in post-stroke aphasia. *Expert Rev. Neurother.* **2017**, *17*, 1091–1107. [CrossRef]
6. Breining, B.L.; Sebastian, R. Neuromodulation in post-stroke aphasia treatment. *Curr. Phys. Med. Rehabil. Rep.* **2020**, *8*, 44–56. [CrossRef]
7. Schwarzer, V.; Bährend, I.; Rosenstock, T.; Dreyer, F.R.; Vajkoczy, P.; Picht, T. Aphasia and cognitive impairment decrease the reliability of rnTMS language mapping. *Acta Neurochir.* **2018**, *160*, 343–356. [CrossRef] [PubMed]
8. Luber, B.; Lisanby, S.H. Enhancement of human cognitive performance using transcranial magnetic stimulation (TMS). *Neuroimage* **2014**, *85*, 961–970. [CrossRef]
9. Meinzer, M.; Darkow, R.; Lindenberg, R.; Flöel, A. Electrical stimulation of the motor cortex enhances treatment outcome in post-stroke aphasia. *Brain A J. Neurol.* **2016**, *139*, 1152–1163. [CrossRef]
10. Jenabi, M.; Peck, K.K.; Young, R.J.; Brennan, N.; Holodny, A.I. Probabilistic fiber tracking of the language and motor white matter pathways of the supplementary motor area (SMA) in patients with brain tumors. *J. Neuroradiol.* **2014**, *41*, 342–349. [CrossRef]
11. Lou, W.; Peck, K.K.; Brennan, N.; Mallela, A.; Holodny, A. Left-lateralization of resting state functional connectivity between the presupplementary motor area and primary language areas. *Neuroreport* **2017**, *28*, 545–550. [CrossRef] [PubMed]
12. Bathla, G.; Gene, M.N.; Peck, K.K.; Jenabi, M.; Tabar, V.; Holodny, A.I. Resting State Functional Connectivity of the Supplementary Motor Area to Motor and Language Networks in Patients with Brain Tumors. *J. Neuroimaging Off. J. Am. Soc. Neuroimaging* **2019**, *29*, 521–526. [CrossRef]
13. Berninger, V.W.; Richards, T.L.; Nielsen, K.H.; Dunn, M.W.; Raskind, M.H.; Abbott, R.D. Behavioral and brain evidence for language by ear, mouth, eye, and hand and motor skills in literacy learning. *Int. J. Sch. Educ. Psychol.* **2019**, *7*, 182–200. [CrossRef]
14. Lefaucheur, J.; André-Obadia, N.; Antal, A.; Ayache, S.S.; Baeken, C.; Benninger, D.H.; Cantello, R.M.; Cincotta, M.; de Carvalho, M.; De Ridder, D.; et al. Evidence-based guidelines on the therapeutic use of repetitive transcranial magnetic stimulation (rTMS). *Clin. Neurophysiol. Off. J. Int. Fed. Clin. Neurophysiol.* **2014**, *125*, 2150–2206. [CrossRef]
15. Lefaucheur, J.; Aleman, A.; Baeken, C.; Benninger, D.H.; Brunelin, J.; Di Lazzaro, V.; Filipović, S.R.; Grefkes, C.; Hasan, A.; Hummel, F.C.; et al. Corrigendum to "Evidence-based guidelines on the therapeutic use of repetitive transcranial magnetic stimulation (rTMS): An update (2014–2018). *Clin. Neurophysiol.* **2020**, *131*, 1168–1169. [CrossRef] [PubMed]
16. Huang, Y.Z.; Edwards, M.J.; Rounis, E.; Bhatia, K.P.; Rothwell, J.C. Theta burst stimulation of the human motor cortex. *Neuron* **2005**, *45*, 201–206. [CrossRef] [PubMed]
17. Suppa, A.; Huang, Y.; Funke, K.; Ridding, M.C.; Cheeran, B.; Di Lazzaro, V.; Ziemann, U.; Rothwell, J.C. Ten Years of Theta Burst Stimulation in Humans: Established Knowledge, Unknowns and Prospects. *Brain Stimul.* **2016**, *9*, 323–335. [CrossRef]
18. Meng, Y.; Zhang, D.; Hai, H.; Zhao, Y.; Ma, Y. Efficacy of coupling intermittent theta-burst stimulation and 1 Hz repetitive transcranial magnetic stimulation to enhance upper limb motor recovery in subacute stroke patients: A randomized controlled trial. *Restor. Neurol. Neuros.* **2020**, *38*, 109–118. [CrossRef]
19. Chen, Y.; Huang, Y.; Chen, C.; Chen, C.; Chen, H.; Wu, C.; Lin, K.; Chang, T. Intermittent theta burst stimulation enhances upper limb motor function in patients with chronic stroke: A pilot randomized controlled trial. *BMC Neurol.* **2019**, *19*, 69. [CrossRef]
20. Li, Y.; Luo, H.; Yu, Q.; Yin, L.; Li, K.; Li, Y.; Fu, J. Cerebral Functional Manipulation of Repetitive Transcranial Magnetic Stimulation in Cognitive Impairment Patients After Stroke: An fMRI Study. *Front. Neurol.* **2020**, *11*, 977. [CrossRef]
21. Wang, J.X.; Rogers, L.M.; Gross, E.Z.; Ryals, A.J.; Dokucu, M.E.; Brandstatt, K.L.; Hermiller, M.S.; Voss, J.L. Targeted enhancement of cortical-hippocampal brain networks and associative memory. *Science* **2014**, *345*, 1054–1057. [CrossRef] [PubMed]
22. Hawco, C.; Armony, J.L.; Daskalakis, Z.J.; Berlim, M.T.; Chakravarty, M.M.; Pike, G.B.; Lepage, M. Differing Time of Onset of Concurrent TMS-fMRI during Associative Memory Encoding: A Measure of Dynamic Connectivity. *Front. Hum. Neurosci.* **2017**, *11*, 404. [CrossRef] [PubMed]
23. Suppa, A.; Fabbrini, A.; Guerra, A.; Petsas, N.; Asci, F.; Di Stasio, F.; Trebbastoni, A.; Vasselli, F.; De Lena, C.; Pantano, P.; et al. Altered speech-related cortical network in frontotemporal dementia. *Brain Stimul.* **2020**, *13*, 765–773. [CrossRef] [PubMed]
24. Romero Lauro, L.J.; Vergallito, A.; Anzani, S.; Vallar, G. Primary motor cortex and phonological recoding: A TMS-EMG study. *Neuropsychologia* **2020**, *139*, 107368. [CrossRef]
25. Vukovic, N.; Shtyrov, Y. Learning with the wave of the hand: Kinematic and TMS evidence of primary motor cortex role in category-specific encoding of word meaning. *Neuroimage* **2019**, *202*, 116179. [CrossRef] [PubMed]
26. Hara, T.; Abo, M.; Kakita, K.; Mori, Y.; Yoshida, M.; Sasaki, N. The Effect of Selective Transcranial Magnetic Stimulation with Functional Near-Infrared Spectroscopy and Intensive Speech Therapy on Individuals with Post-Stroke Aphasia. *Eur. Neurol.* **2017**, *77*, 186–194. [CrossRef]
27. Greicius, M.D.; Supekar, K.; Menon, V.; Dougherty, R.F. Resting-state functional connectivity reflects structural connectivity in the default mode network. *Cereb. Cortex.* **2009**, *19*, 72–78. [CrossRef] [PubMed]
28. Baria, A.T.; Baliki, M.N.; Parrish, T.; Apkarian, A.V. Anatomical and functional assemblies of brain BOLD oscillations. *J. Neurosci.* **2011**, *31*, 7910–7919. [CrossRef]
29. Chen, Y.; Xiang, C.; Liu, W.; Jiang, N.; Zhu, P.; Ye, L.; Li, B.; Lin, Q.; Min, Y.; Su, T.; et al. Application of amplitude of low-frequency fluctuation to altered spontaneous neuronal activity in classical trigeminal neuralgia patients: A resting-state functional MRI study. *Mol. Med. Rep.* **2019**, *20*, 1707–1715. [CrossRef]
30. Hua, J.; Liu, P.; Kim, T.; Donahue, M.; Rane, S.; Chen, J.J.; Qin, Q.; Kim, S. MRI techniques to measure arterial and venous cerebral blood volume. *Neuroimage* **2019**, *187*, 17–31. [CrossRef] [PubMed]

31. Tang, Y.; Jiao, X.; Wang, J.; Zhu, T.; Zhou, J.; Qian, Z.; Zhang, T.; Cui, H.; Li, H.; Tang, X.; et al. Dynamic Functional Connectivity Within the Fronto-Limbic Network Induced by Intermittent Theta-Burst Stimulation: A Pilot Study. *Front. Neurosci.-Switz.* **2019**, *13*, 944. [CrossRef] [PubMed]
32. Hoptman, M.J.; Zuo, X.; Butler, P.D.; Javitt, D.C.; D'Angelo, D.; Mauro, C.J.; Milham, M.P. Amplitude of low-frequency oscillations in schizophrenia: A resting state fMRI study. *Schizophr. Res.* **2010**, *117*, 13–20. [CrossRef]
33. Buckner, R.L.; Sepulcre, J.; Talukdar, T.; Krienen, F.M.; Liu, H.; Hedden, T.; Andrews-Hanna, J.R.; Sperling, R.A.; Johnson, K.A. Cortical hubs revealed by intrinsic functional connectivity: Mapping, assessment of stability, and relation to Alzheimer's disease. *J. Neurosci.* **2009**, *29*, 1860–1873. [CrossRef]
34. Nuzzi, R.; Dallorto, L.; Rolle, T. Changes of Visual Pathway and Brain Connectivity in Glaucoma: A Systematic Review. *Front Neurosci.-Switz.* **2018**, *12*, 363. [CrossRef] [PubMed]
35. Gilmore, N.; Dwyer, M.; Kiran, S. Benchmarks of Significant Change After Aphasia Rehabilitation. *Arch. Phys. Med. Rehabil.* **2019**, *100*, 1131–1139. [CrossRef] [PubMed]
36. Wu, J.; Lyu, Z.; Liu, X.; Li, H.; Wang, Q. Development and Standardization of a New Cognitive Assessment Test Battery for Chinese Aphasic Patients: A Preliminary Study. *Chin. Med. J.-Peking* **2017**, *130*, 2283–2290.
37. Mlinac, M.E.; Feng, M.C. Assessment of Activities of Daily Living, Self-Care, and Independence. *Arch. Clin. Neuropsychol. Off. J. Natl. Acad. Neuropsychol.* **2016**, *31*, 506–516. [CrossRef] [PubMed]
38. Nazarova, M.; Novikov, P.; Ivanina, E.; Kozlova, K.; Dobrynina, L.; Nikulin, V.V. Mapping of multiple muscles with transcranial magnetic stimulation: Absolute and relative test-retest reliability. *Hum. Brain Mapp.* **2021**, *42*, 2508–2528. [CrossRef]
39. Pinot-Monange, A.; Moisset, X.; Chauvet, P.; Gremeau, A.; Comptour, A.; Canis, M.; Pereira, B.; Bourdel, N. Repetitive Transcranial Magnetic Stimulation Therapy (rTMS) for Endometriosis Patients with Refractory Pelvic Chronic Pain: A Pilot Study. *J. Clin. Med.* **2019**, *8*, 508. [CrossRef] [PubMed]
40. Yushkevich, P.A.; Piven, J.; Hazlett, H.C.; Smith, R.G.; Ho, S.; Gee, J.C.; Gerig, G. User-guided 3D active contour segmentation of anatomical structures: Significantly improved efficiency and reliability. *Neuroimage* **2006**, *31*, 1116–1128. [CrossRef] [PubMed]
41. Nachev, P.; Coulthard, E.; Jäger, H.R.; Kennard, C.; Husain, M. Enantiomorphic normalization of focally lesioned brains. *Neuroimage* **2008**, *39*, 1215–1226. [CrossRef]
42. Yan, C.; Wang, X.; Zuo, X.; Zang, Y. DPABI: Data Processing & Analysis for (Resting-State) Brain Imaging. *Neuroinformatics* **2016**, *14*, 339–351.
43. Zou, Q.; Zhu, C.; Yang, Y.; Zuo, X.; Long, X.; Cao, Q.; Wang, Y.; Zang, Y. An improved approach to detection of amplitude of low-frequency fluctuation (ALFF) for resting-state fMRI: Fractional ALFF. *J. Neurosci. Meth.* **2008**, *172*, 137–141. [CrossRef]
44. Alkadhi, H.; Crelier, G.R.; Boendermaker, S.H.; Golay, X.; Hepp-Reymond, M.; Kollias, S.S. Reproducibility of primary motor cortex somatotopy under controlled conditions. *AJNR. Am. J. Neuroradiol.* **2002**, *23*, 1524–1532.
45. Li, H.; Li, L.; Shao, Y.; Gong, H.; Zhang, W.; Zeng, X.; Ye, C.; Nie, S.; Chen, L.; Peng, D. Abnormal Intrinsic Functional Hubs in Severe Male Obstructive Sleep Apnea: Evidence from a Voxel-Wise Degree Centrality Analysis. *PLoS ONE* **2016**, *11*, e0164031.
46. Dang, T.P.; Mattan, B.D.; Kubota, J.T.; Cloutier, J. The ventromedial prefrontal cortex is particularly responsive to social evaluations requiring the use of person-knowledge. *SCI REP-UK* **2019**, *9*, 5054. [CrossRef] [PubMed]
47. Wang, M.; Liao, H.; Shen, Q.; Cai, S.; Zhang, H.; Xiang, Y.; Liu, S.; Wang, T.; Zi, Y.; Mao, Z.; et al. Changed Resting-State Brain Signal in Parkinson's Patients With Mild Depression. *Front. Neurol.* **2020**, *11*, 28. [CrossRef] [PubMed]
48. Mistry, S.; Michou, E.; Rothwell, J.; Hamdy, S. Remote effects of intermittent theta burst stimulation of the human pharyngeal motor system. *Eur. J. Neurosci.* **2012**, *36*, 2493–2499. [CrossRef]
49. Zimerman, M.; Hummel, F.C. Non-invasive brain stimulation: Enhancing motor and cognitive functions in healthy old subjects. *Front. Aging Neurosci.* **2010**, *2*, 149. [CrossRef]
50. Ghaffari, H.; Yoonessi, A.; Darvishi, M.J.; Ahmadi, A. Normal Electrical Activity of the Brain in Obsessive-Compulsive Patients After Anodal Stimulation of the Left Dorsolateral Prefrontal Cortex. *Basic Clin. Neurosci.* **2018**, *9*, 135–146. [CrossRef] [PubMed]
51. Versace, V.; Schwenker, K.; Langthaler, P.B.; Golaszewski, S.; Sebastianelli, L.; Brigo, F.; Pucks-Faes, E.; Saltuari, L.; Nardone, R. Facilitation of Auditory Comprehension After Theta Burst Stimulation of Wernicke's Area in Stroke Patients: A Pilot Study. *Front. Neurol.* **2020**, *10*, 1319. [CrossRef]
52. Sprugnoli, G.; Monti, L.; Lippa, L.; Neri, F.; Mencarelli, L.; Ruffini, G.; Salvador, R.; Oliveri, G.; Batani, B.; Momi, D.; et al. Reduction of intratumoral brain perfusion by noninvasive transcranial electrical stimulation. *Sci. Adv.* **2019**, *5*, u9309. [CrossRef]
53. Du, J.; Hu, J.; Hu, J.; Xu, Q.; Zhang, Q.; Liu, L.; Ma, M.; Xu, G.; Zhang, Y.; Liu, X.; et al. Aberrances of Cortex Excitability and Connectivity Underlying Motor Deficit in Acute Stroke. *Neural. Plast.* **2018**, *2018*, 1318093. [CrossRef] [PubMed]
54. Tatsuno, H.; Hamaguchi, T.; Sasanuma, J.; Kakita, K.; Okamoto, T.; Shimizu, M.; Nakaya, N.; Abo, M. Does a combination treatment of repetitive transcranial magnetic stimulation and occupational therapy improve upper limb muscle paralysis equally in patients with chronic stroke caused by cerebral hemorrhage and infarction?: A retrospective cohort study. *Medicine* **2021**, *100*, e26339. [CrossRef] [PubMed]
55. Kawamura, M.; Takahashi, N.; Kobayashi, Y. Effect of Repetitive Transcranial Magnetic Stimulation on the Right Superior Temporal Gyrus for Severe Aphasia Caused by Damage to the Left Inferior Frontal Gyrus. *Case Rep. Neurol.* **2019**, *11*, 189–198. [CrossRef]

56. Crosson, B.; Moore, A.B.; McGregor, K.M.; Chang, Y.; Benjamin, M.; Gopinath, K.; Sherod, M.E.; Wierenga, C.E.; Peck, K.K.; Briggs, R.W.; et al. Regional changes in word-production laterality after a naming treatment designed to produce a rightward shift in frontal activity. *Brain Lang.* **2009**, *111*, 73–85. [CrossRef] [PubMed]
57. Dodd, K.C.; Nair, V.A.; Prabhakaran, V. Role of the Contralesional vs. Ipsilesional Hemisphere in Stroke Recovery. *Front. Hum. Neurosci.* **2017**, *11*, 469. [CrossRef] [PubMed]
58. Da Silva Júnior, H.B.; Fernandes, M.R.; Souza, Â.M.C. Repetitive Transcranial Magnetic Stimulation Improves Depressive Symptoms and Quality of Life of Poststroke Patients-Prospective Case Series Study. *J. Cent. Nerv. Syst. Dis.* **2019**, *11*, 593269432. [CrossRef] [PubMed]
59. Fecteau, S.; Agosta, S.; Oberman, L.; Pascual-Leone, A. Brain stimulation over Broca's area differentially modulates naming skills in neurotypical adults and individuals with Asperger's syndrome. *Eur. J. Neurosci.* **2011**, *34*, 158–164. [CrossRef]
60. Tadayonnejad, R.; Yang, S.; Kumar, A.; Ajilore, O. Clinical, cognitive, and functional connectivity correlations of resting-state intrinsic brain activity alterations in unmedicated depression. *J. Affect. Disord.* **2015**, *172*, 241–250. [CrossRef]
61. Rao, J.; Hu, G.; Yang, W.; Yu, J.; Xue, C.; Tian, L.; Chen, J. Recovery of aphemia with pure word dumbness after treatment with repetitive transcranial magnetic stimulation: A case report. *J. Neurolinguist* **2021**, *59*, 100991. [CrossRef]
62. Su, F.; Xu, W. Enhancing Brain Plasticity to Promote Stroke Recovery. *Front. Neurol.* **2020**, *11*, 554089. [CrossRef] [PubMed]
63. Chen, H.; Huang, L.; Li, H.; Qian, Y.; Yang, D.; Qing, Z.; Luo, C.; Li, M.; Zhang, B.; Xu, Y. Microstructural disruption of the right inferior fronto-occipital and inferior longitudinal fasciculus contributes to WMH-related cognitive impairment. *CNS Neurosci. Ther.* **2020**, *26*, 576–588. [CrossRef]
64. Velikova, S.; Nordtug, B. Self-guided Positive Imagery Training: Effects beyond the Emotions-A Loreta Study. *Front. Hum. Neurosci.* **2018**, *11*, 644. [CrossRef] [PubMed]

Article

Functional Connectivity States of Alpha Rhythm Sources in the Human Cortex at Rest: Implications for Real-Time Brain State Dependent EEG-TMS

Davide Tabarelli [1,†], Arianna Brancaccio [1,†], Christoph Zrenner [2] and Paolo Belardinelli [1,3,*]

1. Center for Mind/Brain Sciences—CIMeC, University of Trento, I-38123 Trento, Italy; davide.tabarelli@unitn.it (D.T.); arianna.brancaccio@unitn.it (A.B.)
2. Temerty Centre for Therapeutic Brain Intervention, Centre for Addiction and Mental Health, Department of Psychiatry, University of Toronto, Toronto, ON M6J 1H4, Canada; christoph.zrenner@gmail.com
3. Department of Neurology & Stroke, University of Tübingen, D-72070 Tübingen, Germany
* Correspondence: paolo.belardinelli@unitn.it
† These authors contributed equally to this work.

Abstract: Alpha is the predominant rhythm of the human electroencephalogram, but its function, multiple generators and functional coupling patterns are still relatively unknown. In this regard, alpha connectivity patterns can change between different cortical generators depending on the status of the brain. Therefore, in the light of the communication through coherence framework, an alpha functional network depends on the functional coupling patterns in a determined state. This notion has a relevance for brain-state dependent EEG-TMS because, beyond the local state, a network connectivity overview at rest could provide further and more comprehensive information for the definition of 'instantaneous state' at the stimulation moment, rather than just the local state around the stimulation site. For this reason, we studied functional coupling at rest in 203 healthy subjects with MEG data. Sensor signals were source localized and connectivity was studied at the Individual Alpha Frequency (IAF) between three different cortical areas (occipital, parietal and prefrontal). Two different and complementary phase-coherence metrices were used. Our results show a consistent connectivity between parietal and prefrontal regions whereas occipito-prefrontal connectivity is less marked and occipito-parietal connectivity is extremely low, despite physical closeness. We consider our results a relevant add-on for informed, individualized real-time brain state dependent stimulation, with possible contributions to novel, personalized non-invasive therapeutic approaches.

Keywords: alpha oscillations; functional connectivity; source reconstruction; MEG; EEG state-dependent TMS

1. Introduction

The origin of alpha waves and the function they subserve constitute long-lasting scientific issues in neuroscience. Already by 1929, Berger had managed to isolate alpha waves by means of a pioneering EEG set-up using scalp electrodes and described this rhythm as the most prominent in the human electroencephalogram [1]. Recent advances due to brain source reconstruction and invasive electrophysiological recordings have provided evidence that alpha waves originate from several cortical and subcortical sites, with direct evidence suggesting both thalamus and diverse cortical areas as possible origins of such rhythm [2–4].

The functional role of alpha oscillations also is still far from being completely understood. The "pulsed inhibition hypothesis", for instance, proposes that alpha oscillations actively inhibit neuronal firing in a phasic manner, opening and closing interleaved periods of "high" and "low" excitability of the cortex by cyclically producing bouts of inhibition [5,6]. This hypothesis is in line with the idea behind brain-state dependent stimulation, that the outcome of an electro/magnetic perturbation depends on the instantaneous phase state of a specific brain rhythm in a given area. In fact, studies investigating alpha in the occipital cortex demonstrate that the alpha phase at the instant of a visual stimulus predicts high or low probability of detection [7].

However, it is still not clear which is the exact role of alpha oscillations in opening states of cortical excitability. For example, the aforementioned "pulsed inhibition hypothesis" does not seem to apply to findings of brain-state dependent stimulation in the parietal cortex, where the same directionality of inhibition vs. excitation as in the visual cortex does not emerge [8–10]. In this regard, studies investigating μ-alpha in the parietal cortex show that a principle of pulsed facilitation rather than inhibition would better explain the role of alpha in the hand-knob of the sensorimotor cortex. In fact, it has been shown by means of real-time EEG phase-dependent transcranial magnetic stimulation (TMS) that stimulation at μ-alpha troughs results in facilitated motor evoked responses (MEPs, [8]), while at peaks of the same oscillation, inhibition does not seem to occur in a significant proportion of instances [9].

Even if there is no consensus as to whether alpha shapes neuronal recruitment by determining windows of lower and higher excitability, or rather only by opening windows of greater excitability, alpha oscillations appear to have a role in shaping neural recruitment.

Most of the protocols conceived to modulate cortical excitability by means of TMS use a predefined stimulus sequence irrespective of the instantaneous brain-state, as opposed to a real-time brain-state dependent stimulation, which delivers the stimulus in determined phases of the alpha cycle [8]. If the phase of alpha reflects a phasic increase and decrease of cortical excitability (as reported in the occipital cortex by [11], we should find different outcomes depending on the local alpha phase at the instant of stimulation. First attempts in this direction have tried to post-hoc determine the phase of alpha waves at the moment of the stimulus. This has been mainly tested in the occipital cortex, where the conscious visual percept has been linked post-hoc to the alpha phase at the instant of stimulus presentation [7,12]). However, the effects were generally assessed by statistically estimating the probability that the stimulus could be delivered at a given phase. Differently, phase-state dependent stimulation leverages instantaneous EEG phase-states to trigger the TMS pulse exactly at the phase of interest, without post-hoc tracing of it back to the moment of analysis. Therefore, studies using this novel approach have led to more consistent and reproducible results. For example, EEG-TMS has been used to investigate motor excitability depending on mu-alpha in the parietal cortex. This line of research is relatively recent, but several consistent pieces of evidence show that participants with a detectable mu-alpha rhythm show larger motor evoked potentials (MEPs) when TMS stimulation at the motor spot is delivered at mu troughs or at the early rising phase, compared to the conditions where positive peaks or random phases are targeted [8,13,14]. These results are in line with the hypothesis of the alpha rhythm as a mechanism modulating cortical excitability. Attempts to link specific alpha phases to enhanced cortical excitability have also been made in studies targeting the dorsolateral frontal cortex (DLPFC), where alpha-synchronized rTMS at troughs appears to provide for a local alpha power decrease in patients with drug resistant depression disorder. The same result is not obtained by intermittent theta burst or random phase stimulation [15].

To sum up, the role of alpha phase has been investigated in at least three different cortical areas and a relationship between alpha phase and cortical excitability has been found in frontal, parietal and occipital regions [8,11,16]. However, there is still no clear and organic consensus on a generalized functional role of the alpha phase. For example, the alpha phase in the occipital cortex seems to elicit opposite effects with respect to the other cortical areas: trough targeting has been proposed to decrease the possibility of perceiving the stimulus [7]. Moreover, it has been shown that the conscious perception of a visual stimulus not only depends on alpha phase in the occipital cortex but also on that in fronto-central regions at the moment of stimulation [7,12] this would be in line with the idea of the frontal control network allowing access to a conscious perception.

This evidence, together with the assumption that coherence between brain regions underlies integration of information [17], suggests that knowing the functional coupling between stimulated (or post-hoc investigated) regions and other areas potentially connected with the stimulation site is crucial in designing experiments aiming at a trial-based selective perturbation of a given "brain state". In this light, considering the functional coupling patterns of the stimulated site is important in explaining EEG-TMS results and is even more crucial when EEG-TMS protocols are intended, not only in terms of a single stimulation site, but also as a technique for pathway-specific modulation targeting multiple functional hubs: the next horizon of brain-state dependent stimulation. In this regard, this technique has been shown to have the potential to modulate specific pathways [18], for example, by coupling the stimulation to an activity state of that pathway (e.g., [13]). Therefore, reliable metrics are required that can extract extended network states through long-range connectivity. Such solid brain state landmarks would be certainly useful (before and after a neuro-modulatory intervention) to assess whether the brain-stimulation has exerted the desired network state modulation. Most importantly, however, they would be even more crucial when time-resolved connectivity state estimates, describing the connectivity pattern of the network, are used as real-time trigger condition. Nevertheless, gold-standard metrics for connectivity generally consist of pseudo-statistics regarding different phase consistency over several tens of trials and a consensus for a conceptual definition of single-trial instantaneous connectivity is still missing. For this reason, we here try to open the way to addressing the need for brain-state dependent protocols to take into account not only the local state which usually triggers the stimulus, but also a general sensitivity state of the system being modulated in its excitability.

Therefore, this study aims to identify suitable functional pathways between well-known cortical alpha generators, whose connectivity state can be assessed with MEG/EEG at rest. Here, we present a pipeline that effectively determines potential connections from resting MEG/EEG data using Weighted Pairwise Phase Consistency (WPPC; [19]) and Weighted Phase Lag Index (WPLI; [20]). It is worth noticing that the choice of a connectivity metric has to take into account its advantages and disadvantages in the context of application [21,22]. Here we used WPPC and WPLI, exploiting their partial complementarity, for the data under investigation. Both metrics depend on the consistency between the phases of the signal of interest, and are not biased by sample size. WPPC is based on the pairwise phase difference, thus it is still affected by zero-lag correlations introduced by the spatial spread of the inverse solution. As a matter of fact, WPPC also provides positive and statistically significant values when the phase difference is exactly zero. This is a result that does not reflect real synchronization, because the finite (but not null) propagation time of the nervous signal on the pathway connecting the two areas is not taken into account. In contrast, WPLI does not have this problem because it is computed from the sign of the imaginary part of the coherence, which vanishes for zero-lag correlations. However, the WPLI signal to noise ratio is optimal for a phase delay of $\pi/2$, which corresponds at a typical frequency of 10 Hz to an absolute time delay of 25 ms. Since we are investigating the temporally resolved cortico-cortical connectivity, this is a large delay if compared to the typical brain conduction delays within the same hemisphere. Thus WPLI, despite being a less sensitive measure of phase consistency between the brain regions under investigation, is useful to confirm that the connectivity already detected by the more sensitive WPPC is real and not due to the spatial spread of the inverse solution. In this sense, the two metrices are partially complementary. With this methodological set-up, we investigated 203 healthy subjects with several minutes of resting MEG data and found alpha connectivity stronger for parietal-prefrontal areas rather than for occipito-parietal areas. The results of this study will be relevant to design real-time brain state-dependent EEG-TMS experiments: connectivity priors [23] will be highly relevant as off-line acquired a priori information for the development of real-time connectivity state estimation algorithms.

2. Methods

2.1. Dataset and Acquisition

We analyzed magnetoencephalographic (MEG) resting state recordings of 203 healthy participants (age range 18–57 years; see Figure 1a for age distribution) from the Cambridge Centre for Aging and Neuroscience (CamCAN) public dataset. Prior to inclusion in the dataset for neuroimaging measurements, all participants were tested for cognitive decline (MMSE > 24; [24]), for matching vision, hearing and English language inclusion criteria and for absence of serious neurological and psychiatric conditions (see [25] for details about the dataset and acquisition parameters). For each participant, resting state activity (eyes closed; 8 min and 40 s) and empty-room noise background (3 min) were recorded using a 306-channel VectorView MEG system (102 magnetometers; 204 first order planar gradiometers; sampling rate = 1000 Hz; high pass filter 0.03 Hz; low pass filter 330 Hz). Anatomical landmarks (nasion, left and right pre-auricular points) were registered, as well as at least 75 additional (isotrak) points, to model the head surface. For human resting state data, continuous monitoring of the head position (cHPI), electro-ocular (EOHG, EOVG) and electro-cardiac (ECG) recordings were available. Additionally, T1 weighted anatomical data were collected using a 3T Siemens TIM Trio scanner (MPRAGE, TR = 2250 ms, TE = 2.99 s; FOV = 256 × 256 × 192; voxel size 1 × 1 × 1 mm). All the subsequent analysis was performed only on planar gradiometers using Fieldtrip ([26]), SPM ([27]), CAT12 (http://www.neuro.uni-jena.de/cat/ (accessed on 1 October 2021)) and custom MATLAB code.

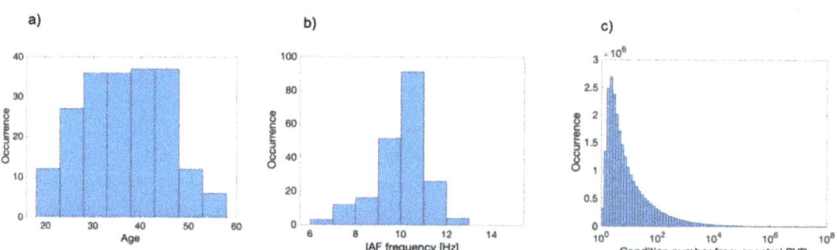

Figure 1. (a) Distribution of sample age. (b) Distribution of detected IAF on the whole sample. (c) Distribution of all condition numbers from SVD, aiming to define optimal dipole direction for spectral data.

2.2. MEG Data Preprocessing

Collected MEG datasets (resting state and empty-room) were inspected and bad channels showing high noise levels and/or SQUID jumps were marked and removed from the data. External magnetic source nuisance was removed by means of temporal Signal Space Separation (tSSS, [28,29]) with a correlation threshold of 0.98 and a sliding time window of 10 s. Line noise contribution was removed by applying a 5th-order Butterworth two-pass filter centered at the line frequency and its first 3 harmonics, and spanning an interval of 2 Hz. Head position of the resting state data was corrected every 200 ms using cHPI data and re-referenced to a common head position. Human MEG data were cleaned from physiological artifacts using an automated procedure. Muscular activity was detected by filtering data between 100 and 140 Hz, transforming each channel data into a z-score relative to the whole channel time series and rejecting segments where the z-score averaged across channels was greater than 5 for at least 200 ms. Then we ran an extended Infomax Independent Component Analysis (ICA; [30,31]) on data filtered between 0.5 and 125 Hz (Butterworth 4th order, two-pass) and resampled to 250 Hz. Data resampling and digital filtering commute with the ICA unmixing matrix estimation, provided the artifactual activity of interest lies in the spared frequency band [31]. For this reason, and for computational efficiency, we computed ICA on resampled data and applied unmixing matrices to unfiltered data at the original sampling rate of 1000 Hz. Correlation coefficients of each extracted component with electro-physiological channels (EHOG, EVOG and ECG) were computed and rejected using a recursive z-score based procedure that robustly discarded all components with correlation more than 2 standard deviations from the coefficient distribution mean of all components. Data cleaned from artifacts were then filtered and segmented. Empty-room data, later used for noise covariance matrix estimation, were filtered broadly between 0.5 and 140 Hz (4th-order Butterworth two-pass filter) and segmented in 1 s epochs. Human resting state recordings were instead filtered around the alpha frequency band (4th-order Butterworth two-pass filter between 5 and 16 Hz) and split into 2 s segments. Finally, all empty-room and resting state segments exceeding a threshold of 10 standard deviations with respect to the channel average, were further discarded in order to deal with potential residual SQUID jumps and/or filter border effects.

2.3. Anatomical Data Processing

Structural T1w MRI scans were processed with the aim of extracting cortical surfaces for forward and inverse model calculation and for common space mapping. We processed T1w images using the CAT 12 toolbox pipeline (http://www.neuro.uni-jena.de/cat/ (accessed on 1 October 2021)). After bias normalization, denoising and skull stripping, volumetric data were automatically co-registered to a template space (IXII 555 MNI space; www.brain-development.org, (accessed on 1 October 2021)) Different tissue was then segmented in native space in order to extract surface models of the head, the brain enclosing surface and the cortical mantle. We modeled the cortical mantle with a tessellation (20,484 vertices) of the mid thickness surface, i.e., the surface in between the pial and the white/gray matter interface. The surface enclosing the brain and a surface model of the head were also modeled as 20,000 vertices meshes. In addition to the extracted surfaces in native subject space, co-registered spheres for projection on the FreeSurfer Average (5th order icosahedron "fsaverage5"; [32]) superficial template space were computed, as well as the corresponding interpolation matrices between template and native coordinate spaces. Interpolation coefficients for each point were defined as inversely weighted average of first neighborhood. The mapping of the native space cortical surface onto the FreeSurfer average allowed us to identify, in template space, vertices belonging to 360 regions of interests of the 'state of the art' multimodal parcellation from the Human Connectome Project [33]. This atlas was generated by combining structural, diffusion and resting state fMRI data from 210 healthy young adults. Being interested in phase consistency between frontal, parietal and occipital areas, we defined three corresponding brain sectors, for each hemisphere, by pooling together correspondent ROIs from the atlas. We selected the ROIs in order to achieve a sufficiently large coverage of the three sectors of interest, while avoiding excessive proximity that might lead to spurious connectivity due to potential spread of the inverse solution. Finally, this led to 16, 7 and 18 regions of interest for the occipital, parietal and frontal sectors, respectively. For the sake of computational efficiency, we refined the sectors by trimming the borders in order to have the same number of dipoles per sector (D = 930). A list of corresponding regions of interest, with MNI coordinates, can be found in Table 1. All the subsequent analyses were performed only on these ROIs.

Table 1. List of Region of Interest (ROI) defining the three sectors. A brief description (from van Essen) and the coordinate of the centroid in mm with respect to the template MNI space is reported.

ROI	Sector	MNI Coordinates of Centroid (mm)		
		x	y	z
V1	Occipital	−13.1	−82.0	1.5
V2	Occipital	−12.4	−81.5	3.6
ProS	Occipital	−18.5	−52.2	0.1
V3	Occipital	−18.3	−86.2	5.4
V4	Occipital	−29.7	−82.5	−3.9
V6	Occipital	−13.9	−78.0	27.2

Table 1. Cont.

ROI	Sector	MNI Coordinates of Centroid (mm)		
		x	y	z
V6A	Occipital	−18.6	−84.3	38.1
V7	Occipital	−23.8	−81.9	26.6
IPS1	Occipital	−22.6	−71.7	33.0
V3A	Occipital	−17.2	−88.4	23.0
V3B	Occipital	−28.2	−78.9	16.3
V3CD	Occipital	−35.3	−85.7	12.3
IP0	Occipital	−30.4	−73.5	25.5
PGp	Occipital	−39.8	−80.1	22.1
LO1	Occipital	−37.8	−82.9	4.2
LO2	Occipital	−42.7	−83.3	−4.9
1	Parietal	−47.1	−24.5	52.3
2	Parietal	−35.4	−34.4	49.7
3a	Parietal	−34.3	−21.8	41.8
3b	Parietal	−36.8	−24.1	51.6
4	Parietal	−26.7	−19.7	53.8
6mp	Parietal	−14.1	−13.2	65.7
6d	Parietal	−34.9	−12.7	61.9
8BL	Frontal	−11.6	35.1	50.8
9p	Frontal	−18.9	44.0	36.4
9m	Frontal	−7.7	51.0	21.8
9a	Frontal	−19.7	53.2	23.8
8Ad	Frontal	−23.3	24.7	41.2
9–46d	Frontal	−28.7	42.1	21.4
8BM	Frontal	−6.3	29.5	43.1
8Av	Frontal	−37.1	18.0	47.4
46	Frontal	−36.6	35.6	28.3
8C	Frontal	−40.3	16.1	35.0
p9–46v	Frontal	−43.3	29.2	26.3
a32pr	Frontal	−10.2	28.1	28.6
d32	Frontal	−10.0	38.5	21.1
a9–46v	Frontal	−37.1	47.7	8.8
10d	Frontal	−12.1	62.9	8.4
p10p	Frontal	−23.6	55.0	5.2
p47r	Frontal	−41.2	40.3	1.5
IFSa	Frontal	−42.0	31.2	13.2

2.4. MEG Source Reconstruction Based on Individual Anatomies

We used Minimum Norm Estimation [34] to solve the inverse problem and compute the projection matrix from sensors to source space. Native space surfaces were co-registered to MEG coordinates in a first step, using correspondence between anatomical landmarks as recorded during the MEG session and in the MRI anatomical image. Second, co-registration was refined by aligning additional head surface points registered during MEG acquisition to the tessellation representing the head surface. In this procedure, MEG isotrak points anterior to the nasion were discarded since the anatomical MRI images were defaced prior to being publicly available. We then computed a forward model solution for planar gradiometers using the singleshell method and the brain enclosing surface as computed before, and depth normalizing lead fields with a factor of 0.5. The source model for MNE was defined as a free orientation set of dipoles uniformly distributed on the cortical surface (mean spacing 3.1 mm) and the inverse solution was computed, with a 1% noise covariance regularization. This procedure leads to a set of three time series of estimated cortical activations for each resting state epoch and for each vertex of the source model mesh, in the three cartesian coordinate system. MEG is blind to the radial component of magnetic field generated by a current dipole in the source model [34,35]. Thus, even for realistic forward solutions, the estimated current in the most radial direction, with respect to the head surface, is almost zero. Therefore, we performed a Singular Value Decomposition (SVD) of the three source time series at each vertex and for each epoch, retaining only the components associated with the first two singular values, while the third was always almost zero. In this way we reduced the current estimate to its projection on the plane orthogonal to the radial direction, resulting in two source activity time series for each vertex and for each dipole represented hereafter by the two-dimensional vectors *x(t)*.

2.5. Spectral Analysis

For each epoch and vertex of the cortical mesh, we computed Fourier coefficients *X(f)* from the time dependent activity vectors *x(t)* in a frequency interval $f = [8{:}14]$ Hz using a Hanning window taper. Having thus two Fourier coefficients at each frequency, dipole and epoch, we decided to reduce spectral data defining an optimal dipole orientation as follows: we computed the cross spectral density matrix between the two Fourier components, performed an SVD and keeping only the direction relative to the first singular value. This resulted in an optimal dipole orientation $\hat{u}$ that depends on the dipole position, the epoch and the frequency of interest, thus optimizing the detection of the brain signal at the frequency of interest. We reduced accordingly the Fourier coefficient vector to a scalar one defined as $X(f) = \mathbf{X}(f) \cdot \hat{u}$. As a measure of the quality of the procedure, we pooled together all condition numbers resulting from SVDs on all subjects: the average condition number was 1.5×10^3. This means that, on average, the dipole orientations were, in time, almost fixed (see also [36]) and thus our optimization procedure captures most of the spectral content of the data. The distribution of the logarithm of all condition numbers can be inspected in Figure 1c, showing the reliability of the assumption. Fourier coefficients in the optimized direction *X(f)* were then used to compute power spectral density at each vertex and for each epoch. Pooling together all spectral densities and detecting the peak in the frequency band of interest, we defined, for each subject, an Individual Alpha Frequency (IAF); a distribution of all the 203 IAFs is reported in Figure 1b. All the subsequent connectivity analysis was then performed at the individual alpha frequency. For this reason, the frequency dependence of Fourier coefficients *X(f)* will be dropped hereafter in the notation.

2.6. Connectivity Analysis and Group Statistical Validation

We are interested in connectivity and phase relationships between region of interest belonging to the frontal, parietal and occipital areas within each hemisphere. For this reason, we computed two different spectral based connectivity metrics between all the combinations of areas belonging to the three different sectors. In particular, given the symmetricity of the connectivity metrics we use, we crossed occipital areas with parietal and frontal, and parietal region of interests with frontal ones. For each resulting combination, we computed the Weighted Pairwise Phase Consistency (WPPC; [20]) and the Weighted Phase Lag Index (WPLI; [19]). The WPPC is based on the distribution of the phase differences between all the pairs of observations (resting state 2 s epochs in our case). WPPC is a robust non-biased measure of phase consistency of brain signals, but it is still affected by zero-lag artificial correlations induced by the imperfection of the inverse model solution [21,22]. Complementary to WPPC, we estimated WPLI as a connectivity measure not affected by the artificial zero-lag correlations. Being based on the imaginary part of the coherence, WPLI is immune to zero-lag connectivity but it has another disadvantage: it achieves maximal Signal to Noise Ratio (SNR) when the two brain signals are in a $\pi/2$ relationship. For the frequency band of interest (around 10 Hz) this means a time delay of ~25 ms, a long time when compared to typical brain signal propagation time. However, both metrics strongly depend on a consistent phase relationship between signal of interest, and then they can complementarily provide information about the phase consistency of alpha rhythmicity between the sectors we are investigating.

Given the combination of two brain sectors (*A* and *B*) from the ones defined above (O = occipital; P = parietal; F = frontal) we computed whole sector connectivity matrices, using both WPLI and WPPC, as follows. Named for brevity as $X = X_A(d_a)$ and $Y = Y_B(d_B)$, the Fourier coefficients at each vertex d_a and d_b of the sectors A and B, respectively, we computed a regional dipole-wise connectivity matrix $C_{AB}^M(d_a, d_b) \in \mathcal{M}(D \times D)$, where M represents the metrics (M = WPPC or WPLI). Henceforth, for the sake of clarity, subscripts and superscripts will be omitted when not necessary. In addition, we computed, for each metrics, correspondent null connectivity matrices $\widetilde{C}_A(d_a, d_r)$ and $\widetilde{C}_B(d_b, d_r)$ from sector of interest to a set of D dipoles $\{d_r\}$ randomly chosen within each subject space among the set of dipoles not belonging to any sector of interest. These null connectivity matrices were then used for group statistical validation, under the null hypothesis that connectivity to random dipoles, differently chosen for each subject, will be distributed according only to the metrics' bias and sensitivity. To this aim we performed a bootstrap procedure at the group level by comparing actual and random connectivity matrices: for each combination (A, B) and for each subject, the actual connectivity matrix C_{AB} and the random ones $\widetilde{C}_A$ and $\widetilde{C}_B$ were permuted 10,000 times in order to estimate the empirical distribution of the Fisher regularized difference of connectivity $dC \equiv \tan^{-1}(C_{AB}) - \tan^{-1}(\widetilde{C}_{A/B})$. The comparison between the real dC with its null empirical distribution provided, for each dipole, a bootstrap p-value. (only positive tailed comparisons were considered). Resulting D^2 p-values were then FDR corrected (q = 0.05) and only significant elements of C_AB were considered in the subsequent analysis. Furthermore, we reduced the information in each connectivity matrix by summarizing connectivity between the two sets of ROIs $\{r_a\}$ and $\{r_b\}$ from the atlas [33] belonging to sectors A and B, respectively. To this aim we defined, for all connectivity metrics, the following two quantities:

$$\Delta(r_a, r_b) \equiv \frac{\#[C_{AB}(d_a \in r_a, d_b \in r_B)]}{D}$$

$$\Gamma(r_a, r_b) \equiv P_{95}\{f^{sg}[C_{AB}(d_a \in r_a, d_b \in r_B)]\}$$

where $\#[\cdot]$ and $f^{sg}[\cdot]$ represents the count and the distribution of significant connectivity values, respectively, while P_95 [·] represents the 95% percentile. The quantities Γ and Δ were computed between regions for each combination {r_a,r_b} and from a single region of interest to all the target sector {r_a,B}. Finally, bi-hemispheric results were collapsed by averaging the contribution of both hemispheres. We defined the quantities Δ and Γ to summarize the connectivity between ROIs and/or sectors, given that inspecting he whole dipole by dipole connectivity matrices would have been confusing and not clear to the reader. The two quantities are conceptually derived from standard graph theory analysis. The connectivity degree Δ(r_a,r_b) simply counts, from the connectivity matrix, the number of statistically significant connections between dipoles of the two ROIs, disregarding the strength of the connectivity and normalizing the count to the total number of dipoles in the sector. This is conceptually analogous to the node degree in the context of network analysis. The mean significant connectivity Γ(r_a,r_b) provides a refinement of the connectivity degree, including the information about the connectivity (or phase consistency) strength. This is achieved by selecting from the connectivity matrix only the statistically significant connections between the dipoles of the two ROIs, extracting the 95% percentile of the resulting distribution instead of just counting them. It is worth noticing that choosing the 95% percentile is a conservative choice, given that usually connectivity values bounded between 0 and 1, even when Fisher regularized, can give rise to non-symmetrical distributions.

3. Results

Degree and mean connectivity between each ROI combination ($\Delta(r_a, r_b)$ and $\Gamma(r_a, r_b)$) and from each ROI to the other sectors as a whole ($\Delta(r_a, B)$ and $\Gamma(r_a, B)$) for WPPC and WPLI are shown in Figures 2 and 3 as color coded source maps and connecto-grams. All extracted values are listed in Tables 2 and 3. In general, we can notice that the connectivity estimated from both metrics, each one differently depending on a consistent phase relationship, is predominant in the parietal–frontal connection. While the connection between occipital and frontal sectors appears to be still relevant, the weakest phase consistency has been found between the occipital and the parietal areas. The predominance of phase consistency in parieto–frontal connections is also confirmed by the value of $\Gamma(P, F) = 0.06$, to be compared with $\Gamma(O, P) = 0.03$ and $\Gamma(O, F) = 0.03$ obtained by WPPC considering the whole area as a single ROI in the analysis. When considering the same values but extracted from WPLI, the pattern is less evident ($\Gamma(O, P) = 0.012$; $\Gamma(O, F) = 0.015$ and $\Gamma(P, F) = 0.013$), since the mean connectivity between sectors as detected by the imaginary part of the coherence is almost the same. However, this has to be interpreted in the light of the different sensitivity of the two metrics with respect to the absolute value of phase shift at IAF between the source activity in the two areas, which is ultimately related to the propagation time of the nervous signal on the pathway connecting the ROIs under consideration. As a matter of fact, comparing values of connectivity parameters as extracted from WPPC and WPLI, we can see that the former are in general larger than the first ones. The pattern of a predominant fronto-parietal connectivity with a less consistent pareto-occipital connectivity is also less marked in the WPLI case. We interpret this, methodologically, by considering the different sensitivity of WPPC and WPLI with respect to phase delay absolute values. Being based on the imaginary part of the coherence, WPLI is more sensitive to phase consistent signals whose time delay corresponds to values close to $\phi = \pi/2$, which translate in the alpha band into a propagation time delay of signals of ~25 ms. This time is relatively large if compared to typical brain conduction delays, therefore we expect the phase delay value giving rise to significant connectivity measures to be more toward a phase difference of $\phi = 0$, thus expecting WPPC to be more sensitive. However, because of the finite conduction delays in the white matter, and in general in the brain, we expect more distant areas having larger propagation time. Thus, we expect phase consistency between distant areas to be more evident in the WPLI.

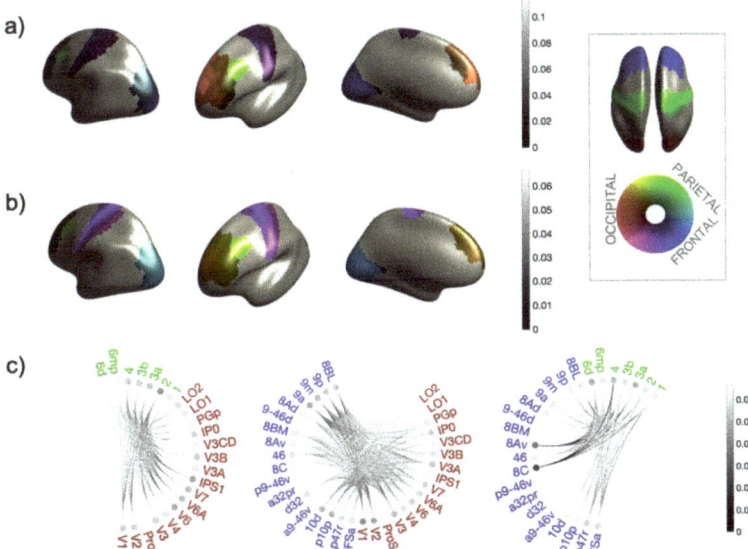

Figure 2. Results for WPPC connectivity. (**a**) Values of $\Delta(r_a, B)$, for all sector combinations in color code. (**b**) Values of $\Gamma(r_a, B)$, for all sector combinations in color code. (**c**) Connectome plots for the three relevant sector combinations computed from G. Red/green/blue ROI names belongs to occipital/parietal/frontal sectors respectively. The gray level of lines encodes the magnitude of $\Gamma(r_a, r_b)$ while the gray level of the dot encodes $\Gamma(r_a, B)$. In the inset the color code is explained: assigning red/green/blue colors to occipital/parietal/frontal sectors, respectively, the intensity of each area represents the overall magnitude of the parameter of interest, while the color encodes the proportion of the magnitude due to connectivity to the other sectors. So, for example, a parietal area

more towards blue has more consistent phase relationship to the frontal sectors, while, if more towards red, the most relevant connection is to the occipital sector.

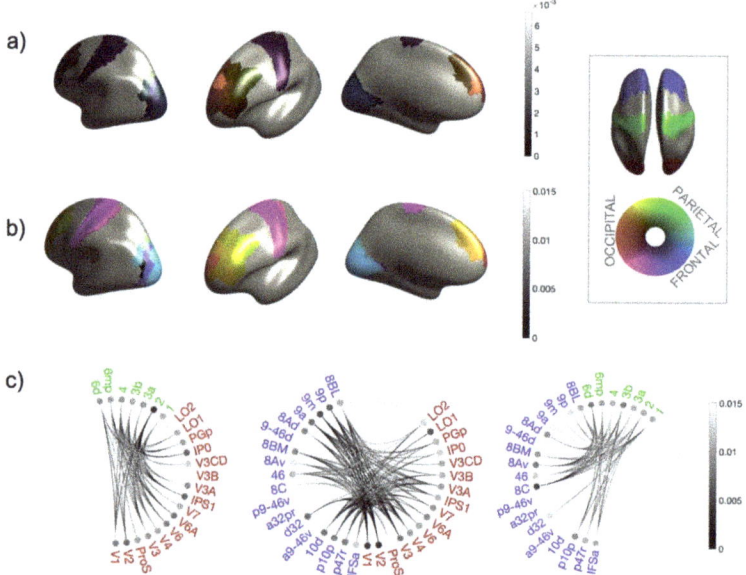

Figure 3. Results for WPLI connectivity. (**a**) Values of $\Delta(r_a, B)$, for all sector combinations in color code. (**b**) Values of $\Gamma(r_a, B)$, for all sector combinations in color code. (**c**) Connectome plots for the three relevant sector combinations computed from G. Red/green/blue ROI names belong to occipital/parietal/frontal sectors, respectively. The gray level of lines encodes the magnitude of $\Gamma(r_a, r_b)$ while the gray level of the dot encodes $\Gamma(r_a, B)$. In the inset the color code is explained: assigning red/green/blue colors to occipital/parietal/frontal sectors, respectively, the intensity of each area represents the overall magnitude of the parameter of interest, while the color encodes the proportion of the magnitude due to connectivity to the other sectors. So, for example, a parietal area more towards blue has more consistent phase relationship to the frontal sectors, while, if more towards red, the most relevant connection is to the occipital sector.

Table 2. Connectivity values for all ROIS towards each of the three sectors. Values of $\Gamma(r_a, b)$ and $\Delta(r, B)$ computed from WPPC are shown.

	Occipital	Parietal	Frontal	
V1	0.00005	0.01813	0.04801	0.03330
V2	0.00026	0.02336	0.03238	0.03052
ProS				
V3	0.00054	0.02260	0.01536	0.02715
V4	0.00007	0.01750	0.00167	0.01769
V6	0.01140	0.03235	0.01202	0.02595
V6A	0.00591	0.02071	0.01167	0.01895
V7	0.00787	0.02024	0.01866	0.02350
IPS1	0.01689	0.03146	0.00290	0.01733
V3A	0.00240	0.02260	0.01683	0.02353
V3B	0.00526	0.02268	0.00203	0.01546
V3CD	0.00017	0.01639	0.00426	0.01660

Table 2. Cont.

	Occipital		Parietal		Frontal	
IP0			0.00256	0.02089	0.00459	0.01869
PGp			0.00036	0.01662	0.00583	0.01964
LO1					0.00054	0.01236
LO2			0.00007	0.01632	0.00008	0.01108
1	0.00147	0.01513			0.01000	0.02590
2	0.01323	0.03352			0.00150	0.02274
3a	0.00182	0.02163			0.01169	0.06285
3b	0.00166	0.01980			0.00928	0.04439
4	0.00153	0.01990			0.01354	0.06664
6mp	0.00026	0.01459			0.00653	0.03106
6d	0.00037	0.01261			0.02895	0.05629
8BL	0.09001	0.02671	0.00231	0.01952		
9p	0.08669	0.02861	0.00182	0.01623		
9m	0.10542	0.03381	0.00006	0.01323		
9a	0.11954	0.02970	0.00015	0.01537		
8Ad	0.03460	0.01914	0.00374	0.01783		
9–46d	0.05946	0.02154	0.00038	0.01336		
8BM	0.04143	0.02319	0.00015	0.01470		
8Av	0.00697	0.01461	0.06058	0.04594		
46	0.01625	0.01679	0.00248	0.01423		
8C	0.00071	0.01432	0.10138	0.05798		
p9–46v	0.00158	0.01213	0.01557	0.01848		
a32pr	0.00488	0.01591	0.00003	0.00649		
d32	0.02711	0.02249	0.00003	0.00840		
a9–46v	0.04548	0.01925	0.00283	0.01968		
10d	0.07100	0.02592	0.00028	0.01980		
p10p	0.05559	0.02151	0.00038	0.01635		
p47r	0.01739	0.01740	0.01290	0.02046		
IFSa	0.00425	0.01535	0.00776	0.01874		

Table 3. Connectivity values for all ROIS towards each of the three sectors. Values of $\Gamma(r_a, b)$ and $\Delta(r, B)$ computed from WPLI are shown.

	Occipital		Parietal		Frontal	
V1			0.00004	0.01007	0.00103	0.01402
V2			0.00002	0.01189	0.00103	0.01511
ProS					0.00008	0.00814
V3			0.00004	0.01173	0.00041	0.01337
V4					0.00012	0.01177
V6			0.00007	0.01102	0.00013	0.01063
V6A			0.00014	0.00900	0.00013	0.01001
V7					0.00019	0.01040
IPS1			0.00023	0.01053		
V3A			0.00005	0.01201	0.00014	0.01081
V3B					0.00016	0.01120
V3CD						
IP0			0.00016	0.01141	0.00003	0.00919
PGp			0.00009	0.01005	0.00009	0.01133
LO1						
LO2					0.00017	0.00917
1	0.00010	0.00946			0.00006	0.01010
2	0.00022	0.01309			0.00006	0.01011
3a	0.00008	0.01093			0.00022	0.01174
3b	0.00013	0.01032			0.00011	0.01131
4	0.00015	0.01078			0.00017	0.01302
6mp	0.00007	0.00970			0.00009	0.00931

Table 3. Cont.

	Occipital		Parietal		Frontal	
6d	0.00018	0.01071			0.00027	0.01096
8BL	0.00266	0.01300	0.00008	0.01005		
9p	0.00664	0.01340	0.00006	0.00831		
9m	0.00505	0.01418	0.00003	0.00788		
9a	0.00261	0.01296				
8Ad	0.00063	0.01018	0.00010	0.00953		
9–46d	0.00206	0.01211	0.00002	0.00974		
8BM	0.00133	0.01180	0.00002	0.00744		
8Av	0.00011	0.00886	0.00074	0.01142		
46	0.00043	0.01072	0.00010	0.00949		
8C	0.00003	0.00789	0.00090	0.01296		
p9–46v	0.00017	0.00939	0.00025	0.01076		
a32pr	0.00051	0.01047	0.00002	0.00623		
d32	0.00134	0.01224	0.00003	0.00763		
a9–46v	0.00119	0.01163	0.00005	0.00786		
10d	0.00269	0.01193				
p10p	0.00150	0.01111	0.00012	0.00804		
p47r	0.00051	0.01068	0.00007	0.00797		
IFSa	0.00029	0.00949	0.00011	0.00918		

Inspecting more in detail $\Delta(r_a, B)$ and $\Gamma(r_a, B)$), as we can see from Figure 2a,b, the overall higher values for both the quantities extracted from WPPC correspond to the pareto–frontal connections, giving frontal area 8C the overall highest values ($\Delta(8C, P) = 0.1$ and $\Gamma(8C, P) = 0.06$)). This predominance can be also appreciated in the connecto-grams (Figure 2c), where connections between frontal areas 8C and 8av appear to be the highest in this sector's combination and with respect to the other sector combinations. This result is further confirmed by values obtained from WPLI, as shown in Figure 3. In this case also, referring to the parietal-frontal connection, area 8C is the most connected, with $\Delta(8C, P) = 0.0009$ and $\Gamma(8C, P) = 0.013$. We can thus conclude that this pattern is not due to the proximity of area 8C with respect to the frontal sector, WPLI being insensitive to zero-lag connections by design. However, in this case, the degree and the mean connectivity of areas 8c and 8av are not the highest with respect to other sector combinations, as can be appreciated in the connecto-grams of Figure 3c and in the color-coded source maps of mean connectivity in Figure 3b.

As regards the occipito–frontal connections, the highest WPPC connectivity and degree values has been found from frontal area 9a to the occipital sector, being $\Delta(9a, O) = 0.12$ and $\Gamma(9a, O) = 0.03$. The corresponding values for WPLI are $\Delta(9a, O) = 0.003$ and $\Gamma(9a, O) = 0.013$ while, in this case, areas 9m and 9p have also been found among the most connected to the occipital sector ($\Delta(9m, O) = 0.007$ and $\Gamma(9m, O) = 0.015$; $\Delta(9p, O) = 0.007$ and $\Gamma(9p, O) = 0.013$). This pattern is confirmed also by inspecting the connecto-grams in Figure 2c for WPPC and Figure 3c for WPLI, where, in the latter case, connectivity from frontal areas to occipital is more evident than in the WPPC case. As we pointed out before, we interpret this as a byproduct of the different sensitivity of WPPC and WPLI to coherent signals whose phase delay is more towards $\phi = 0$ or $\phi = \pi/2$.

Finally, also in the single ROI connectivity values inspection, the occipito–parietal connectivity seems to be less relevant. This can be appreciated in the connecto-grams for WPPC in Figure 2c, where the highest values, in this case, can be found in frontal area 2 ($\Gamma(2, O) = 0.034$ and in occipital area V6 ($\Gamma(V6, F) = 0.033$) for the WPPC. This is confirmed also from WPLI results ($\Gamma(2, O) = 0.013$ and ($\Gamma(V6, F) = 0.012$). However, in this case the predominance of these two areas in the occipito-parietal connectivity is less marked as can be noticed from the connecto-grams in Figure 3c.

4. Discussion

Our aim was to provide the brain-state dependent EEG-TMS community with a work describing functional phase-dependent relationships between frontal, parietal and occipital areas. In these regions, alpha phase-dependent cortical excitability has been studied both in healthy controls and patients via brain-state dependent stimulation [8,13–15]. (Since the attempts at manipulating the effects of stimulation triggered by the phase detected on a region different from the target have no clear effectiveness [8], we believe that taking into account off-line phase-dependent connectivity patterns between the areas of interest would be of substantial help in predicting results from these studies.

Furthermore, consideration of the general phase-dependent connectivity state of the target region, (either trigger-region or not), could be essential for explaining changes in connectivity patterns for the time after application of plasticity protocols up to the moment where the behavioral measure has returned to baseline. Considering not only the TMS trigger-state (phase of the frequency of interest) but also the general connectivity state of the brain between the regions of interest would in fact be useful both for predicting plasticity protocols' efficacy and for interpreting results. For these reasons, we investigated spectral-resolved connectivity metrics that depend on a consistent phase relationship at IAF between frontal, parietal and occipital regions at rest. Results showed greater functional coupling between frontal and parietal areas, compared to very low functional coupling values between frontal and occipital and parietal and occipital pairs (Figure 2). These results are consistent with previous evidence of a fronto-parietal network at rest (e.g., [37]). We used MEG data and three different phase-coherence measures to show that a clear coupling emerges at IAF between frontal and parietal parcels at rest.

The ability to assess the strength of frontoparietal coupling from resting state data before and after an intervention makes this a suitable pathway for the investigation of pathway-specific plasticity. The alpha frequency band is relevant because this is the frequency band where phase-specific modulation of cortical excitability was previously demonstrated both in the motor system [8,13,14] and in the prefrontal cortex [15]. Therefore, in frontal and sensorimotor regions, state-dependent stimulation based on the phase of alpha frequency appears effective in modulating cortical excitability at rest. However, when TMS is applied to the motor cortex in response to the phase of alpha recorded from the occipital cortex at rest, no modulation of corticospinal excitability was found [8].

For future development of real-time estimators of instantaneous connectivity states, we propose that connectivity metrics which depend on phase relationship between areas, and therefore describe whether two regions share trial-averaged functional coupling, can provide off-line connectivity priors in order to enable more accurate time-resolved connectivity estimates. This broad state of functional coupling will, in fact, constitute a novel level of a priori information to assess whether real-time EEG-TMS paradigms will be effective in modulating cortical excitability in one area when the trigger is recorded from another area. In this light, for example, our finding of the limited functional coupling at rest between occipital and parietal areas can help elaborate on the reason why no effect was found when the hand knob of the motor cortex was stimulated on the basis of the occipital alpha phase [8].

Therefore, we propose that, in order to improve the efficacy of brain-state dependent stimulation protocols, the definition of "brain state" should not only refer to the instantaneous phase of the considered frequency which is triggering the stimulus, but also to the predisposition of the network to be globally perturbed by a stimulus triggered by a given phase of a determined frequency. In this regard, measures of functional coupling of regions at rest or during tasks that involve different functional coupling patterns would allow the obtaining of "off-line" connectivity priors.

In essence, beside the targeting of a local oscillation phase, brain-state dependent stimulation might leverage the off-line acquired, a priori knowledge regarding functional patterns in order to causally test the functional coupling between the brain regions involved. As an example of potential relevance of this approach, it has been shown that a network of prefrontal and occipito–parietal areas is involved in visual target detection and that alpha phase in this network is essential for a correct visuospatial processing [38,39]. Standard brain-state dependent stimulation could be used in order to further test the relationship between these cortical areas. In fact, brain-state dependent stimulation could potentially modulate cortical excitability of one area given the phase of alpha from another region. However, according to our approach, this brain-state dependent stimulation could lead to more effective results only when the two regions are embedded in a functional pattern that describes a state of sensitivity to be modulated [13,21]. A relevant part of cortical excitability could derive therefore from functional coupling of the target areas at rest (or during a task) at a given frequency.

This new perspective would also be useful for developing new and more effective individualized approaches for rehabilitation via brain-state dependent stimulation. In this regard, individualized stimulation rehabilitative protocols have been proposed in stroke patients for whom specific adaptive connectivity patterns between specific cortical regions should be reinforced in order to achieve recovery [40]. In this light, our approach combining brain-state dependent stimulation with phase-dependent connectivity measures would nicely suit this purpose.

Finally, it must be noticed that our work is the first attempt at measuring coherence at IAF during rest between these three regions of the cerebral cortex. Most of the resting state literature consists of fMRI studies, which lack the temporal resolution that MEG measures provide [41–43]. It is also worth noticing that, depending on the connectivity metric employed, we obtain slightly different results in the coupling between frontal and occipital areas, with the WPLI showing some functional coupling between the two regions compared to other measures. The WPLI is a measure based on the imaginary part of coherence. This metric is not sensitive to zero-lag spurious correlations. However, its maximum SNR spuriously coincides with $\pi/4$. Since we are investigating IAF which, on average, refers to around 10 Hz, the maximum SNR of WPLI measures has a phase-lag of about 25 ms. For these reasons, the functional relationship between distant areas like the frontal and occipital cortices might be captured thanks to this longer phase-lag that allows pick-up of the coupling between signals with longer propagation time.

To summarize, our findings suggest that brain-state dependent stimulation could benefit from taking into account a broader concept of "state" of the system. Depending on whether brain-state dependent stimulation is practiced at rest or during a task, one should consider the functional coupling patterns between the areas of interest in the frequency band whose phase is used as a trigger. Furthermore, we suggest that the physical distance between the regions could also be taken into account when choosing the coherence metric to assess functional coupling based on a consistent phase relationship in the frequency band of interest.

5. Conclusions

Brain-state dependent brain-stimulation has the potential to reliably modulate specific neural pathways using a "Closed-Loop" approach [44]. How to achieve a time-resolved real-time estimation of EEG-derived brain connectivity states remains a key challenge. In this study, we addressed the off-line estimation of phase-based connectivity patterns between different regions of interest, presenting a pipeline to determine candidate pathways for real-time paradigms, for instance where the trigger-signal is extracted from one cortical site, and the stimulation targets a different site, or considering the target region and downstream effects to distal cortical regions.

In summary, offline connectivity measures have a twofold purpose: first, the estimate of connectivity patterns before and after can represent a biomarker of cortico-cortical changes after single and repetitive state-dependent TMS. In fact, for mu-alpha, stimulation at troughs was found to enhance Transcranial evoked potentials (TEPs) in both ipsi and contralateral hemisphere at 100 ms after stimulus even when the stimulus was provided at 90% of motor threshold [10]. Moreover, the same kind of stimulation has been found to be responsible for an enhancement of connectivity between the two sensorimotor cortices [45]. Connectivity joined with TEP analysis could be even more relevant when analyzing non-motor areas such as the dorsolateral prefrontal cortex, where one cannot rely on as reliable and consistent an outcome as that of the MEPs.

Second, an off-line connectivity analysis can be used as prior information for future online estimates, to determine a correlation between areas oscillating at the same frequency or in an anticorrelation, a phase state which was previously suggested as a possible marker of functional segregation [46], or no correlation at all. The offline measures can serve as a benchmark for determining the optimal trade-off between the temporal resolution (window-length) and the estimator error variance for a given experimental condition, noting that the window-length must be short enough to capture physiological transitions [47].

Author Contributions: Conceptualization, D.T., A.B. and P.B.; Formal Analysis, D.T.; Visualization: D.T.; Writing—Original Draft Preparation, A.B., D.T. and P.B.; Writing—Review & Editing, A.B., D.T., P.B. and C.Z. All authors have read and agreed to the published version of the manuscript.

Funding: This research received no direct external funding. CamCAN funding was provided by the UK Biotechnology and Biological Sciences Research Council (grant number BB/H008217/1), together with support from the UK Medical Research Council and University of Cambridge, UK.

Institutional Review Board Statement: Data analyzed in this work were obtained from the CamCAN repository. Data were collected in compliance with the Helsinki Declaration (Cambridgeshire 2 Research Ethics Committee; reference: 10/H0308/50). See [24] for details.

Informed Consent Statement: Data analyzed in this manuscript are publicly available from of the Cambridge Centre for Aging and Neuroscience (CamCAN) public dataset [25]. No data were collected by the authors for this study. For information about original informed consent see the CamCAN dataset documentation [25].

Data Availability Statement: Data analyzed in this study are part of the Cambridge Centre for Aging and Neuroscience (CamCAN) public dataset [25] and are publicly available.

Conflicts of Interest: The authors declare no conflict of interest.

References

1. Berger, H. Über das elektroenkephalogramm des menschen. *Arch. Psychiatr. Nervenkrankh.* **1929**, *87*, 527–570. [CrossRef]
2. Da Silva, F.L.; Vos, J.E.; Mooibroek, J.; Van Rotterdam, A. Relative contributions of intracortical and thalamo-cortical processes in the generation of alpha rhythms, revealed by partial coherence analysis. *Electroencephalogr. Clin. Neurophysiol.* **1980**, *50*, 449–456. [CrossRef]
3. Vijayan, S.; Kopell, N.J. Thalamic model of awake alpha oscillations and implications for stimulus processing. *Proc. Natl. Acad. Sci. USA* **2012**, *109*, 18553–18558. [CrossRef] [PubMed]
4. Stolk, A.; Brinkman, L.; Vansteensel, M.J.; Aarnoutse, E.; Leijten, F.S.; Dijkerman, C.H.; Knight, R.T.; de Lange, F.P.; Toni, I. Electrocorticographic dissociation of alpha and beta rhythmic activity in the human sensorimotor system. *eLife* **2019**, *8*, e48065. [CrossRef]
5. Klimesch, W.; Sauseng, P.; Hanslmayr, S. EEG alpha oscillations: The inhibition–timing hypothesis. *Brain Res. Rev.* **2007**, *53*, 63–88. [CrossRef]
6. Jensen, O.; Mazaheri, A. Shaping functional architecture by oscillatory alpha activity: Gating by inhibition. *Front. Hum. Neurosci.* **2010**, *4*, 186. [CrossRef]
7. Mathewson, K.E.; Gratton, G.; Fabiani, M.; Beck, D.M.; Ro, T. To see or not to see: Prestimulus α phase predicts visual awareness. *J. Neurosci.* **2009**, *29*, 2725–2732. [CrossRef]
8. Zrenner, C.; Desideri, D.; Belardinelli, P.; Ziemann, U. Real-time EEG-defined excitability states determine efficacy of TMS-induced plasticity in human motor cortex. *Brain Stimul.* **2018**, *11*, 374–389. [CrossRef]
9. Bergmann, T.O.; Lieb, A.; Zrenner, C.; Ziemann, U. Pulsed facilitation of corticospinal excitability by the sensorimotor μ-alpha rhythm. *J. Neurosci.* **2019**, *39*, 10034–10043. [CrossRef]
10. Desideri, D.; Zrenner, C.; Ziemann, U.; Belardinelli, P. Phase of sensorimotor μ-oscillation modulates cortical responses to transcranial magnetic stimulation of the human motor cortex. *J. Physiol.* **2019**, *597*, 5671–5686. [CrossRef]
11. Mazaheri, A.; Jensen, O. Rhythmic pulsing: Linking ongoing brain activity with evoked responses. *Front. Hum. Neurosci.* **2010**, *4*, 177. [CrossRef] [PubMed]
12. Dugué, L.; Marque, P.; VanRullen, R. The phase of ongoing oscillations mediates the causal relation between brain excitation and visual perception. *J. Neurosci.* **2011**, *31*, 11889–11893. [CrossRef] [PubMed]

13. Stefanou, M.I.; Desideri, D.; Belardinelli, P.; Zrenner, C.; Ziemann, U. Phase synchronicity of μ-rhythm determines efficacy of interhemispheric communication between human motor cortices. *J. Neurosci.* **2018**, *38*, 10525–10534. [CrossRef] [PubMed]
14. Schaworonkow, N.; Gordon, P.C.; Belardinelli, P.; Ziemann, U.; Bergmann, T.O.; Zrenner, C. μ-rhythm extracted with personalized EEG filters correlates with corticospinal excitability in real-time phase-triggered EEG-TMS. *Front. Neurosci.* **2018**, *12*, 954. [CrossRef]
15. Zrenner, B.; Zrenner, C.; Gordon, P.C.; Belardinelli, P.; McDermott, E.J.; Soekadar, S.R.; Fallgatter, A.J.; Ziemann, U.; Müller-Dahlhaus, F. Brain oscillation-synchronized stimulation of the left dorsolateral prefrontal cortex in depression using real-time EEG-triggered TMS. *Brain Stimul.* **2020**, *13*, 197–205. [CrossRef]
16. Gordon, P.C.; Dörre, S.; Belardinelli, P.; Stenroos, M.; Zrenner, B.; Ziemann, U.; Zrenner, C. Prefrontal theta-phase synchronized brain stimulation with real-time EEG-Triggered TMS. *Front. Hum. Neurosci.* **2021**, 335. [CrossRef]
17. Fries, P. Rhythms for cognition: Communication through coherence. *Neuron* **2015**, *88*, 220–235. [CrossRef]
18. Nieminen, J.O.; Sinisalo, H.; Souza, V.H.; Malmi, M.; Yuryev, M.; Tervo, A.E.; Stenroos, M.; Milardovich, D.; Korhonen, J.T.; Koponen, L.M.; et al. Multi-locus transcranial magnetic stimulation system for electronically targeted brain stimulation. *Brain Stimul.* **2022**, *15*, 116–124. [CrossRef]
19. Vinck, M.; van Wingerden, M.; Womelsdorf, T.; Fries, P.; Pennartz, C.M.A. The pairwise phase consistency: A bias-free measure of rhythmic neuronal synchronization. *Neuroimage* **2010**, *51*, 112–122. [CrossRef]
20. Vinck, M.; Oostenveld, R.; van Wingerden, M.; Battaglia, F.; Pennartz, C.M.A. An improved index of phase-synchronization for electrophysiological data in the presence of volume-conduction, noise and sample-size bias. *Neuroimage* **2011**, *55*, 1548–1565. [CrossRef]
21. Bastos, A.M.; Schoffelen, J.M. A tutorial review of functional connectivity analysis methods and their interpretational pitfalls. *Front. Syst. Neurosci.* **2016**, *9*, 175. [CrossRef] [PubMed]
22. Gross, J.; Kluger, D.S.; Abbasi, O.; Chalas, N.; Steingräber, N.; Daube, C.; Schoffelen, J.-M. Comparison of undirected frequency-domain connectivity measures for cerebro-peripheral analysis. *Neuroimage* **2021**, *245*, 118660. [CrossRef] [PubMed]
23. Pezzulo, G.; Zorzi, M.; Corbetta, M. The secret life of predictive brains: What's spontaneous activity for? *Trends Cogn. Sci.* **2021**, *25*, 730–743. [CrossRef] [PubMed]
24. Folstein, M.F.; Folstein, S.E.; McHugh, P.R. 'Mini-mental state' a practicalmethod for grading the cognitive state of patients for the clinician. *J. Psychiatr. Res.* **1975**, *12*, 189–198. [CrossRef]
25. Taylor, J.R.; Williams, N.; Cusack, R.; Auer, T.; Shafto, M.A.; Dixon, M.; Tyler, L.K.; Henson, R.N. The Cambridge Centre for ageing and neuroscience (Cam-CAN) data repository: Structural and functional MRI, MEG, and cognitive data from a cross-sectional adult lifespan sample. *Neuroimage* **2017**, *144*, 262–269.
26. Oostenveld, R.; Fries, P.; Maris, E.; Schoffelen, J.-M. FieldTrip: Open source software for advanced analysis of MEG, EEG, and invasive electrophysiological data. *Comput. Intell. Neurosci.* **2011**, *2011*, 868305. [CrossRef] [PubMed]
27. Friston, K.J. *Statistical Parametric Mapping: The Analysis of Funtional Brain Images*, 1st ed.; Elsevier: Amsterdam, The Netherlands; Academic Press: Boston, MA, USA, 2007.
28. Taulu, S.; Simola, J. Spatiotemporal signal space separation method for rejecting nearby interference in MEG measurements. *Phys. Med. Biol.* **2006**, *51*, 1759–1768. [CrossRef]
29. Taulu, S.; Simola, J.; Kajola, M. Applications of the signal space separation method. *IEEE Trans. Signal Process.* **2005**, *53*, 3359–3372. [CrossRef]
30. Lee, T.-W.; Girolami, M.; Sejnowski, T.J. Independent component analysis using an extended infomax algorithm for mixed subgaussian and supergaussian sources. *Neural Comput.* **1999**, *11*, 417–441. [CrossRef]
31. Hyvärinen, A.; Karhunen, J.; Oja, E. *Independent Component Analysis*; Wiley: Hoboken, NJ, USA, 2001.
32. Fischl, B.; Sereno, M.I.; Tootell, R.B.H.; Dale, A.M. High-resolution intersubject averaging and a coordinate system for the cortical surface. *Hum. Brain Mapp.* **1999**, *8*, 272–284. [CrossRef]
33. Glasser, M.F.; Coalson, T.S.; Robinson, E.C.; Hacker, C.D.; Harwell, J.; Yacoub, E.; Ugurbil, K.; Andersson, J.; Beckmann, C.F.; Jenkinson, M.; et al. A multi-modal parcellation of human cerebral cortex. *Nature* **2016**, *536*, 171–178. [CrossRef] [PubMed]
34. Ilmoniemi, R.; Sarvas, J. *Brain Signals: Physics and Mathematics of MEG and EEG*; MIT Press: Cambridge, MA, USA, 2019.
35. Hämäläinen, M.; Hari, R.; Ilmoniemi, R.J.; Knuutila, J.; Lounasmaa, O.V. Magnetoencephalography—Theory, instrumentation, and applications to noninvasive studies of the working human brain. *Rev. Mod. Phys.* **1993**, *65*, 413–497. [CrossRef]
36. Gross, J.; Kujala, J.; Hamalainen, M.; Timmermann, L.; Schnitzler, A.; Salmelin, R. Dynamic imaging of coherent sources: Studying neural interactions in the human brain. *Proc. Natl. Acad. Sci. USA* **2001**, *98*, 694–699. [CrossRef] [PubMed]
37. Markett, S.; Reuter, M.; Montag, C.; Voigt, G.; Lachmann, B.; Rudorf, S.; Elger, C.E.; Weber, B. Assessing the function of the fronto-parietal attention network: Insights from resting-state fMRI and the attentional network test. *Hum. Brain Mapp.* **2014**, *35*, 1700–1709. [CrossRef]
38. Capotosto, P.; Babiloni, C.; Romani, G.L.; Corbetta, M. Frontoparietal cortex controls spatial attention through modulation of anticipatory alpha rhythms. *J. Neurosci.* **2009**, *29*, 5863–5872. [CrossRef]
39. Hamm, J.P.; Dyckman, K.A.; McDowell, J.E.; Clementz, B.A. Pre-cue fronto-occipital alpha phase and distributed cortical oscillations predict failures of cognitive control. *J. Neurosci.* **2012**, *32*, 7034–7041. [CrossRef]
40. Brancaccio, A.; Tabarelli, D.; Belardinelli, P. A new framework to interpret individual inter-hemispheric compensatory communication after stroke. *J. Pers. Med.* **2022**, *12*, 59. [CrossRef]

41. Fox, M.D.; Corbetta, M.; Snyder, A.Z.; Vincent, J.L.; Raichle, M.E. Spontaneous neuronal activity distinguishes human dorsal and ventral attention systems. *Proc. Natl. Acad. Sci. USA* **2006**, *103*, 10046–10051. [CrossRef]
42. Mantini, D.; Perrucci, M.G.; Del Gratta, C.; Romani, G.L.; Corbetta, M. Electrophysiological signatures of resting state networks in the human brain. *Proc. Natl. Acad. Sci. USA* **2007**, *104*, 13170–13175. [CrossRef]
43. Smitha, K.A.; Akhil Raja, K.; Arun, K.M.; Rajesh, P.G.; Thomas, B.; Kapilamoorthy, T.R.; Kesavadas, C. Resting state fMRI: A review on methods in resting state connectivity analysis and resting state networks. *Neuroradiol. J.* **2017**, *30*, 305–317. [CrossRef]
44. Zrenner, C.; Belardinelli, P.; Müller-Dahlhaus, F.; Ziemann, U. Closed-loop neuroscience and non-invasive brain stimulation: A tale of two loops. *Front. Cell. Neurosci.* **2016**, *10*, 92. [CrossRef] [PubMed]
45. Momi, D.; Ozdemir, R.A.; Tadayon, E.; Boucher, P.; Shafi, M.M.; Pascual-Leone, A.; Santarnecchi, E. Network-level macroscale structural connectivity predicts propagation of transcranial magnetic stimulation. *Neuroimage* **2021**, *229*, 117698. [CrossRef] [PubMed]
46. Maris, E.; Fries, P.; van Ede, F. Diverse phase relations among neuronal rhythms and their potential function. *Trends Neurosci.* **2016**, *39*, 86–99. [CrossRef] [PubMed]
47. Ermolova, M.; Metsomaa, J.; Zrenner, C.; Kozák, G.; Marzetti, L.; Ziemann, U. Spontaneous phase-coupling within cortico-cortical networks: How time counts for brain-state-dependent stimulation. *Brain Stimul.* **2021**, *14*, 404–406. [CrossRef] [PubMed]

Article

Effects of Transcranial Magnetic Stimulation Therapy on Evoked and Induced Gamma Oscillations in Children with Autism Spectrum Disorder

Manuel F. Casanova [1,2], Mohamed Shaban [3], Mohammed Ghazal [4], Ayman S. El-Baz [5], Emily L. Casanova [1], Ioan Opris [6] and Estate M. Sokhadze [1,2,*]

1. Department of Biomedical Sciences, University of South Carolina School of Medicine-Greenville, 701 Grove Rd., Greenville, SC 29605, USA; Manuel.Casanova@prismahealth.org (M.F.C.); Emily.Casanova@prismahealth.org (E.L.C.)
2. Department of Psychiatry & Behavioral Sciences, University of Louisville, 401 E Chestnut Str., #600, Louisville, KY 40202, USA
3. Department of Electrical and Computer Engineering, University of South Alabama, Mobile, AL 36688, USA; mshaban@southalabama.edu
4. BioImaging Research Lab, Electrical and Computer Engineering Abu Dhabi University, Abu Dhabi 59911, UAE; mohammed.ghazal@adu.ac.ae
5. Department of Bioengineering, University of Louisville, Louisville, KY 40202, USA; ayman.elbaz@louisville.edu
6. School of Medicine, University of Miami, Miami, FL 33136, USA; ixo82@miami.edu
* Correspondence: tato.sokhadze@louisville.edu; Tel.: +1-(502)-294-6522

Received: 30 May 2020; Accepted: 30 June 2020; Published: 3 July 2020

Abstract: Autism spectrum disorder (ASD) is a behaviorally diagnosed neurodevelopmental condition of unknown pathology. Research suggests that abnormalities of elecltroencephalogram (EEG) gamma oscillations may provide a biomarker of the condition. In this study, envelope analysis of demodulated waveforms for evoked and induced gamma oscillations in response to Kanizsa figures in an oddball task were analyzed and compared in 19 ASD and 19 age/gender-matched neurotypical children. The ASD group was treated with low frequency transcranial magnetic stimulation (TMS), (1.0 Hz, 90% motor threshold, 18 weekly sessions) targeting the dorsolateral prefrontal cortex. In ASD subjects, as compared to neurotypicals, significant differences in evoked and induced gamma oscillations were evident in higher magnitude of gamma oscillations pre-TMS, especially in response to non-target cues. Recordings post-TMS treatment in ASD revealed a significant reduction of gamma responses to task-irrelevant stimuli. Participants committed fewer errors post-TMS. Behavioral questionnaires showed a decrease in irritability, hyperactivity, and repetitive behavior scores. The use of a novel metric for gamma oscillations. i.e., envelope analysis using wavelet transformation allowed for characterization of the impedance of the originating neuronal circuit. The results suggest that gamma oscillations may provide a biomarker reflective of the excitatory/inhibitory balance of the cortex and a putative outcome measure for interventions in autism.

Keywords: Autism spectrum disorder; evoked and induced gamma oscillations; EEG; TMS; oddball task; reaction time; aberrant and repetitive behaviors

---

1. Introduction

Rhythmic patterns of neural activity, manifested in the electroencephalogram (EEG) as voltage oscillations, have been linked to varied cognitive functions such as perception, attention, memory, and consciousness. The reciprocal interaction between excitation (pyramidal cells) and inhibition (interneurons) during cortical activation provides the genesis for brainwave oscillations [1]. Those brainwaves with the highest frequency, between 30 and 90 Hz, comprise the gamma bandwidth [1,2]. Fast-spiking interneurons that provide for the perisomatic inhibition of pyramidal cells, control the rhythm (clockwork) of these high frequency oscillations [1]. Immunocytochemical characterization of these cells reveals that they express the calcium-binding albumin protein parvalbumin (PV). The high metabolic activity of PV cells, which comprise the largest subgroup of cortical interneurons, makes them highly susceptible to oxidative injury. This pathoclisis helps explain their putative relationship to abnormalities of gamma aminobutyric acidergic (GABAergic) neurotransmission in many psychiatric disorders [3].

Reduced numbers of PV-expressing cells have been reported in human postmortem brain samples [4] and animal models of autism spectrum disorder (ASD) (e.g., *Fmr1*, VPA, *Nlgn3*, R451C, and *Cntnap2*) [5]. More significantly, the reduced levels of PV expression correlate with ASD-like behavioral deficits (e.g., sociability, vocalization) and, curiously enough, with symptoms usually ascribed to ASD comorbidities (e.g., pain sensitivity, seizures) [6,7]. Long-lasting reversal of PV (GABAergic) deficits by pharmacologic or cell type-specific gene rescue, normalizes, or at least diminishes, cognitive dysfunction, and social deficits in these animal models [5,8,9]. It is therefore unsurprising that ASD researchers have proposed using gamma-band-based metrics, both a putative "electrophysiological endophenotype" [6] of PV pathology and a metric indicative of the cortical balance between inhibition and excitation [2], as an outcome measure for interventions aimed at targeting the underlying pathology of ASD [10–12].

Gamma band activity is thought to reflect the mechanism for the integration of information in neural networks within and between brain regions (for reviews see [12,13]). Gamma rhythm is normally defined as EEG band in the frequency range between 30 to 90 Hz (or even higher), although there is an opinion [14] that different frequency sub-bands (e.g., 30–35 Hz, 40–48 Hz, etc.) may have distinct functional significance. Our study focuses on gamma sub-band within 35–45 Hz (so-called 40 Hz-centered gamma [15,16]). Oscillatory activity in the 40 Hz-centered gamma range has been related to Gestalt perception and to cognitive functions such as attention, learning, and memory [17,18]. Binding of widely distributed cell assemblies by synchronization of gamma frequency activity is thought to underlie cohesive stimulus representation in the brain [19,20]. It has been proposed that "weak central coherence" in autism could result from a reduction in the integration of specialized local networks in the brain caused by a deficit in temporal binding that depends on gamma synchronization [21–23]. It is important to emphasize that there are distinct functional differences between spontaneous gamma, evoked gamma power and coherence, and event-related induced gamma power and coherence [14]. Sensory evoked gamma coherences reflect the property of modality-specific networks activated by a sensory stimulation. Event-related (or cognitive) induced gamma and its coherences manifest coherent activity of sensory and cognitive networks triggered by and governed by requirements of a cognitive task. In autism, synchronization between these neural networks is abnormal and reflects an imbalance of the excitation/inhibition bias of the cerebral cortex (vide supra, [2]). Studies have shown that resting gamma power appears to be inversely correlated to ASD severity as measured by the Social Responsiveness Scale (SRS) [24].

Illusory contour or illusory figure (e.g., Kanizsa figure [25]) perception is a very useful model to study the integration of local image features into a coherent percept, and tests based on several illusory figures were productively used to investigate the impairment of such integration in children with ASD. Brown et al. [22] tested adolescents with autism in an experiment that presented Kanizsa shapes with visual illusions and reported excessive evoked gamma at 80 and 120 ms post-stimulus, in addition to enhanced induced gamma (200–400 ms). Inability to reduce gamma activity would lead to the inability to decide which event requires attention when there are multiple choices. In autism, uninhibited gamma activity suggests that none of the circuits in the brain can come to dominance because too many are active simultaneously [21–23]. Abnormalities of gamma synchrony can result in significant cognitive deficits, such as reduced attentional control, and other dysfunctions present in ASD. In addition,

EEG recordings during a Kanizsa figure task have shown an overall increase in gamma oscillatory activity in ASD as compared to neurotypicals [22]. These findings are thought to reflect a reduction in the "signal to noise" level due to diminished inhibitory processing [22]. These observations are of clinical significance as several studies have now reported in ASD that abnormalities in gamma oscillations are normalized by low frequency repetitive transcranial stimulation (TMS). This neuromodulatory therapy also provides improvements in both repetitive behaviors and executive functions [10,26–29].

In a recent study comparing ASD and neurotypical controls, spectral analysis of the outer envelope joining the upper peaks of gamma oscillations allowed researchers to characterize the settling time after peak voltage amplitude [30]. At baseline, with no active treatment instituted, the latency of the ringing decay assessed using frequency analysis was significantly diminished in ASD as compared to control subjects. A short ringing time indicates a system whose efficiency of operation or sensitivity is diminished [31]. The oscillations induced by tasks involving the integration of features, as for example in a reaction time tasks using Kanizsa illusory features. Our group has used oddball task paradigms of target classification and discrimination which required a response to target Kanizsa squares among non-target Kanizsa triangles and other non-Kanizsa distractor figures in order to examine event-related potentials (ERP) and amplitude of gamma-band EEG activity [10,12,29]. We reported differences between neurotypical children and children with ASD diagnosis in reaction time and ERP measures as well as amplitude of gamma responses. Furthermore, we reported normalization of ERP responses and improved behavioral symptoms in children with ASD following repetitive transcranial magnetic stimulation (rTMS) treatment [28]. The current study was focused on more advanced analysis of evoked and induced gamma oscillations using the same illusory figure task. The gamma waveforms elicited by this task exhibit a characteristic dampening after peak amplitude in which the outer envelope of successive peaks traces a decay curve that persists until baseline.

The study used demodulation of gamma oscillations allowing to examine both the envelope of a signal as well as the periodic waveform that carries the same suggesting that resonance behavior, exemplified in the carrier wave may tie neural populations operating at the same frequency. Analysis of the envelope of gamma oscillations can be used to investigate the impedance of involved circuits and the excitatory inhibitory balance of the cerebral cortex [30]. We propose that the metrics of gamma oscillations, ingrained in both its carrier and its envelope, may provide important information contributing to better understanding of functional significance of EEG gamma waveforms.

Despite all of the evidence, the utility of gamma-band related variables as diagnostic biomarkers is currently unexplored, suggesting an urgent need for using gamma oscillation measures as functional markers of response to interventions such as rTMS or other types of neuromodulation. This sensitivity is what allows a system to respond selectively to a given frequency while eliminating others. Brainwave oscillations are not a finely tuned process; amplitude, frequency and phase all vary across individual gamma cycles [32]. In autism, the low sensitivity makes the synchronization between neuronal networks imprecise. The end result is a distortion in the ability to form cohesive perceptual experiences and a reduction in the brain's ability to provide for nuanced responses to both environmental and social exigencies.

The findings described in the previous paragraphs led the authors to study and compare, in ASD and a neurotypical control population, the metrics for gamma oscillations that describe its envelope. This study expands on previous findings by analyzing the evoked and induced components of gamma oscillations using wavelet transformation. Given the many reports in the literature which translate gamma oscillations to possible behavioral states we also analyzed for possible correlates to aberrant and/or repetitive behaviors in our ASD population post-TMS treatment.

2. Methods and Materials

2.1. Subjects

Children and adolescents with ASD diagnosis were recruited through referrals from several pediatric clinics. All patients (N = 19 mean age, 14.4 ± 3.61 years old, 5 females) were diagnosed according to the Diagnostic and Statistical Manual of Mental Disorders (DSM-IVTR) and/or DSM-5 [33,34]. Diagnosis of autism was further ascertained with the Autism Diagnostic Interview-Revised (ADI-R) [35]). A developmental pediatrician evaluated the patients, ascertained them to be in good health, had normal hearing, and were willing to participate in lab testing. Participants were excluded if they had a history of seizures, impairment of vision, genetic disorders, and/or brain abnormalities based on neuroimaging studies. Exclusionary criteria for this group were as follows: (a) current diagnosis of any Axis I psychiatric disorder, such as psychosis, bipolar disorder, and schizophrenia; (b) current psychiatric symptoms requiring medication other than those for attention deficit/hyperactivity disorder (ADHD); (c) severe medical, cognitive or psychiatric impairments that would preclude from cooperation with the study protocol; and (d) inability to read, write, or speak English. The EEG test procedures also required the following exclusionary criteria: (1) impaired, non-correctable vision or hearing; (2) significant neurological disorder (epilepsy, encephalitis) or head injury. Subjects enrolled in the study were high-functioning children or adolescents with a full-scale Intelligence Quotient (IQ) of more than 80 according to evaluations using the Wechsler Intelligence Scale for Children, Fourth Edition (WISC-IV, [36]) or the Wechsler Abbreviated Scale of Intelligence (WASI, [37]). Children with an ASD diagnosis who were on stimulant medication were included in this study only if they were taken off medication on the day of the lab visit for testing. Four participants had ADHD diagnosed before they were diagnosed with either autistic disorder or Asperger Syndrome by DSM-IV [33], while one subject was diagnosed by DSM-5 [34] as ASD comorbid with ADHD.

Typically developing children (i.e., control subjects, CNT group, N = 19, 14.8 ± 3.67 years old, 6 females) were recruited through advertisements in the local media. All control participants were free of neurological or significant medical disorders, had normal hearing and vision, and were free of psychiatric, learning, or developmental disorders based on self- and parental reports. Subjects were screened for a history of psychiatric or neurological diagnosis using the Structured Clinical Interview for DSM-IV Non-Patient Edition (SCID-NP [38]). Participants within the control and ASD groups were matched by age, gender, full scale IQ, and socioeconomic status (based on parental level of education and annual household income) of their family. The age of participants was in 9 to 17 years range, with majority of them being adolescents within 11–15 years range, but since according to the National Institutes of Health (NIH) definition the participants under 18 years old are still are categorized as children they are further referred to for convenience as "children" rather than "children and adolescents".

The study was conducted in accordance with relevant national regulations and institutional policies and complied with the Helsinki declaration. The protocol of the study including informed consent and assent forms that were reviewed and received approval of University of Louisville (Louisville, KY, USA) Institutional Review Board (IRB) (Ethical approval protocol#006.07). Children and their parents or legal guardians received detailed information about the research study specifics, including its purpose, responsibilities, reimbursement rate, risk vs. benefits evaluation, etc. The participants were reimbursed only for oddball tests ($25 for each procedure), and did not receive any reimbursement for the TMS treatment. Investigators provided consent and assent forms to all families who expressed interest in participation in this treatment research study and answered all questions related to the project. If the child and his family member confirmed their commitment and agreed to be part of it, both child and parent signed and dated the consent and assent forms and received a copy co-signed by the study investigator.

2.2. Experimental Task: Visual Oddball with Illusory Kanizsa Figures

The test used in the study was a three-stimuli oddball task with rare illusory Kanizsa squares (target, 25%), rare Kanizsa triangle (non-target Kanizsa, 25%) and frequent non-Kanizsa stimuli (standards, 50%) [25]. Non-target Kanizsa and non-Kanizsa standard stimuli are further referred to as task-irrelevant stimuli. Visual stimuli were presented for 250 ms with inter-trial interval in the 1100–1300 ms range. Subjects were instructed to press a button on a keypad with their right index finger when a target appeared, and not to respond to any of the task-irrelevant stimuli. This task required the processing of both stimulus features (shape and collinearity) for discrimination of targets. Before the test, all subjects had a brief practice block to get familiar with the specifics of the task, make sure that they understood the test requirements, and that they could recognize the target stimulus correctly. There was a total of 240 trials in the study and 20 trials in the practice block. The test took approximately 20–25 min to complete. Participants with ASD diagnosis had at least one lab visit before the test to ensure habituation to the experimental setting and lab environment as well as conditioning to the EEG sensor net.

2.3. Event-Related Gamma Oscillations Recording

The dense-array (128 channel) electroencephalogram (EEG) was recorded with an Electrical Geodesics Inc. Netstation system (EGI-Philips, Eugene, OR, USA). Experimental control (e.g., stimulus presentation, reaction time) was executed using E-prime software (version 1.1, Psychological Software Tools (PST), Inc., Pittsburg, PA, USA). Visual stimuli were presented on a monitor located in front of the subject, while motor responses were recorded with a 4-button keypad (PST's Serial Box). EEG was recorded with 512 Hz sampling rate, analog Notch (60 Hz, IIR, 5th order) filter and analog bandpass elliptical filters set at 0.1 to 100 Hz range. Electrodes impedance was kept under 40 Ω as recommended by the EGI Netstation manual. Raw EEG recordings were segmented off-line spanning 200 ms pre-stimulus baseline and 800 ms epoch post-stimulus. EEG data was screened for artifacts and all trials that had eye blinks, gross movements and other artifacts were removed using Netstation artifact rejection tools [39]. Other details of our experimental procedure and EEG data acquisition, pre-processing and analysis can be found in our prior studies using the same methodology [26–29,40,41]. Stimulus-locked dependent EEG variables for the frontal (F3, F4, F7, F8) and parietal (P3, P4, P7, P8) sites-of-interest referenced to vertex (Cz) and nasion as a ground were used for gamma oscillation analysis. Analysis of gamma oscillations was performed on trial-by-trial basis. Data set was not re-referenced for average reference frame but rather accepted trials were left with initial vertex reference to avoid distortion of gamma waves. Evoked gamma was analyzed with 40–160 ms window and induced gamma within 180-400 ms window post-stimulus following ranges recommended by prior studies using similar designs and comparable stimuli [12,14,18–20,22].

2.4. EEG Analysis

Method Description

EEG signal $x(t)$ represented by 500 time samples per trial was analyzed using a Continuous Wavelet Transform (CWT) where the Morlet wavelet is used as an analysis wavelet. The CWT for the EEG signal (i.e., $X(\tau,s)$) is defined as follows:

$$X(\tau,s) = \frac{1}{\sqrt{s}} \int_0^\infty x(t)\left(\frac{t-\tau}{s}\right)dt \qquad (1)$$

where τ is the Morlet wavelet that is a continuous function in both time and frequency, is the time shift of the wavelet, and s is scale of the wavelet. In fact, the scale reveals crucial information about the signal. The smaller the scale, the more compressed the wavelet in the time domain, and the more focus on high frequencies (i.e., rapidly changing details). However, the larger the scale, the more stretched the wavelet in the time domain, and the more focus on low frequencies (i.e., slow changing coarse features). $X(\tau,s)$ is then calculated at 500 different scales spanning various frequencies constituting the EEG signal.

Further, in order to obtain the gamma wave, a frequency-localized inverse CWT is used where the coefficients $X(\tau,s)$ corresponding to the frequency range (35–45 Hz) are extracted and an inverse CWT is applied to the extracted coefficients $X(\tau,s)$ as follows:

$$G(t) = \frac{1}{C} \int_{s1}^{s2} \int_0^\infty \frac{1}{s^{2.5}} X(\tau,s) \varphi\left(\frac{t-\tau}{s}\right) d\tau ds \qquad (2)$$

where $G(t)$ is the continuous time gamma wave, φ is the dual function of such that both functions are orthonormal and C is a constant calculated from the mother wavelet. Also, s_1 corresponds to the highest frequency in the gamma wave frequency band while s_2 corresponds to the lowest frequency in the frequency band.

Next, we will define several characteristics of the discrete time representation of the gamma wave (i.e., $G(n)$) such as zero crossings, peaks, major peak, latencies, and areas of the left and right halves of the positive envelope of the gamma wave.

Zero crossings are defined as the time location where the sign of the gamma wave changes from the positive to the negative and vice versa. Zero crossings can be classified into two types; upward zero crossings and downward zero crossings. An upward zero crossing is found when the gamma wave changes from a negative to a positive value while a downward zero crossing is located when the gamma wave changes from a positive to a negative value. Both upward zero crossing (Z_{upward}) and downward zero crossing ($Z_{downward}$) are defined as follows:

$$Z_{upward} = n_0 \text{ s.t. } G(n_0) < 0 \text{ and } G(n_0 + 1) > 0 \qquad (3)$$

$$Z_{downward} = n_1 \text{ s.t. } G(n_1) > 0 \text{ and } G(n_1 + 1) < 0 \qquad (4)$$

A peak is defined as the maximum value of the set of values within the time interval defined between an upward zero crossing and a downward zero crossings. The amplitude of the peaks ($P(\bar{n})$) can be represented using the following mathematical equation:

$$P(\bar{n}) = \max G(n) | Z_{upward} < n < Z_{downward} \qquad (5)$$

Further, the amplitude of the major peak (P_M) of the evoked and induced gamma waves can be defined as the maximum value of all the peaks ($P(\bar{n})$) located within the corresponding timeframe.

$$P_M = \max \overrightarrow{P(n)} | \bar{n}_1 < \bar{n} < \bar{n}_2 \qquad (6)$$

where $\bar{n}_1$ and $\bar{n}_2$ are time location of the starting and ending peaks of the evoked or induced gamma waves.

Slopes are calculated between the major peak and the time location of the lowest peak values mentioned above. A positive slope (S_+) is defined between the lowest peak at the beginning of the gamma wave time interval ($\bar{n}_1$) and the major peak while a negative slope (S_-) is defined between the major peak and the lowest peak at the end of the gamma wave time interval ($\bar{n}_2$).

$$S_+ = \frac{P_M - P(\bar{n}_1)}{\bar{n} - \bar{n}_1} \qquad (7)$$

$$S_- = -\frac{P_M - P(\bar{n}_2)}{\bar{n} - \bar{n}_2} \qquad (8)$$

Latencies are defined as the time difference between the location of the major peak and the location of the lowest-amplitude peaks at the beginning and at the end of the gamma wave time intervals (i.e., $\bar{n}_1$ and $\bar{n}_2$ respectively).

$$Latency_1 = \bar{n} - \bar{n}_1 \qquad (9)$$

$$Latency_2 = -(\bar{n} - \bar{n}_2) \qquad (10)$$

The areas of the left and right halves for the positive envelope of the gamma wave can be approximated using the following equation:

$$A_L = \frac{1}{2}(Latency_1 \times P_M) \qquad (11)$$

$$A_R = \frac{1}{2}(Latency_2 \times P_M) \qquad (12)$$

2.5. Transcranial Magnetic Stimulation

Repetitive TMS was administration using a Magstim Rapid device (Magstim Co., Sheffield, UK) with a 70-mm wing span figure-eight coil. For the identification of resting motor threshold (MT) for each hemisphere the output of the magnetic stimulator was increased by 5% steps until a 50 µV deflection of electromyogram (EMG) or a visible twitch in the First Dorsal Interosseous (FDI) muscle was detected in at least 2 or 3 trials of TMS delivered over the motor cortex controlling the contralateral FDI. Electromyogram was recorded with a psychophysiological monitor C-2 J&J Engineering Inc. with USE-3 software (version 3, J&J Engineering Inc., Poulsbo, WA, USA) and Physiodata applications (Physiodata, Inc., Winslow, WA, USA).

The rTMS was administered on a weekly basis with the following stimulation parameters: 1.0 Hz frequency, 90% MT, 180 pulses per session with 9 trains of 20 pulses each with 20–30 s intervals between the trains. Initial six weekly rTMS session were administered over the left dorsolateral prefrontal cortex (DLPFC) and the next 6 were over the right DLPFC, while the additional 6 treatments were done bilaterally (evenly over the left and right DLPFC). The procedure for stimulation placed the TMS coil 5 cm anterior, and in a parasagittal plane, to the site of maximal FDI response as judged by the FDI EMG response. To ensure better positioning of the TMS coil a swimming cap was used on a head of subject. The location for TMS stimulation was performed with anatomical landmarks [42,43] that approximate the scalp region used for F3 and F4 EEG electrode placements in the 10-20 International System. Motor threshold was detected for the left hemisphere during session 1, for the right hemisphere at session 7, and for both hemispheres at session 13.

We selected 90% of the MT based on reports from prior studies where low frequency rTMS was used for the stimulation of DLPFC in whole range of neurological and psychiatric disorders [44–47]. we decided to have stimulation power below MT as a safety precaution meant to lower the probability of seizures in this study population. The decision to use low frequency (below or equal 1 Hz) magnetic stimulation was based on the finding that at this frequency range rTMS exerts an inhibitory influence on the stimulated cortex [48]. Visual oddball tests in the ASD group were conducted with a week before the 18 session-long rTMS course and within a week after the completion of the course of intervention.

2.6. Behavioral and Social Functioning Evaluation

For evaluation of social and behavioral functioning caregiver (parent or guardian) reports were used. Participants in the ASD group were evaluated before TMS course and within a week following treatment. Aberrant Behavior Checklist (ABC, [49]) is a rating scale to assess Irritability, Lethargy/Social Withdrawal, Stereotypy, Hyperactivity, and Inappropriate Speech based on parent/caregiver report. Repetitive Behavior Scale—Revised (RBS-R [50]) is a caregiver completed rating scale to assess stereotyped, self-injurious, compulsive, ritualistic, sameness, and restricted range.

2.7. Statistical Analysis

Repeated measure ANOVA was the primary model for statistical analyses of subject-averaged evoked and induced gamma oscillation metrics (positive and negative areas of gamma oscillation envelope), motor response, and behavioral questionnaires data. Dependent behavioral variables were RT, omission and commission response rate, and total accuracy. Dependent stimulus-locked evoked and induced gamma variables were positive/ascending and negative/descending areas values at pre-determined frontal (F3, F4, F7, F8) and parietal (P3, P4, P7, P8) EEG sites of interest. The within-participant factors for analysis of TMS effects were the following: *Stimulus*

(Target Kanizsa (TRG), Non-target Kanizsa (NTG), non-Kanizsa (NOK)), *Hemisphere* (Left, Right), and *Time* (ASD Baseline, ASD Post-TMS). Comparisons for ASD and CNT groups used also *Stimulus × Hemisphere × Group* (CNT, ASD pre-(ASD), ASD post-TMS (TMS)) factors. Post-hoc analyses were conducted where appropriate using one-way ANOVA. For behavioral rating scores, a *Treatment* (pre- vs. post-TMS) factor was used. Histograms with normal distribution curves along with skewness and kurtosis data were obtained for each dependent variable to determine normality of distribution and appropriateness of data for ANOVA tests. All dependent variables in the study had normal distribution. Greenhouse-Geisser (GG) corrected p-values were employed where appropriate in all ANOVAs. For the estimation of the effect size, we used a Partial Eta Squared (η_p^2). IBM SPSS (version 26.0, Armonk, NY, USA) and Sigma Stat statistical packages (version 9.0, Systat Software, Inc. San Jose, CA, USA) were used for data analysis.

3. Results

3.1. Behavioral Responses (Reaction Time and Accuracy)

There were no group differences in reaction time (RT) between ASD and neurotypical (CNT) children groups. The ASD group had more commission errors (11.22 ± 15.48 % in ASD vs. 1.55 ± 3.48% in CNT, $F_{1,36} = 7.06$, $p = 0.012$) and more omission errors (2.12 ± 2.72% vs. 0.55 ± 1.18%; $F_{1,36} = 5.31$, $p = 0.027$).

Effects of TMS on RT to targets were not significant ($p = 0.51$, n.s.). There were post-TMS group differences in accuracy, namely in total percentage of errors (13.35 ± 16.84% pre- vs. 3.16 ± 3.09% post-TMS, $F_{1,36} = 6.73$, $p = 0.014$). Group differences in commission error percentage were also statistically significant (11.22 ± 15.48 vs. 2.23 ± 2.51% post-TMS, $F_{1,36} = 6.24$, $p = 0.017$).

3.2. Behavior Evaluations Post-TMS

Repetitive Behavior Scale Outcomes: Repetitive behavior subscales (RBS-R, [50]) showed group difference for Stereotype Behavior and T-score. Stereotypy scores decreased from 5.53 ± 3.85 to 3.26 ± 2.57 ($F_{1,36} = 4.53$, $p = 0.040$); Compulsive Behavior scores decreased from 3.95 ± 2.82 to 2.00 ± 1.82, $F1,36 = 6.39$, $p = 0.016$), and Total Repetitive Behaviors T-score decreased from 23.74 ± 14.54 to 5.21 ± 9.84 ($F_{1,36} = 4.47$, $p = 0.041$).

Aberrant Behavior Checklist Outcomes: Two of the ABC [49] subscales showed significant post-TMS differences. Irritability scores decreased from 11.74 ± 8.66 to 6.63 ± 5.65 ($F_{1,36} = 4.62$, $p = 0.038$); and Hyperactivity scores from 17.47 ± 15.60 to 8.79 ± 7.34 ($F_{1,36} = 4.79$, $p = 0.035$). Lethargy/Social Withdrawal scores showed trend to statistically non-significant decrease from 7.89 ± 4.89 to 5.89 ± 4.45, $F_{1,36} = 1.60$, $p = 0.241$ (n.s.).

3.3. Evoked and Induced Gamma Oscillations

Evoked gamma. Positive slope area (ascending half-envelope area): *Stimulus* had main effect at the frontal sites F7/F8 ($F_{2,55} = 17.63$, $p < 0.001$, $\eta_p^2 = 0.391$), *Stimulus* (TRG, NTG, NOK) × *Group* (CNT, ASD, TMS) interaction was significant ($F_{2,55} = 4.56$, $p = 0.002$, $\eta_p^2 = 0.139$). Main effects of *Stimulus* and *Stimulus × Group* interactions at F3/F4 did not reach a significant level. At the parietal P3/P4, main effect stimulus was significant ($F_{2,55} = 13.85$, $p < 0.001$, $\eta_p^2 = 0.201$). *Stimulus × Group* effects was statistically significant both at P3/P4 ($F_{2,55} = 2.45$, $p = 0.044$, $\eta_p^2 = 0.085$) and at P7/P8 ($F_{2,55} = 3.87$, $p = 0.027$, $\eta_p^2 = 0.123$). Gamma waveforms in 2 groups (CNT and ASD) are depicted in Figure 1.

Positive area of evoked gamma oscillation to targets (TRG) was higher in ASD as compared to CNT at P7 ($F_{1,36} = 7.77$, $p < 0.008$; to non-target Kanizsa (NTG) at F7, F8, P3, and P4 (all $p < 0.05$); and to standard non-Kanizsa (NOK) stimuli at F8 and P7 (all $p < 0.05$). Descriptors of differences and statistical metrics are shown in Table 1.

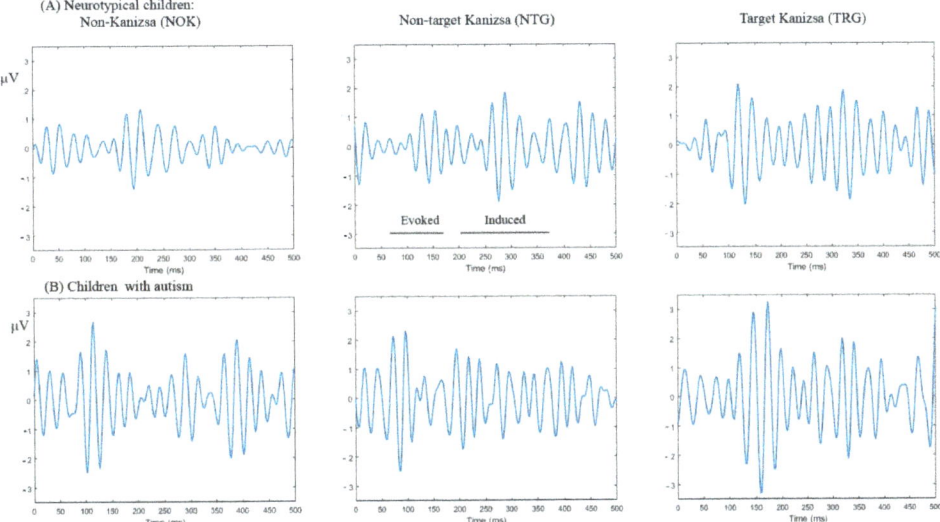

Figure 1. Evoked and induced gamma oscillation waveforms in response to target Kanizsa, non-target Kanizsa, and non-Kanizsa stimuli in neurotypical children (**A**) and children with autism spectrum disorder (ASD) (**B**). Both evoked and induced gamma oscillations have higher magnitude in the ASD group at the frontal sites (F3, F4).

Table 1. Evoked and induced gamma oscillations positive and negative areas in response to rare target Kanizsa (Target), rare non-target Kanizsa (Non-target) and frequent non-Kanizsa standard (Non-Kanizsa) stimuli at the frontal and parietal electroencephalogram (EEG) sites in 19 children with autism (ASD) and in neurotypicals (CNT). Means ± SD. * $p < 0.05$; ** $p < 0.001$; *** $p < 0.001$.

Stimulus/EEG Site	CNT	ASD	F(1,36)	p
Positive/ascending area, evoked				
Target at F7	50.08 ± 22.61	69.22 ± 19.62	7.77	0.008 **
Non-target at F7	44.47 ± 13.13	59.84 ± 20.54	7.55	0.009 **
Non-target at F8	48.10 ± 15.77	60.57 ± 16.37	5.73	0.022 *
Non-target at P3	53.27 ± 19.20	73.72 ± 23.48	8.64	0.006 **
Non-Kanizsa at F7	51.56 ± 21.35	69.82 ± 27.74	5.17	0.029 *
Non-Kanizsa at P7	61.09 ± 19.23	75.92 ± 21.13	5.12	0.030*
Negative/descending area, evoked				
Target at F7	55.23 ± 4.43	49.51 ± 4.42	15.86	<0.001 ***
Non-target at F7	53.86 ± 4.55	49.21 ± 4.85	9.29	0.004 **
Non-Kanizsa at F7	61.63 ± 22.54	47.01 ± 3.12	7.84	0.008 **
Non-Kanizsa at P7	53.06 ± 21.58	68.35 ± 24.21	4.22	0.047 *
Negative/descending area, induced				
Target at P7	53.71 ± 22.95	70.30 ± 22.90	4.97	0.032 *
Non-target at P7	52.96 ± 27.07	73.67 ± 26.53	5.67	0.023 *

The TMS treatment (Figure 2) significantly decreased positive areas of evoked gamma responses to TRG at F7, F8, P7, P8; NTG at P3, P4, and P8; and NOK F7, F8, P7, P8 (all $p < 0.05$).

Negative slope area (descending half-envelope area): *Stimulus* type had main effect at the frontal F3/F4 ($F_{2,55} = 3.88$, $p = 0.023$). *Stimulus × Group* interaction at the same site was also significant ($F_{2,55} = 4.22$, $p = 0.019$, $\eta_p^2 = 0.092$). *Stimulus* had main effect at F7/F8 sites ($F_{2,55} = 7.82$, $p = 0.001$, $\eta_p^2 = 0.180$) and *Stimulus × Group* interaction was also statistically significant ($F_{2,55} = 6.36$, $p = 0.012$, $\eta_p^2 = 0.115$). Similar interaction effects were observed at the inferior parietal sites P7/P8 ($F_{2,55} = 2.12$, $p = 0.039$, $\eta_p^2 = 0.087$). Figures 3 and 4 illustrate these effects.

Post-hoc analysis showed following group differences: Negative area of evoked gamma oscillations at the F7 site was higher in ASD as compared to CNT to TRG ($F_{1,36} = 15.85, p < 0.001$); to NTG ($F_{1,36} = 9.23, p < 0.001$); and to NOK ($F_{1,36} = 12.02, p = 0.001$). TMS treatment decreased negative area to targets at F8, to NTG at F7, F8, P7, P8, (all $p < 0.005$) and to NOK stimuli at F7, F8, P7, P8 (all $p < 0.05$). Descriptive statistics are presented in Table 2.

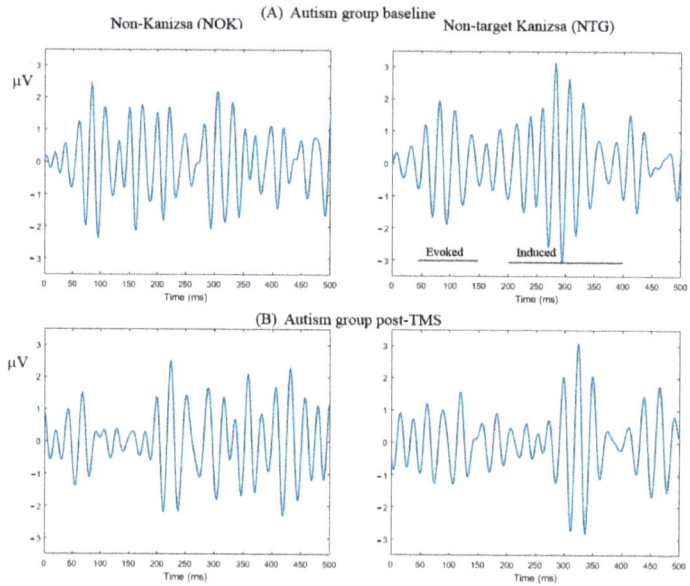

Figure 2. Evoked and induced gamma oscillations in response to task-irrelevant non-target Kanizsa and non-Kanizsa stimuli in children with ASD at the baseline (**A**, upper raw) and post-TMS treatment (**B**, lower raw) at the parietal sites (P7, P8). Magnitude of both evoked and induced gamma oscillations decreased post-transcranial magnetic stimulation (TMS).

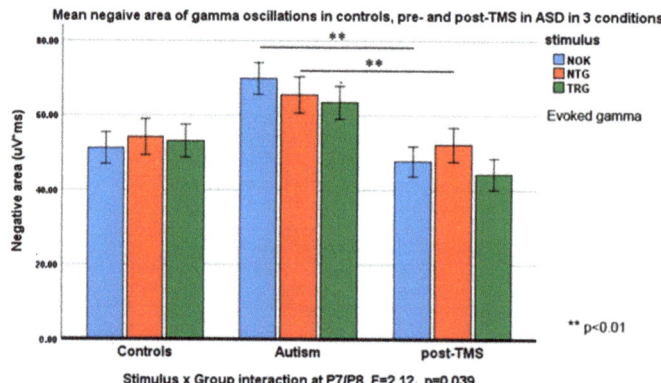

Figure 3. Negative slope areas (Mean ± SE) of evoked gamma oscillations in response to target Kanizsa (TRG), non-target Kanizsa (NTG), and non-Kanizsa (NOK) stimuli in typical children, children with ASD pre- and post-TMS treatment. *Stimulus* (TRG, NTG, NOK) × *Group* (CNT, ASD pre-, ASD post-TMS) interaction was significant at the frontal sites (F3, F4). Note higher area of responses at baseline and a decrease of areas in response to non-target items post-TMS.

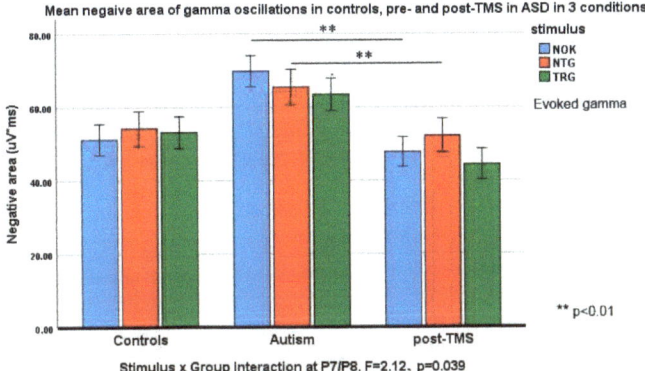

Figure 4. Negative slope area of evoked gamma oscillations in response to target Kanizsa (TRG), non-target Kanizsa (NTG), and non-Kanizsa (NOK) stimuli in typical children, children with ASD pre- and post-TMS treatment. *Stimulus* (TRG, NTG, NOK) × *Group* (CNT, ASD pre-, ASD post-TMS) interaction was significant at the parietal sites (P7, P8). Note higher area of responses at baseline and a decrease of areas in response to non-target items post-TMS similar to effects at the frontal sites.

Table 2. Effect of rTMS course on evoked and induced gamma oscillations' positive and negative areas in response to rare target Kanizsa (Target), rare non-target Kanizsa (Non-target) and frequent non-Kanizsa standard (Non-Kanizsa) stimuli at the frontal and parietal EEG sites in 19 participants with ASD (Means ± SD). * $p < 0.05$; ** $p < 0.001$; *** $p < 0.001$.

Stimulus/EEG Site	ASD Pre-TMS	ASD Post-TMS	F(1,36)	p
	Positive/ascending area, evoked			
Target at F7	67.25 ± 20.79	51.67 ± 18.62	6.25	0.017 *
Target at F8	74.46 ± 25.45	51.26 ± 20.97	9.97	0.003 **
Target at P7	69.22 ± 19.62	47.59 ± 19.68	12.08	0.001 **
Target at P8	56.37 ± 15.22	44.11 ± 16.69	5.82	0.021 **
Non-target at P3	73.72 ± 23.48	53.97 ± 19.95	8.27	0.007 **
Non-target at P4	76.16 ± 21.59	56.52 ± 22.92	7.74	0.008 **
Non-target at P8	66.49 ± 22.90	45.90 ± 17.67	10.25	0.003 **
Non-Kanizsa at F7	75.81 ± 25.23	48.48 ± 16.71	16.62	<0.001 ***
Non-Kanizsa at F8	75.92 ± 21.13	50.91 ± 21.58	13.66	0.001 **
Non-Kanizsa at P7	69.82 ± 27.74	47.01 ± 18.62	9.49	0.004 **
Non-Kanizsa at P8	59.99 ± 25.29	44.28 ± 19.43	4.91	0.033 *
	Negative/descending area, evoked			
Target at F8	48.76 ± 4.57	41.01 ± 15.70	4.29	0.045 *
Non-target at F7	51.29 ± 3.70	39.68 ± 15.57	10.03	0.003 **
Non-target at F8	54.69 ± 4.79	39.87 ± 15.38	16.17	<0.001 ***
Non-target at P7	52.46 ± 6.38	41.45 ± 15.13	8.65	0.006 **
Non-target at P8	52.07 ± 5.63	43.60 ± 16.97	4.29	0.045 *
Non-Kanizsa at F7	51.34 ± 5.25	38.80 ± 14.94	12.02	0.001 **
Non-Kanizsa at F8	51.94 ± 4.85	39.48 ± 15.63	11.07	0.002 **
Non-Kanizsa at P7	53.71 ± 5.19	40.77 ± 16.00	12.02	0.001 **
Non-Kanizsa at P8	51.13 ± 5.32	40.48 ± 15.68	7.92	0.008 **
	Negative/descending area, induced			
Non-target at P3	56.80 ± 20.26	42.71 ± 16.14	5.98	0.019 **
Non-target at P4	56.37 ± 19.48	43.05 ± 18.45	4.93	0.032 *
Non-target at P7	70.30 ± 22.90	49.33 ± 19.63	9.73	0.003 **
Non-target at P8	65.66 ± 24.02	46.18 ± 22.20	7.11	0.011*
Non-Kanizsa at P3	61.07 ± 20.00	44.24 ± 20.33	6.94	0.012 *
Non-Kanizsa at P7	70.99 ± 22.65	46.82 ± 20.50	12.23	0.001 **

Induced gamma: <u>Positive area</u>: No *Stimulus* × *Group* interactions were found for any frontal or parietal topographies. Negative area of induced gamma oscillations, on the other hand, showed statistical effects and group differences: *Stimulus* type had main effect both at P3/P4 ($F_{2,55} = 13.8$,

$p = 0.016$, $\eta_p^2 = 0.071$) and at P7/P8 sites ($F_{2,55} = 9.06$, $p < 0.001$, $\eta_p^2 = 0.251$). *Stimulus × Group* interaction was significant at P3/P4 ($F_{2,55} = 2.54$, $p = 0.044$, $\eta_p^2 = 0.085$) and at P7/P8 sites ($F_{2,55} = 3.87$, $p = 0.027$, $\eta_p^2 = 0.123$). Some of these interactions are depicted in Figures 5 and 6.

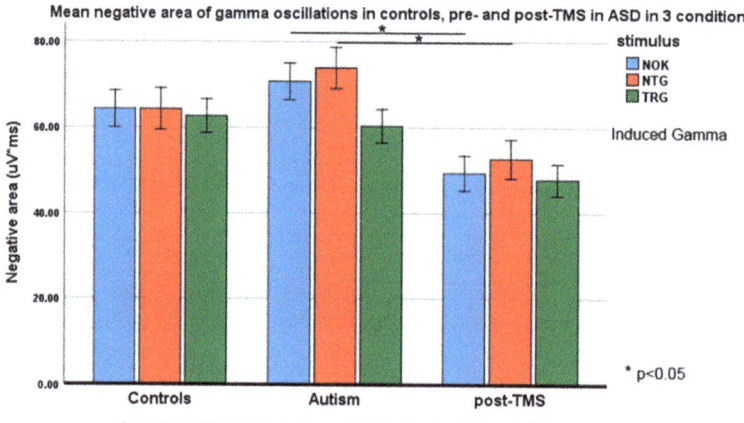

Figure 5. Negative slope area (Mean ± SE) of induced gamma oscillations in response to target Kanizsa (TRG), non-target Kanizsa (NTG) and non-Kanizsa (NOK) stimuli in typical children, children with ASD pre- and post-TMS treatment. *Stimulus* (TRG, NTG, NOK) × *Group* (CNT, ASD pre-, ASD post-TMS) interaction was significant at the inferior frontal sites (F7, F8). Note decrease of areas in response to non-target items post-TMS.

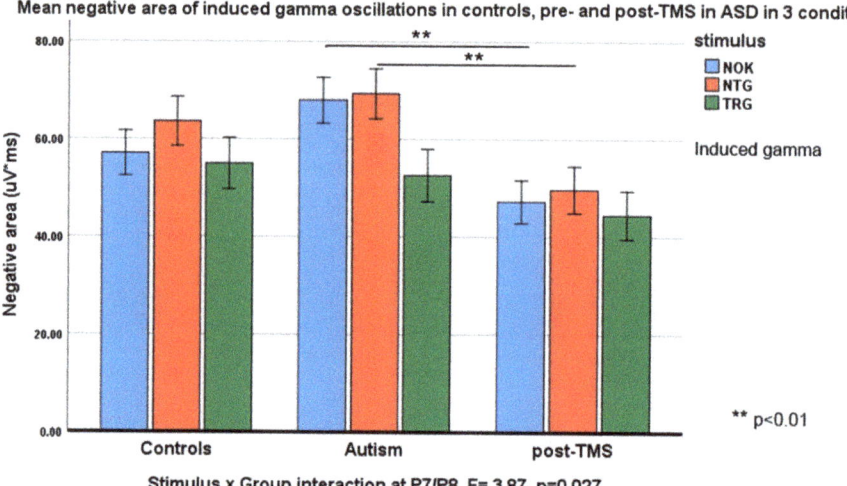

Figure 6. Negative slope area (Mean ± SE) of induced gamma oscillations in response to target Kanizsa (TRG), non-target Kanizsa (NTG) and non-Kanizsa (NOK) stimuli in typical children, children with ASD pre- and post-TMS treatment. *Stimulus* (TRG, NTG, NOK) × *Group* (CNT, ASD pre-, ASD post-TMS) interaction was significant at the inferior parietal sites (P7, P8). Note higher area of response at baseline and a decrease of areas in response to non-target items post-TMS similar to effects at the frontal sites.

Negative area of induced gamma to targets was higher in ASD as compared to CNT only at P7 site ($\overline{F_{1,36} = 4.97}$, $p = 0.032$). TMS decrease negative area of the induced gamma to NTG at P3, P4, P7, and P8 (all $p < 0.05$), and to NOK stimuli at P3 and P7 (both $p < 0.05$).

4. Discussion

The results of our envelope analysis show that autistic subjects as compared to neurotypicals exhibit significant differences at baseline (pre-TMS) in both evoked and induced gamma oscillations. This difference was most salient in response to non-target cues as recorded from frontal and parietal recording sites (F3, F4, F7, F8, P3, P7, P8). Post-TMS treatment, our autistic subjects exhibited a significant reduction of gamma responses to task-irrelevant stimuli. Significant changes were also observed in both the positive and negative slope areas of evoked gamma responses. The rise of the modulated waveform being higher for our ASD subjects as compared to controls and diminishing towards control levels after TMS treatment. Normalization of gamma oscillations occurred concomitantly to changes in behaviors; more specifically questionnaires (ABC, RBS-R) showed a decrease in irritability, hyperactivity, and repetitive scores. In addition, post-TMS. participants committed fewer errors (total percentage of errors and commission error percentage).

According to pioneers of 40 Hz neurofeedback training [16] gamma response as such represents attention-related processes as it is basically a stimulus evaluation and response selection mechanism activity reflection at different stages of information processing. The early (evoked) gamma oscillation could be considered as an attention–trigger process that gives information about the arrival of a stimulus and need for more detailed processing that is occurring later in time and is reflected in cognitive event-related potentials (e.g., N200 and P300) and late (induced) gamma responses. Decrease magnitude of the evoked gamma responses to non-target stimuli post-TMS facilitates processing of targets at the later stages of the responses and is reflected in a decreased induced gamma magnitude.

Demodulation of EEG frequency bands has been used in the analysis of brainwave activity during altered states of consciousness [51]. Mathematically, it has been proven that the integration of the amplitude of the modulated EEG waveform accurately describes EEG cortical activation patterns of event-related desynchronization (ERD) [52]. Amplitude modulation usually decreases or disappears in the period prior to and at the very beginning of task performance [53]. In ASD patients, with a high autism spectrum quotient [54], ERD (ergo the integration of the amplitude modulated EEG waveform) serves to discriminate clinically significant differences such as the processing of human facial expression [55].

EEG sub-band modulation has been proposed as an automated non-invasive diagnostic tool for neuropsychiatric disorders such as Alzheimer's disease wherein the slowly-varying EEG amplitudes of modulated waveforms appears to be most affected [56]. This feature, which measures the rate at which EEG sub-bands are modulated, has been called the "spectro-temporal modulation energy" [56]. TMS has been shown to change EEG activity while complex demodulation has been used to extract the power of its different bandwidths [57]. As early as the first postnatal year, abnormalities in the power of brain oscillations (delta and gamma sub-bands) help distinguish infants with ASD diagnoses from others [58].

In the present study, demodulation and integration of the area under the upper envelope of the gamma waveform, showed a similar effect for both the ascending and descending train of oscillations. The spectro-temporal modulation energy for both evoked and induced gamma was larger in autism than in neurotypicals and diminished towards neurotypical levels after TMS treatment. In modeling membrane potentials, a depolarizing pulse causes the interaction of parasitic properties of inductance and capacitance from some of its anatomical components. The output of the circuit is a smooth periodic sinusoidal oscillation that resonates at a particular frequency. In the absence of a driving force, the involved circuit has a tendency to vibrate around its natural frequency. It is speculated that the ability of neurons to resonate allows them to preferentially respond to those inputs arriving at the same frequency. The end result of this effect is the promotion of synchronized activity within and between neuronal populations [59]. It is believed that evoked gamma activity represents the binding of information within a confined cortical field while the induced component represents the binding across networks of different cortical regions [12,13,28]. Our results showed significant differences in both the evoked and induced components of gamma oscillations in our ASD patients as compared to neurotypicals. The findings may help explain the atypical perceptual processing symptoms observed in ASD as information processing is impeded both within and between brain regions [21].

In our initial study, using envelope analysis of gamma oscillations in ASD [30], a short period of ringing decay at baseline (pre-TMS) was thought to indicate a system of low sensitivity [31]; that is, one where the resonator's center frequency in relation to its bandwidth is broadened. TMS increases the sensitivity of the system by transforming the envelope of gamma oscillations from one akin to a ramp or sawtooth waveform (baseline) to a symmetrical configuration (post-treatment). In ASD, prior to active treatment, the envelope of gamma oscillations ramps upward to a peak amplitude which is higher than in controls. This waveform is characterized by a longer latency to peak amplitude and a shorter settling time [30]. The non-sinusoidal nature of the envelope thus engendered makes it difficult to identify a dominant frequency while its sharper inflection procreates harmonic distortion. The harmonically related oscillations create additional signals that are injected back to the system and makes high-frequency EEG recordings in this patient population unreliable. TMS treatment helps transform the envelope to one, more in keeping with a sinusoidal waveform, one where spectral analysis would more easily identify a fundamental frequency that facilitates neural binding. We believe that, in ASD, the increased spectro-temporal modulation energy (vide supra) at baseline may be the result of an increased contribution of excitation to an oscillatory system that is trying to define its fundamental frequency.

It may be reasonable to assume that abnormalities of gamma oscillation may be manifested more fully in those regions of the brain, like the prefrontal lobes, that exhibit rich and varied connectivity. Gamma oscillations play a crucial role in the binding of information between neural networks. The exceptional richness in connectivity of the frontal lobes is an important anatomical feature that may support our unique cognitive abilities as humans. Researchers describe it as the "traffic hub" of the nervous system [60]. The dorsolateral prefrontal cortex (DLPFC), a region of the prefrontal lobes, projects to a large number of cortical areas involved in visuospatial processing, auditory information and sensory integration, motor response, reward and punishment, memory, and error detection [61]. This region of the brain is most typically associated with executive functions including working memory, selective attention, mental flexibility, response initiation, impulse control, and action monitoring [62,63]. The connectivity that characterizes this area of the frontal lobes, therefore provides a low functional breakdown threshold expressed as gamma oscillation abnormalities and executive dysfunction [64].

Previous studies have shown that TMS targeting the frontal lobes improves executive functions [65]. These improvements have been noted on tests of cognitive flexibility, conceptual tracking, attention and working memory [65–67]. In ASD, TMS can change some of the core deficits, more specifically, those impairments in self-monitoring that constitute our supervisory attentional system [10,29,40,41,68]. Higher amplitude of early evoked gamma activity has been correlated to faster behavioral responses, which could constitute impulsive behaviors [69]. Indeed, results from previous trials suggest that individuals with ASD have a reduced sensitivity for monitoring errors (i.e., a diminished response time after committing an error) and instituting corrective actions. Researchers believe that a deficit in monitoring errors might manifest itself as perseverative behaviors typical of ASD [10,29,40,41,68].

In keeping with these findings TMS targeting the DLPFC in the present study provided for improvement in behaviors as noted in caregivers' reports. The most notable change was a decrease in the T-score of the Repetitive Behavior Scale-Revised [50], along with decreased irritability and hyperactivity rating scores in the Aberrant Behavior Checklist questionnaire [49].

In analogy to electronic communication systems, demodulating brain oscillations allows us to examine both the envelope of a signal as well as the periodic waveform that carries the same. Gutfreund et al. [59] suggested that resonance behavior, exemplified in the carrier wave, helps tie together those neural populations that find themselves working at the same frequency. By way of contrast, the envelope of gamma oscillations could provide information regarding the impedance of involved circuits and the excitatory–inhibitory balance of the cerebral cortex [30]. Indeed, the amplitude of oscillations in the gamma frequency range, that helps define its envelope, have been shown to be dependent on the GABAergic tone [70–73]. We suggest that these metrics of gamma oscillations, those ingrained in both its carrier and its envelope, provide complementary information that allows for the better interpretation of EEG waveforms. In future studies, we hope to use these new metrics of gamma oscillations in clinical designs using different parameters of stimulation. Changes in the pulse width of the magnetic field, that better influence the recruitment of excitatory and inhibitory cells within the cerebral cortex [74], could better normalize gamma oscillation abnormalities while improving clinical outcomes of our ASD patients.

5. Conclusions

In this exploratory study we employed envelope analysis of demodulated waveforms for evoked and induced gamma oscillations in response to Kanizsa figures in a three-stimuli visual oddball task and compared outcomes in children and adolescents with ASD and in neurotypical peers. In ASD group, as compared to neurotypicals, significant differences in evoked and induced gamma oscillations were evident in larger gamma oscillations reflected in a higher area of gamma oscillation envelopes. Low frequency rTMS treatment in ASD resulted in a significant reduction of gamma responses to non-target stimuli and in improved accuracy of performance in the oddball task. Several rating scores of behavioral questionnaires also showed improvement post-TMS. Application of a novel metric for gamma oscillations based on envelope analysis using wavelet transformation allowed to suggest that gamma oscillations may provide a viable biomarker reflective of the excitatory/inhibitory balance of the cortex, a useful diagnostic index and a putative outcome measure of rTMS intervention in autism.

Author Contributions: All authors contributed substantially to the work reported in this paper. M.F.C., A.S.E.-B., M.S. and E.M.S. conceived and designed the experiment and data analysis strategy, participated in oddball test data collection, statistical analysis and interpretation of results. E.M.S. administered visual oddball test to patients and collected raw data. M.S. and M.G. participated in EEG data extraction, preprocessing and statistical analysis and also contributed in preparation of figures; E.L.C. and I.O. participated in reaction time and accuracy data processing and statistical analysis; A.S.E. assisted in preparation of methodological part related to EEG data handling, analysis and interpretation; M.F.C. participated in patient clinical records analysis and in interpretation of data and discussion, he also reviewed the manuscript and edited the final version. All authors have read and agreed to the published version of the manuscript.

Funding: This research was funded by the National Institutes of Health Eureka R01 Grant MH86784 to Manuel F. Casanova.

Acknowledgments: Funding for this study was provided by the NIH grant R01 MH086784 to Manuel F. Casanova.

Conflicts of Interest: The authors declare no conflict of interest.

References

1. Buzsáki, G.; Wang, X.J. Mechanisms of gamma oscillations. *Annu. Rev. Neurosci.* **2012**, *35*, 203–225. [CrossRef]
2. Snijders, T.M.; Milivojevic, B.; Kemner, C. Atypical excitation-inhibition balance in autism captured by the gamma response to contextual modulation. *Neuroimage Clin.* **2013**, *3*, 65–72. [CrossRef]
3. Steullet, P.; Cabungcal, J.-H.; Coyle, J.; Didriksen, M.; Gill, K.; Grace, A.A.; Hensch, T.K.; LaMantia, A.-S.; Lindemann, L.; Maynard, T.M.; et al. Oxidative stress-driven parvalbumin interneuron impairment as a common mechanism in models of schizophrenia. *Mol. Psychiatry* **2017**, *22*, 936–943. [CrossRef]
4. Hashemi, E.; Ariza, J.; Rogers, H.; Noctor, S.C.; Martinez-Cerdeno, V. The number of parvalbumin-expressing interneurons is decreased in the prefrontal cortex in autism. *Cereb. Cortex* **2017**, *27*, 1931–1943. [PubMed]
5. Lee, E.; Lee, J.; Kim, E. Excitation/inhibition imbalance in animal models of autism spectrum disorders. *Biol. Psychiatry* **2017**, *81*, 838–847. [CrossRef] [PubMed]
6. Saunders, J.A.; Tatard-Leitman, V.M.; Suh, J.; Billingslea, E.N.; Roberts, T.P.; Siegel, S.J. Knockout of NMDA receptors in parvalbumin interneurons recreates autism-like phenotypes. *Autism Res.* **2013**, *6*, 69–77. [CrossRef] [PubMed]
7. Wöhr, M.; Orduz, D.; Gregory, P.; Moreno, H.; Khan, U.; Vörckel, K.J.; Wolfer, D.P.; Welzl, H.; Gall, D.; Schiffmann, S.N.; et al. Lack of parvalbumin in mice leads to behavioral deficits relevant to all human autism core symptoms and related neural morphofunctional abnormalities. *Transl. Psychiatry* **2015**, *5*, e525. [CrossRef]

8. Selimbeyoglu, A.; Kim, C.K.; Inoue, M.; Lee, S.Y.; Hong, A.S.O.; Kauvar, I.; Ramakrishnan, C.; Fenno, L.E.; Davidson, T.J.; Wright, M.; et al. Modulation of prefrontal cortex excitation/inhibition balance rescues social behavior in CNTNAP2-deficient mice. *Sci. Transl. Med.* **2017**, *9*, eaah6733. [CrossRef] [PubMed]
9. Mukherjee, A.; Carvalho, F.; Eliez, S.; Caroni, P. Long-lasting rescue of network and cognitive dysfunction in a genetic schizophrenia model. *Cell* **2019**, *178*, 1387–1402. [CrossRef]
10. Sokhadze, E.M.; El-Baz, A.; Baruth, J.; Mathai, G.; Sears, L.; Casanova, M.F. Effects of low frequency repetitive transcranial magnetic stimulation (rTMS) on gamma frequency oscillations and event-related potentials during processing of illusory figures in autism. *J. Autism Dev. Disord.* **2009**, *39*, 619–634. [CrossRef]
11. Rojas, D.C.; Wilson, L.B. Gamma-band abnormalities as markers of autism spectrum disorders. *Biomark Med.* **2014**, *8*, 353–368. [CrossRef] [PubMed]
12. Casanova, M.F.; Baruth, J.; El-Baz, A.S.; Sokhadze, G.E.; Hensley, M.; Sokhadze, E.S. Evoked and induced gamma frequency oscillations in autism. In *Imaging the Brain in Autism*; Casanova, M.F., El-Baz, A.S., Suri, J.S., Eds.; Springer: New York, NY, USA, 2013; pp. 87–106.
13. Rippon, G. Gamma abnormalities in autism spectrum disorders. In *Autism Imaging and Devices*; Casanova, M.F., El-Baz, A., Suri, J.S., Eds.; CRC Press, Taylor and Francis Group: Boca Raton, FL, USA, 2017; pp. 457–496.
14. Başar, E.; Tülay, E.; Güntekin, B. Multiple gamma oscillations in the brain: A new strategy to differentiate functional correlates and P300 dynamics. *Int. J. Psychophysiol.* **2015**, *95*, 406–420. [CrossRef] [PubMed]
15. Galambos, R.A. Comparison of certain gamma band 40 Hz brain rhythms in cat and man. In *Induced Rhythms in the Brain*; Basar, E., Bullock, T.H., Eds.; Birkhauser: Boston, MA, USA, 1992; pp. 201–216.
16. Sheer, D.E. Focused arousal and 40 Hz EEG. In *The Neuropsychology of Learning Disorders*; Knight, R.M., Baker, D.J., Eds.; University Park Press: Baltimore, MD, USA, 1976.
17. Kahana, M.J. The cognitive correlates of human brain oscillations. *J. Neurosci.* **2006**, *26*, 1669–1672. [CrossRef] [PubMed]
18. Kaiser, J.; Lutzenberger, W. Induced gamma-band activity and human brain function. *Neuroscientist* **2003**, *9*, 475–484. [CrossRef]
19. Tallon-Baudry, C.; Bertrand, O. Oscillatory gamma activity in humans and its role in object representation. *Trends Cogn. Sci.* **1999**, *3*, 151–162. [CrossRef]
20. Tallon-Baudry, C.; Bertrand, O.; Henaff, M.A.; Isnard, J.; Fischer, C. Attention modulates gamma-band oscillations differently in the human lateral occipital cortex and fusiform gyrus. *Cereb. Cortex* **2005**, *15*, 654–662. [CrossRef]
21. Brock, J.; Brown, C.C.; Boucher, J.; Rippon, G. The temporal binding deficit hypothesis of autism. *Dev. Psychopathol.* **2002**, *14*, 209–224. [CrossRef]
22. Brown, C.; Gruber, T.; Boucher, J.; Rippon, G.; Brock, J. Gamma abnormalities during perception of illusory figures in autism. *Cortex* **2005**, *41*, 364–376. [CrossRef]
23. Rippon, G.; Brock, J.; Brown, C.; Boucher, J. Disordered connectivity in the autistic brain: Challenges for the "new psychophysiology". *Int. J. Psychophysiol.* **2007**, *63*, 164–172. [CrossRef]
24. Maxwell, C.R.; Villalobos, M.E.; Schultz, R.T.; Herpertz-Dahlmann, B.; Konrad, K.; Kohls, G. Atypical laterality of resting gamma oscillations in autism spectrum disorders. *J. Autism Dev. Disord.* **2015**, *45*, 292–297. [CrossRef]
25. Kanizsa, G. Subjective contours. *Sci. Am.* **1976**, *234*, 48–52. [CrossRef] [PubMed]
26. Baruth, J.M.; Casanova, M.F.; El-Baz, A.; Horrell, T.; Mathai, G.; Sears, L.; Sokhadze, E. Low-frequency repetitive transcranial magnetic stimulation (rTMS) modulates evoked-gamma frequency oscillations in autism spectrum disorder (ASD). *J. Neurother* **2010**, *14*, 179–194. [CrossRef]
27. Baruth, J.M.; Williams, E.L.; Sokhadze, E.M.; El-Baz, A.S.; Casanova, M.F. Beneficial effects of repetitive transcranial magnetic stimulation (rTMS) on behavioral outcome measures in autism spectrum disorder. *Autism Sci. Dig.* **2011**, *1*, 52–57.
28. Casanova, M.F.; Baruth, J.M.; El-Baz, A.; Tasman, A.; Sears, L.; Sokhadze, E. Repetitive transcranial magnetic stimulation (rTMS) modulates event-related potential (ERP) indices of attention in autism. *Transl. Neurosci.* **2012**, *3*, 170–180. [CrossRef] [PubMed]
29. Sokhadze, E.M.; El-Baz, A.S.; Sears, L.L.; Opris, I.; Casanova, M.F. rTMS neuromodulation improves electrocortical functional measures of information processing and behavioral responses in autism. *Front. Syst. Neurosci.* **2014**, *8*, 134. [CrossRef] [PubMed]

30. Casanova, M.F.; Shaban, M.; Ghazal, M.; Sokhadze, E.; Casanova, E.L.; El-Baz, A.S. Ringing decay of gamma oscillations and transcranial magnetic stimulation therapy in autism spectrum disorder. *Appl. Psychophysiol. Biofeedback* **2020**. under review.
31. Silver, M.L.; Tiedemann, D. *Dynamic Geotechnical Testing*; ASTM International: West Conshohocken, PA, USA, 1978.
32. Attallah, B.V.; Scanziani, M. Instantaneous modulation of gamma oscillation frequency by balancing excitation with inhibition. *Neuron* **2009**, *62*, 566–577. [CrossRef]
33. American Psychiatric Association. *Diagnostic and Statistical Manual of Mental Disorders (DSM-IV)*, 4th ed.; American Psychiatric Publishing, Inc.: Washington, DC, USA, 2000.
34. American Psychiatric Association. *Diagnostic and Statistical Manual of Mental Disorders (DSM-5)*, 5th ed.; American Psychiatric Publishing, Inc.: Washington, DC, USA, 2013.
35. LeCouteur, A.; Lord, C.; Rutter, M. *The Autism Diagnostic Interview—Revised (ADI-R)*; Western Psychological Services: Los Angeles, CA, USA, 2003.
36. Wechsler, D. *Wechsler Intelligence Scale for Children (WISC-IV)*, 4th ed.; Harcourt Assessment, Inc.: San Antonio, TX, USA, 2003.
37. Wechsler, D. *Wechsler Abbreviated Scale of Intelligence (WASI)*; Harcourt Assessment, Inc.: San Antonio, TX, USA, 1999.
38. First, M.B.; Spitzer, R.L.; Gibbon, M.; Williams, J.B.W. *Structured Clinical Interview for DSM-IV-TR Axis I Disorders, Research Version (SCID-I/NP)*; Biometrics Research, New York State Psychiatric Institute: New York, NY, USA, 2002.
39. Luu, P.; Tucker, D.M.; Englander, R.; Lockfeld, A.; Lutsep, H.; Oken, B. Localizing acute stroke-related EEG changes: Assessing the effects of spatial undersampling. *J. Clin. Neurophysiol.* **2001**, *18*, 302–317. [CrossRef]
40. Sokhadze, E.M.; Baruth, J.M.; Sears, L.; Sokhadze, G.E.; El-Baz, A.; Casanova, M.F. Prefrontal neuromodulation using rTMS improves error monitoring and correction function in autism. *Appl. Psychophysiol. Biofeedback* **2012**, *37*, 91–102. [CrossRef]
41. Sokhadze, E.M.; Lamina, E.V.; Casanova, E.L.; Kelly, D.P.; Opris, I.; Tasman, A.; Casanova, M.F. Exploratory study of rTMS neuromodulation effects on electrocortical functional measures of performance in an oddball test and behavioral symptoms in autism. *Front. Syst. Neurosci.* **2018**, *12*, 20. [CrossRef]
42. Mir-Moghtadaei, A.; Caballero, R.; Fried, P.; Fox, M.D.; Lee, K.; Giacobbe, P.; Daskalakis, Z.J.; Blumberger, D.M.; Downar, J. Concordance between BeamF3 and MRI-neuronavigated target sites for repetitive transcranial magnetic stimulation of the left dorsolateral prefrontal cortex. *Brain Stimul.* **2015**, *8*, 965–973. [CrossRef] [PubMed]
43. Pommier, B.; Vassal, F.; Boutet, C.; Jeannin, S.; Peyron, R.; Faillenot, I. Easy methods to make the neuronavigated targeting of DLPFC accurate and routinely accessible for rTMS. *Neurophysiol. Clin.* **2017**, *7*, 35–46. [CrossRef] [PubMed]
44. Pascual-Leone, A.; Walsh, V.; Rothwell, J. Transcranial magnetic stimulation in cognitive neuroscience–virtual lesion, chronometry, and functional connectivity. *Curr. Opin. Neurobiol.* **2000**, *10*, 232–237. [CrossRef]
45. Oberman, L.M.; Enticott, P.G.; Casanova, M.F.; Rotenberg, A.; Pascual-Leone, A.; McCracken, J.T.; TMS in ASD Consensus Group. Transcranial magnetic stimulation in autism spectrum disorder: Challenges, promise, and roadmap for future research. *Autism Res.* **2016**, *9*, 184–203. [CrossRef]
46. Wassermann, E.M.; Lisanby, S.H. Therapeutic application of repetitive transcranial magnetic stimulation: A review. *Clin. Neurophysiol.* **2001**, *112*, 1367–1377. [CrossRef]
47. Wassermann, E.M.; Zimmermann, T. Transcranial magnetic stimulation: Therapeutic promises and scientific gaps. *Pharm. Ther.* **2012**, *133*, 98–107. [CrossRef]
48. Maeda, F.; Keenan, J.P.; Tormos, J.M.; Topka, H.; Pascual-Leone, A. Modulation of corticospinal excitability by repetitive transcranial magnetic stimulation. *Clin. Neurophysiol.* **2000**, *111*, 800–805. [CrossRef]
49. Aman, M.G.; Singh, N.N. *Aberrant Behavior Checklist-Community. Supplementary Manual*; Slosson Educational Publications: East Aurora, NY, USA, 1994.
50. Bodfish, J.W.; Symons, F.J.; Lewis, J. *Repetitive Behavior Scale (Western Carolina Center Research Reports)*; Western Carolina Center: Morganton, NC, USA, 1999.
51. Jovanov, E.; Rakovic, D.; Radivojevic, V.; Kusic, D. Band Power Envelope Analysis: A New Method in Quantitative EEG. In Proceedings of the 17th International Conference of the Engineering in Medicine and Biology Society, Montréal, QC, Canada, 20–23 September 1995.

52. Clochon, P.; Fontbonne, J.; Lebrun, N.; Etevenon, P. A new method for quantifying EEG event-related desynchronization: Amplitude envelope analysis. *Electroencephalogr. Clin. Neurophysiol.* **1996**, *98*, 126–129. [CrossRef]
53. Draganova, R.; Popivanov, D. Assessment of EEG frequency dynamics using complex demodulation. *Physiol. Res.* **1999**, *48*, 157–165.
54. Baron-Cohen, S.; Wheelwright, S.; Skinner, R.; Martin, J.; Clubley, E. The autism-spectrum quotient (AQ): Evidence from Asperger syndrome/high-functioning autism, males and females, scientists and mathematicians. *J. Autism Dev. Disord.* **2001**, *31*, 5–17. [CrossRef]
55. Cooper, N.R.; Simpson, A.; Till, A.; Simmons, K.; Puzzo, I. Beta event-related desynchronization as an index of individual differences in processing human facial expression: Further investigations of autistic traits in typically developing adults. *Front. Hum. Neurosci.* **2013**, *7*, 159. [CrossRef] [PubMed]
56. Trambaiolli, L.R.; Falk, T.H.; Fraga, F.J.; Anghinah, R.; Lorena, A.C. EEG spectro-temporal modulation energy: A new feature for automated diagnosis of Alzheimer's disease. *Conf. Proc. IEEE Eng. Med. Biol. Soc.* **2011**, *2011*, 3828–3831. [PubMed]
57. Griskova, I.; Ruksenas, O.; Dapsys, K.; Herpertz, S.; Hoppner, J. The effects of 10 Hz repetitive transcranial magnetic stimulation on resting EEG power spectrum in healthy subjects. *Neurosci. Lett.* **2007**, *419*, 162–167. [CrossRef] [PubMed]
58. Gabard-Durnam, L.J.; Wilkinson, C.; Kapur, K.; Tager-Flusberg, H.; Levin, A.R.; Nelson, C.A. Longitudinal EEG power in the first postnatal year differentiates autism outcomes. *Nat. Commun.* **2019**, *10*, 4188. [CrossRef]
59. Gutfreund, Y.; Yarom, Y.; Segev, I. Subthreshold oscillations and resonant frequency in guinea-pig cortical neurons: Physiology and modelling. *J. Physiol.* **1995**, *483*, 621–640. [CrossRef] [PubMed]
60. Baars, B.J.; Gage, N.M. *Cognition, Brain, and Consciousness*, 2nd ed.; Academic Press: Cambridge, MA, USA, 2010.
61. Arnsten, A.F.T. The neurobiology of thought: The groundbreaking discoveries of Patricia Goldman-Rakic 1937–2003. *Cereb. Cortex* **2013**, *23*, 2269–2281. [CrossRef] [PubMed]
62. Curtis, C.E.; D'Esposito, M. Persistent activity in the prefrontal cortex during working memory. *Trends Cogn. Sci.* **2003**, *7*, 415–423. [CrossRef]
63. Panerai, S.; Tasca, D.; Ferri, R.; D'Arrigo, V.G.; Elia, M. Executive functions and adaptive behavior in autism spectrum disorder with and without intellectual disability. *Psychiatry J.* **2014**, *2014*, 941809. [CrossRef]
64. Hill, E.L. Executive dysfunction in autism. *Trends Cogn. Sci.* **2004**, *8*, 26–32. [CrossRef]
65. Moser, D.J.; Jorge, R.E.; Manes, F.; Paradiso, S.; Benjamin, M.L.; Robinson, R.G. Improved executive functioning following repetitive transcranial magnetic stimulation. *Neurology* **2002**, *58*, 1288–1290. [CrossRef]
66. Palaus, M.; Marron, E.M.; Viejo-Sobera, R.; Redolar-Ripoll, D. Cognitive Enhancement by Means of TMS: Memory and Executive Functions. In Proceedings of the 5th Conference of the European Societies of Neuropsychology, Tampere, Finland, 9–11 September 2015.
67. Hauer, L.; Sellner, J.; Brigo, F.; Trinka, E.; Sebastianelli, L.; Saltuari, L.; Versace, V.; Höller, Y.; Nardone, R. Effects of repetitive transcranial magnetic stimulation over prefrontal cortex on attention in psychiatric disorders: A systematic review. *J. Clin. Med.* **2019**, *8*, 416. [CrossRef]
68. Sokhadze, E.; Baruth, J.; El-Baz, A.; Horrell, T.; Sokhadze, G.; Carroll, T.; Tasman, A.; Sears, L.; Casanova, M.F. Impaired error monitoring and correction function in autism. *J. Neurother* **2010**, *14*, 79–95. [CrossRef] [PubMed]
69. van Es, M.W.J.; Schoffelen, J.M. Stimulus-induced gamma power predicts the amplitude of the subsequent visual evoked response. *Neuroimage* **2019**, *186*, 703–712. [CrossRef]
70. Brunel, N.; Wang, X.J. What determines the frequency of fast network oscillations with irregular neural discharges? I. Synaptic dynamics and excitation-inhibition balance. *J. Neurophysiol.* **2003**, *90*, 415–430. [CrossRef] [PubMed]
71. Edden, R.A.E.; Muthukumaraswamy, S.D.; Freeman, T.C.A.; Singh, K.D. Orientation discrimination performance is predicted by GABA concentration and gamma oscillation frequency in human primary visual cortex. *J. Neurosci.* **2009**, *29*, 15721–15726. [CrossRef] [PubMed]

72. Muthukumaraswamy, S.D.; Edden, R.A.E.; Jones, D.K.; Swettenham, J.B.; Singh, K.D. Resting GABA concentration predicts peak gamma frequency and fMRI amplitude in response to visual stimulation in humans. *Proc. Natl. Acad. Sci. USA* **2009**, *106*, 8356–8361. [CrossRef] [PubMed]
73. Saxena, N.; Muthukumaraswamy, S.D.; Diukova, A.; Singh, K.; Hall, J.; Wise, R. Enhanced stimulus-induced gamma activity in humans during propofol-induced sedation. *PLoS ONE* **2013**, *8*, e57685. [CrossRef]
74. Hannah, R.; Rocchi, L.; Tremblay, S.; Wilson, E.; Rothwell, J.C. Pulse width biases the balance of excitation and inhibition recruited by transcranial magnetic stimulation. *Brain Stimul.* **2020**, *13*, 536–538. [CrossRef]

© 2020 by the authors. Licensee MDPI, Basel, Switzerland. This article is an open access article distributed under the terms and conditions of the Creative Commons Attribution (CC BY) license (http://creativecommons.org/licenses/by/4.0/).

Article

Downregulation of CD73/A_{2A}R-Mediated Adenosine Signaling as a Potential Mechanism of Neuroprotective Effects of Theta-Burst Transcranial Magnetic Stimulation in Acute Experimental Autoimmune Encephalomyelitis

Milorad Dragić [1,*], Milica Zeljković [1], Ivana Stevanović [2,3], Marija Adžić [1], Anđela Stekić [1], Katarina Mihajlović [1], Ivana Grković [4], Nela Ilić [5,6], Tihomir V. Ilić [3], Nadežda Nedeljković [1] and Milica Ninković [2,3]

1. Department for General Physiology and Biophysics, Faculty of Biology, University of Belgrade, 11000 Belgrade, Serbia; milica.zeljkovic@bio.bg.ac.rs (M.Z.); amarija@bio.bg.ac.rs (M.A.); andjela.stekic@bio.bg.ac.rs (A.S.); katarina.mihajlovic@bio.bg.ac.rs (K.M.); nnedel@bio.bg.ac.rs (N.N.)
2. Institute for Medical Research, Military Medical Academy, 11000 Belgrade, Serbia; ivanav13@yahoo.ca (I.S.); ninkovic7@gmail.com (M.N.)
3. Medical Faculty of Military Medical Academy, University of Defense, 11000 Belgrade, Serbia; tihoilic@gmail.com
4. Department of Molecular Biology and Endocrinology, Vinča Institute of Nuclear Sciences-National Institute of the Republic of Serbia, University of Belgrade, 11000 Belgrade, Serbia; istanojevic@vinca.rs
5. Medical Faculty, University of Belgrade, 11000 Belgrade, Serbia; nelavilic@gmail.com
6. Clinic of Physical Medicine and Rehabilitation, Clinical Center of Serbia, 11000 Belgrade, Serbia
* Correspondence: milorad.dragic@bio.bg.ac.rs

Abstract: Multiple sclerosis (MS) is a chronic neurodegenerative disease caused by autoimmune-mediated inflammation in the central nervous system. Purinergic signaling is critically involved in MS-associated neuroinflammation and its most widely applied animal model—experimental autoimmune encephalomyelitis (EAE). A promising but poorly understood approach in the treatment of MS is repetitive transcranial magnetic stimulation. In the present study, we aimed to investigate the effect of continuous theta-burst stimulation (CTBS), applied over frontal cranial bone, on the adenosine-mediated signaling system in EAE, particularly on CD73/A_{2A}R/A_1R in the context of neuroinflammatory activation of glial cells. EAE was induced in two-month-old female DA rats and in the disease peak treated with CTBS protocol for ten consecutive days. Lumbosacral spinal cord was analyzed immunohistochemically for adenosine-mediated signaling components and pro- and anti-inflammatory factors. We found downregulated IL-1β and NF-κB-ir and upregulated IL-10 pointing towards a reduction in the neuroinflammatory process in EAE animals after CTBS treatment. Furthermore, CTBS attenuated EAE-induced glial eN/CD73 expression and activity, while inducing a shift in A_{2A}R expression from glia to neurons, contrary to EAE, where tight coupling of eN/CD73 and A_{2A}R on glial cells is observed. Finally, increased glial A_1R expression following CTBS supports anti-inflammatory adenosine actions and potentially contributes to the overall neuroprotective effect observed in EAE animals after CTBS treatment.

Keywords: CD73; adenosine; A_{2A}R; A_1R; neuroinflammation; theta-burst stimulation; rTMS; purinergic signaling

1. Introduction

Multiple sclerosis (MS) is a progressive demyelinating and neurodegenerative disorder driven by the adaptive immune response [1,2] and inflicts primary damage to the myelin sheath [3]. Succeeding inflammation and glial cell activation result in diffuse plaques of demyelination and axonal loss in multiple areas of the brain and spinal cord, which are the main cause of progressive neurological disability and motor dysfunctions in MS [4]. The histopathological deteriorations create many other symptoms, including pain, depression, spasticity, and cognitive deficits, which also progress over time [5]. The most common form of MS is relapse-remitting MS (RRMS), occurring in 85% of patients, characterized by symptomatic loss-of-function periods (relapses), followed by complete or partial remissions. Even though a compelling improvement is made regarding the introduction of new disease-modifying treatments, a significant number of patients will still develop a secondary progressive form of MS (SPMS) [6]. Accordingly, new therapeutic, neuroprotective, and myelination supportive approaches able to ameliorate neuroinflammation and neurotoxic reactive phenotype of astrocytes and microglia are major unmet clinical needs in MS [7].

Neuroinflammation driven by astrocytes and microglia in pathological conditions including MS—and its most widely used animal model experimental autoimmune encephalomyelitis (EAE)—is closely regulated by purinergic signaling. Specifically, the neuroinflammatory responses of glial cells begin with an emergence of danger-associated molecular patterns (DAMP), among which adenosine triphosphate (ATP) plays a particular role. Under the pathological conditions, damaged or dying neurons release large amounts of ATP [8,9], which acts at nucleotide-responsive purinoreceptors, P2X or P2Y, to initiate pro-inflammatory actions of glial cells [8]. The action of extracellular ATP at purinoreceptors is ceased by its sequential hydrolysis, mediated by the ectonucleotidase enzyme chain (CD39/NTPDase1, NTPDase2, and CD73). The last step is the hydrolysis of AMP, mediated via ecto-5′-nucleotidase (eN/Cluster of differentiation 73 (CD73)), resulting in the production of adenosine [10,11]. One longitudinal study in MS patients [12] showed that impaired metabolism of extracellular ATP and drop of adenosine in the cerebrospinal fluid were associated with significantly faster disability progression in MS patients over time. Similarly, reduced production of adenosine in blood serum and increased production in the spinal cord tissue [13] were registered along with a strong upregulation of CD73 by reactive astrocytes during the symptomatic phase of EAE [14].

Adenosine, generated by the catalytic action of CD73, acts at P1 receptor subtypes, A_1, A_{2A}, A_{2B}, and A_3, which are abundantly expressed in the CNS [15]. Adenosine plays a critical role in the regulation and complex modal changes in glial cells during neuroinflammation [16]. Although it is generally considered that adenosine, unlike ATP, elicits anti-inflammatory and immunosuppressive effects [17], its effects critically depend on a particular P1 receptor subtype(s), which mediates the adenosine [18]. Thus, concerning MS/EAE, evidence suggests that potentiation of A_1R and blockade of $A_{2A}R$-mediated adenosine actions induce strong neuroprotective actions via the attenuation of glial cells' reactivity [15,18–22].

One promising but poorly exploited clinical approach in MS is repetitive transcranial magnetic stimulation (rTMS). rTMS refers to a non-invasive and painless stimulation protocol designed to modulate excitability and activity in several brain systems, by applying magnetic pulses delivered in predefined administration patterns [23–25]. Theta-burst stimulation protocol (TBS) is a highly effective version of rTMS, which affords a short stimulation time, low stimulus intensity, and improved reliability of rTMS [26]. Over the past decade, several studies have shown that rTMS stimulation induces measurable clinical outcomes in several neurological disorders, including depression, schizophrenia, stroke, Alzheimer's disease, and Parkinson's disease, and significant improvement of motor and cognitive functions in healthy subjects [27–32]. So far, studies have demonstrated limited clinical value of rTMS in MS patients [33,34], particularly regarding motor dysfunction [35,36].

Despite numerous positive neurological outcomes in several neurological disorders, mechanisms underlying the plasticity induced by TBS are poorly understood, which urges the need for preclinical animal testing. Up to date, the efficacy of different TBS protocols has been explored in animal studies using EAE, as an experimental paradigm of RRMS. The studies have demonstrated reduced oxidative stress [37,38], attenuation of gliosis [39], and increased expression of brain-derived neurotrophic factor (BDNF) [40]. Therefore, the present study aims to explore the effect of continuous theta-burst stimulation (CTBS) protocol on purinergic system activity in the context of neuroinflammation associated with experimental autoimmune encephalomyelitis in Dark Agouti rats. If proven effective, these data could incite translation into clinical practice as an early/add-on non-invasive therapeutic intervention.

2. Material and Methods

2.1. Ethical Statement

All experimental procedures were approved by Ethics Comity of Military Medical Academy (Application No. 323-07-00622/2017-05). Care was taken to minimize the pain and discomfort of the experimental animals in accordance with EU Directive 2010/63/EU.

2.2. Animals

This study was performed on two months old female Dark Agouti (DA) rats (150–200 g) acquired from Military Medical Academy local colony. All animas were housed under standardized conditions (constant humidity 55 ± 3%, temperature 23 ± 2 oC, 13/11 h light/dark regime) in polyethylene cages (3 animals per cage) with food and water ad libitum.

2.3. Induction of Experimental Autoimmune Encephalomyelitis

Acute experimental autoimmune encephalomyelitis was induced as previously described [39]. Briefly, animals were anesthetized with sodium pentobarbital (45 mg/kg, Trittay, Germany) and s.c. injected with 0.1 mL of encephalogenic emulsion comprising complete Freund's Adjuvant (CFA, 1 mg/mL *Mycobacterium tuberculosis*, Sigma, St. Louis, MO, USA) and rat spinal cord tissue homogenate (50% *w/v* in saline) in right hind foot.

The animals were weighed and daily scored for neurological signs of EAE for 24 days post-injection (dpi) using the standard EAE scoring scale (0–5): 0 = unaffected/no sign of illness; 0.5 = reduced tail tone; 1 = tail atony; 1.5 = slightly clumsy gait, impaired righting ability or combination; 2 = hind limb paresis; 2.5 = partial hind limb paralysis; 3 = complete hind limb paralysis; 3.5 = complete hind limb paralysis accompanied with forelimb weakness; 4 = tetraplegic; 5 = morbidus state or death [41]. Daily score was averaged taking into account all animals within the experimental group.

2.4. Theta-Burst Stimulation Protocol

In the present study, theta-burst stimulation (TBS) was applied in the form of continuous protocol (CTBS), as previously described [39,40,42]. Briefly, the stimulation was performed using MagStim Rapid2 device via 25 mm figure-of-eight coil (The MagStim Company, Whitland, UK). Continuous protocol was applied according to [43]. The CTBS block was administered as a single 40 s train of bursts repeated at a frequency of 5 Hz, each block containing 600 pulses. Stimulation intensity was set at 30% of maximal output, just below a motor threshold value. The stimulation was applied by holding the center of the coil directly above the frontal cranial bone in close contact with the scalp of a manually immobilized animal. Given that a coil size is larger than cranium of an animal, application over the frontal cranial bone provides equally distributed whole brain stimulation.

2.5. Experimental Groups and Treatment

All animals were divided into four experimental groups: naïve, healthy animals ($n = 8$), EAE animals (sacrificed on day 24, $n = 8$), EAE animals subjected to CTBS protocol ($n = 8$), and animals subjected to sham CTBS noise artifact ($n = 8$). Animals were subjected to either CTBS or noise artifact for 10 consecutive days, starting at 14 dpi, when clinical scoring showed disease peak (Figure 1). The next day, animals were decapitated using Harvard Apparatus, and spinal cord tissue was processed for immunohistochemistry. Given that sham groups did not produce any qualitative/quantitative change when compared to non-treated animals, those images were not shown.

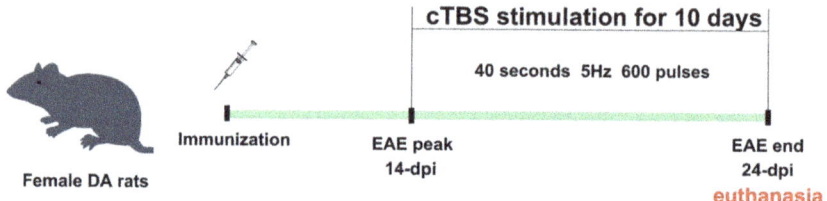

Figure 1. CTBS treatments of EAE rats. Rats were immunized for EAE at day 0 and scored and weighed every day until day 24. The first symptoms appeared around 10 dpi and peaked around 14 dpi. The animals were subjected to CTBS or sham noise artifact for 10 consecutive days from disease peak and euthanized.

2.6. Enzyme Histochemistry

Ectonucleotidase enzyme histochemistry based on the AMP-hydrolyzing activities of and eN/CD73 has been applied, as previously described [44]. Briefly, cryosections were preincubated for 30 min at RT in TRIS-maleate sucrose buffer (TMS), containing 0.25 M sucrose, 50 mM TRIS-maleate, 2 mM $MgCl_2$ (pH 7.4), and 2 mM levamisole, to inhibit tissue non-specific alkaline phosphatase. The enzyme reaction was carried out at 37 °C/90 min, in TMS buffer, containing 2 mM $Pb(NO_3)_2$, 5 mM $MnCl_2$, 3% dextran T250, and 1 mM substrate (ATP, ADP, or AMP), as substrate. After thorough washing, slides were immersed in 1% (v/v) $(NH_4)_2S$, and the product of enzyme reaction was visualized as an insoluble brown precipitate at a site of the enzyme activity. After dehydration in graded ethanol solutions (70–100% EtOH and 100% xylol), slides were mounted with a DPX-mounting medium (Sigma Aldrich, Saint Louise, MO, USA). The sections were examined under LEITZ DM RB light microscope (Leica Mikroskopie and Systems GmbH, Wetzlar, Germany), equipped with LEICA DFC320 CCD camera (Leica Microsystems Ltd., Heerbrugg, Switzerland) and analyzed using LEICA DFC Twain Software (Leica, Wetzlar, Germany).

2.7. Immunofluorescence and Confocal Microscopy

Lumbar areas of the spinal cords (3–4 animals per group) were removed from decapitated animals and fixed in 4% paraformaldehyde (0.1 M PBS, pH 7.4, 12 h at 4 °C) and dehydrated in graded sucrose solution (10–30% in 0.1 M PBS, pH 7.4). After dehydration, 25 μm sections were cut on crytome and collected serially, mounted on supefrost glass slides, air-dried for 1–2 h at room temperature, and stored at 20 °C until staining. After rehydration and washing steps in PBS, sections were blocked with 5% normal donkey serum at room temperature for 1 h, followed by incubation with primary antibodies (Table 1). Slides were then probed with appropriate secondary antibodies (Table 1) for 2 h at room temperature in the dark chamber. Slides were covered using the Mowiol medium (Sigma Aldrich, USA) and left to dry at 4 °C over night. Slides were examined using a confocal laser-scanning microscope (LSM 510, Carl Zeiss, GmbH, Jena, Germany) using Ar multi-line (457, 478, 488, and 514 nm), HeNe (543 nm), HeNe (643 nm) lasers using 63× (×2 digital zoom) DIC oil, 40× and monochrome camera AxioCam ICm1 camera (Carl Zeiss, GmbH, Germany).

2.8. Quantification of Immunofluorescence and Multi-Image Colocalization Analysis

All image quantification and analysis were performed using ImageJ software (free download from https://imagej.net/Dowloads, accessed on 10 April 2021). In order to evaluate a degree of overlap and correlation between multiple channels, we performed multi-image colocalization analysis using the JACoP ImageJ plugin. A degree of overlap and correlation between channels was estimated by calculating Pearson's correlation coefficient (PCC) and Manders' correlation coefficient (MCC). We captured 7–9 images/animal of the white matter under the same conditions (1024 × 1024, laser gain and exposure) and performed PCC and MCC analysis. Analysis was performed on 40× magnification for PCC and 63× magnifications for MCC analysis. Given that astrocytes and microglia were closely related and often intermingled without clear borders, especially in EAE group, whole images were used for analysis rather than single cell [45]. PCC is a statistical parameter that reflects co-occurrence and correlation of analyzed channels. On the other hand, MCC measures fractional overlap between two signals, signal 1 and signal 2. MCC_1 quantifies the fraction of signal 1 that co-localizes with signal 2, while MCC_2 represents the fraction of signal 2 that overlaps with signal 1 [45].

Table 1. Antibodies used for immunohistochemistry.

Antibody	Source and Type	Used Dilution	Manufacturer
Iba-1	Goat, polyclonal	1:400	Abcam ab5076, RRID:AB_2224402
CD73, rNu-9L(I4,I5)	Rabbit, polyclonal	1:300	Ectonucleotidases-ab.com
GFAP	Rabbit, polyclonal	1:500	DAKO, Agilent Z0334, RRID:AB_10013382
IL-10	Goat, polyclonal	1:100	Santa Cruz Biotechnology, sc-1783, RRID: AB_2125115
NF-kB	Rabbit, polyclonal	1:100	Santa Cruz Biotechnology, sc-109, RRID: AB_632039
IL-1β/IL-1F2	Goat, polyclonal	1:100	R&D Systems, AF-501-NA, RRID: AB_354508
A2AR	Rabbit, polyclonal	1:300	Abcam, ab3461, RRID: AB_303823
A1R	Rabbit, polyclonal	1:200	Novus Biologicals, NB300-549, RRID: AB_10002337
Anti-mouse IgG Alexa Fluor 488	Donkey, polyclonal	1:400	Invitrogen A21202, RRID:AB_141607
Anti-goat IgG Alexa Fluor 488	Donkey, polyclonal	1:400	Invitrogen A-11055, RRID:AB_142672
Anti-rabbit IgG Alexa Fluor 555	Donkey, polyclonal	1:400	Invitrogen A-21428, RRID:AB_141784
Anti-mouse IgG Alexa Fluor 647	Donkey, polyclonal	1:400	Thermo Fisher Scientific A-31571, RRID:AB_162542

2.9. Statistical Analysis

The values are presented either as mean ± SD or SEM, as indicated. Data were first assessed for normality using Shapiro–Wilk followed by adequate parametric test. One-way ANOVA followed by Tuckey post hoc test were used in GraphPad Prism v. 6.03. The $p < 0.05$ was considered to be significant (Table 2).

Table 2. Results of ANOVA analysis performed for results obtained from image analysis.

Analysis Performed	ANOVA Results	p Values
PCC GFAP–CD73	$F_{(2, 28)} = 0.7792$	$p = 0.4736$
PCC IBA1–CD73	$F_{(2, 31)} = 33.48$	$p < 0.0001$
MCC1 GFAP–CD73	$F_{(2, 28)} = 3.228$	$p = 0.0648$
MCC2 CD73-GFAP	$F_{(2, 28)} = 4.975$	$p < 0.05$
MCC1 IBA1–CD73	$F_{(2, 27)} = 5.482$	$p < 0.05$
MCC2 CD73–IBA1	$F_{(2, 27)} = 17.05$	$p < 0.0001$
PCC A_1R–CD73	$F_{(2, 30)} = 22.19$	$p < 0.0001$
PCC IBA1–A_1R	$F_{(2, 28)} = 9.155$	$p < 0.01$
MCC1 GFAP–A_1R	$F_{(2, 27)} = 24.45$	$p < 0.0001$
MCC2 A_1R-GFAP	$F_{(2, 27)} = 9.217$	$p < 0.01$
MCC1 IBA1–A_1R	$F_{(2, 28)} = 9.502$	$p < 0.01$
MCC2 A_1R–IBA1	$F_{(2, 28)} = 5.458$	$p < 0.05$
PCC GFAP–$A_{2A}R$	$F_{(2, 33)} = 12.74$	$p < 0.001$
PCC IBA1–$A_{2A}R$	$F_{(2, 32)} = 25.23$	$p < 0.0001$
MCC1 GFAP–$A_{2A}R$	$F_{(2, 26)} = 20.86$	$p < 0.0001$
MCC2 $A_{2A}R$-GFAP	$F_{(2, 27)} = 8.629$	$p < 0.01$
MCC1 IBA1–$A_{2A}R$	$F_{(2, 29)} = 10.93$	$p < 0.001$
MCC2 $A_{2A}R$–IBA1	$F_{(2, 29)} = 31.472$	$p < 0.0001$

3. Results

3.1. The Effect of Continuous Theta-Burst Stimulation on the Disease Course

Injection of the encephalitogenic emulsion in susceptible DA rats resulted in a typical acute disease, characterized by gradual neurological deterioration and significant weight loss followed by a spontaneous recovery (Figure 2), as previously reported [39]. Briefly, in the non-treated group (EAE), the first clinical signs of EAE appeared at ~10 post-injection (dpi), peaked at 14 dpi, and withdrew at ~24 dpi. In the group subjected to the CTBS protocol (EAE+CTBS), the stimulation was applied to start from 14 dpi for 10 consecutive days. The effect of the CTBS noise artifact was explored in the sham group of animals (EAE+CTBSpl), which were subjected to the noise artifact according to the same experimental scheme. Significant reduction in duration, disability, and weight loss were observed after CTBS treatment, compared to both sham and naïve animals, as previously published (Figure 2) [37,39].

3.2. CTBS Promotes Anti-Inflammatory Milieu in EAE

One of the critical pathological features of EAE/MS is the invasion of peripheral immune cells into the CNS parenchyma and the release of pro-inflammatory mediators, which initiate the neuroinflammatory response of astrocytes and microglia. Therefore, we first examined the effect of CTBS on the inflammatory milieu induced by EAE. IL-1β is a master inflammatory cytokine and the effector molecule in MS/EAE [46]. While control tissue did not express IL-1β-immunoreactive (*ir*) signal (Figure 3A,D), conspicuous IL-1β-*ir*, mostly residing at GFAP-*ir* astrocytes and IBA-1-*ir* microglial cells, were observed in the gray (Figure 3B) and white matter (Figure 3E) of EAE animals, respectively. Prominent IL-1β-*ir* was also observed at neuronal cell bodies in both ventral and dorsal gray matter (Figure 3B). However, the upregulation of IL-1β was completely prevented in EAE animals subjected to CTBS, together with the GFAP-*ir* and Iba-1-*ir* lowered to the level seen in healthy control (Figure 3C,F). The downstream signaling cascade of IL-1β initiates nuclear factor kappa-light-chain-enhancer of activated B cells (NF-κB) family of transcription factors, which trigger the transcription of proinflammatory genes [47]. Strong NF-κB-*ir*, mostly residing at GFAP-*ir* astrocytes cells in EAE animals (Figure 4B, arrowhead), was attenuated to a control level after CTBS treatment protocol (Figure 4C). The cytokine IL-10, on the other hand, exhibits immune response downregulatory properties, which include suppression of the synthesis and release of pro-inflammatory cytokines (PMID: 10320650). Basal IL-10-*ir* in control sections (Figure 5A) was attenuated in EAE (Figure 5B), while CTBS protocol enhanced the intensity of IL-10-*ir* in comparison to control (Figure 5C). The IL-10-*ir* mostly resided at GFAP-*ir* astrocytes (Figure 5C). Sham-treated animals did not show any observable changes when compared to EAE (not shown).

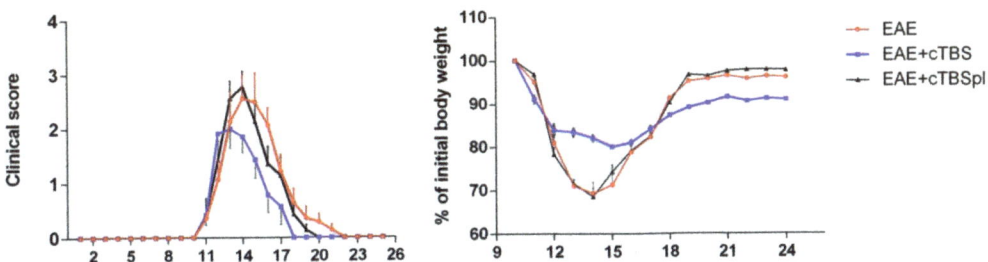

Figure 2. Effects of CTBS treatment on the clinical score of EAE and weight of DA rats. Clinical score and weight of EAE (red circles) in DA rats treated with CTBS protocol (blue square) and CTBS sham noise artifact (black triangles). Animals were monitored from 0 dpi when EAE was induced until 24 dpi when animals were sacrificed.

3.3. CTBS Attenuates EAE-Induced Expression of CD73

The main objective of the present study was to evaluate the effects of CTBS on purinergic system activity in the context of neuroinflammatory activation of astrocytes and microglia. Hence, we first examined the level of expression and cellular localization of CD73 in the spinal cord tissue in control, non-treated, and CTBS-treated EAE animals (Figure 6). The degree of overlap between CD73 and selected fluorescence signals was determined by calculating PCC and MCC coefficients, which reflect the co-occurrence of selected signals and the fraction of pixels with positive values for selected signals, respectively. In control sections, faint CD73-*ir* was mainly associated with quiescent GFAP-*ir* cells and only sporadically with IBA-1-*ir* microglia (Figure 6A,a). A prominent increase in CD73-*ir* in EAE was mainly associated with IBA-1-*ir* microglia (Figure 6B), which is reflected with the increase in both PCC and MCC$_2$ for the two signals, and only marginally with GFAP-*ir* ($p < 0.05$; Figure 6D). The increase in CD73-*ir* was completely reversed by the CTBS treatment (Figure 6C,c), which was reflected with a decrease in MCC$_2$ value primarily for CD73-IBA-1, but also for CD73-GFAP overlap ($p < 0.05$, Figure 6E). The occurrence of CD73-*ir* with both fluorescence tracers for astrocytes and microglia was confirmed with the Z-stack imaging (Figure 6F). Interestingly, the fraction of the CD73-*ir* in control and CTBS sections was found without association with GFAP- and IBA-1-*ir* (Figure 6c, arrowheads).

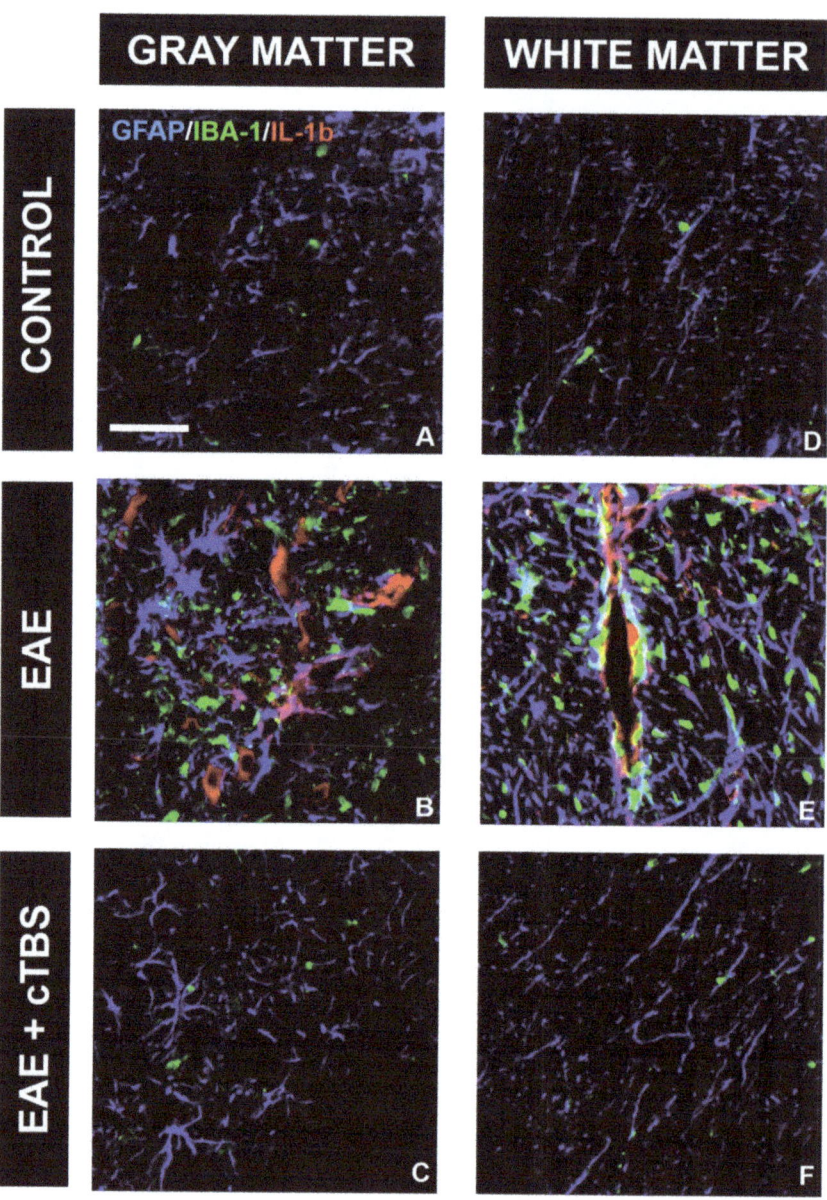

Figure 3. Effect of CTBS treatment on IL-1β expression in gray and white matter of EAE rats. Triple immunofluorescence labeling directed to astrocyte marker GFAP (blue), microglial marker IBA-1 (green), and pro-inflammatory cytokine IL-1β (red). Expression of IL-1β was not detected in control sections (**A**,**D**). In EAE sections, increased IL-1β immunostaining in gray (**B**) and white matter (**E**), colocalizing with both GFAP and IBA-1 cells. After CTBS treatment, no IL-1β-*ir* was observed (**C**,**F**). Scale bar corresponds to 50 μm.

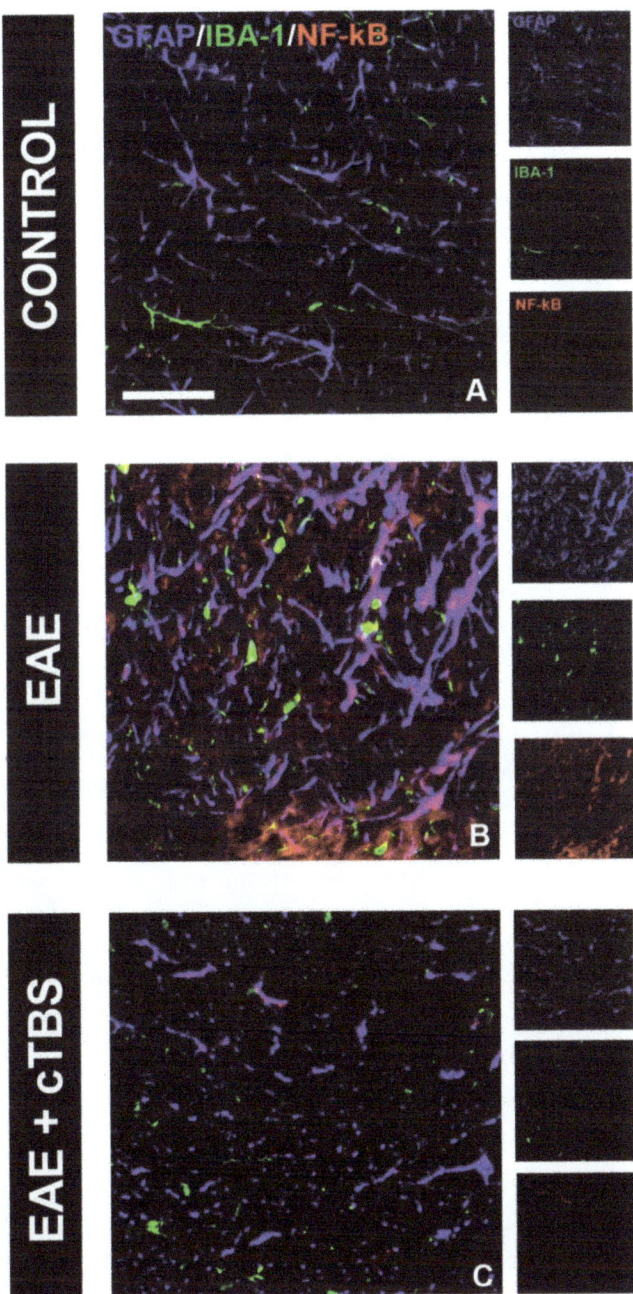

Figure 4. Effects of CTBS treatment on NF-κB expression in EAE rats. Triple immunofluorescence labeling directed to astrocyte marker GFAP (blue), microglial marker IBA-1 (green), and NF-κB (red). Faint colocalization of NF-κB-*ir* and GFAP was observed in control sections (**A**). In EAE sections, a marked increase in NF-κB-*ir* was observed predominantly colocalizing with GFAP+ cells (**B**). CTBS treatment decreased immunostaining of NF-κB, and only scattered NF-κB+/GFAP+ cells were observed (**C**). Scale bar corresponds to 50 μm.

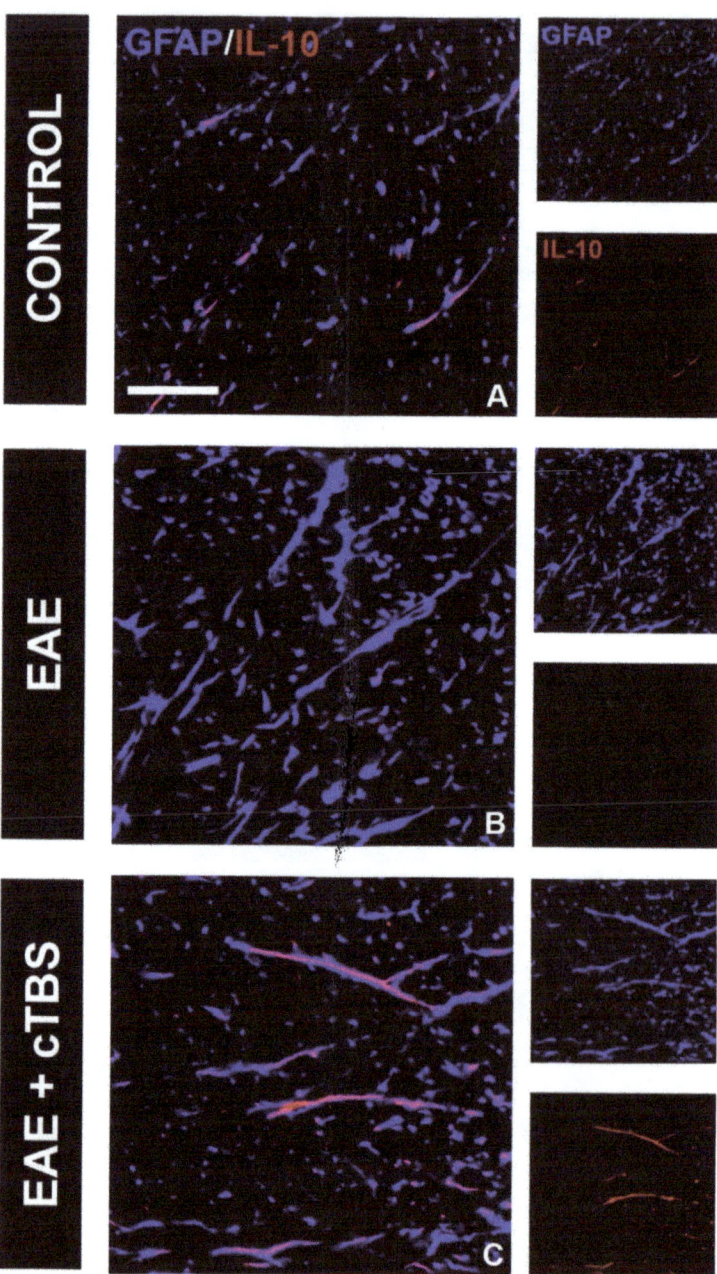

Figure 5. Effects of CTBS treatment on IL-10 expression in EAE rats. Double immunofluorescence labeling directed to astrocyte marker GFAP (blue) and anti-inflammatory cytokine IL-10 (red). Control sections revealed modest colocalization of IL-10 and GFAP (**A**), which was barely detectable in EAE animas (**B**). CTBS treatment led to marked increase in immunostaining of IL-10, which was confined to quiescent GFAP+ cells (**C**). Scale bar corresponds to 50 μm.

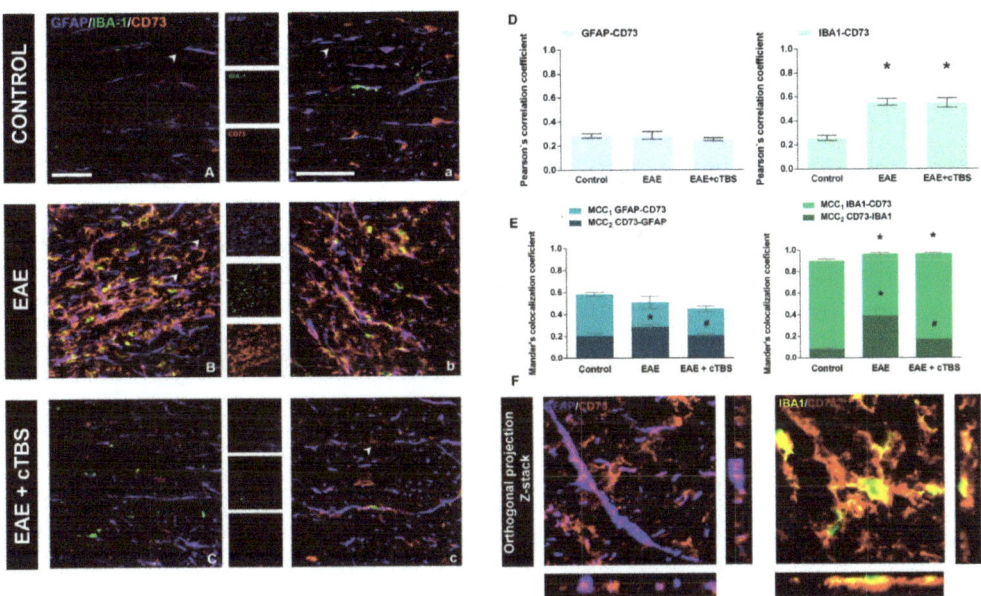

Figure 6. Effects of CTBS treatment on eN/CD73 expression in EAE rats. Triple immunofluorescence labeling directed to astrocyte marker GFAP (blue), microglial marker IBA-1 (green), and eN/CD73 (red). In control section, faint staining of eN/CD73 was observed colocalizing dominantly with GFAP+ cells (**A,a**). In EAE sections, a marked increase in eN/CD73 staining was observed colocalizing with GFAP+ and IBA-1+ cells (**B,b**). After CTBS treatment, a significant reduction in eN/CD73-*ir* was observed (**C,c**). Pearson correlation coefficients (PCC) indicating the level of signal overlap between GFAP-*ir* and eN/CD73-*ir* and IBA-1-*ir* and eN/CD73-*ir*. Bars show mean PCC ± SEM, from 7–9 images/animal (**D**). Mander's colocalization coefficient (MCC) indicating level of signal colocalization between GFAP/CD73 (MCC$_1$, light blue), CD73/GFAP (MCC$_2$, dark blue), IBA-1/CD73 (MCC$_1$, light green), and CD73/IBA1 (MCC$_2$, dark green) (**E**). Orthogonal Z-stack projection of GFAP/CD73 and IBA-1/CD73 (**F**). Level of significance: * $p < 0.05$ or less when compared to control, # $p < 0.05$ when compared to EAE. Scale bar corresponds to 50 μm.

3.4. CTBS Attenuates EAE-Induced Upregulation of CD73 and Shift in A_1R-to-$A_{2A}R$ Expression

Altered immunofluorescence imaging directed to CD73 pointed to significant alterations of CD73 expression, both in EAE and after CTBS treatment. Therefore, the expression of the CD73 enzyme activity was shown by AMP-based enzyme histochemistry (Figure 7). The diffuse histochemical reaction produced by CD73-catalyzed hydrolysis of AMP was dominantly observed in the control spinal cord gray matter (Figure 7A,B), whereas the white matter was faintly stained (Figure 7A,C). In EAE sections, an increased reaction was observed in both gray (Figure 7D,E) and white matter (Figure 7D,F), with numerous amoeboid CD73-reactive cells (Figure 7E). Again, CTBS treatment resulted in histochemical staining almost identical to the control (Figure 7G–I). Diffuse staining dominated the ventral and dorsal gray matter (Figure 7G), whereas no infiltrations of amoeboid cells could be found in the white matter (Figure 7H,I).

Signaling actions of adenosine in the CNS are mostly mediated via high-affinity inhibitory A_1R and excitatory $A_{2A}R$ receptors, differentially involved in neuroinflammatory processes [15,18]. In physiological conditions, the expression is dominated by A_1R mostly found in association with the gray and white matter parenchyma (Figure 8A,a). The induction of EAE is associated with marked loss of A_1R-*ir*, particularly from the white matter projection pathways (Figure 8B,b). However, CTBS treatment restored and even enhanced the intensity of A_1R-*ir* (Figure 8C,c). The determination of PCC and MCC had shown that CTBS increases the proportion of both GFAP-*ir* astrocytes and IBA-1-*ir* cells, which expressed A_1R-*ir*, whereas the overall fraction of A_1R-*ir* is expressed by the glial cells (Figure 8D,E; $p < 0.05$), also confirmed by Z-stack imaging (Figure 8F). Therefore, EAE is associated with the significant axonal loss of A_1R-*ir*, whereas CTBS restores the expression and even potentiates it at responsive glial cells.

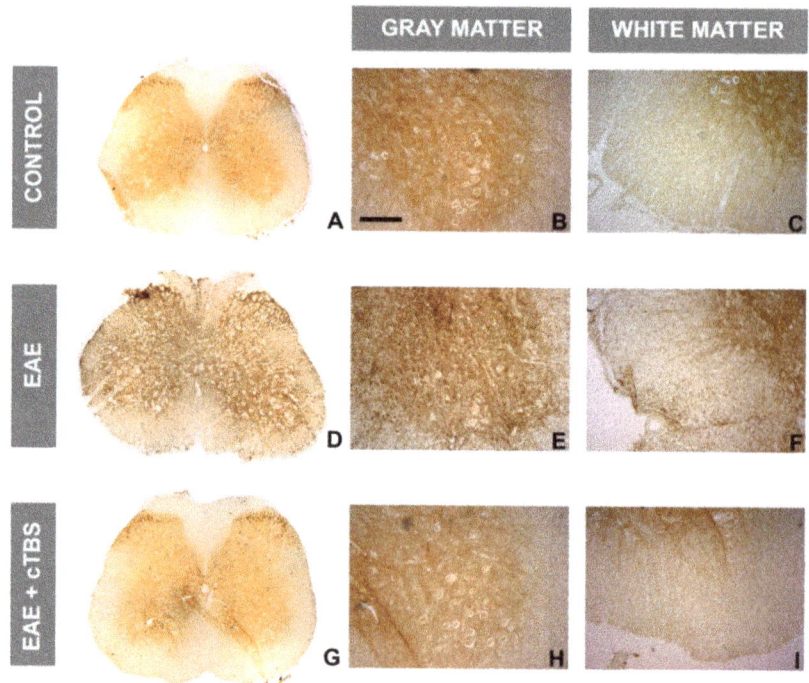

Figure 7. Effects of CTBS treatment on AMP-based enzyme histochemistry in lumbar spinal cords of EAE rats. Enzyme histochemistry in the presence of AMP as a substrate labeling structures that exhibit eN/CD73 activity in the spinal cord of control, EAE, and CTBS-treated EAE sections. Control sections (**A**) exhibited diffuse staining patterns localized mainly in gray matter (**B**), while white matter was devoid of staining. (**C**) EAE sections reveled (**D**) a marked increase in eN/CD73 activity localized in gray (**E**) and white matter (**F**). After CTBS protocol (**G**), faint activity was observed in both gray (**H**) and white matter (**I**), similarly to control sections. Scale bar corresponds to 50 μm.

Concerning the $A_{2A}R$, the intensity of *ir* was weak in control sections, and no significant co-localization was observed with either GFAP-*ir* or IBA-1 (Figure 9A,a). EAE was associated with significant enhancement of $A_{2A}R$-*ir*, particularly co-localized with GFAP- and IBA1-*ir* (Figure 9B,b), reflected through a significant increase in PCC for the association of $A_{2A}R$ with GFAP and IBA-1 (Figure 9D). Again, CTBS treatment markedly decreased the intensity of $A_{2A}R$-*ir* and induced massive dissociation between GFAP- and IBA-1-*ir*. A significant part of $A_{2A}R$-*ir* after CTBS resided at 5–7 μm in diameter ovoid structures, probably axon fibers (Figure 9c, arrowhead). Combined immunofluorescence directed to $A_{2A}R$ and neurofilament H protein showed a strong association of $A_{2A}R$ with neuronal cell bodies in the gray matter and with axonal fibers in the white matter (Figure 10A,B). The CTBS treatment reduced $A_{2A}R$ expression on glial cells and increased it on spinal cord neurons.

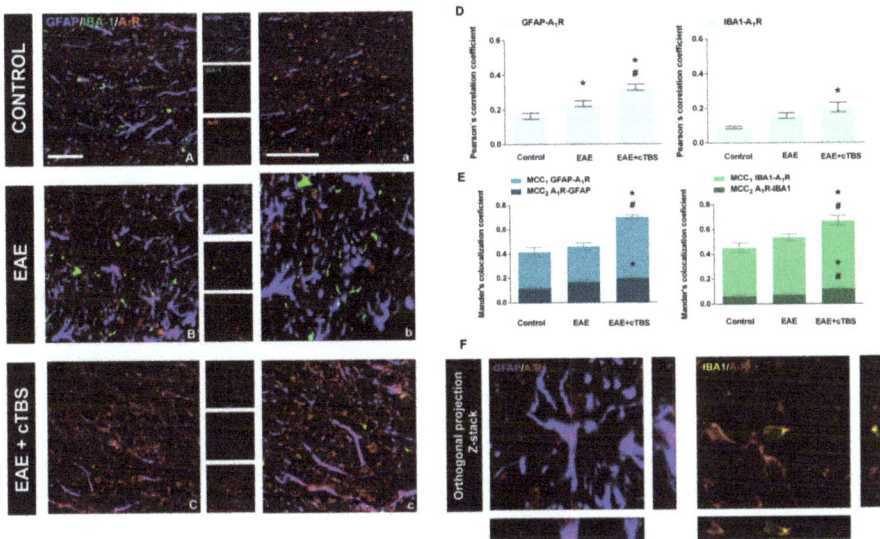

Figure 8. Effects of CTBS treatment on A_1R expression in lumbar spinal cords of EAE rats. Triple immunofluorescence labeling directed to astrocyte marker GFAP (blue), microglial marker IBA-1 (green), and $A_{2A}R$ (red). In control sections, moderate staining of A_1R-*ir* was observed mostly confined to what appeared to be neuronal elements (**A,a**). In EAE sections, no apparent change in A_1R-*ir* was observed compared to control (**B,b**). After CTBS treatment A_1R-*ir* was significantly increased on glial cells (**C,c**). Pearson correlation coefficients (PCC) indicating the level of signal overlap between GFAP-*ir* and A_1R-*ir* and IBA-1-*ir* and A_1R-*ir*. Bars show mean PCC ± SEM, from 7–9 images/animal (**D**). Mander's colocalization coefficient (MCC) indicating level of signal colocalization between GFAP/A_1R (MCC_1, light blue), A_1R/GFAP (MCC_2, dark blue), IBA-1/A_1R (MCC_1, light green), and A_1R/IBA1 (MCC_2, dark green). Bars show mean MCC ± SEM, from 7–9 images/animal (**E**). Orthogonal Z-stack projection of GFAP/A_1R and IBA-1/A_1R (**F**). Level of significance: * $p < 0.05$ or less when compared to control, # $p < 0.05$ when compared to EAE. Scale bar corresponds to 50 µm.

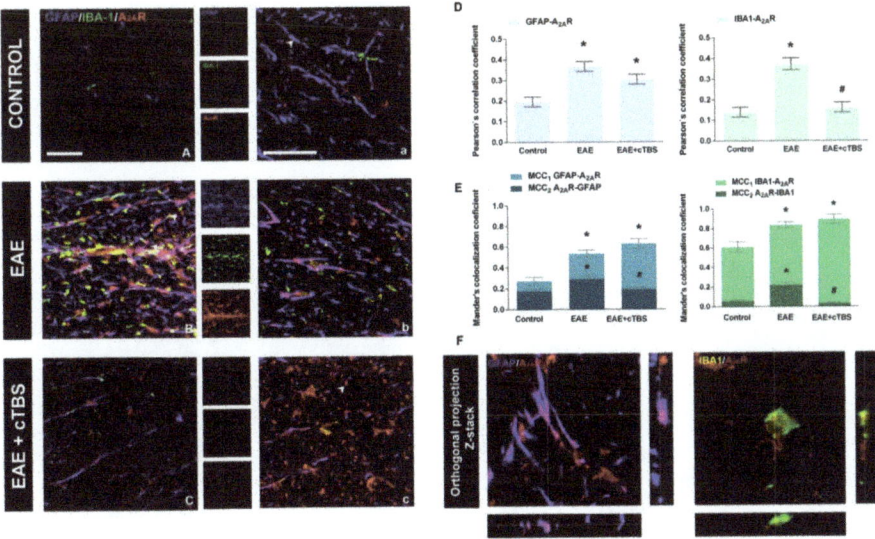

Figure 9. Effects of CTBS treatment on $A_{2A}R$ expression in lumbar spinal cords of EAE rats. Triple immunofluorescence

labeling directed to astrocyte marker GFAP (blue), microglial marker IBA-1 (green), and $A_{2A}R$ (red). In control sections, faint staining of $A_{2A}R$-*ir* was observed (**A,a**). A prominent increase in $A_{2A}R$-*ir* was observed in EAE, both in association with GFAP- and IBA-1-*ir* (**B,b**). The CTBS treatment decreased the overall intensity of $A_{2A}R$-*ir* in the gray matter and was reduced on glial cells, but an increase in staining was detected in non-glial elements (**C,c**). Pearson correlation coefficients (PCC) indicating the level of signal overlap between GFAP-*ir* and $A_{2A}R$-*ir* and IBA-1-*ir* and $A_{2A}R$-*ir*. Bars show mean PCC ± SEM, from 7–9 images/animal (**D**). Mander's colocalization coefficient (MCC) indicating level of signal colocalization between GFAP/$A_{2A}R$ (MCC_1, light blue), $A_{2A}R$/GFAP (MCC_2, dark blue), IBA-1/$A_{2A}R$ (MCC_1, light green), and $A_{2A}R$/IBA1 (MCC_2, dark green) Bars show mean MCC ± SEM, from 7–9 images/animal (**E**). Orthogonal Z-stack projection of GFAP/$A_{2A}R$ and IBA-1/$A_{2A}R$ (**F**). Level of significance: * $p < 0.05$ or less when compared to control, # $p < 0.05$ when compared to EAE. Scale bar corresponds to 50 µm.

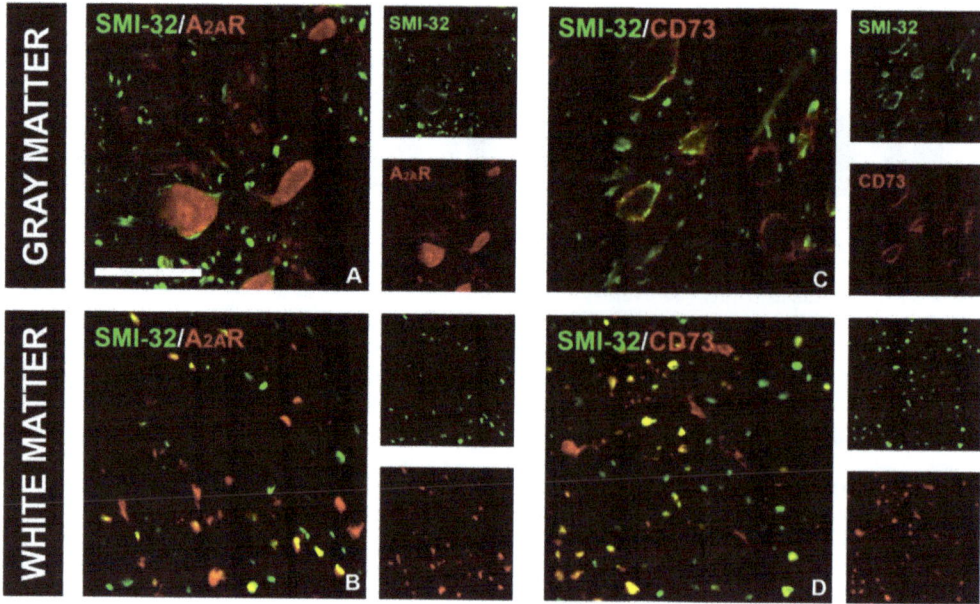

Figure 10. $A_{2A}R$ and CD73 expression in gray and white matter in lumbar spinal cords of CTBS-treated rats. $A_{2A}R$ signal was colocalized with SMI-32: in gray matter, co-staining was observed in neuronal soma (**A**), whereas in white matter, neuronal axons showed $A_{2A}R$ immunoreactivity (**B**). CD73 signal was colocalized with SMI-32: in gray matter (**C**), whereas in white matter, neuronal axons showed SMI-32/CD73 colocalization (**D**). Scale bar corresponds to 50 µm.

4. Discussion

EAE is a widely used experimental model of the autoimmune neurodegenerative pathology driven by an intertwined network of adaptive immune and CNS resident cells and their inflammatory mediators, which reproduce all the critical events in MS. According to current understanding, pro-inflammatory mediator IL-1β and its main downstream target, NF-κB, are critically involved in the pathogenesis of MS/EAE [48], while the induction of anti-inflammatory cytokine IL-10 correlates with the clinical recovery [49]. The involvement of extracellular ATP, adenosine, and their respective P2 and P1 purinoceptors in the neurodegenerative processes associated with MS/EAE is established as well [50]. Several recent reports emphasize the contribution of ectonucleotidases and ATP/ADP- [41,51,52] and adenosine-mediated signaling in the neuroinflammatory process in EAE pathology (Safarzadeh et al., 2016 [53]; Nedeljkovic, 2019 [18], Lavrnja et al., 2015 [14]; Zhou et al., 2019 [54]). Accordingly, the present study shows that the neuroinflammatory process in EAE is associated with prominent upregulation of CD73 in lumbosacral spinal cord tissue, mostly by reactive microglia and astrocytes activated in response to immune cell invasion to the CNS. Given that CD73 is the only adenosine-producing enzyme in the extracellular milieu [55], the strong induction of CD73 corroborates the finding of the substantial accumulation of adenosine in the extracellular space during EAE (Lavrnja et al., 2009 [13]; 2015 [14]). Although adenosine is generally considered a powerful anti-inflammatory and immunosuppressive molecule [56,57], it exerts pleiotropic actions depending on the functional coupling with particular P1 receptor subtype [15,18,20]. Thus, in physiological conditions, extracellular adenosine, present in low micromolar concentrations, mainly activates inhibitory a A_1R receptor subtype ubiquitously present in the CNS cell types. However, in neuroinflammatory conditions, the actions of adenosine are mediated largely via excitatory $A_{2A}R$ and low-affinity $A_{2B}R$ receptor subtypes. Indeed, the upregulation of $A_{2A}R$ and its tight spatial coupling with CD73 is another common feature of inflamed tissue in several brain pathologies, including EAE/MS [15,58,59]. Our present study, thus, corroborates the view that the gain-of-function in CD73/$A_{2A}R$ and enhanced adenosine signaling drives neuroinflammation and directs the course of EAE.

By using the pathological context of EAE, the principal goal of our study was to show the ability and efficiency of the CTBS protocol to revert the EAE-induced alterations in adenosine signaling and, thus, to point to potential merit of TBS as a therapeutic approach in MS/EAE. Beneficial and anti-inflammatory actions of TBS have been demonstrated in several neurological and psychiatric disorders and animal models, so far [60–65]. In the current study, we have observed that animals subjected to CTBS experienced milder neurological dysfunctions for a shorter time than in the group of non-treated EAE. At the histopathological level, the CTBS protocol prevented the release of IL-1β and reduced NF-κB signaling, while increasing the expression of anti-inflammatory IL-10. These effects altogether suggested that CTBS exerted neuroimmune downregulating properties. Indeed, animals subjected to CTBS exhibited significantly lower numbers of reactive microglial cells and hypertrophied astrocytes, which are the typical histological hallmark of the spinal cord tissue injury in EAE [14,39]. The treatment also decreased both the levels of CD73 enzyme activity and the protein expression, particularly by microglia and astrocytes, suggesting a decrease in the extracellular level of adenosine. Given that CD73 itself is necessary for the peripheral T cells entry and the induction of EAE [66,67], altered expression of CD7 by microglia and astrocytes may be seen as the critical factor of the reduced peripheral immune cell entry and local neuroinflammation [18,66].

Besides CD73, the CTBS treatment completely reverted the expression of adenosine receptors, at least the dominant A_1R and $A_{2A}R$ subtypes. Specifically, CTBS prevented the exclusion of A_1R-mediated signaling observed in EAE and even enhanced the purinoceptors expression in respect to naïve animals. The enhanced expression was mainly observed at astrocytes and microglia, at which the A_1R receptor activation decreases proinflammatory cytokines and chemokines, thus reducing astrocyte ability to interact with autoreactive CD4$^+$ lymphocytes (Liu et al., 2018 [68]; Cunha, 2005 [58]; Liu et al., 2018 [68]; Bijelić et al., 2020 [52]). Furthermore, the CTBS treatment prevented excessive $A_{2A}R$ signaling and decreased the co-occurrence of both the $A_{2A}R$ and CD73 with the glial cells markers. Instead, CTBS induced neuronal expression of $A_{2A}R$, which is known to regulate the tonic expression and synaptic actions of BDNF [40], thus promoting neuronal survival [69–71]. Namely, neuronal $A_{2A}R$-mediated signaling increases BDNF synthesis and the resulting synaptic efficiency and LTD-induced plasticity [72,73], which may be one of the possible mechanisms of the CTBS-induced protective actions in EAE.

In the end, we would like to point out some limitations of our study. Due to size of the TBS stimulation coil, when applied, the whole brain of DA rats is being stimulated, and therefore, we could not ascribe observed beneficial effects to a specific brain region. The beneficial effects observed in this study are most likely mediated via various descending cerebro-spinal tracts. It is possible that focal stimulation of a specific region would yield even better effects; therefore, further research is required in this direction. Another potential limitation would be the selected time of stimulation, since we chose to stimulate animals in the peak of disease and monitor them until the end of disease. Even though it is more common practice to start treatment in the onset of acute EAE, we wanted to examine beneficial effects that could translate to more real situation, since MS patients seek medical attention usually during the peak of their symptoms, which corresponds to the peak of acute EAE in experimental animals.

5. Conclusions

Our study convincingly demonstrates that the applied CTBS protocol efficiently counteracts the EAE-induced effects on adenosine signaling and attenuates the reactive state of microglia and astrocytes at histological and biochemical levels, thus providing powerful protective and reparative potential in EAE. Given the paucity of effective treatments in MS, the TBS protocols could be a safe and effective complementary therapeutic approach, together with other disease-modifying treatments, that could provide better clinical outcome in MS.

Author Contributions: M.D.: conceptualization, methodology, validation, investigation, writing—original draft; M.Z.: methodology, validation, formal analysis, writing—original draft; I.S.: methodology, validation, investigation, formal analysis, writing—review and editing; M.A.: methodology, validation, formal analysis, writing—review and editing; A.S.: methodology, validation, formal analysis, investigation; K.M.: methodology, validation, formal analysis, investigation; I.G.: methodology, resources, formal analysis, writing—review and editing; N.I.: methodology, investigation, writing—review and editing; T.V.I.: conceptualization, methodology, resources, formal analysis, writing—review and editing; N.N.: methodology, resources, formal analysis, writing—review and editing; M.N.: conceptualization, resources, formal analysis, writing—review and editing. All authors have read and agreed to the published version of the manuscript.

Funding: This work was supported by the Ministry of Education, Science and Technological development, the Republic of Serbia (Grant No. 451-93-9/2021-14/200178) and the University of Defense (Grant No. MFVMA/04/19-21).

Institutional Review Board Statement: The study was conducted according to the guidelines of the Declaration of Helsinki, and approved by The Ethical Community of the Military Medical Academy (Belgrade, Serbia) license no. 323-07-00622/2017-05 (May 2017).

Informed Consent Statement: Not applicable.

Data Availability Statement: The datasets generated during and/or analyzed during the current study are available from the corresponding author on reasonable request.

Conflicts of Interest: The authors declare that they have no conflict of interest.

References

1. Lassmann, H.; Brück, W.; Lucchinetti, C.F. The immunopathology of multiple sclerosis: An overview. *Brain Pathol.* **2007**, *17*, 210–218. [CrossRef] [PubMed]
2. Filippi, M.; Preziosa, P.; Langdon, D.; Lassmann, H.; Paul, F.; Rovira, À.; Schoonheim, M.M.; Solari, A.; Stankoff, B.; Rocca, M.A. Identifying Progression in Multiple Sclerosis: New Perspectives. *Ann. Neurol.* **2020**, *88*, 438–452. [CrossRef] [PubMed]
3. Brambilla, R. The contribution of astrocytes to the neuroinflammatory response in multiple sclerosis and experimental autoimmune encephalomyelitis. *Acta Neuropathol.* **2019**, *137*, 757–783. [CrossRef] [PubMed]
4. Duffy, S.S.; Lees, J.G.; Moalem-Taylor, G. The contribution of immune and glial cell types in experimental autoimmune encephalomyelitis and multiple sclerosis. *Mult. Scler. Int.* **2014**, *2014*, 285245. [CrossRef] [PubMed]
5. Compston, A.; Coles, A. Multiple sclerosis. *Lancet* **2002**, *359*, 1221–1231. [CrossRef]
6. Fletcher, J.M.; Lalor, S.J.; Sweeney, C.M.; Tubridy, N.; Mills, K.H.G. T cells in multiple sclerosis and experimental autoimmune encephalomyelitis. *Clin. Exp. Immunol.* **2010**, *162*, 1–11. [CrossRef] [PubMed]
7. Sättler, M.B.; Bähr, M. Future neuroprotective strategies. *Exp. Neurol.* **2010**, *225*, 40–47. [CrossRef] [PubMed]
8. di Virgilio, F.; Ceruti, S.; Bramanti, P.; Abbracchio, M.P. Purinergic signalling in inflammation of the central nervous system. *Trends Neurosci.* **2009**, *32*, 79–87. [CrossRef] [PubMed]
9. Giuliani, A.L.; Sarti, A.C.; di Virgilio, F. Ectonucleotidases in Acute and Chronic Inflammation. *Front. Pharmacol.* **2020**, *11*, 619458. [CrossRef]
10. Yegutkin, G.G. Nucleotide- and nucleoside-converting ectoenzymes: Important modulators of purinergic signalling cascade. *Biochim. Biophys. Acta* **2008**, *1783*, 673–694. [CrossRef]

11. Adzic, M.; Nedeljkovic, N. Unveiling the Role of Ecto-5′-Nucleotidase/CD73 in Astrocyte Migration by Using Pharmacological Tools. *Front. Pharmacol.* **2018**, *9*, 153. [CrossRef] [PubMed]
12. Lazzarino, G.; Amorini, A.M.; Eikelenboom, M.; Killestein, J.; Belli, A.; Di Pietro, V.; Tavazzi, B.; Barkhof, F.; Polman, C.; Uitdehaag, B.; et al. Cerebrospinal fluid ATP metabolites in multiple sclerosis. *Mult. Scler.* **2010**, *16*, 549–554. [CrossRef] [PubMed]
13. Lavrnja, I.; Bjelobaba, I.; Stojiljkovic, M.; Pekovic, S.; Mostarica-Stojkovic, M.; Stošić-Grujičić, S.; Nedeljkovic, N. Time-course changes in ectonucleotidase activities during experimental autoimmune encephalomyelitis. *Neurochem. Int.* **2009**, *55*, 193–198. [CrossRef]
14. Lavrnja, I.; Laketa, D.; Savić, D.; Bozic, I.; Bjelobaba, I.; Pekovic, S.; Nedeljkovic, N. Expression of a second ecto-5′-nucleotidase variant besides the usual protein in symptomatic phase of experimental autoimmune encephalomyelitis. *J. Mol. Neurosci. MN* **2015**, *55*, 898–911. [CrossRef]
15. Cunha, R.A. How does adenosine control neuronal dysfunction and neurodegeneration? *J. Neurochem.* **2016**, *139*, 1019–1055. [CrossRef]
16. Abbracchio, M.P.; Ceruti, S. Roles of P2 receptors in glial cells: Focus on astrocytes. *Purinergic Signal.* **2006**, *2*, 595–604. [CrossRef]
17. di Virgilio, F.; Vuerich, M. Purinergic signaling in the immune system. *Auton. Neurosci. Basic Clin.* **2015**, *191*, 117–123. [CrossRef]
18. Nedeljkovic, N. Complex regulation of ecto-5′-nucleotidase/CD73 and A(2A)R-mediated adenosine signaling at neurovascular unit: A link between acute and chronic neuroinflammation. *Pharmacol. Res.* **2019**, *144*, 99–115. [CrossRef]
19. Antonioli, L.; Pacher, P.; Vizi, E.S.; Haskó, G. CD39 and CD73 in immunity and inflammation. *Trends Mol. Med.* **2013**, *19*, 355–367. [CrossRef]
20. Fredholm, B.B.; Chen, J.-F.; Cunha, R.A.; Svenningsson, P.; Vaugeois, J.-M. Adenosine and brain function. *Int. Rev. Neurobiol.* **2005**, *63*, 191–270.
21. Smith, G.E.; Pankratz, V.S.; Negash, S.; Machulda, M.M.; Petersen, R.C.; Boeve, B.F.; Knopman, D.S.; Lucas, J.A.; Ferman, T.J.; Graff-Radford, N.; et al. A plateau in pre-Alzheimer memory decline: Evidence for compensatory mechanisms? *Neurology* **2007**, *69*, 133–139. [CrossRef] [PubMed]
22. Blackburn, M.R.; Vance, C.O.; Morschl, E.; Wilson, C.N. Adenosine receptors and inflammation. *Handb. Exp. Pharmacol.* **2009**, *193*, 215–269.
23. Barker, A.T.; Jalinous, R.; Freeston, I.L. Non-invasive magnetic stimulation of human motor cortex. *Lancet* **1985**, *1*, 1106–1107. [CrossRef]
24. Hallett, M. Transcranial magnetic stimulation: A primer. *Neuron* **2007**, *55*, 187–199. [CrossRef]
25. Lefaucheur, J.-P.; Aleman, A.; Baeken, C.; Benninger, D.H.; Brunelin, J.; Di Lazzaro, V.; Filipović, S.R.; Grefkes, C.; Hasan, A.; Hummel, F.C.; et al. Evidence-based guidelines on the therapeutic use of repetitive transcranial magnetic stimulation (rTMS): An update (2014-2018). *Clin. Neurophysiol. Off. J. Int. Fed. Clin. Neurophysiol.* **2020**, *131*, 474–528. [CrossRef]
26. Huang, Y.-Z.; Rothwell, J.C.; Chen, R.-S.; Lu, C.-S.; Chuang, W.-L. The theoretical model of theta burst form of repetitive transcranial magnetic stimulation. *Clin. Neurophysiol. Off. J. Int. Fed. Clin. Neurophysiol.* **2011**, *122*, 1011–1018. [CrossRef]
27. Koch, G.; Brusa, L.; Caltagirone, C.; Peppe, A.; Oliveri, M.; Stanzione, P.; Centonze, D. rTMS of supplementary motor area modulates therapy-induced dyskinesias in Parkinson disease. *Neurology* **2005**, *65*, 623–625. [CrossRef]
28. George, M.S.; Lisanby, S.H.; Avery, D.; McDonald, W.M.; Durkalski, V.; Pavlicova, M.; Anderson, B.; Nahas, Z.; Bulow, P.; Zarkowski, P.; et al. Daily left prefrontal transcranial magnetic stimulation therapy for major depressive disorder: A sham-controlled randomized trial. *Arch. Gen. Psychiatry* **2010**, *67*, 507–516. [CrossRef]
29. Khedr, E.M.; Fetoh, N.A.-E. Short- and long-term effect of rTMS on motor function recovery after ischemic stroke. *Restor. Neurol. Neurosci.* **2010**, *28*, 545–559. [CrossRef]
30. Ridding, M.C.; Ziemann, U. Determinants of the induction of cortical plasticity by non-invasive brain stimulation in healthy subjects. *J. Physiol.* **2010**, *588*, 2291–2304. [CrossRef]
31. Hasan, A.; Guse, B.; Cordes, J.; Wölwer, W.; Winterer, G.; Gaebel, W.; Langguth, B.; Landgrebe, M.; Eichhammer, P.; Frank, E.; et al. Cognitive Effects of High-Frequency rTMS in Schizophrenia Patients With Predominant Negative Symptoms: Results From a Multicenter Randomized Sham-Controlled Trial. *Schizophr. Bull.* **2016**, *42*, 608–618. [CrossRef]
32. Rabey, J.M.; Dobronevsky, E. Repetitive transcranial magnetic stimulation (rTMS) combined with cognitive training is a safe and effective modality for the treatment of Alzheimer's disease: Clinical experience. *J. Neural Transm.* **2016**, *123*, 1449–1455. [CrossRef]
33. Nasios, G.; Messinis, L.; Dardiotis, E.; Papathanasopoulos, P. Repetitive Transcranial Magnetic Stimulation, Cognition, and Multiple Sclerosis: An Overview. *Behav. Neurol.* **2018**, *2018*, 8584653. [CrossRef]
34. Liu, M.; Fan, S.; Xu, Y.; Cui, L. Non-invasive brain stimulation for fatigue in multiple sclerosis patients: A systematic review and meta-analysis. *Mult. Scler. Relat. Disord.* **2019**, *36*, 101375. [CrossRef]
35. Centonze, D.; Petta, F.; Versace, V.; Rossi, S.; Torelli, F.; Prosperetti, C.; Marfia, G.; Bernardi, G.; Koch, G.; Miano, R.; et al. Effects of motor cortex rTMS on lower urinary tract dysfunction in multiple sclerosis. *Mult. Scler.* **2007**, *13*, 269–271. [CrossRef]
36. Mori, T.; Koyama, N.; Guillot-Sestier, M.-V.; Tan, J.; Town, T. Ferulic acid is a nutraceutical β-secretase modulator that improves behavioral impairment and alzheimer-like pathology in transgenic mice. *PLoS ONE* **2013**, *8*, e55774. [CrossRef]
37. Stevanovic, I.; Ninkovic, M.; Mancic, B.; Milivojevic, M.; Stojanovic, I.; Ilic, T.; Vujovic, M.; Djukic, M. Compensatory Neuro-protective Response of Thioredoxin Reductase against Oxidative-Nitrosative Stress Induced by Experimental Autoimmune Encephalomyelitis in Rats: Modulation by Theta Burst Stimulation. *Molecules* **2020**, *25*, 3922. [CrossRef]

38. Medina-Fernandez, F.J.; Escribano, B.M.; Agüera, E.; Aguilar-Luque, M.; Feijoo, M.; Luque, E.; Garcia-Maceira, F.I.; Pascual-Leone, A.; Drucker-Colin, R.; Tunez, I.; et al. Effects of transcranial magnetic stimulation on oxidative stress in experimental autoimmune encephalomyelitis. *Free Radic. Res.* **2017**, *51*, 460–469. [CrossRef]
39. Dragic, M.; Zeljkovic, M.; Stevanovic, I.; Ilic, T.; Ilic, N.; Nedeljkovic, N.; Ninkovic, M. Theta burst stimulation ameliorates symptoms of experimental autoimmune encephalomyelitis and attenuates reactive gliosis. *Brain Res. Bull.* **2020**, *162*, 208–217. [CrossRef]
40. Stevanovic, I.; Mancic, B.; Ilic, T.; Milosavljevic, P.; Lavrnja, I.; Stojanovic, I.; Ninkovic, M. Theta burst stimulation influence the expression of BDNF in the spinal cord on the experimental autoimmune encephalomyelitis. *Folia Neuropathol.* **2019**, *57*, 129–145. [CrossRef]
41. Jakovljevic, M.; Lavrnja, I.; Bozic, I.; Savic, D.; Bjelobaba, I.; Pekovic, S.; Sévigny, J.; Nedeljkovic, N.; Laketa, D. Down-regulation of NTPDase2 and ADP-sensitive P2 Purinoceptors Correlate with Severity of Symptoms during Experimental Autoimmune Encephalomyelitis. *Front. Cell. Neurosci.* **2017**, *11*, 333. [CrossRef] [PubMed]
42. Mancic, B.; Stevanovic, I.; Ilic, T.V.; Djuric, A.; Stojanovic, I.; Milanovic, S.; Ninkovic, M. Transcranial theta-burst stimulation alters GLT-1 and vGluT1 expression in rat cerebellar cortex. *Neurochem. Int.* **2016**, *100*, 120–127. [CrossRef] [PubMed]
43. Huang, Y.Z.; Chen, R.S.; Rothwell, J.C.; Wen, H.Y. The after-effect of human theta burst stimulation is NMDA receptor dependent. *Clin. Neurophysiol.* **2007**, *118*, 1028–1032. [CrossRef] [PubMed]
44. Dragić, M.; Zarić, M.; Mitrović, N.; Nedeljković, N.; Grković, I. Application of Gray Level Co-Occurrence Matrix Analysis as a New Method for Enzyme Histochemistry Quantification. *Microsc. Microanal.* **2019**, *25*, 690–698. [CrossRef]
45. Dunn, K.W.; Kamocka, M.M.; McDonald, J.H. A practical guide to evaluating colocalization in biological microscopy. *Am. J. Physiol. Physiol.* **2011**, *300*, C723–C742. [CrossRef]
46. Lin, C.-C.; Edelson, B.T. New Insights into the Role of IL-1β in Experimental Autoimmune Encephalomyelitis and Multiple Sclerosis. *J. Immunol.* **2017**, *198*, 4553–4560. [CrossRef]
47. Lin, W.; Yue, Y.; Stone, S. Role of nuclear factor κB in multiple sclerosis and experimental autoimmune encephalomyelitis. *Neural Regen. Res.* **2018**, *13*, 1507–1515. [CrossRef]
48. van Loo, G.; De Lorenzi, R.; Schmidt, H.; Huth, M.; Mildner, A.; Schmidt-Supprian, M.; Lassmann, H.; Prinz, M.R.; Pasparakis, M. Inhibition of transcription factor NF-kappaB in the central nervous system ameliorates autoimmune encephalitis in mice. *Nat. Immunol.* **2006**, *7*, 954–961. [CrossRef]
49. Ozenci, V.; Kouwenhoven, M.; Huang, Y.M.; Xiao, B.; Kivisäkk, P.; Fredrikson, S.; Link, H. Multiple sclerosis: Levels of interleukin-10-secreting blood mononuclear cells are low in untreated patients but augmented during interferon-beta-1b treatment. *Scand. J. Immunol.* **1999**, *49*, 554–561. [CrossRef]
50. Burnstock, G. An introduction to the roles of purinergic signalling in neurodegeneration, neuroprotection and neuroregeneration. *Neuropharmacology* **2016**, *104*, 4–17. [CrossRef]
51. Jakovljevic, M.; Lavrnja, I.; Bozic, I.; Milosevic, A.; Bjelobaba, I.; Savic, D.; Sévigny, J.; Pekovic, S.; Nedeljkovic, N.; Laketa, D. Induction of NTPDase1/CD39 by Reactive Microglia and Macrophages Is Associated With the Functional State During EAE. *Front. Neurosci.* **2019**, *13*, 410. [CrossRef]
52. Bijelić, D.D.; Milićević, K.D.; Lazarević, M.N.; Miljković, D.M.; Pristov, J.J.B.; Savić, D.Z.; Petković, B.B.; Andjus, P.R.; Momčilović, M.B.; Nikolić, L.M. Central nervous system-infiltrated immune cells induce calcium increase in astrocytes via astroglial purinergic signaling. *J. Neurosci. Res.* **2020**, *98*, 2317–2332. [CrossRef]
53. Safarzadeh, E.; Jadidi-Niaragh, F.; Motallebnezhad, M.; Yousefi, M. The role of adenosine and adenosine receptors in the immunopathogenesis of multiple sclerosis. *Inflamm. Res.* **2016**, *65*, 511–520. [CrossRef]
54. Zhou, S.; Liu, G.; Guo, J.; Kong, F.; Chen, S.; Wang, Z. Pro-inflammatory Effect of Downregulated CD73 Expression in EAE Astrocytes. *Front. Cell. Neurosci.* **2019**, *13*, 233. [CrossRef]
55. Zimmermann, H.; Zebisch, M.; Sträter, N. Cellular function and molecular structure of ecto-nucleotidases. *Purinergic Signal.* **2012**, *8*, 437–502. [CrossRef]
56. Haskó, G.; Cronstein, B. Regulation of inflammation by adenosine. *Front. Immunol.* **2013**, *4*, 85. [CrossRef]
57. Morandi, F.; Horenstein, A.L.; Rizzo, R.; Malavasi, F. The Role of Extracellular Adenosine Generation in the Development of Autoimmune Diseases. *Mediat. Inflamm.* **2018**, *2018*, 1–10. [CrossRef]
58. Cunha, R.A. Neuroprotection by adenosine in the brain: From A1 receptor activation to A2A receptor blockade. *Purinergic Signal.* **2005**, *1*, 111–134. [CrossRef]
59. Borroto-Escuela, D.O.; Hinz, S.; Navarro, G.; Franco, R.; Müller, C.E.; Fuxe, K. Understanding the Role of Adenosine A2AR Heteroreceptor Complexes in Neurodegeneration and Neuroinflammation. *Front. Neurosci.* **2018**, *12*, 43. [CrossRef]
60. George, M.S.; Wassermann, E.M.; Williams, W.A.; Callahan, A.; Ketter, T.A.; Basser, P.; Hallett, M.; Post, R.M. Daily repetitive transcranial magnetic stimulation (rTMS) improves mood in depression. *NeuroReport* **1995**, *6*, 1853–1856. [CrossRef]
61. Siebner, H.R.; Mentschel, C.; Auer, C.; Conrad, B. Repetitive transcranial magnetic stimulation has a beneficial effect on bradykinesia in Parkinson's disease. *Neuroreport* **1999**, *10*, 589–594. [CrossRef]
62. Downar, J.; Daskalakis, Z.J. New Targets for rTMS in Depression: A Review of Convergent Evidence. *Brain Stimul.* **2013**, *6*, 231–240. [CrossRef]

63. Okada, K.; Matsunaga, K.; Yuhi, T.; Kuroda, E.; Yamashita, U.; Tsuji, S. The long-term high-frequency repetitive transcranial magnetic stimulation does not induce mRNA expression of inflammatory mediators in the rat central nervous system. *Brain Res.* **2002**, *957*, 37–41. [CrossRef]
64. Sasso, V.; Bisicchia, E.; Latini, L.; Ghiglieri, V.; Cacace, F.; Carola, V.; Molinari, M.; Viscomi, M.T. Repetitive transcranial magnetic stimulation reduces remote apoptotic cell death and inflammation after focal brain injury. *J. Neuroinflamm.* **2016**, *13*, 150. [CrossRef]
65. Clarke, D.; Beros, J.; Bates, K.A.; Harvey, A.R.; Tang, A.D.; Rodger, J. Low intensity repetitive magnetic stimulation reduces expression of genes related to inflammation and calcium signalling in cultured mouse cortical astrocytes. *Brain Stimul.* **2021**, *14*, 183–191. [CrossRef]
66. Mills, J.H.; Thompson, L.F.; Mueller, C.; Waickman, A.T.; Jalkanen, S.; Niemela, J.; Airas, L.; Bynoe, M.S. CD73 is required for efficient entry of lymphocytes into the central nervous system during experimental autoimmune encephalomyelitis. *Proc. Natl. Acad. Sci. USA* **2008**, *105*, 9325–9330. [CrossRef]
67. Mills, J.H.; Alabanza, L.M.; Mahamed, D.A.; Bynoe, M.S. Extracellular adenosine signaling induces CX3CL1 expression in the brain to promote experimental autoimmune encephalomyelitis. *J. Neuroinflamm.* **2012**, *9*, 193. [CrossRef] [PubMed]
68. Liu, G.; Zhang, W.; Guo, J.; Kong, F.; Zhou, S.; Chen, S.; Wang, Z.; Zang, D. Adenosine binds predominantly to adenosine receptor A1 subtype in astrocytes and mediates an immunosuppressive effect. *Brain Res.* **2018**, *1700*, 47–55. [CrossRef] [PubMed]
69. Ramirez, S.H.; Fan, S.; Maguire, C.A.; Perry, S.; Hardiek, K.; Ramkumar, V.; Gelbard, H.A.; Dewhurst, S.; Maggirwar, S.B. Activation of adenosine A2A receptor protects sympathetic neurons against nerve growth factor withdrawal. *J. Neurosci. Res.* **2004**, *77*, 258–269. [CrossRef] [PubMed]
70. Wiese, S.; Jablonka, S.; Holtmann, B.; Orel, N.; Rajagopal, R.; Chao, M.V.; Sendtner, M. Adenosine receptor A2A-R contributes to motoneuron survival by transactivating the tyrosine kinase receptor TrkB. *Proc. Natl. Acad. Sci. USA* **2007**, *104*, 17210–17215. [CrossRef] [PubMed]
71. Jeon, S.J.; Bak, H.; Seo, J.; Han, S.M.; Lee, S.H.; Han, S.-H.; Kwon, K.J.; Ryu, J.H.; Cheong, J.H.; Ko, K.H.; et al. Oroxylin A Induces BDNF Expression on Cortical Neurons through Adenosine A2A Receptor Stimulation: A Possible Role in Neuroprotection. *Biomol. Ther.* **2012**, *20*, 27–35. [CrossRef]
72. Kopec, B.M.; Kiptoo, P.; Zhao, L.; Rosa-Molinar, E.; Siahaan, T.J. Noninvasive Brain Delivery and Efficacy of BDNF to Stimulate Neuroregeneration and Suppression of Disease Relapse in EAE Mice. *Mol. Pharm.* **2019**, *17*, 404–416. [CrossRef]
73. Costenla, A.R.; Diógenes, M.J.; Canas, P.M.; Rodrigues, R.J.; Nogueira, C.; Maroco, J.; Agostinho, P.M.; Ribeiro, J.A.; Cunha, R.A.; De Mendonça, A. Enhanced role of adenosine A2A receptors in the modulation of LTP in the rat hippocampus upon ageing. *Eur. J. Neurosci.* **2011**, *34*, 12–21. [CrossRef]

Review

Contribution of TMS and TMS-EEG to the Understanding of Mechanisms Underlying Physiological Brain Aging

Andrea Guerra [1], Lorenzo Rocchi [2,3], Alberto Grego [4], Francesca Berardi [4], Concetta Luisi [4] and Florinda Ferreri [4,5,*]

1. IRCCS Neuromed, 86077 Pozzilli (IS), Italy; andrea.guerra@uniroma1.it
2. Department of Clinical and Movements Neurosciences, UCL Queen Square Institute of Neurology, University College London, London WC1N 3BG, UK; l.rocchi@ucl.ac.uk
3. Department of Medical Sciences and Public Health, University of Cagliari, 09124 Cagliari, Italy
4. Department of Neuroscience, University of Padua, 35122 Padua, Italy; alberto.grego@aopd.veneto.it (A.G.); francesca.berardi@unipd.it (F.B.); concetta.luisi@unipd.it (C.L.)
5. Department of Clinical Neurophysiology, Kuopio University Hospital, University of Eastern Finland, 70210 Kuopio, Finland
* Correspondence: florinda.ferreri@unipd.it

Abstract: In the human brain, aging is characterized by progressive neuronal loss, leading to disruption of synapses and to a degree of failure in neurotransmission. However, there is increasing evidence to support the notion that the aged brain has a remarkable ability to reorganize itself, with the aim of preserving its physiological activity. It is important to develop objective markers able to characterize the biological processes underlying brain aging in the intact human, and to distinguish them from brain degeneration associated with many neurological diseases. Transcranial magnetic stimulation (TMS), coupled with electromyography or electroencephalography (EEG), is particularly suited to this aim, due to the functional nature of the information provided, and thanks to the ease with which it can be integrated with behavioral manipulation. In this review, we aimed to provide up to date information about the role of TMS and TMS-EEG in the investigation of brain aging. In particular, we focused on data about cortical excitability, connectivity and plasticity, obtained by using readouts such as motor evoked potentials and transcranial evoked potentials. Overall, findings in the literature support an important potential contribution of TMS to the understanding of the mechanisms underlying normal brain aging. Further studies are needed to expand the current body of information and to assess the applicability of TMS findings in the clinical setting.

Keywords: aging; transcranial magnetic stimulation; EEG; excitability; connectivity; plasticity

Citation: Guerra, A.; Rocchi, L.; Grego, A.; Berardi, F.; Luisi, C.; Ferreri, F. Contribution of TMS and TMS-EEG to the Understanding of Mechanisms Underlying Physiological Brain Aging. *Brain Sci.* **2021**, *11*, 405. https://doi.org/10.3390/brainsci11030405

Academic Editors: Nico Sollmann and Petro Julkunen

Received: 10 February 2021
Accepted: 19 March 2021
Published: 22 March 2021

Publisher's Note: MDPI stays neutral with regard to jurisdictional claims in published maps and institutional affiliations.

Copyright: © 2021 by the authors. Licensee MDPI, Basel, Switzerland. This article is an open access article distributed under the terms and conditions of the Creative Commons Attribution (CC BY) license (https://creativecommons.org/licenses/by/4.0/).

1. Introduction

Physiological aging is a finely controlled process which entails biological alterations at different dimensional scales, from molecules, to cells and integrated systems [1,2]. Some of these changes likely represent the undesirable outcome of exposure to stressors [3], whereas others are physiological and might subtend the attempt of the organism to maintain its function. The latter is particularly true for the central nervous system, whose functional remodeling during aging supports the preservation of activity performance in daily tasks [4]. Identifying specific patterns of neuronal activity linked to aging is a major challenge in neuroscience. In the motor system, for instance, neuromuscular degeneration, alterations in motoneuronal properties, changes in the activity and connectivity of multiple cortical circuits and modifications in the efficiency of synaptic plasticity mechanisms can be observed in older people [5–9]. Some of these abnormalities overlap those occurring in the early stages of patients with neurodegenerative diseases, including movement disorders or dementia [10–20]. Thus, understanding the various changes underlying physiological aging may help in discriminating between normal and pathological conditions. Neurophysiological techniques, in particular transcranial magnetic stimulation (TMS) and

TMS-electroencephalography (EEG) coregistration, are very useful to this aim. Indeed, they are able to assess the activity of several brain circuits with a very high-temporal resolution, thus allowing to identify even subtle changes in the functionality of mechanisms controlling movement, learning and cognitive performances [21,22].

In this review, we will describe changes in cortical excitability, connectivity and plasticity occurring during physiological aging in humans, as assessed by TMS and TMS-EEG studies. Since TMS allows to explore mainly neurophysiological functions of the primary motor cortex (M1), the majority of studies discussed in this article will focus on the motor system. In the first section of the review we report research providing evidence for spatially-restricted (i.e., limited to M1) neurophysiological changes during aging. Indeed, over the last decade, a large number of studies found that aging is associated with changes in global corticospinal excitability and function of different neurotransmitter systems within M1, including GABA-A-ergic, GABA-B-ergic, cholinergic and glutamatergic. In the second section we discuss age-related modifications occurring in large-scale sensorimotor networks, as investigated by TMS and TMS-EEG connectivity measures. Lastly, in the third section of the article, we review non-invasive brain stimulation (NIBS) studies which assessed possible alterations in synaptic plasticity and metaplasticity during aging. Indeed, cortical plastic changes occur throughout the normal lifespan in response to the numerous events that represent everyday experiences, and synaptic plasticity mechanisms, such as long-term potentiation (LTP) and long-term depression (LTD), must be tightly regulated to prevent saturation, which would impair learning and memory [23–28].

2. Neurophysiological Changes in Local Motor Circuits during Aging
2.1. TMS Studies

In the last decades, TMS has been extensively used to investigate the physiology of brain aging in a safe and non-invasive way, particularly with regards to the sensorimotor system. It is well known that a single TMS pulse results in multiple descending volleys, i.e., an early direct (D) followed by indirect (I) waves [29]. The physiological mechanisms underlying the generation of I-waves are still unclear and different hypotheses have been made, ranging from oscillating activation of corticospinal tract (CST) neurons, to reverberation of activity within interneuronal circuits and converging pyramidal output cells [30–32]. The motor-evoked potential (MEP) arises from the temporal summation of these descending volleys at the spinal level, but its basic characteristics (i.e., latency and amplitude) also reflect the combination of excitatory and inhibitory events occurring in a more complex synaptic network, at different levels of the motor pathway [31,33]. The easiest and most reproducible single-pulse (sp) TMS measure, used to probe the excitability of M1, is the resting motor threshold (RMT), which is defined as the stimulation intensity required to elicit MEPs in resting muscles of at least 50 µV in 5 out of 10 trials [30,34]. The RMT can be considered as an estimate of global cortical excitability, it reflects axonal excitability, likely of the intracortical elements activated by TMS, and it has been reported to have high test-retest reliability in healthy older adults, in particular when using monophasic rather than biphasic TMS pulses [35]. It is sensitive not only to cortical excitability, but also to the scalp-to-cortex distance, as this alters the amount of energy required to bring CST neurons to threshold [36]. Given that age-related cortical atrophy can increase the coil-to-cortex distance, the RMT can be used as a measure to explore brain aging [37,38]. Indeed, several studies reported an increase in RMT in older adults [39,40]; however, this has not been replicated by other authors [41–43]. This inconsistency can be due to the possibly different degree of individual cortical atrophy in elderly subjects and to the lack of a structural neuroimaging assessment in the various studies, which would allow a coil-to-cortex distance measurement in all participants and, accordingly, an ad-hoc RMT normalization. Another hypothesis proposed to explain the lack of consistency in age-related differences in RMT is that this variable changes in discrete stages during the life span. Supporting this idea, Shibuya and colleagues (2016) recently evaluated the RMT in 113 subjects with an age ranging from 20 to 83 years; they found that the RMT increases from age 20 to 50 and then decreases, roughly following a quadratic function [44]. Up to date, a long-term longitudinal RMT follow up is missing; this could potentially have more value, compared to cross-sectional measurements, in clarifying electrophysiological correlates of the gray matter volume loss in the aging brain.

In addition to RMT, the stimulus-response relationship may be intrinsically altered in the aged motor cortex. It is well known that the MEP amplitude increases sigmoidally with stimulation intensity, and that voluntary activation of the target muscle shifts this input-output curve to the left [45,46]. Pitcher and colleagues (2003) explored the input-output curve in young and elderly subjects and demonstrated a rightward shift in the latter group, due to the higher stimulation intensity necessary to achieve the maximum MEP. The slope of the curve and the RMT were not different between the two groups [47]. These data have been interpreted as reflecting a reduction in the amount of spinal motoneurons activated by TMS or in the stimulus-induced neuronal synchronization with aging [47,48].

Besides spTMS, more complex TMS protocols, which consist in delivering two stimuli via the same coil (paired-pulse TMS; ppTMS) [49,50] or by coupling a TMS pulse with peripheral electric stimulation [51] have been also applied in older subjects. While spTMS measures generally reflect the degree of the overall corticospinal excitability [52], these protocols allow to investigate the physiology of intracortical circuits which rely on different neurotransmitter classes [30,53,54]. More in detail, in ppTMS a conditioning stimulus (CS) precedes a test stimulus (TS) by an interstimulus interval (ISI) which varies according to the protocol used, allowing the evaluation of intra-cortical inhibitory and facilitatory circuits. Short-interval intracortical inhibition (SICI) is elicited when a subthreshold CS is followed by a suprathreshold TS with an ISI of 1–6 ms and likely reflects inhibition mediated by GABA-A receptors, while intracortical facilitation (ICF) entails longer ISIs (6–30 ms) and is likely mediated by NMDA receptors [30,49,53,55]. The investigation of ppTMS measures in physiological aging has produced inconsistent results. SICI has been reported to be either decreased [41,56–59], normal [42,43,60,61] or even increased [39,62] in elderly. Differently, ICF was found to be generally normal in the majority of [41,44,58,62], but not all [39,61], studies, suggesting intact NMDA-related glutamatergic neurotransmission in these subjects. Long-interval intracortical inhibition (LICI) is another ppTMS measure, which refers to inhibition of a test MEP using a suprathreshold CS applied 50–200 ms before, and likely depends on GABA-B receptors [53,63,64]. Similar to what observed with SICI, the effectiveness of LICI was found to be variable with aging, being either increased [39] or decreased [43,58] in different studies. Thus, ppTMS researches have not yet clarified whether GABA-ergic neurotransmission is generally preserved or impaired in physiological aging. Finally, short interval intracortical facilitation (SICF) can be elicited by a suprathreshold first stimulus followed by a second stimulus at about RMT [65,66], by an ISI compatible with I-waves period (i.e., 1.3–1.5 ms). In this protocol, the observed increase in MEP is likely due to summation of different I-waves at the M1 level and mainly reflects the activity of non-NMDA glutamatergic neurotransmission [17,31]. Clark and colleagues (2011) described a higher SICF in elderly than young subjects at ISI 1.5 ms, while the opposite result was present at ISI 2.5 ms. These apparently contrasting data confirm that physiological mechanisms underlying the first and later SICF peaks differ [17,31,67,68] and may suggest that they have different sensitivity to the aging process [69]. More recently, however, Opie and colleagues tested SICF in young and older adults in more detail and compared I-wave intervals that were optimal for I-waves summation in each group. Although they confirmed the reduction in the second SICF peak in the elderly, a similar change was also found in the other peaks, thus not corroborating the previous hypothesis and leaving the question open [70,71]. Additionally, these studies found a delayed latency of the third SICF peak with aging, which influences both plasticity induction and motor function [70,71].

Discrepancies in the effects of ppTMS protocols across studies can have several explanations. The first could be the choice of the ISI due the fact that paired-pulse protocol is associated with a relatively wide range of possible ISIs, and the timing optimal to obtain the predicted changes in MEP may be different across subjects [30]. While some of ppTMS studies have used multiple ISIs [41,42,44,58], the majority employed only one ISI for all subjects [39,43,56,57,59–61]. As different subjects will probably manifest optimal responses at different ISI, the outcome of future experimental studies could be improved by the use of paired-pulse curves which may decrease the variability and improve the outcome in future studies. Furthermore, concerning SICI, many studies tested this measure only using an ISI of 3 ms, which, according to some reports [72], may be influenced by SICF; however, this has not been universally confirmed [66]. Another important factor contributing to the inconsistency of ppTMS protocols is the choice of CS intensity, which can be calibrated either based on the RMT, or the active motor threshold (AMT). The intensities used vary considerably, usually ranging from 50% to 95% RMT [39,41,42,44,55,56] or from 70% to 90% AMT [73–75], and this variability may contribute to divergent findings in SICI studies. Another point to take into account is that ppTMS measures may have different sensitivity to coil orientation [66,76], and that this pattern of susceptibility to current direction might change according to the age of subjects. In this regard, Sale and colleagues (2016) assessed SICI in young and older adults with TMS coil in posterior-to-anterior (PA) and, then, anterior-to-posterior (AP) position. Interestingly, they found no age-related differences in SICI with PA-directed TMS, while SICI was more effective in older than young subjects using AP-directed TMS [77]. Unfortunately, both the CS and TS were delivered with the same current direction, thus making it difficult to understand which one drives the effect. A further important issue is that ppTMS measures are known to be modulated by different brain states, including motor tasks [78,79], and that this change may be dependent on age. For instance, the amount of SICI during a reaction time task was related to manual dexterity in aging subjects [57]. Moreover, while SICI increases during motor stopping in young subjects, this phenomenon does not occur in older adults [80]. Analyzing task-related changes in ppTMS measures may thus increase their sensitivity in investigating possible age-related decline in intracortical motor circuits. Overall, future studies should standardize the methodology of ppTMS to help comparisons across different age groups.

Differently from ppTMS, short-afferent inhibition (SAI) has shown more consistent alterations in the aging brain. SAI is elicited by electrical stimulation of a peripheral nerve preceding contralateral M1 TMS by an ISIs slightly longer than the latency of the N20 component of the somatosensory evoked potential. This form of inhibition likely reflects intracortical sensorimotor interaction and involves cholinergic as well as GABA-A-ergic circuits [51,81–83]. Although a first study found that SAI was comparable between older and younger adults [42], later studies showed decreased SAI in the elderly [40,84–86], which was also correlated with the age of subjects [87]. These data overall point to a progressive impairment of cholinergic activity during aging.

2.2. TMS-EEG Studies

TMS-EEG allows to measure the perturbation induced by a TMS pulse on cortical activity, both at the local and network level. Thanks to its high temporal resolution, TMS-EEG can provide meaningful clues on the functional properties of cerebral circuits in physiological and pathological conditions [88–92]. Importantly, TMS–EEG allows the assessment of cortical physiology by discerning causal interactions from pure temporal correlations [93,94]. TMS-evoked EEG responses were first measured in the late nineties [95] and, subsequently, scalp topography and possible generator sources of TMS-evoked potentials (TEPs) have been described. It is now clear that TEPs represent genuine responses of the cerebral cortex to TMS, provided that indirect sources of brain activation, such as somatosensory and auditory stimulation, are appropriately reduced or suppressed [96–99]. TEPs occurring in the first 20–40 ms after the TMS pulse most likely reflect the responses of local cortical circuits, whereas longer-latency responses involve more distributed networks [100]. TMS-EEG studies which aimed to investigate age-related modifications in M1 generally showed decreased excitability, a finding which supports the outcomes of MEP investigations [42,47,101]. However TMS-EEG yields complex results, based on the metric used for analysis [102]. For instance, the global-mean field power (GMFP), which provides a measure of the TMS-induced brain response on the whole scalp [103,104], has been shown to be smaller in the elderly, pointing towards a generalized cortical hypoexcitability. This is in agreement with analysis of single TEP peaks, which showed a decrease in the local P30 amplitude after M1 stimulation; however, an opposite trend was found in the ipsilateral prefrontal areas, reflected in an increased TEP amplitude. This has been interpreted as a true hyperexcitability of the prefrontal cortex, rather than a compensatory phenomenon to M1 hypoexcitability [102]. The result on P30 has not been confirmed by another study, which failed to identify an amplitude difference between groups, suggesting instead a decreased P30 latency in elderly subjects [105].

TMS-EEG allows the investigation of specific cortical neurotransmitter systems, with the assumption that different TEP components reflect selective activity of receptor classes. Data obtained in humans and non–human primates point to age-dependent changes in GABA-A and GABA-B receptor density and subunit composition, particularly in frontal cortices [106,107]. In particular, while GABA-ergic neurotransmission seems to become more efficient during the first years of postnatal development in the dorsolateral prefrontal cortex (DLPFC) [108], there is evidence for GABA-ergic neurons activity impairment in animal and human motor areas with aging [57–59,109–111]. Previous TMS-EEG data have found that the P30 and N45 peaks are related to GABA-A neurotransmission, whereas activity of GABA-B receptors shows a stronger link with later peaks, both in healthy and in diseased brains [82,90,100,112,113]. The amplitude of the N45 has been shown to be modulated by aging, albeit in opposite directions in two studies. It was reduced in Ferreri et al. [102], a data which supported a GABA-A-ergic neurotransmission deficit in M1 in older subjects, and increased in a following study [105]. As a further confounding factor, both P30 and N45 amplitude did not change when SICI, which measures GABA-A activity, was obtained by stimulation of DLPFC [114]. While the cause of these contradictory findings is currently unclear, it may reflect the variability observed for SICI in TMS studies in aging (see Section 2.1). Importantly, to date, no study has assessed SICI in M1 with concurrent EEG recording in older adults; this would help in better elucidating possible age-related changes in GABA-A mediated inhibition in the motor system. Only few studies used other ppTMS protocols during EEG recording in the elderly. Noda and colleagues (2017) tested NMDA receptor-mediated glutamatergic activity in the DLPFC using ICF; when comparing the results with a group of healthy controls, they found a reduction in N45 amplitude. The authors speculated that this result would reflect a functional decline in glutamatergic excitatory neurotransmission in older adults [114]. Finally, a recent study recorded TEPs during M1 LICI and demonstrated a greater inhibition of the N45 wave in older, compared to younger subjects; this may suggest that this component is particularly sensitive in detecting modulation of GABA-B-ergic inhibitory circuits which occurs during aging [105].

3. Neurophysiological Changes in Wide-Range Networks during Aging
3.1. TMS Studies

Long-range connectivity between different brain areas can be tested with TMS by delivering pulses with two different coils at specific ISIs. The most robust connectivity protocol is interhemispheric inhibition (IHI), which is obtained by conditioning a MEP with a preceding suprathreshold TMS pulse on the contralateral M1. IHI probably generated by local inhibitory interneurons within M1, which activate GABA-B receptors on principal cells [115] and are activated by interhemispheric excitatory pathways passing through the corpus callosum [116–118]. IHI is most pronounced at ISIs of 8–10 ms and 40–50 ms, which are referred to as short and long latency IHI, respectively (SIHI and LIHI). Most studies, but not all [119,120], testing IHI at rest, revealed that both SIHI [121,122] and LIHI [123–125] did not differ between older and young subjects. By contrast, there is evidence that elderly subjects have less lateralized cortical activation during various motor tasks, including hand grip [126]. In neuroimaging studies, young subjects demonstrate a decreased activity in the

M1 ipsilateral to the moving hand (iM1) during task, while this pattern is cancelled or even reversed in older subjects [127]. TMS research have further explored this phenomenon and found that LIHI from the contralateral M1 (cM1) to iM1 decreased during motor tasks in older compared to young participants, suggesting a reduced inhibitory drive from cM1 to iM1 with aging [121,128]. By combining neurophysiological and neuroimaging data, the authors clarified that the reduced inhibition of the iM1 underlies a progressive involvement of this area during simple motor tasks with aging [128]. Another study demonstrated that the iM1 also contributes to motor preparation in elderly. Indeed, TMS was delivered over iM1 before an index finger movement, and this caused delayed motor responses in older, but not in young, adults [129]. Besides IHI, the interaction between the two M1 can be functionally assessed by evaluating the ipsilateral silent period (ISP), defined as the interruption of voluntary electromyographic (EMG) activity induced by suprathreshold TMS applied in the iM1. ISP is thought to be mediated by transcallosal inhibition between the stimulated and the pre-activated cM1 [116,130]. A number of studies have reported changes in ISP in older subjects, such as delayed onset and decreased depth or area [131–134]. Taken together, findings on IHI and ISP indicate a decline in interhemispheric inhibition with increasing age, with recent evidence that this decline is uniform across the lifespan [134].

In addition to the interactions between the homologous M1, IHI can be extended to a widespread inhibitory system projecting from various cortical areas, including the dorsolateral prefrontal cortex, dorsal premotor cortex and somatosensory cortex, to cM1 [118]. In this regard, one study tested whether physiological aging alters the functional connectivity between the left dorsal premotor cortex (PMd) and right M1 [124]. Paired-pulse TMS was delivered immediately before a simple left index finger movement in young and older subjects, and the data showed that only in the latter group there was a facilitatory interaction between left PMd and right M1. Moreover, the degree of modulation was associated with faster responses [124]. A more recent study combined LIHI with neuroimaging and behavioral measures to assess interhemispheric connectivity between DLPFC and cM1 during the preparation of a complex bimanual coordination task in aging [125]. Interestingly, it was found that the ability to disinhibit functional connectivity between DLPFC and cM1 was impaired in older subjects, and this alteration was paralleled by decreased bimanual performance.

Overall, based on TMS data, it may be argued that the increased activity in ipsilateral M1 and premotor regions before/during simple movements reflects the involvement of additional areas as an attempt to preserve normal motor performances despite advancing age [128,135,136]. Conversely, the altered modulatory activity of DLPFC to contralateral M1 may underlie the decline of bimanual performance in older subjects [125].

3.2. TMS-EEG Studies

As previously stated, recording of EEG activity during TMS provides the possibility to non-invasively and directly probe brain connectivity [89]. Particularly, it has been suggested that the first part of the TMS-evoked EEG response reflects local excitation of the stimulated cortex (see Section 2.2), whereas the spatiotemporal distribution of later TEP components over the scalp reflect the activation of distant cortical areas, either via cortico-cortical connections or projections from subcortical structures [137–140]. Recent evidence has suggested that age and neurodegeneration influence late TEP components [12,102,105,114]. Among the late TEPs described in the EEG signals evoked by M1 stimulation, the N100 is the dominant negative peak and it has been related to the GABA-B-ergic neurotransmission [100,113]. In the elderly, its scalp distribution and source activation has been demonstrated to be significantly different from younger subjects, suggesting hypoexcitability in prefrontal and premotor cortices of the stimulated hemisphere, coupled with hyperexcitability in the median anterior EEG channels [102]. The neural generators of late TEP components are not entirely clarified. In previous studies [100], it had been suggested that they could be related to the activity of reverberant cortico-cortical as well as cortico-subcortical circuits driven by GABA-B neurotransmission, and finally re-engaging the stimulated M1. Age-related differences in spatial distribution of the N100 have been confirmed by a more recent study: here, by using a LICI protocol, the authors demonstrated that the paired-pulse inhibitory effects on N100 wave are increased in older adults, thus suggesting a potentiation of pre- and post-synaptic GABA-B-mediated inhibition [105].

Previous TMS-EEG studies showed that TMS-evoked EEG signals strongly depend on the brain state at the time of stimulus delivery [94]; this can be determined by excitability of local circuits, or by the activity of diffuse neuromodulatory systems [141,142]. In agreement with this notion, features of EEG rhythms preceding a TMS pulse applied on M1 have been shown to influence MEP amplitude [143,144], and this process changes with aging [142,145]. It is known that MEP amplitude shows a degree of inter-trial variability, which depends on several factors, including fluctuations in excitability of cortical and spinal neurons [146,147] and corticospinal connectivity [148]. TMS-EEG allows to verify whether MEP amplitude variability also depends on cortico-cortical connectivity changes. In this regard, Ferreri and colleagues (2014) found that, in young subjects, MEPs are significantly larger when the ipsilateral M1-prefrontal cortex coherence in the beta−2 band and the ipsilateral M1-parietal cortex coherence in the delta band are high. However, elderly subjects showed higher M1-parietal cortex delta coherence than young participants, and this measure was unrelated to MEP size variations [145]. Since the delta rhythm may underlie functional disconnection between areas [149,150], the results of this study possibly reflect functional unbinding of M1 from the somatosensory cortices' inhibitory control. This mechanism may be compensatory to age-related decrease in cortical excitability and motor functions [145].

4. Neurophysiological Changes in Plasticity and Metaplasticity Processes during Aging

Physiological aging is characterized by a weakening of different brain functions, mainly linked to neuroplasticity processes, such as learning and memory [1]. In humans, various TMS-based protocols allow the assessment of synaptic plasticity mechanisms in a non-invasive way; the most widely used are paired associative stimulation (PAS) and theta-burst stimulation (TBS). PAS is based on associative synaptic plasticity [151] and its effects probably reflect spike-timing dependent plasticity, where the precise timing of pre- and post-synaptic neurons firing is crucial for the direction of long-lasting changes. If the inter-spike interval is positive (pre- before post-synaptic action potential), LTP occurs, whereas if the interval is negative (post- before pre-synaptic action potential), LTD is elicited [152–155]. In humans, the PAS protocol implies the combination of repetitive cortical TMS and peripheral nerve stimulation [156,157]. If M1 stimulation occurs around 25 ms after the electric median nerve stimulation at the wrist (PAS_{25}), MEP amplitude is increased for 30–60 min (LTP-like effects), whereas if the inter-stimulus interval is shorter, i.e., 10 ms (PAS_{10}), MEP amplitude is decreased (LTD-like effects) [156–158]. TBS is based on evidence in animal models demonstrating that high-frequency bursts of stimuli rhythmically delivered in the theta frequency range transiently modulate hippocampal neuronal firing [159], and that LTP/LTD-like changes can be recorded by measuring changes in post-synaptic responses following the stimulation [155,160,161]. The classical TBS paradigm consists of bursts of three TMS pulses at 50 Hz, repeated at 5 Hz [162,163]. If the pattern of stimulation is intermittent (iTBS), i.e., short trains of 2 s given every 10 s, cortical excitability is enhanced and MEP amplitude increases up to 30 min after stimulation (LTP-like effects). If the pattern is continuous (cTBS), i.e., bursts given continuously for 40 s, cortical excitability is inhibited and MEP amplitude decreases for 20–60 min (LTD-like effects) [155,163–166].

A recent meta-analysis on NIBS studies in the aged population suggested that there is a general trend towards decrease in motor cortex plasticity, with a certain degree of variability between different studies and different plasticity-inducing protocols [8]. In 2008, two reports by different groups showed, for the first time, that PAS-induced LTP-like plasticity of M1 may deteriorate with physiological aging [167,168]. Müller-Dahlhaus and colleagues verified the effects of PAS in a cohort of 27 subject with variable age (range: 22–71 years) and found that the magnitude of PAS effects was negatively correlated with age, with a smaller

MEP facilitation in elderly subjects [167]. A direct comparison of PAS effects between young and aged subjects was provided by Tecchio and colleagues (2008). Although it was somewhat confirmed that the long-lasting increase of M1 excitability after PAS is weaker in older than young participants, this effect was clearly driven by the female population [168]. The authors pointed to a possible impairment in intracortical excitatory network activity due to hormonal changes during menopause, a hypothesis which was confirmed in a later research [169]. In a following study, a larger number of healthy subjects was enrolled and divided in three groups based on age (young: 21–39 years, middle: 40–59 years, elderly: 60–79 years). The expected PAS-induced facilitation of MEP amplitude was observed in the young and middle groups, but not in the elderly group, further confirming the impaired LTP-like plasticity in M1 with aging [170]. Interestingly, age-related decline in response to PAS has been demonstrated to be restored by L-dopa [171], a finding that suggests that the alteration in PAS response observed in the elderly might have a functional, rather than a structural substrate. In contrast with PAS, only few studies investigated the influence of aging on M1 plasticity induced by TBS. Dickins and colleagues compared MEP amplitude changes after iTBS over the dominant M1 between 20 young (18–28 years) and older subjects (65–76 years). In contrast to PAS studies, M1 excitability increased in a comparable way between the two groups after iTBS [172]. Similar findings were also obtained in more recent research, indicating that iTBS-induced LTP-like plasticity of M1 is not affected by aging [173,174].

The different neurophysiological findings on age-related synaptic plasticity changes obtained using PAS and TBS may depend on the different types of plasticity mechanisms elicited by the different protocols, as described above in this section [155,157,163,175,176]. Moreover, PAS acts through a combination of sensory input and direct cortical stimulation activating the same M1 neurons, a process which reflects heterosynaptic plasticity due to sensorimotor interaction [155,157,177,178]. Conversely, TBS operates through the repetitive activation of the same synapses by M1 stimulation alone, which reflects homosynaptic plasticity mechanisms [155,163,176]. Impaired PAS-induced and normal TBS-induced effects may therefore suggest that synaptic plasticity processes are not diffusely impaired by physiological aging. Rather, the alteration is restricted to processes requiring the activation of specific intracortical circuits and/or sensorimotor interaction mechanisms.

The direction and amount of synaptic plasticity can be influenced by neuronal activity occurring immediately before or during the induction of plasticity [179,180]. These features of plasticity can be framed in the context of metaplasticity, which can be shortly defined as "the plasticity of synaptic plasticity"; this involves a wide range of mechanisms and, from a behavioral point of view, has an important role in the regulation of important brain functions, including memory and learning [27,180,181]. Metaplasticity in humans can be explored through different protocols, from priming an exogenous or endogenous plasticity-inducing protocol with NIBS, to delivering plasticity-inducing protocols with longer duration [64,73,75]. For instance, Opie and colleagues (2017) delivered iTBS 10 min after the application of sham TBS (sham TBS + iTBS), cTBS (cTBS + iTBS), or iTBS (iTBS + iTBS) in young and older participants. The results showed that, whereas priming iTBS with either cTBS or iTBS boosted M1 plasticity in young subjects, MEP facilitation after sham TBS + iTBS did not differ from iTBS + iTBS, and was even larger than cTBS + iTBS, in the older group [173]. Similar findings were also found using PAS: priming the stimulation with $PAS_{N20+2ms}$ caused enhancement of plasticity in young but not in older subjects [182]. In another study by the same group, a visuo-motor training task was performed after facilitatory, inhibitory or sham PAS in young and older adults. While the baseline level of motor skill did not differ between sessions in young subjects, priming with PAS had a detrimental effect on skill acquisition in older ones [183]. Recent evidence suggests that the amount of NIBS-induced plasticity of M1 changes by concurrently modulating cortical gamma oscillations through transcranial alternating current stimulation (tACS), and this effect has been interpreted as reflecting gating phenomena [18,184–186]. However, gamma-tACS has been demonstrated to boost LTP-like plasticity induced by iTBS to a larger extent in young than older adults and, in the latter group, the effect of gamma-tACS decreased with increasing age [174]. In summary, unlike synaptic plasticity mechanisms, which seem to be altered only in part by the aging process, the existing evidence point to a higher susceptibility of metaplasticity processes by physiological aging.

5. Limitations, Perspectives and Conclusions

The present review offers a summary of data obtained with TMS, either coupled with EMG or EEG, about physiological brain aging. Overall, the findings suggest that TMS can offer valuable insight into several functional derangements occurring throughout the life span, including a trend of decrease in brain excitability, altered long-range cortico-cortical connectivity and impaired associative plasticity and metaplasticity processes (Figure 1). There are, however, a number of caveats pertaining to acquisition of TMS-EMG and TMS-EEG variables, as well as their interpretation, which we deem important to discuss in this final section. The first, general issue is that, despite the relatively high number of studies investigating cortical excitability and plasticity during aging, their sample sizes are generally small. This factor would determine low statistical power and may, at least in part, explain the variability of results observed in the literature [8]. A second problem is related to the degree of cerebral atrophy which accompanies physiological aging. As this causes an increase in the distance between scalp and coil, TMS measures are necessarily affected by it [37,38]. Even the most reliable among them, such as the RMT [34], could yield limited information in a context where anatomical data are not available. Therefore, when possible, it is important to obtain structural brain information, along with neurophysiological assessment. In absence of the former, spTMS measures could still be able to track within-subject longitudinal changes, but the pathophysiological/clinical value of such follow up still need to be established. Paired-pulse TMS measures partially solve the confound represented by decreased brain volume, since they are usually calculated as ratios between conditioned and unconditioned MEPs. However, ppTMS protocols are affected by variability and reliability issues of experimental paradigms and output measures [175,187]. For instance, SICI considerably varies between individuals, even if the same CS intensities and ISIs are tested [66,188]. Therefore, to compare results across studies, it would be useful to explore a range of ISIs/CS intensities in large cohorts of subjects, so as to obtain data about maximal effects and thresholds. Moreover, since the putative circuits tested by different ppTMS protocols can interact [72], the interpretation of the effects of CS is not always straightforward. For instance, since SICI acts by suppressing I waves [189], an hypersynchronized and/or hyperexcitable state of excitatory M1 interneurons may secondarily result in SICI decrease [70,71,190]. This is possibly the case in amyotrophic lateral sclerosis [191] and Parkinson's Disease [13,17,192–194]; whether a similar scenario occurs in physiological aging is not known and is probably worth exploring.

The possibility of adding simultaneous EEG recording has substantially increased the range of variables that can be tested with TMS, thus expanding the amount of information that is possible to obtain. However, the nature of TMS-EEG signals has not been completely clarified yet. For instance, the information given by TEPs obtained from M1 TMS may not completely overlap with those provided by MEPs; indeed, the latter arise from excitation of PTN and associated circuitry, whereas the former probably reflect activity of a larger ensemble of cortical cells [89,100,195]. Therefore, caution should be used when comparing conclusions drawn by the two variables, especially for ppTMS protocols, which were devised for MEPs, and are still of uncertain interpretation in the TMS-EEG setting [112,196,197]. Another issue is related to the use of TEPs to measure brain connectivity. This is usually performed by measuring the spatial distribution of specific TEP components; however, this alone does not take into account volume conduction, and has not been assessed in conjunction with more common measures of EEG connectivity [198]. In addition, TEPs, especially in their late components (N100, P180), can be contaminated by EEG responses generated by indirect brain activation due to the somatosensory and auditory stimulation intrinsic to TMS, if adequate countermeasures are not properly taken (e.g., suppression of the TMS click by the use of a masking noise, use of ear defenders, application of a foam layer under the coil) [99,199]. This should lead to careful review of past studies where sensory input by TMS was not properly masked; in particular, it should be noted that effective masking procedures were only seldom used in studies involving older adults. This should prompt to strict control of these confounding factors in future work.

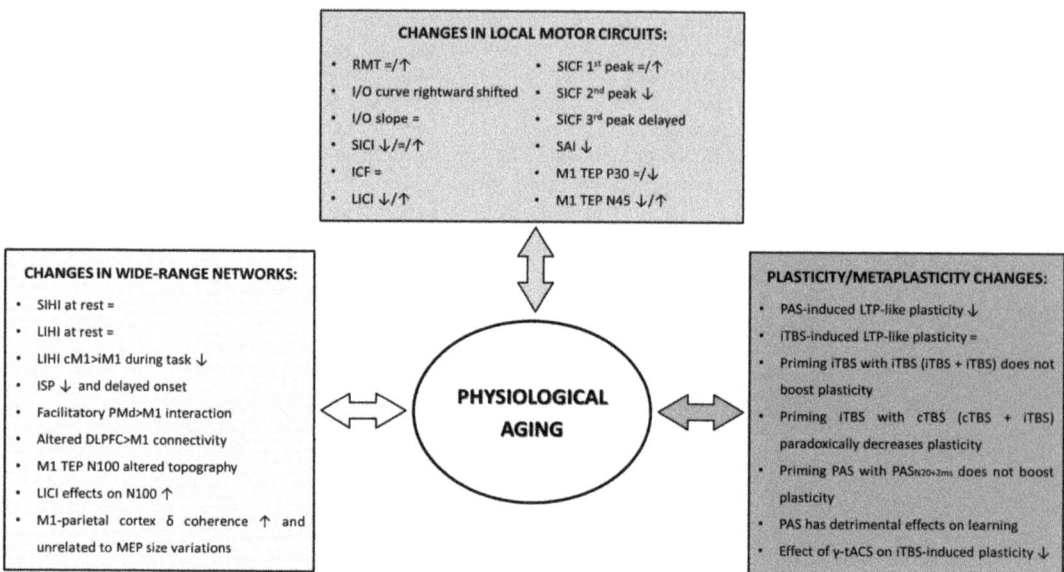

Figure 1. Age-related changes in local motor circuits, wide-range networks and plasticity processes RMT: resting motor threshold; I/O curve: input-output curve; SICI: short-interval intracortical inhibition; ICF: intracortical facilitation; LICI: long-interval intracortical inhibition; SICF: short-interval intracortical facilitation; SAI: short-latency afferent inhibition; M1: primary motor cortex; TEP: transcranial evoked potential; SIHI: short-latency interhemispheric inhibition; LIHI: long-latency interhemispheric inhibition; cM1: contralateral M1; iM1: ipsilateral M1; ISP: ipsilateral silent period; PMd: dorsal premotor cortex; DLPFC: dorsolateral prefrontal cortex; PAS: paired-associative stimulation; iTBS: intermittent theta burst stimulation; cTBS: continuous theta burst stimulation; tACS: transcranial alternating current stimulation.

In conclusion, we believe that TMS and TMS-EEG can give an important contribution to the understanding of the mechanisms underlying physiological brain aging, provided that technical pitfalls and interpretation biases are considered. Future studies should seek to integrate electrophysiological and structural data and clarify how these relate to impairment of daily activities in the elderly population, with the ultimate goal of reliably distinguishing physiological and compensatory processes from disease.

Funding: This work was partially supported by the Italian Institute of Health, Grant GR-2016-02361802.

Institutional Review Board Statement: Not applicable.

Informed Consent Statement: Not applicable.

Data Availability Statement: Not applicable.

Conflicts of Interest: The authors declare no conflict of interest.

References

1. Burke, S.N.; Barnes, C.A. Neural plasticity in the ageing brain. *Nat. Rev. Neurosci.* **2006**, *7*, 30–40. [CrossRef]
2. Yeoman, M.; Scutt, G.; Faragher, R. Insights into CNS ageing from animal models of senescence. *Nat. Rev. Neurosci.* **2012**, *13*, 435–445. [CrossRef]
3. Enzinger, C.; Fazekas, F.; Matthews, P.M.; Ropele, S.; Schmidt, H.; Smith, S.; Schmidt, R. Risk factors for progression of brain atrophy in aging: Six-year follow-up of normal subjects. *Neurology* **2005**, *64*, 1704–1711. [CrossRef]
4. Rossini, P.M.; Ferilli, M.A.N.; Rossini, L.; Ferreri, F. Clinical neurophysiology of brain plasticity in aging brain. *Curr. Pharm. Des.* **2013**, *19*, 6426–6439. [CrossRef] [PubMed]
5. Doherty, T.J. Invited review: Aging and sarcopenia. *J. Appl. Physiol. (1985)* **2003**, *95*, 1717–1727. [CrossRef] [PubMed]
6. Semmler, J.G.; Kornatz, K.W.; Meyer, F.G.; Enoka, R.M. Diminished task-related adjustments of common inputs to hand muscle motor neurons in older adults. *Exp. Brain Res.* **2006**, *172*, 507–518. [CrossRef] [PubMed]
7. Faulkner, J.A.; Larkin, L.M.; Claflin, D.R.; Brooks, S.V. Age-related changes in the structure and function of skeletal muscles. *Clin. Exp. Pharmacol. Physiol.* **2007**, *34*, 1091–1096. [CrossRef]
8. Bhandari, A.; Radhu, N.; Farzan, F.; Mulsant, B.H.; Rajji, T.K.; Daskalakis, Z.J.; Blumberger, D.M. A meta-analysis of the effects of aging on motor cortex neurophysiology assessed by transcranial magnetic stimulation. *Clin. Neurophysiol.* **2016**, *127*, 2834–2845. [CrossRef] [PubMed]
9. Damoiseaux, J.S. Effects of aging on functional and structural brain connectivity. *Neuroimage* **2017**, *160*, 32–40. [CrossRef]
10. Ferreri, F.; Pauri, F.; Pasqualetti, P.; Fini, R.; Dal Forno, G.; Rossini, P.M. Motor cortex excitability in Alzheimer's disease: A transcranial magnetic stimulation study. *Ann. Neurol.* **2003**, *53*, 102–108. [CrossRef]
11. Ferreri, F.; Pasqualetti, P.; Määttä, S.; Ponzo, D.; Guerra, A.; Bressi, F.; Chiovenda, P.; Del Duca, M.; Giambattistelli, F.; Ursini, F.; et al. Motor cortex excitability in Alzheimer's disease: A transcranial magnetic stimulation follow-up study. *Neurosci. Lett.* **2011**, *492*, 94–98. [CrossRef] [PubMed]
12. Ferreri, F.; Vecchio, F.; Vollero, L.; Guerra, A.; Petrichella, S.; Ponzo, D.; Määtta, S.; Mervaala, E.; Könönen, M.; Ursini, F.; et al. Sensorimotor cortex excitability and connectivity in Alzheimer's disease: A TMS-EEG Co-registration study. *Hum. Brain Mapp.* **2016**, *37*, 2083–2096. [CrossRef]
13. Bologna, M.; Guerra, A.; Paparella, G.; Giordo, L.; Alunni Fegatelli, D.; Vestri, A.R.; Rothwell, J.C.; Berardelli, A. Neurophysiological correlates of bradykinesia in Parkinson's disease. *Brain* **2018**, *141*, 2432–2444. [CrossRef] [PubMed]
14. Bologna, M.; Guerra, A.; Colella, D.; Cioffi, E.; Paparella, G.; Di Vita, A.; D'Antonio, F.; Trebbastoni, A.; Berardelli, A. Bradykinesia in Alzheimer's disease and its neurophysiological substrates. *Clin. Neurophysiol.* **2020**, *131*, 850–858. [CrossRef] [PubMed]
15. Guerra, A.; Costantini, E.M.; Maatta, S.; Ponzo, D.; Ferreri, F. Disorders of consciousness and electrophysiological treatment strategies: A review of the literature and new perspectives. *Curr. Pharm. Des.* **2014**, *20*, 4248–4267. [CrossRef]
16. Guerra, A.; Petrichella, S.; Vollero, L.; Ponzo, D.; Pasqualetti, P.; Määttä, S.; Mervaala, E.; Könönen, M.; Bressi, F.; Iannello, G.; et al. Neurophysiological features of motor cortex excitability and plasticity in Subcortical Ischemic Vascular Dementia: A TMS mapping study. *Clin. Neurophysiol.* **2015**, *126*, 906–913. [CrossRef]
17. Guerra, A.; Suppa, A.; D'Onofrio, V.; Di Stasio, F.; Asci, F.; Fabbrini, G.; Berardelli, A. Abnormal cortical facilitation and L-dopa-induced dyskinesia in Parkinson's disease. *Brain Stimul.* **2019**, *12*, 1517–1525. [CrossRef]
18. Guerra, A.; Asci, F.; D'Onofrio, V.; Sveva, V.; Bologna, M.; Fabbrini, G.; Berardelli, A.; Suppa, A. Enhancing Gamma Oscillations Restores Primary Motor Cortex Plasticity in Parkinson's Disease. *J. Neurosci.* **2020**, *40*, 4788–4796. [CrossRef]
19. Benussi, A.; Alberici, A.; Ferrari, C.; Cantoni, V.; Dell'Era, V.; Turrone, R.; Cotelli, M.S.; Binetti, G.; Paghera, B.; Koch, G.; et al. The impact of transcranial magnetic stimulation on diagnostic confidence in patients with Alzheimer disease. *Alzheimer's Res. Ther.* **2018**, *10*, 94. [CrossRef]
20. Colella, D.; Guerra, A.; Paparella, G.; Cioffi, E.; Di Vita, A.; Trebbastoni, A.; Berardelli, A.; Bologna, M. Motor dysfunction in mild cognitive impairment as tested by kinematic analysis and transcranial magnetic stimulation. *Clin. Neurophysiol.* **2021**, *132*, 315–322. [CrossRef]
21. Hill, A.T.; Rogasch, N.C.; Fitzgerald, P.B.; Hoy, K.E. TMS-EEG: A window into the neurophysiological effects of transcranial electrical stimulation in non-motor brain regions. *Neurosci. Biobehav. Rev.* **2016**, *64*, 175–184. [CrossRef]
22. Rawji, V.; Latorre, A.; Sharma, N.; Rothwell, J.C.; Rocchi, L. On the Use of TMS to Investigate the Pathophysiology of Neurodegenerative Diseases. *Front. Neurol.* **2020**, *11*, 584664. [CrossRef] [PubMed]
23. Cardin, J.A.; Abel, T. Memory suppressor genes: Enhancing the relationship between synaptic plasticity and memory storage. *J. Neurosci. Res.* **1999**, *58*, 10–23. [CrossRef]
24. Rosenzweig, E.S.; Barnes, C.A. Impact of aging on hippocampal function: Plasticity, network dynamics, and cognition. *Prog. Neurobiol.* **2003**, *69*, 143–179. [CrossRef]
25. Disterhoft, J.F.; Oh, M.M. Learning, aging and intrinsic neuronal plasticity. *Trends Neurosci.* **2006**, *29*, 587–599. [CrossRef] [PubMed]
26. Pauwels, L.; Chalavi, S.; Swinnen, S.P. Aging and brain plasticity. *Aging* **2018**, *10*, 1789–1790. [CrossRef] [PubMed]
27. Abraham, W.C.; Richter-Levin, G. From Synaptic Metaplasticity to Behavioral Metaplasticity. *Neurobiol. Learn. Mem.* **2018**, *154*, 1–4. [CrossRef] [PubMed]
28. Dunn, A.R.; Kaczorowski, C.C. Regulation of intrinsic excitability: Roles for learning and memory, aging and Alzheimer's disease, and genetic diversity. *Neurobiol. Learn. Mem.* **2019**, *164*, 107069. [CrossRef]

29. Amassian, V.E.; Stewart, M.; Quirk, G.J.; Rosenthal, J.L. Physiological basis of motor effects of a transient stimulus to cerebral cortex. *Neurosurgery* **1987**, *20*, 74–93. [CrossRef] [PubMed]
30. Rossini, P.M.; Burke, D.; Chen, R.; Cohen, L.G.; Daskalakis, Z.; Di Iorio, R.; Di Lazzaro, V.; Ferreri, F.; Fitzgerald, P.B.; George, M.S.; et al. Non-invasive electrical and magnetic stimulation of the brain, spinal cord, roots and peripheral nerves: Basic principles and procedures for routine clinical and research application. An updated report from an I.F.C.N. Committee. *Clin. Neurophysiol.* **2015**, *126*, 1071–1107. [CrossRef] [PubMed]
31. Ziemann, U. I-waves in motor cortex revisited. *Exp. Brain Res.* **2020**, *238*, 1601–1610. [CrossRef]
32. Guerra, A.; Ranieri, F.; Falato, E.; Musumeci, G.; Di Santo, A.; Asci, F.; Di Pino, G.; Suppa, A.; Berardelli, A.; Di Lazzaro, V. Detecting cortical circuits resonant to high-frequency oscillations in the human primary motor cortex: A TMS-tACS study. *Sci. Rep.* **2020**, *10*, 7695. [CrossRef]
33. Di Lazzaro, V.; Rothwell, J.; Capogna, M. Noninvasive Stimulation of the Human Brain: Activation of Multiple Cortical Circuits. *Neuroscientist* **2017**, *24*, 246–260. [CrossRef]
34. Brown, K.E.; Lohse, K.R.; Mayer, I.M.S.; Strigaro, G.; Desikan, M.; Casula, E.P.; Meunier, S.; Popa, T.; Lamy, J.-C.; Odish, O.; et al. The reliability of commonly used electrophysiology measures. *Brain Stimul.* **2017**, *10*, 1102–1111. [CrossRef]
35. Davila-Pérez, P.; Jannati, A.; Fried, P.J.; Cudeiro Mazaira, J.; Pascual-Leone, A. The Effects of Waveform and Current Direction on the Efficacy and Test-Retest Reliability of Transcranial Magnetic Stimulation. *Neuroscience* **2018**, *393*, 97–109. [CrossRef]
36. Gomes-Osman, J.; Indahlastari, A.; Fried, P.J.; Cabral, D.L.F.; Rice, J.; Nissim, N.R.; Aksu, S.; McLaren, M.E.; Woods, A.J. Non-invasive Brain Stimulation: Probing Intracortical Circuits and Improving Cognition in the Aging Brain. *Front. Aging Neurosci.* **2018**, *10*, 177. [CrossRef] [PubMed]
37. Kozel, F.A.; Nahas, Z.; deBrux, C.; Molloy, M.; Lorberbaum, J.P.; Bohning, D.; Risch, S.C.; George, M.S. How coil-cortex distance relates to age, motor threshold, and antidepressant response to repetitive transcranial magnetic stimulation. *J. Neuropsychiatry Clin. Neurosci.* **2000**, *12*, 376–384. [CrossRef]
38. List, J.; Kübke, J.C.; Lindenberg, R.; Külzow, N.; Kerti, L.; Witte, V.; Flöel, A. Relationship between excitability, plasticity and thickness of the motor cortex in older adults. *Neuroimage* **2013**, *83*, 809–816. [CrossRef] [PubMed]
39. McGinley, M.; Hoffman, R.L.; Russ, D.W.; Thomas, J.S.; Clark, B.C. Older adults exhibit more intracortical inhibition and less intracortical facilitation than young adults. *Exp. Gerontol.* **2010**, *45*, 671–678. [CrossRef] [PubMed]
40. Young-Bernier, M.; Davidson, P.S.R.; Tremblay, F. Paired-pulse afferent modulation of TMS responses reveals a selective decrease in short latency afferent inhibition with age. *Neurobiol. Aging* **2012**, *33*, 835.e1–835.e11. [CrossRef]
41. Peinemann, A.; Lehner, C.; Conrad, B.; Siebner, H.R. Age-related decrease in paired-pulse intracortical inhibition in the human primary motor cortex. *Neurosci Lett.* **2001**, *313*, 33–36. [CrossRef]
42. Oliviero, A.; Profice, P.; Tonali, P.A.; Pilato, F.; Saturno, E.; Dileone, M.; Ranieri, F.; Di Lazzaro, V. Effects of aging on motor cortex excitability. *Neurosci. Res.* **2006**, *55*, 74–77. [CrossRef]
43. Opie, G.M.; Semmler, J.G. Age-related differences in short- and long-interval intracortical inhibition in a human hand muscle. *Brain Stimul.* **2014**, *7*, 665–672. [CrossRef] [PubMed]
44. Shibuya, K.; Park, S.B.; Geevasinga, N.; Huynh, W.; Simon, N.G.; Menon, P.; Howells, J.; Vucic, S.; Kiernan, M.C. Threshold tracking transcranial magnetic stimulation: Effects of age and gender on motor cortical function. *Clin. Neurophysiol.* **2016**, *127*, 2355–2361. [CrossRef] [PubMed]
45. Hess, C.W.; Mills, K.R.; Murray, N.M. Responses in small hand muscles from magnetic stimulation of the human brain. *J. Physiol.* **1987**, *388*, 397–419. [CrossRef]
46. Bologna, M.; Rocchi, L.; Paparella, G.; Nardella, A.; Li Voti, P.; Conte, A.; Kojovic, M.; Rothwell, J.C.; Berardelli, A. Reversal of Practice-related Effects on Corticospinal Excitability has no Immediate Effect on Behavioral Outcome. *Brain Stimul.* **2015**, *8*, 603–612. [CrossRef] [PubMed]
47. Pitcher, J.B.; Ogston, K.M.; Miles, T.S. Age and sex differences in human motor cortex input-output characteristics. *J. Physiol.* **2003**, *546*, 605–613. [CrossRef] [PubMed]
48. Pitcher, J.B.; Doeltgen, S.H.; Goldsworthy, M.R.; Schneider, L.A.; Vallence, A.-M.; Smith, A.E.; Semmler, J.G.; McDonnell, M.N.; Ridding, M.C. A comparison of two methods for estimating 50% of the maximal motor evoked potential. *Clin. Neurophysiol.* **2015**, *126*, 2337–2341. [CrossRef] [PubMed]
49. Kujirai, T.; Caramia, M.D.; Rothwell, J.C.; Day, B.L.; Thompson, P.D.; Ferbert, A.; Wroe, S.; Asselman, P.; Marsden, C.D. Corticocortical inhibition in human motor cortex. *J. Physiol.* **1993**, *471*, 501–519. [CrossRef]
50. Ganos, C.; Rocchi, L.; Latorre, A.; Hockey, L.; Palmer, C.; Joyce, E.M.; Bhatia, K.P.; Haggard, P.; Rothwell, J. Motor cortical excitability during voluntary inhibition of involuntary tic movements. *Mov. Disord.* **2018**, *33*, 1804–1809. [CrossRef] [PubMed]
51. Tokimura, H.; Di Lazzaro, V.; Tokimura, Y.; Oliviero, A.; Profice, P.; Insola, A.; Mazzone, P.; Tonali, P.; Rothwell, J.C. Short latency inhibition of human hand motor cortex by somatosensory input from the hand. *The Journal of Physiology* **2000**, *523*, 503–513. [CrossRef] [PubMed]
52. Ibáñez, J.; Hannah, R.; Rocchi, L.; Rothwell, J.C. Premovement Suppression of Corticospinal Excitability may be a Necessary Part of Movement Preparation. *Cereb. Cortex* **2020**, *30*, 2910–2923. [CrossRef] [PubMed]
53. Ziemann, U.; Reis, J.; Schwenkreis, P.; Rosanova, M.; Strafella, A.; Badawy, R.; Müller-Dahlhaus, F. TMS and drugs revisited 2014. *Clin. Neurophysiol.* **2015**, *126*, 1847–1868. [CrossRef]

54. Rocchi, L.; Latorre, A.; Ibanez Pereda, J.; Spampinato, D.; Brown, K.E.; Rothwell, J.; Bhatia, K. A case of congenital hypoplasia of the left cerebellar hemisphere and ipsilateral cortical myoclonus. *Mov. Disord.* **2019**, *34*, 1745–1747. [CrossRef]
55. Paparella, G.; Rocchi, L.; Bologna, M.; Berardelli, A.; Rothwell, J. Differential effects of motor skill acquisition on the primary motor and sensory cortices in healthy humans. *J. Physiol.* **2020**, *598*, 4031–4045. [CrossRef]
56. Marneweck, M.; Loftus, A.; Hammond, G. Short-interval intracortical inhibition and manual dexterity in healthy aging. *Neurosci. Res.* **2011**, *70*, 408–414. [CrossRef] [PubMed]
57. Heise, K.-F.; Zimerman, M.; Hoppe, J.; Gerloff, C.; Wegscheider, K.; Hummel, F.C. The aging motor system as a model for plastic changes of GABA-mediated intracortical inhibition and their behavioral relevance. *J. Neurosci.* **2013**, *33*, 9039–9049. [CrossRef]
58. Hermans, L.; Levin, O.; Maes, C.; van Ruitenbeek, P.; Heise, K.-F.; Edden, R.A.E.; Puts, N.A.J.; Peeters, R.; King, B.R.; Meesen, R.L.J.; et al. GABA levels and measures of intracortical and interhemispheric excitability in healthy young and older adults: An MRS-TMS study. *Neurobiol. Aging* **2018**, *65*, 168–177. [CrossRef]
59. Cuypers, K.; Verstraelen, S.; Maes, C.; Hermans, L.; Hehl, M.; Heise, K.-F.; Chalavi, S.; Mikkelsen, M.; Edden, R.; Levin, O.; et al. Task-related measures of short-interval intracortical inhibition and GABA levels in healthy young and older adults: A multimodal TMS-MRS study. *Neuroimage* **2020**, *208*, 116470. [CrossRef]
60. Rogasch, N.C.; Dartnall, T.J.; Cirillo, J.; Nordstrom, M.A.; Semmler, J.G. Corticomotor plasticity and learning of a ballistic thumb training task are diminished in older adults. *J. Appl. Physiol. (1985)* **2009**, *107*, 1874–1883. [CrossRef]
61. Bashir, S.; Perez, J.M.; Horvath, J.C.; Pena-Gomez, C.; Vernet, M.; Capia, A.; Alonso-Alonso, M.; Pascual-Leone, A. Differential effects of motor cortical excitability and plasticity in young and old individuals: A Transcranial Magnetic Stimulation (TMS) study. *Front. Aging Neurosci.* **2014**, *6*, 111. [CrossRef]
62. Smith, A.E.; Ridding, M.C.; Higgins, R.D.; Wittert, G.A.; Pitcher, J.B. Age-related changes in short-latency motor cortex inhibition. *Exp. Brain Res.* **2009**, *198*, 489–500. [CrossRef]
63. Nakamura, H.; Kitagawa, H.; Kawaguchi, Y.; Tsuji, H. Intracortical facilitation and inhibition after transcranial magnetic stimulation in conscious humans. *J. Physiol.* **1997**, *498 Pt 3*, 817–823. [CrossRef]
64. Erro, R.; Rocchi, L.; Antelmi, E.; Liguori, R.; Tinazzi, M.; Berardelli, A.; Rothwell, J.; Bhatia, K.P. High frequency somatosensory stimulation in dystonia: Evidence fordefective inhibitory plasticity. *Mov. Disord.* **2018**, *33*, 1902–1909. [CrossRef] [PubMed]
65. Ziemann, U.; Tergau, F.; Wassermann, E.M.; Wischer, S.; Hildebrandt, J.; Paulus, W. Demonstration of facilitatory I wave interaction in the human motor cortex by paired transcranial magnetic stimulation. *J. Physiol.* **1998**, *511 Pt 1*, 181–190. [CrossRef]
66. Hannah, R.; Rocchi, L.; Tremblay, S.; Wilson, E.; Rothwell, J.C. Pulse width biases the balance of excitation and inhibition recruited by transcranial magnetic stimulation. *Brain Stimul.* **2020**, *13*, 536–538. [CrossRef] [PubMed]
67. Ortu, E.; Deriu, F.; Suppa, A.; Tolu, E.; Rothwell, J.C. Effects of volitional contraction on intracortical inhibition and facilitation in the human motor cortex. *J. Physiol.* **2008**, *586*, 5147–5159. [CrossRef] [PubMed]
68. Shirota, Y.; Hamada, M.; Terao, Y.; Matsumoto, H.; Ohminami, S.; Furubayashi, T.; Nakatani-Enomoto, S.; Ugawa, Y.; Hanajima, R. Influence of short-interval intracortical inhibition on short-interval intracortical facilitation in human primary motor cortex. *J. Neurophysiol.* **2010**, *104*, 1382–1391. [CrossRef]
69. Clark, J.; Loftus, A.; Hammond, G. Age-related changes in short-interval intracortical facilitation and dexterity. *Neuroreport* **2011**, *22*, 499–503. [CrossRef]
70. Opie, G.M.; Cirillo, J.; Semmler, J.G. Age-related changes in late I-waves influence motor cortex plasticity induction in older adults. *J. Physiol.* **2018**, *596*, 2597–2609. [CrossRef]
71. Opie, G.M.; Hand, B.J.; Semmler, J.G. Age-related changes in late synaptic inputs to corticospinal neurons and their functional significance: A paired-pulse TMS study. *Brain Stimul.* **2020**, *13*, 239–246. [CrossRef]
72. Peurala, S.H.; Müller-Dahlhaus, J.F.M.; Arai, N.; Ziemann, U. Interference of short-interval intracortical inhibition (SICI) and short-interval intracortical facilitation (SICF). *Clin. Neurophysiol.* **2008**, *119*, 2291–2297. [CrossRef]
73. Erro, R.; Antelmi, E.; Bhatia, K.P.; Latorre, A.; Tinazzi, M.; Berardelli, A.; Rothwell, J.C.; Rocchi, L. Reversal of Temporal Discrimination in Cervical Dystonia after Low-Frequency Sensory Stimulation. *Mov. Disord.* **2020**. [CrossRef] [PubMed]
74. Gövert, F.; Becktepe, J.; Balint, B.; Rocchi, L.; Brugger, F.; Garrido, A.; Walter, T.; Hannah, R.; Rothwell, J.; Elble, R.; et al. Temporal discrimination is altered in patients with isolated asymmetric and jerky upper limb tremor. *Mov. Disord.* **2020**, *35*, 306–315. [CrossRef]
75. Latorre, A.; Cocco, A.; Bhatia, K.P.; Erro, R.; Antelmi, E.; Conte, A.; Rothwell, J.C.; Rocchi, L. Defective Somatosensory Inhibition and Plasticity Are Not Required to Develop Dystonia. *Mov. Disord.* **2020**. [CrossRef]
76. Hanajima, R.; Okabe, S.; Terao, Y.; Furubayashi, T.; Arai, N.; Inomata-Terada, S.; Hamada, M.; Yugeta, A.; Ugawa, Y. Difference in intracortical inhibition of the motor cortex between cortical myoclonus and focal hand dystonia. *Clin. Neurophysiol.* **2008**, *119*, 1400–1407. [CrossRef] [PubMed]
77. Sale, M.V.; Lavender, A.P.; Opie, G.M.; Nordstrom, M.A.; Semmler, J.G. Increased intracortical inhibition in elderly adults with anterior-posterior current flow: A TMS study. *Clin. Neurophysiol.* **2016**, *127*, 635–640. [CrossRef]
78. Ibáñez, J.; Spampinato, D.A.; Paraneetharan, V.; Rothwell, J.C. SICI during changing brain states: Differences in methodology can lead to different conclusions. *Brain Stimul.* **2020**, *13*, 353–356. [CrossRef] [PubMed]
79. Neubert, F.-X.; Mars, R.B.; Olivier, E.; Rushworth, M.F.S. Modulation of short intra-cortical inhibition during action reprogramming. *Exp. Brain Res.* **2011**, *211*, 265–276. [CrossRef]

80. Hermans, L.; Maes, C.; Pauwels, L.; Cuypers, K.; Heise, K.-F.; Swinnen, S.P.; Leunissen, I. Age-related alterations in the modulation of intracortical inhibition during stopping of actions. *Aging* **2019**, *11*, 371–385. [CrossRef]
81. Di Lazzaro, V.; Pilato, F.; Dileone, M.; Tonali, P.A.; Ziemann, U. Dissociated effects of diazepam and lorazepam on short-latency afferent inhibition. *J. Physiol.* **2005**, *569*, 315–323. [CrossRef] [PubMed]
82. Ferreri, F.; Ponzo, D.; Hukkanen, T.; Mervaala, E.; Könönen, M.; Pasqualetti, P.; Vecchio, F.; Rossini, P.M.; Määttä, S. Human brain cortical correlates of short-latency afferent inhibition: A combined EEG-TMS study. *J. Neurophysiol.* **2012**, *108*, 314–323. [CrossRef]
83. Hwang, Y.T.; Rocchi, L.; Hammond, P.; Hardy, C.J.; Warren, J.D.; Ridha, B.H.; Rothwell, J.; Rossor, M.N. Effect of Donepezil on Transcranial Magnetic Stimulation Parameters in Alzheimer's Disease. Available online: https://pubmed.ncbi.nlm.nih.gov/29560413/ (accessed on 26 January 2021).
84. Young-Bernier, M.; Kamil, Y.; Tremblay, F.; Davidson, P.S.R. Associations between a neurophysiological marker of central cholinergic activity and cognitive functions in young and older adults. *Behav. Brain Funct.* **2012**, *8*, 17. [CrossRef]
85. Young-Bernier, M.; Tanguay, A.N.; Davidson, P.S.R.; Tremblay, F. Short-latency afferent inhibition is a poor predictor of individual susceptibility to rTMS-induced plasticity in the motor cortex of young and older adults. *Front. Aging Neurosci.* **2014**, *6*, 182. [CrossRef]
86. Brown, K.E.; Neva, J.L.; Feldman, S.J.; Staines, W.R.; Boyd, L.A. Sensorimotor integration in healthy aging: Baseline differences and response to sensory training. *Exp. Gerontol.* **2018**, *112*, 1–8. [CrossRef] [PubMed]
87. Di Lorenzo, F.; Ponzo, V.; Bonnì, S.; Motta, C.; Negrão Serra, P.C.; Bozzali, M.; Caltagirone, C.; Martorana, A.; Koch, G. Long-term potentiation-like cortical plasticity is disrupted in Alzheimer's disease patients independently from age of onset: Cortical Plasticity in Alzheimer's Disease. *Ann. Neurol.* **2016**, *80*, 202–210. [CrossRef] [PubMed]
88. Ziemann, U. Transcranial magnetic stimulation at the interface with other techniques: A powerful tool for studying the human cortex. *Neuroscientist* **2011**, *17*, 368–381. [CrossRef] [PubMed]
89. Ferreri, F.; Rossini, P.M. TMS and TMS-EEG techniques in the study of the excitability, connectivity, and plasticity of the human motor cortex. *Rev. Neurosci.* **2013**, *24*, 431–442. [CrossRef]
90. Ferreri, F.; Ponzo, D.; Vollero, L.; Guerra, A.; Di Pino, G.; Petrichella, S.; Benvenuto, A.; Tombini, M.; Rossini, L.; Denaro, L.; et al. Does an intraneural interface short-term implant for robotic hand control modulate sensorimotor cortical integration? An EEG-TMS co-registration study on a human amputee. *Restor. Neurol. Neurosci.* **2014**, *32*, 281–292. [CrossRef]
91. Hannah, R.; Rocchi, L.; Tremblay, S.; Rothwell, J. Controllable Pulse Parameter TMS and TMS-EEG As Novel Approaches to Improve Neural Targeting with rTMS in Human Cerebral Cortex. Available online: https://pubmed.ncbi.nlm.nih.gov/27965543/ (accessed on 26 January 2021).
92. Tremblay, S.; Rogasch, N.C.; Premoli, I.; Blumberger, D.M.; Casarotto, S.; Chen, R.; Di Lazzaro, V.; Farzan, F.; Ferrarelli, F.; Fitzgerald, P.B.; et al. Clinical utility and prospective of TMS-EEG. *Clin. Neurophysiol.* **2019**, *130*, 802–844. [CrossRef]
93. Bonato, C.; Miniussi, C.; Rossini, P.M. Transcranial magnetic stimulation and cortical evoked potentials: A TMS/EEG co-registration study. *Clin. Neurophysiol.* **2006**, *117*, 1699–1707. [CrossRef]
94. Ilmoniemi, R.J.; Kicić, D. Methodology for combined TMS and EEG. *Brain Topogr.* **2010**, *22*, 233–248. [CrossRef]
95. Ilmoniemi, R.J.; Virtanen, J.; Ruohonen, J.; Karhu, J.; Aronen, H.J.; Näätänen, R.; Katila, T. Neuronal responses to magnetic stimulation reveal cortical reactivity and connectivity. *Neuroreport* **1997**, *8*, 3537–3540. [CrossRef] [PubMed]
96. Gosseries, O.; Sarasso, S.; Casarotto, S.; Boly, M.; Schnakers, C.; Napolitani, M.; Bruno, M.-A.; Ledoux, D.; Tshibanda, J.-F.; Massimini, M.; et al. On the cerebral origin of EEG responses to TMS: Insights from severe cortical lesions. *Brain Stimul.* **2015**, *8*, 142–149. [CrossRef]
97. Mäki, H.; Ilmoniemi, R.J. Projecting out muscle artifacts from TMS-evoked EEG. *Neuroimage* **2011**, *54*, 2706–2710. [CrossRef]
98. Ilmoniemi, R.J.; Hernandez-Pavon, J.C.; Makela, N.N.; Metsomaa, J.; Mutanen, T.P.; Stenroos, M.; Sarvas, J. Dealing with artifacts in TMS-evoked EEG. In Proceedings of the 2015 37th Annual International Conference of the IEEE Engineering in Medicine and Biology Society (EMBC), Milan, Italy, 25–29 August 2015; Volume 2015, pp. 230–233.
99. Rocchi, L.; Di Santo, A.; Brown, K.; Ibáñez, J.; Casula, E.; Rawji, V.; Di Lazzaro, V.; Koch, G.; Rothwell, J. Disentangling EEG responses to TMS due to cortical and peripheral activations. *Brain Stimul.* **2020**, *14*, 4–18. [CrossRef] [PubMed]
100. Ferreri, F.; Pasqualetti, P.; Määttä, S.; Ponzo, D.; Ferrarelli, F.; Tononi, G.; Mervaala, E.; Miniussi, C.; Rossini, P.M. Human brain connectivity during single and paired pulse transcranial magnetic stimulation. *Neuroimage* **2011**, *54*, 90–102. [CrossRef] [PubMed]
101. Rossini, P.M.; Desiato, M.T.; Caramia, M.D. Age-related changes of motor evoked potentials in healthy humans: Non-invasive evaluation of central and peripheral motor tracts excitability and conductivity. *Brain Res.* **1992**, *593*, 14–19. [CrossRef]
102. Ferreri, F.; Guerra, A.; Vollero, L.; Ponzo, D.; Maatta, S.; Mervaala, E.; Iannello, G.; Di Lazzaro, V. Age-related changes of cortical excitability and connectivity in healthy humans: Non-invasive evaluation of sensorimotor network by means of TMS-EEG. *Neuroscience* **2017**, *357*, 255–263. [CrossRef]
103. Lehmann, D.; Skrandies, W. Reference-free identification of components of checkerboard-evoked multichannel potential fields. *Electroencephalogr. Clin. Neurophysiol.* **1980**, *48*, 609–621. [CrossRef]
104. Casula, E.P.; Rocchi, L.; Hannah, R.; Rothwell, J.C. Effects of pulse width, waveform and current direction in the cortex: A combined cTMS-EEG study. *Brain Stimul.* **2018**, *11*, 1063–1070. [CrossRef] [PubMed]
105. Opie, G.M.; Sidhu, S.K.; Rogasch, N.C.; Ridding, M.C.; Semmler, J.G. Cortical inhibition assessed using paired-pulse TMS-EEG is increased in older adults. *Brain Stimul.* **2018**, *11*, 545–557. [CrossRef]

106. Hashimoto, T.; Arion, D.; Unger, T.; Maldonado-Avilés, J.G.; Morris, H.M.; Volk, D.W.; Mirnics, K.; Lewis, D.A. Alterations in GABA-related transcriptome in the dorsolateral prefrontal cortex of subjects with schizophrenia. *Mol. Psychiatry* **2008**, *13*, 147–161. [CrossRef]
107. Fillman, S.G.; Duncan, C.E.; Webster, M.J.; Elashoff, M.; Weickert, C.S. Developmental co-regulation of the beta and gamma GABAA receptor subunits with distinct alpha subunits in the human dorsolateral prefrontal cortex. *Int. J. Dev. Neurosci.* **2010**, *28*, 513–519. [CrossRef]
108. Duncan, C.E.; Webster, M.J.; Rothmond, D.A.; Bahn, S.; Elashoff, M.; Shannon Weickert, C. Prefrontal GABA(A) receptor alpha-subunit expression in normal postnatal human development and schizophrenia. *J. Psychiatr. Res.* **2010**, *44*, 673–681. [CrossRef]
109. Matsumura, M.; Sawaguchi, T.; Oishi, T.; Ueki, K.; Kubota, K. Behavioral deficits induced by local injection of bicuculline and muscimol into the primate motor and premotor cortex. *J. Neurophysiol.* **1991**, *65*, 1542–1553. [CrossRef]
110. Matsumura, M.; Sawaguchi, T.; Kubota, K. GABAergic inhibition of neuronal activity in the primate motor and premotor cortex during voluntary movement. *J. Neurophysiol.* **1992**, *68*, 692–702. [CrossRef] [PubMed]
111. Gao, F.; Edden, R.A.E.; Li, M.; Puts, N.A.J.; Wang, G.; Liu, C.; Zhao, B.; Wang, H.; Bai, X.; Zhao, C.; et al. Edited magnetic resonance spectroscopy detects an age-related decline in brain GABA levels. *Neuroimage* **2013**, *78*, 75–82. [CrossRef] [PubMed]
112. Premoli, I.; Castellanos, N.; Rivolta, D.; Belardinelli, P.; Bajo, R.; Zipser, C.; Espenhahn, S.; Heidegger, T.; Müller-Dahlhaus, F.; Ziemann, U. TMS-EEG signatures of GABAergic neurotransmission in the human cortex. *J. Neurosci.* **2014**, *34*, 5603–5612. [CrossRef] [PubMed]
113. Premoli, I.; Rivolta, D.; Espenhahn, S.; Castellanos, N.; Belardinelli, P.; Ziemann, U.; Müller-Dahlhaus, F. Characterization of GABAB-receptor mediated neurotransmission in the human cortex by paired-pulse TMS-EEG. *Neuroimage* **2014**, *103*, 152–162. [CrossRef]
114. Noda, Y.; Zomorrodi, R.; Cash, R.F.H.; Barr, M.S.; Farzan, F.; Rajji, T.K.; Chen, R.; Daskalakis, Z.J.; Blumberger, D.M. Characterization of the influence of age on GABAA and glutamatergic mediated functions in the dorsolateral prefrontal cortex using paired-pulse TMS-EEG. *Aging* **2017**, *9*, 556–572. [CrossRef] [PubMed]
115. Irlbacher, K.; Brocke, J.; Mechow, J.V.; Brandt, S.A. Effects of GABA(A) and GABA(B) agonists on interhemispheric inhibition in man. *Clin. Neurophysiol.* **2007**, *118*, 308–316. [CrossRef]
116. Ferbert, A.; Priori, A.; Rothwell, J.C.; Day, B.L.; Colebatch, J.G.; Marsden, C.D. Interhemispheric inhibition of the human motor cortex. *J. Physiol.* **1992**, *453*, 525–546. [CrossRef]
117. Wahl, M.; Lauterbach-Soon, B.; Hattingen, E.; Jung, P.; Singer, O.; Volz, S.; Klein, J.C.; Steinmetz, H.; Ziemann, U. Human motor corpus callosum: Topography, somatotopy, and link between microstructure and function. *J. Neurosci.* **2007**, *27*, 12132–12138. [CrossRef]
118. Ni, Z.; Gunraj, C.; Nelson, A.J.; Yeh, I.-J.; Castillo, G.; Hoque, T.; Chen, R. Two phases of interhemispheric inhibition between motor related cortical areas and the primary motor cortex in human. *Cereb. Cortex* **2009**, *19*, 1654–1665. [CrossRef] [PubMed]
119. Plow, E.B.; Cunningham, D.A.; Bonnett, C.; Gohar, D.; Bayram, M.; Wyant, A.; Varnerin, N.; Mamone, B.; Siemionow, V.; Hou, J.; et al. Neurophysiological correlates of aging-related muscle weakness. *J. Neurophysiol.* **2013**, *110*, 2563–2573. [CrossRef]
120. Mooney, R.A.; Cirillo, J.; Byblow, W.D. Adaptive threshold hunting reveals differences in interhemispheric inhibition between young and older adults. *Eur. J. Neurosci.* **2018**, *48*, 2247–2258. [CrossRef]
121. Talelli, P.; Waddingham, W.; Ewas, A.; Rothwell, J.C.; Ward, N.S. The effect of age on task-related modulation of interhemispheric balance. *Exp. Brain Res.* **2008**, *186*, 59–66. [CrossRef] [PubMed]
122. Hinder, M.R.; Schmidt, M.W.; Garry, M.I.; Summers, J.J. Unilateral contractions modulate interhemispheric inhibition most strongly and most adaptively in the homologous muscle of the contralateral limb. *Exp. Brain Res.* **2010**, *205*, 423–433. [CrossRef]
123. Talelli, P.; Ewas, A.; Waddingham, W.; Rothwell, J.C.; Ward, N.S. Neural correlates of age-related changes in cortical neurophysiology. *Neuroimage* **2008**, *40*, 1772–1781. [CrossRef]
124. Hinder, M.R.; Fujiyama, H.; Summers, J.J. Premotor-motor interhemispheric inhibition is released during movement initiation in older but not young adults. *PLoS ONE* **2012**, *7*, e52573. [CrossRef] [PubMed]
125. Fujiyama, H.; Van Soom, J.; Rens, G.; Gooijers, J.; Leunissen, I.; Levin, O.; Swinnen, S.P. Age-Related Changes in Frontal Network Structural and Functional Connectivity in Relation to Bimanual Movement Control. *J. Neurosci.* **2016**, *36*, 1808–1822. [CrossRef] [PubMed]
126. Cabeza, R. Cognitive neuroscience of aging: Contributions of functional neuroimaging. *Scand. J. Psychol.* **2001**, *42*, 277–286. [CrossRef]
127. Ward, N.S.; Swayne, O.B.C.; Newton, J.M. Age-dependent changes in the neural correlates of force modulation: An fMRI study. *Neurobiol. Aging* **2008**, *29*, 1434–1446. [CrossRef]
128. Boudrias, M.-H.; Gonçalves, C.S.; Penny, W.D.; Park, C.; Rossiter, H.E.; Talelli, P.; Ward, N.S. Age-related changes in causal interactions between cortical motor regions during hand grip. *Neuroimage* **2012**, *59*, 3398–3405. [CrossRef]
129. Fujiyama, H.; Hinder, M.R.; Summers, J.J. Functional role of left PMd and left M1 during preparation and execution of left hand movements in older adults. *J. Neurophysiol.* **2013**, *110*, 1062–1069. [CrossRef]
130. Chen, R.; Yung, D.; Li, J.-Y. Organization of ipsilateral excitatory and inhibitory pathways in the human motor cortex. *J. Neurophysiol.* **2003**, *89*, 1256–1264. [CrossRef]

131. Davidson, T.; Tremblay, F. Age and hemispheric differences in transcallosal inhibition between motor cortices: An ispsilateral silent period study. *BMC Neurosci.* **2013**, *14*, 62. [CrossRef]
132. Petitjean, M.; Ko, J.Y.L. An age-related change in the ipsilateral silent period of a small hand muscle. *Clin. Neurophysiol.* **2013**, *124*, 346–353. [CrossRef] [PubMed]
133. Coppi, E.; Houdayer, E.; Chieffo, R.; Spagnolo, F.; Inuggi, A.; Straffi, L.; Comi, G.; Leocani, L. Age-related changes in motor cortical representation and interhemispheric interactions: A transcranial magnetic stimulation study. *Front. Aging Neurosci.* **2014**, *6*, 209. [CrossRef]
134. Strauss, S.; Lotze, M.; Flöel, A.; Domin, M.; Grothe, M. Changes in Interhemispheric Motor Connectivity Across the Lifespan: A Combined TMS and DTI Study. *Front. Aging Neurosci.* **2019**, *11*, 12. [CrossRef]
135. Ward, N.S. Compensatory mechanisms in the aging motor system. *Ageing Res. Rev.* **2006**, *5*, 239–254. [CrossRef]
136. Seidler, R.D.; Bernard, J.A.; Burutolu, T.B.; Fling, B.W.; Gordon, M.T.; Gwin, J.T.; Kwak, Y.; Lipps, D.B. Motor control and aging: Links to age-related brain structural, functional, and biochemical effects. *Neurosci. Biobehav. Rev.* **2010**, *34*, 721–733. [CrossRef]
137. Lee, L.; Harrison, L.M.; Mechelli, A. A report of the functional connectivity workshop, Dusseldorf 2002. *Neuroimage* **2003**, *19*, 457–465. [CrossRef]
138. Komssi, S.; Kähkönen, S. The novelty value of the combined use of electroencephalography and transcranial magnetic stimulation for neuroscience research. *Brain Res. Rev.* **2006**, *52*, 183–192. [CrossRef]
139. Casula, E.P.; Maiella, M.; Pellicciari, M.C.; Porrazzini, F.; D'Acunto, A.; Rocchi, L.; Koch, G. Novel TMS-EEG indexes to investigate interhemispheric dynamics in humans. *Clin. Neurophysiol.* **2020**, *131*, 70–77. [CrossRef]
140. Casula, E.P.; Pellicciari, M.C.; Bonnì, S.; Spanò, B.; Ponzo, V.; Salsano, I.; Giulietti, G.; Martino Cinnera, A.; Maiella, M.; Borghi, I.; et al. Evidence for interhemispheric imbalance in stroke patients as revealed by combining transcranial magnetic stimulation and electroencephalography. *Hum. Brain Mapp.* **2021**, *42*, 1343–1358. [CrossRef]
141. Massimini, M.; Ferrarelli, F.; Huber, R.; Esser, S.K.; Singh, H.; Tononi, G. Breakdown of cortical effective connectivity during sleep. *Science* **2005**, *309*, 2228–2232. [CrossRef] [PubMed]
142. Ferreri, F.; Vecchio, F.; Ponzo, D.; Pasqualetti, P.; Rossini, P.M. Time-varying coupling of EEG oscillations predicts excitability fluctuations in the primary motor cortex as reflected by motor evoked potentials amplitude: An EEG-TMS study. *Hum. Brain Mapp.* **2014**, *35*, 1969–1980. [CrossRef] [PubMed]
143. Fu, L.; Rocchi, L.; Hannah, R.; Xu, G.; Rothwell, J.C.; Ibáñez, J. Corticospinal excitability modulation by pairing peripheral nerve stimulation with cortical states of movement initiation. *J. Physiol.* **2019**. [CrossRef] [PubMed]
144. Desideri, D.; Zrenner, C.; Ziemann, U.; Belardinelli, P. Phase of sensorimotor μ-oscillation modulates cortical responses to transcranial magnetic stimulation of the human motor cortex. *J. Physiol.* **2019**, *597*, 5671–5686. [CrossRef] [PubMed]
145. Ferreri, F.; Vecchio, F.; Guerra, A.; Miraglia, F.; Ponzo, D.; Vollero, L.; Iannello, G.; Maatta, S.; Mervaala, E.; Rossini, P.M.; et al. Age related differences in functional synchronization of EEG activity as evaluated by means of TMS-EEG coregistrations. *Neurosci. Lett.* **2017**, *647*, 141–146. [CrossRef] [PubMed]
146. Kiers, L.; Cros, D.; Chiappa, K.H.; Fang, J. Variability of motor potentials evoked by transcranial magnetic stimulation. *Electroencephalogr. Clin. Neurophysiol.* **1993**, *89*, 415–423. [CrossRef]
147. Thickbroom, G.W.; Byrnes, M.L.; Mastaglia, F.L. A model of the effect of MEP amplitude variation on the accuracy of TMS mapping. *Clin. Neurophysiol.* **1999**, *110*, 941–943. [CrossRef]
148. Schulz, H.; Ubelacker, T.; Keil, J.; Müller, N.; Weisz, N. Now I am ready-now i am not: The influence of pre-TMS oscillations and corticomuscular coherence on motor-evoked potentials. *Cereb. Cortex* **2014**, *24*, 1708–1719. [CrossRef]
149. Gloor, P.; Ball, G.; Schaul, N. Brain lesions that produce delta waves in the EEG. *Neurology* **1977**, *27*, 326–333. [CrossRef]
150. Giovanni, A.; Capone, F.; di Biase, L.; Ferreri, F.; Florio, L.; Guerra, A.; Marano, M.; Paolucci, M.; Ranieri, F.; Salomone, G.; et al. Oscillatory Activities in Neurological Disorders of Elderly: Biomarkers to Target for Neuromodulation. *Front. Aging Neurosci.* **2017**, *9*, 189.
151. Hebb, D. *The Organization of Behavior*; Wiley: New York, NY, USA, 1949.
152. Dan, Y.; Poo, M.-M. Spike timing-dependent plasticity: From synapse to perception. *Physiol. Rev.* **2006**, *86*, 1033–1048. [CrossRef]
153. Caporale, N.; Dan, Y. Spike timing-dependent plasticity: A Hebbian learning rule. *Annu. Rev. Neurosci.* **2008**, *31*, 25–46. [CrossRef]
154. Feldman, D.E. The spike-timing dependence of plasticity. *Neuron* **2012**, *75*, 556–571. [CrossRef]
155. Suppa, A.; Asci, F.; Guerra, A. TMS as a Tool to Induce and Explore Plasticity in Humans. *Handb. Clin. Neurol.* **2021**.
156. Stefan, K.; Kunesch, E.; Cohen, L.G.; Benecke, R.; Classen, J. Induction of plasticity in the human motor cortex by paired associative stimulation. *Brain* **2000**, *123 Pt 3*, 572–584. [CrossRef]
157. Suppa, A.; Quartarone, A.; Siebner, H.; Chen, R.; Di Lazzaro, V.; Del Giudice, P.; Paulus, W.; Rothwell, J.C.; Ziemann, U.; Classen, J. The associative brain at work: Evidence from paired associative stimulation studies in humans. *Clin. Neurophysiol.* **2017**, *128*, 2140–2164. [CrossRef]
158. Müller-Dahlhaus, F.; Ziemann, U.; Classen, J. Plasticity resembling spike-timing dependent synaptic plasticity: The evidence in human cortex. *Front. Synaptic. Neurosci.* **2010**, *2*, 34. [CrossRef]
159. Larson, J.; Lynch, G. Theta pattern stimulation and the induction of LTP: The sequence in which synapses are stimulated determines the degree to which they potentiate. *Brain Res.* **1989**, *489*, 49–58. [CrossRef]
160. Bliss, T.V.; Lomo, T. Long-lasting potentiation of synaptic transmission in the dentate area of the anaesthetized rabbit following stimulation of the perforant path. *J. Physiol.* **1973**, *232*, 331–356. [CrossRef] [PubMed]

161. Cooke, S.F.; Bliss, T.V.P. Plasticity in the human central nervous system. *Brain* **2006**, *129*, 1659–1673. [CrossRef]
162. Huang, Y.-Z.; Edwards, M.J.; Rounis, E.; Bhatia, K.P.; Rothwell, J.C. Theta burst stimulation of the human motor cortex. *Neuron* **2005**, *45*, 201–206. [CrossRef] [PubMed]
163. Suppa, A.; Huang, Y.-Z.; Funke, K.; Ridding, M.C.; Cheeran, B.; Di Lazzaro, V.; Ziemann, U.; Rothwell, J.C. Ten Years of Theta Burst Stimulation in Humans: Established Knowledge, Unknowns and Prospects. *Brain Stimul.* **2016**, *9*, 323–335. [CrossRef]
164. Georgiev, D.; Rocchi, L.; Tocco, P.; Speekenbrink, M.; Rothwell, J.C.; Jahanshahi, M. Continuous Theta Burst Stimulation Over the Dorsolateral Prefrontal Cortex and the Pre-SMA Alter Drift Rate and Response Thresholds Respectively During Perceptual Decision-Making. *Brain Stimul.* **2016**, *9*, 601–608. [CrossRef]
165. Méndez, J.C.; Rocchi, L.; Jahanshahi, M.; Rothwell, J.; Merchant, H. Probing the timing network: A continuous theta burst stimulation study of temporal categorization. *Neuroscience* **2017**, *356*, 167–175. [CrossRef]
166. Dumitru, A.; Rocchi, L.; Saini, F.; Rothwell, J.C.; Roiser, J.P.; David, A.S.; Richieri, R.M.; Lewis, G.; Lewis, G. Influence of theta-burst transcranial magnetic stimulation over the dorsolateral prefrontal cortex on emotion processing in healthy volunteers. *Cogn. Affect. Behav. Neurosci.* **2020**, *20*, 1278–1293. [CrossRef]
167. Müller-Dahlhaus, J.F.M.; Orekhov, Y.; Liu, Y.; Ziemann, U. Interindividual variability and age-dependency of motor cortical plasticity induced by paired associative stimulation. *Exp. Brain Res.* **2008**, *187*, 467–475. [CrossRef]
168. Tecchio, F.; Zappasodi, F.; Pasqualetti, P.; Gennaro, L.D.; Pellicciari, M.C.; Ercolani, M.; Squitti, R.; Rossini, P.M. Age dependence of primary motor cortex plasticity induced by paired associative stimulation. *Clin. Neurophysiol.* **2008**, *119*, 675–682. [CrossRef]
169. Polimanti, R.; Simonelli, I.; Zappasodi, F.; Ventriglia, M.; Pellicciari, M.C.; Benussi, L.; Squitti, R.; Rossini, P.M.; Tecchio, F. Biological factors and age-dependence of primary motor cortex experimental plasticity. *Neurol. Sci.* **2016**, *37*, 211–218. [CrossRef]
170. Fathi, D.; Ueki, Y.; Mima, T.; Koganemaru, S.; Nagamine, T.; Tawfik, A.; Fukuyama, H. Effects of aging on the human motor cortical plasticity studied by paired associative stimulation. *Clin. Neurophysiol.* **2010**, *121*, 90–93. [CrossRef]
171. Kishore, A.; Popa, T.; James, P.; Yahia-Cherif, L.; Backer, F.; Varughese Chacko, L.; Govind, P.; Pradeep, S.; Meunier, S. Age-related decline in the responsiveness of motor cortex to plastic forces reverses with levodopa or cerebellar stimulation. *Neurobiol. Aging* **2014**, *35*, 2541–2551. [CrossRef]
172. Dickins, D.S.E.; Sale, M.V.; Kamke, M.R. Plasticity Induced by Intermittent Theta Burst Stimulation in Bilateral Motor Cortices Is Not Altered in Older Adults. *Neural Plast.* **2015**, *2015*, 323409. [CrossRef] [PubMed]
173. Opie, G.M.; Vosnakis, E.; Ridding, M.C.; Ziemann, U.; Semmler, J.G. Priming theta burst stimulation enhances motor cortex plasticity in young but not old adults. *Brain Stimul.* **2017**, *10*, 298–304. [CrossRef]
174. Guerra, A.; Asci, F.; Zampogna, A.; D'Onofrio, V.; Berardelli, A.; Suppa, A. The effect of gamma oscillations in boosting primary motor cortex plasticity is greater in young than older adults. *Clin. Neurophysiol.* **2021**. [CrossRef]
175. Guerra, A.; López-Alonso, V.; Cheeran, B.; Suppa, A. Variability in non-invasive brain stimulation studies: Reasons and results. *Neurosci. Lett.* **2020**, *719*, 133330. [CrossRef]
176. Huang, Y.-Z.; Lu, M.-K.; Antal, A.; Classen, J.; Nitsche, M.; Ziemann, U.; Ridding, M.; Hamada, M.; Ugawa, Y.; Jaberzadeh, S.; et al. Plasticity induced by non-invasive transcranial brain stimulation: A position paper. *Clin. Neurophysiol.* **2017**, *128*, 2318–2329. [CrossRef]
177. Suppa, A.; Li Voti, P.; Rocchi, L.; Papazachariadis, O.; Berardelli, A. Early visuomotor integration processes induce LTP/LTD-like plasticity in the human motor cortex. *Cereb. Cortex* **2015**, *25*, 703–712. [CrossRef]
178. Suppa, A.; Rocchi, L.; Li Voti, P.; Papazachariadis, O.; Casciato, S.; Di Bonaventura, C.; Giallonardo, A.T.; Berardelli, A. The Photoparoxysmal Response Reflects Abnormal Early Visuomotor Integration in the Human Motor Cortex. *Brain Stimul.* **2015**, *8*, 1151–1161. [CrossRef]
179. Abraham, W.C. Metaplasticity: Tuning synapses and networks for plasticity. *Nat. Rev. Neurosci.* **2008**, *9*, 387. [CrossRef]
180. Müller-Dahlhaus, F.; Ziemann, U. Metaplasticity in human cortex. *Neuroscientist* **2015**, *21*, 185–202. [CrossRef]
181. Hulme, S.R.; Jones, O.D.; Abraham, W.C. Emerging roles of metaplasticity in behaviour and disease. *Trends Neurosci.* **2013**, *36*, 353–362. [CrossRef]
182. Opie, G.M.; Post, A.K.; Ridding, M.C.; Ziemann, U.; Semmler, J.G. Modulating motor cortical neuroplasticity with priming paired associative stimulation in young and old adults. *Clin. Neurophysiol.* **2017**, *128*, 763–769. [CrossRef]
183. Opie, G.M.; Hand, B.J.; Coxon, J.P.; Ridding, M.C.; Ziemann, U.; Semmler, J.G. Visuomotor task acquisition is reduced by priming paired associative stimulation in older adults. *Neurobiol. Aging* **2019**, *81*, 67–76. [CrossRef]
184. Guerra, A.; Suppa, A.; Bologna, M.; D'Onofrio, V.; Bianchini, E.; Brown, P.; Di Lazzaro, V.; Berardelli, A. Boosting the LTP-like plasticity effect of intermittent theta-burst stimulation using gamma transcranial alternating current stimulation. *Brain Stimul.* **2018**, *11*, 734–742. [CrossRef]
185. Guerra, A.; Suppa, A.; Asci, F.; De Marco, G.; D'Onofrio, V.; Bologna, M.; Di Lazzaro, V.; Berardelli, A. LTD-like plasticity of the human primary motor cortex can be reversed by γ-tACS. *Brain Stimul.* **2019**, *12*, 1490–1499. [CrossRef] [PubMed]
186. Guerra, A.; Asci, F.; Zampogna, A.; D'Onofrio, V.; Petrucci, S.; Ginevrino, M.; Berardelli, A.; Suppa, A. Gamma-transcranial alternating current stimulation and theta-burst stimulation: Inter-subject variability and the role of BDNF. *Clin. Neurophysiol.* **2020**, *131*, 2691–2699. [CrossRef] [PubMed]
187. Guerra, A.; López-Alonso, V.; Cheeran, B.; Suppa, A. Solutions for managing variability in non-invasive brain stimulation studies. *Neurosci. Lett.* **2020**, *719*, 133332. [CrossRef]

188. Stinear, C.M.; Byblow, W.D. Elevated threshold for intracortical inhibition in focal hand dystonia. *Mov. Disord.* **2004**, *19*, 1312–1317. [CrossRef]
189. Di Lazzaro, V.; Profice, P.; Ranieri, F.; Capone, F.; Dileone, M.; Oliviero, A.; Pilato, F. I-wave origin and modulation. *Brain Stimul.* **2012**, *5*, 512–525. [CrossRef]
190. Opie, G.M.; Semmler, J.G. Preferential Activation of Unique Motor Cortical Networks With Transcranial Magnetic Stimulation: A Review of the Physiological, Functional, and Clinical Evidence. *Neuromodulation* **2020**. [CrossRef]
191. Van den Bos, M.A.J.; Higashihara, M.; Geevasinga, N.; Menon, P.; Kiernan, M.C.; Vucic, S. Imbalance of cortical facilitatory and inhibitory circuits underlies hyperexcitability in ALS. *Neurology* **2018**, *91*, e1669–e1676. [CrossRef]
192. Ni, Z.; Bahl, N.; Gunraj, C.A.; Mazzella, F.; Chen, R. Increased motor cortical facilitation and decreased inhibition in Parkinson disease. *Neurology* **2013**, *80*, 1746–1753. [CrossRef]
193. Shirota, Y.; Ohminami, S.; Tsutsumi, R.; Terao, Y.; Ugawa, Y.; Tsuji, S.; Hanajima, R. Increased facilitation of the primary motor cortex in de novo Parkinson's disease. *Parkinsonism. Relat. Disord.* **2019**, *66*, 125–129. [CrossRef]
194. Ammann, C.; Dileone, M.; Pagge, C.; Catanzaro, V.; Mata-Marín, D.; Hernández-Fernández, F.; Monje, M.H.G.; Sánchez-Ferro, Á.; Fernández-Rodríguez, B.; Gasca-Salas, C.; et al. Cortical disinhibition in Parkinson's disease. *Brain* **2020**, *143*, 3408–3421. [CrossRef]
195. Rocchi, L.; Ibáñez, J.; Benussi, A.; Hannah, R.; Rawji, V.; Casula, E.; Rothwell, J. Variability and Predictors of Response to Continuous Theta Burst Stimulation: A TMS-EEG Study. *Front. Neurosci.* **2018**, *12*, 400. [CrossRef] [PubMed]
196. Premoli, I.; Király, J.; Müller-Dahlhaus, F.; Zipser, C.M.; Rossini, P.; Zrenner, C.; Ziemann, U.; Belardinelli, P. Short-interval and long-interval intracortical inhibition of TMS-evoked EEG potentials. *Brain Stimul.* **2018**, *11*, 818–827. [CrossRef] [PubMed]
197. Rawji, V.; Kaczmarczyk, I.; Rocchi, L.; Rothwell, J.C.; Sharma, N. Short interval intracortical inhibition as measured by TMS-EEG. *Sci. Rep.* **2021**. [CrossRef]
198. Rossini, P.M.; Di Iorio, R.; Bentivoglio, M.; Bertini, G.; Ferreri, F.; Gerloff, C.; Ilmoniemi, R.J.; Miraglia, F.; Nitsche, M.A.; Pestilli, F.; et al. Methods for analysis of brain connectivity: An IFCN-sponsored review. *Clin. Neurophysiol.* **2019**, *130*, 1833–1858. [CrossRef]
199. Belardinelli, P.; Biabani, M.; Blumberger, D.M.; Bortoletto, M.; Casarotto, S.; David, O.; Desideri, D.; Etkin, A.; Ferrarelli, F.; Fitzgerald, P.B.; et al. Reproducibility in TMS-EEG studies: A call for data sharing, standard procedures and effective experimental control. *Brain Stimul.* **2019**, *12*, 787–790. [CrossRef]

Review

Stimulating Memory: Reviewing Interventions Using Repetitive Transcranial Magnetic Stimulation to Enhance or Restore Memory Abilities

Connor J. Phipps, Daniel L. Murman and David E. Warren *

Department of Neurological Sciences, University of Nebraska Medical Center, Omaha, NE 68198, USA; connor.phipps@unmc.edu (C.J.P.); dlmurman@unmc.edu (D.L.M.)
* Correspondence: david.warren@unmc.edu

Abstract: Human memory systems are imperfect recording devices that are affected by age and disease, but recent findings suggest that the functionality of these systems may be modifiable through interventions using non-invasive brain stimulation such as repetitive transcranial magnetic stimulation (rTMS). The translational potential of these rTMS interventions is clear: memory problems are the most common cognitive complaint associated with healthy aging, while pathological conditions such as Alzheimer's disease are often associated with severe deficits in memory. Therapies to improve memory or treat memory loss could enhance independence while reducing costs for public health systems. Despite this promise, several important factors limit the generalizability and translational potential of rTMS interventions for memory. Heterogeneity of protocol design, rTMS parameters, and outcome measures present significant challenges to interpretation and reproducibility. However, recent advances in cognitive neuroscience, including rTMS approaches and recent insights regarding functional brain networks, may offer methodological tools necessary to design new interventional studies with enhanced experimental rigor, improved reproducibility, and greater likelihood of successful translation to clinical settings. In this review, we first discuss the current state of the literature on memory modulation with rTMS, then offer a commentary on developments in cognitive neuroscience that are relevant to rTMS interventions, and finally close by offering several recommendations for the design of future investigations using rTMS to modulate human memory performance.

Keywords: TMS; rTMS; memory; hippocampus; brain networks; non-invasive brain stimulation; mild cognitive impairment; Alzheimer's disease

1. Introduction

Human memory systems are understood to be imperfect recording devices, and the performance of these systems is negatively impacted by age and disease. Memory loss is the most common cognitive complaint in older adults, while clinically significant memory deficits exaggerate age-related trends and are often attributable to neurodegenerative disease. The most common form of pathological memory decline is dementia due to Alzheimer's disease (AD) [1]. Unfortunately, current pharmacological interventions for AD-related memory impairment, such as cholinesterase inhibitors, offer limited benefit for memory loss [2,3]; this is also true of other interventions for AD such as lifestyle changes [4–6]. The lack of effective treatments for memory loss, AD-related or not, leaves a significant need unmet: memory loss has negative consequences for independence, autonomy, and identity. Efficacious treatments for memory loss could preserve these faculties [7–9]. Fortunately, recent findings suggest that targeted non-invasive brain stimulation (NBS) may offer meaningful opportunities for treatment [10]. Specifically, transcranial magnetic stimulation (TMS), a form of NBS, has been reported to improve memory in healthy younger adults, healthy older adults, and individuals with AD [11–14]. TMS may therefore

Citation: Phipps, C.J.; Murman, D.L.; Warren, D.E. Stimulating Memory: Reviewing Interventions Using Repetitive Transcranial Magnetic Stimulation to Enhance or Restore Memory Abilities. *Brain Sci.* **2021**, *11*, 1283. https://doi.org/10.3390/brainsci11101283

Academic Editors: Nico Sollmann and Petro Julkunen

Received: 31 August 2021
Accepted: 25 September 2021
Published: 28 September 2021

Publisher's Note: MDPI stays neutral with regard to jurisdictional claims in published maps and institutional affiliations.

Copyright: © 2021 by the authors. Licensee MDPI, Basel, Switzerland. This article is an open access article distributed under the terms and conditions of the Creative Commons Attribution (CC BY) license (https://creativecommons.org/licenses/by/4.0/).

hold promise as a potential symptomatic treatment for memory loss. Still, a review of the current literature reveals substantial variability in methods, outcome measures, and populations. Consistent with the methodological variability, findings from interventions using repetitive TMS (rTMS) to enhance memory have been inconsistent. To address this, our review seeks to summarize the results of recent rTMS studies in patients on the AD continuum, discuss potential sources of heterogeneity, and provide suggestions on how the field could enhance rigor and reproducibility in future work.

2. Review of Prior Work

2.1. Organization of the Review

We performed a narrative review based on searches of commonly used databases indexing scholarly works (PubMed, google scholar). Following the primary identification of literature, forward and backward citation searches were carried out for studies in this initially identified body. We observed that investigations testing TMS as a tool for memory enhancement or a treatment for memory loss have varied widely in their approaches. Acknowledging this heterogeneity, we identified two key independent variables that were used to organize our review: first *stimulation site* and then *target population*. As with the independent variables, we noted that outcome measures could similarly be divided into *changes in cognitive abilities (memory, executive functions, etc.)* and *changes in brain variables (structure and/or function)*. For a summary of rTMS studies organized and annotated according to these attributes, please refer to Table 1.

Regarding *stimulation sites*, investigators have most often selected rTMS targets within frontal or parietal association areas. Importantly, these regions are located immediately beneath the skull and thus within the limited range of typical TMS systems (~2–3 cm beneath the scalp) [15]. Within the brain's frontal lobe, studies have frequently targeted dorsolateral prefrontal cortex (dlPFC). This popularity of dlPFC as a stimulation target may be attributable to its known contributions to many cognitive processes including working memory [16–19]. dlPFC is, of course, also a common rTMS target for clinical treatment of psychiatric disorders such as major depression [20]. More broadly, dlPFC is generally acknowledged as a brain region that is both feasible and safe for rTMS [21]. A less common alternative for rTMS has been the parietal lobe, and within it, rTMS studies have most frequently targeted posterolateral parietal cortex or angular gyrus (AG) [13,22]. In studies of rTMS and memory, AG has frequently been targeted because of its connectivity, both structural and functional, to medial temporal lobe regions which support declarative/relational memory processes. Further, AG is thought to be part of a large-scale intrinsic brain network, the default mode network (DMN) which has been implicated in normal memory function [23–27]. Additionally, the DMN is particularly impacted by AD [25,28,29], making modulation of DMN by rTMS a potentially intriguing therapy. For information on TMS mechanisms, refer to Box 1.

Regarding *target populations*, while many studies of rTMS effects on memory have focused on healthy younger and healthy older individuals, there are an increasing number of studies investigating the potential for rTMS to treat memory loss within clinical populations (e.g. [30–32]). Studies using rTMS have also recruited individuals with clinical conditions that often precede AD, including (amnestic or non-amnestic) mild cognitive impairment (aMCI/MCI) [33–35].

Table 1. Properties of included studies.

Authors	Target	Intensity	Frequency	Sessions	Session Spacing (Days to Complete)	Cognitive Changes ([+/N/−] for rMTS) [Score Change]	Functional Connectivity Changes ([+/N/-] Area1:Area2)	Target Population	N
Frontal Lobe rTMS									
Cui et al. [36]	(R) dlPFC	90	10	10	CD(WD)	[+]AVLT	[+]PCC:(R)Fusiform Gyrus, [+]PCC:(L)Anterior Cingulate Gyrus	aMCI	25
Schluter et al. [37]	(R) dlPFC	110	10	1	NA	NA	[+]Salience network connectivity	H	15
Bagattini et al. [38]	(L) dlPFC	100	20	20	CD(WD)	[+]Paired-associate learning	NA	AD	50
Bakulin et al. [39]	(L) dlPFC	100	10	1	NA	[+]n-back	NA	HY	12
Beynel et al. [40]	(L) dlPFC	100 / 50	5	4	11	[+]Memory Manipulation / [N]n-back	[N]EEG	H	85
Chung et al. [41]	(L) dlPFC	75 / 100	iTBS	1	NA / NA	[+]n-back / [N]n-back	[N]EEG / [N]EEG	H	16
Davis et al. [42]	(L) dlPFC	120	1 / 5	1	NA / NA	[N]Source Memory / [N]Source Memory	[−]Changes in success related activity / [+]Changes in success related activity	HO	15
Fitzsimmons et al. [43]	(L) dlPFC	110	1	1	NA	[−]Set-shifting	[−]Task-based betweenness centrality of dlPFC	H	16
Li et al. [14]	(L) dlPFC	100	20	30	CD(WD)	[+]MMSE[2.03], [+]ADAS-Cog[-2.89]	[+]Plasticity Response at M1	AD	37
Drumond Marra et al. [35]	(L) dlPFC	110	10	10	CD(WD)	[+]Rivermead Behavioral Memory Test, [−]Logical Memory II, [+]Letter-number sequencing, [−]Trails B	NA	MCI	34

Table 1. Cont.

Authors	Target	Intensity	Frequency	Sessions	Session Spacing (Days to Complete)	Cognitive Changes ([+/N/−] for rMTS) [Score Change]	Functional Connectivity Changes ([+/N/−] Area1:Area2)	Target Population	N
Schluter et al. [37]	(L) dlPFC	110	10	1	NA	NA	[−]Salience network connectivity	H	15
W.-C. Wang et al. [44]	(L) dlPFC	120	1	1	NA	[N]Associative memory	[N]Encoding and retrieval similarity	HO	14
			5		NA	[N]Associative memory	[+]Encoding and retrieval similarity		
Wu et al. [45]	(L) dlPFC	70	iTBS	14	CD(D)	[+]Association memory, [+]Recognition, [+]Logical Memory Test, [+]AVLT	[−](L) dlPFC:(R)Precuneus	AD	13
Xue et al. [46]	(L) dlPFC	90	20	1	NA	NA	[+]low-frequency fluctuation in Rostral Anterior Cingulate Cortex, [+]Rostral Anterior Cingulate Cortex:(L)Temporal Cortex	HY	38
Yuan et al. [34]	(L) dlPFC	80	10	20	CD(WD)	[+]MoCA	[+]ALFF for (R)Inferior Frontal Gyrus, (R)Precuneus, (L)AG, (R)Supramarginal gyrus	aMCI	12
Rutherford et al. [11]	(B) dlPFC	100	20	10(+3)	CD(WD)	[+]MoCA, [+]Word/image Association	NA	AD	10
Lynch et al. [47]	(R) Middle Frontal Gyrus	80	cTBS	1	NA	[−]n-back	NA	HY	24
H. Wang et al. [48]	(R) Middle Frontal Gyrus 1	100	10	2	CD(D)	[+]Face/word Pairs	NA	HY	8
	(L) Middle Frontal Gyrus 2				CD(D)	[+]Face/word Pairs	NA		

Table 1. *Cont.*

Authors	Target	Intensity	Frequency	Sessions	Session Spacing (Days to Complete)	Cognitive Changes ([+]/N/−] for rMTS) [Score Change]	Functional Connectivity Changes ([+/N/−] Area1:Area2)	Target Population	N
Jung et al. [49]	(L/R) Precentral Gyrus	100	1	1	NA	NA	[−]DMN activity when at rest	H	36
Riedel et al. [50]	(R) Medial Frontopolar cortex	100	1	1	NA	NA	[−](R)Medial Frontopolar cortex:Amygdala	HY	55
			20		NA	NA	[+](R)Medial Frontopolar cortex:Amygdala		
Parietal Lobe rTMS
Freedberg et al. [51]	(L) AG	100	20	4	CD(D)	NA	[+](L)AG:(L)Hippocampal Network	HY	6
Hendrikse et al. [52]	(L) AG	100	20	4	CD(D)	[N]Associative Memory	[−]Connectivity within (L)Hippocampal Network,	H	36
Hermiller, et al. [53]	(L) AG	80	cTBS	1	NA	[+]Word Recognition Memory	[+]Hipp:PCC, [+]Hipp:Left medial frontal Gyrus, [+]Hipp:Right Medialfrontal Gyrus	H	24
		80	iTBS		NA	[N]Word Recognition Memory	N		
		100	20		NA	[N]Word Recognition Memory	N		
Hermiller, et al. [54]	(L) AG	100	20	5	CD(D)	[+]Paired-associate learning, [N]Long-term forgetting	NA	HY	16
Kim et al. [55]	(L) AG	100	20	5	CD(D)	[N]Item recognition, [+]Contextual recollection	[+]Posterior-medial network activity	HY	16
Nilakantan et al. [56]	(L) AG	100	20	5	CD(D)	[N]Recollection Success, [+]Recollection Precision	[−]Late-positive evoked potential amplitude, [−]Theta-alpha oscillatory power	HY	12

Table 1. *Cont.*

Authors	Target	Intensity	Frequency	Sessions	Session Spacing (Days to Complete)	Cognitive Changes ([+/N/−] for rMTS) [Score Change]	Functional Connectivity Changes ([+/N/−] Area1:Area2)	Target Population	N
Nilakantan et al. [13]	(L) AG	100	20	5	CD(D)	[N]Recollection Success, [+]Recollection Precision	[+]Recollection signals throughout the hippocampal-cortical network	HO	15
J.X. Wang Voss [12]	(L) AG	100	20	5	CD(D)	[+]Paired-associate learning	[+]Hipp:Posteior Hipp-cortical network	HY	16 *
Velioglu et al. [57]	(L) AG	100	20	10	14	[+]Wechsler Memory Scale-Visual	[−]Activity in Occipito-fusiform Gyrus, [−]Fusiform Gyrus:Precuneus, [−]Lateral Occipital Cortex:Precuneus, [+]Fusiform Gyrus:Frontal Opercular Cortex, [+]Lateral Occipital Cortex: Frontal Opercular Cortex	AD	11
J.X. Wang et al. [58]	(L) AG	100	20	5	CD(D)	[+]Paired-associate learning	[+]Cortical-hipp network connectivity	HY	16 *
Wynn et al. [59]	(L) AG	90	1	1	NA	[+]Delayed Recall Confidence	NA	H	25
Freedberg et al. [60]	(L) AG	100	20	3	CD(D)	NA	[+]Hipp:Precuneus, [+]Hipp:Fusiform Area, [+]Hipp:Lateral Parietal Area, [+]Hipp:Superior Parietal Area	HY	8
Tambini et al. [61]	(R) AG	80	cTBS	1	NA	[+]Associative memory success and confidence	Response was dependent on AG and Hippocampus connectivity	HY	25
Bonni et al. [62]	Precuneus	100	cTBS	1	NA	[−]Source Memory Errors	NA	HY	30
Chen et al. [63]	Precuneus	100	10	10	CD(WD)	[+]JAVLT	[−](L)Parahippocampal gyrus:Hipp memory network, [−](L)Middle temporal gyrus:Hipp memory Network	SCD	38
Koch et al. [64]	Precuneus	100	20	10	CD(WD)	[+]JAVLT Delayed Recall[0.8]	[+]Beta band oscillations	PAD	14

Table 1. Cont.

Authors	Target	Intensity	Frequency	Sessions	Session Spacing (Days to Complete)	Cognitive Changes ([+/N/−] for rMTS) [Score Change]	Functional Connectivity Changes ([+/N/-] Area1:Area2)	Target Population	N
Riberio et al. [65]	Superior Parietal Cortex	80	1	1	NA	[−]Spatial Working Memory	NA	HY	20
H. Wang et al. [48]	Superior Parietal Cortex	100	10	2	2	[+]Face/word Pairs	NA	HY	8
Addicott et al. [66]	(R) Postcentral Gyrus	100	10	5	CD(D)	NA	[+](R)Postcentral gyrus:(L)Insula	H	28
Multisite rTMS									
Leocani et al. [67]	(B) Frontal, Parietal, Temporal	120	10	12(+4)	3 sessions a week for 4 weeks	[+]ADAS-Cog[−1.01]	NA	AD	16
Rabey et al. [68]	neuroAD	90–110	10	30(+24)	CD(WD)	[+]ADAS-Cog[3.76]	NA	AD	15
Nguyen et al. [69]	neuroAD	100	10	30	CD(WD)	[+]ADAS-Cog	NA	MCI, AD	10
Sabbagh et al. [70]	neuroAD	110	10	30	CD(WD)	[+]ADAS-Cog[−0.32]	NA	AD	59

Information from studies reviewed here including authors, TMS target, stimulation intensity, stimulation frequency, number of rTMS sessions, whether cognitive changes were present, whether functional connectivity changes were present, the target population, and the number of subjects. Sessions within parentheses indicated maintenance sessions following intervention. Abbreviations: AD, Alzheimer's disease; ADAS-Cog, The Alzheimer's Disease Assessment Scale-Cognitive Subscale; AG, angular gyrus; aMCI, amnestic mild cognitive impairment; AVLT, Rey Auditory Verbal Learning Test; B, bilateral; CD, rTMS sessions on consecutive days; cTBS, continuous theta-burst stimulation; D, rTMS sessions took place daily; dlPFC, dorsolateral prefrontal cortex; EEG, significant EEG changes present; H, healthy; HO, healthy old; Hip, Hippocampus; HY, healthy young; iTBS, intermittent theta-burst stimulation; L, left; MCI, mild cognitive impairment; MMSE, Mini-Mental State Examination; MoCA, Montreal Cognitive Assessment; N, no change; NA, not applicable PAD, prodromal AD; R, right; SCD, subjective cognitive decline; WD, rTMS sessions took place on week days only; *, same set of participants; +, change associated with better cognition or positive change in resting-state functional connectivity; −, change associated with poorer cognition or negative change in resting-state functional connectivity.

In our review of published work, rTMS interventions for memory most frequently involved *frontal lobe* stimulation targets and *healthy individuals*, so we begin by summarizing findings from those studies.

2.2. rTMS of Frontal Lobe Sites

2.2.1. rTMS of dlPFC: Healthy Young and Healthy Old

In the current literature, dlPFC has been stimulated with a variety of rTMS parameters and some reports suggest left and right dlPFC may respond differently to rTMS. Within studies targeting dlPFC in healthy adults considered in this review, the left hemisphere has been more frequently targeted. rTMS of left dlPFC has produced moderately consistent effects on resting-state functional connectivity (RSFC) but less consistent cognitive outcomes. Regarding cognitive changes associated with left dlPFC rTMS, eight of twelve studies reviewed here reported significant cognitive improvements associated with high-frequency stimulation [14,34,35,38,39,41,42,44–46,71,72]. However, and exemplifying the heterogeneity of outcomes in this domain, one study using low-frequency rTMS reported acute cognitive impairment [43].

Heterogeneity in rTMS methods and outcomes can be observed even in the limited domain of rTMS of left dlPFC of healthy adults. In one study, Chung and colleagues applied intermittent theta-burst stimulation (iTBS) rTMS targeting left dlPFC [41]. rTMS at 50%, 75%, and 100% motor threshold (MT) was associated with different results for each intensity. Their study observed a response similar to an inverted U-shaped curve, with no significant results at 50%, cognitive enhancement at 75%, and intermediate enhancement at 100%. In a similar study, Davis and colleagues applied 5 Hz rTMS at 120% MT to left dlPFC but observed no significant change in cognitive ability [42]. Together, these studies suggest that the greater the rTMS stimulation intensity does not strictly correspond to improved outcomes, and that there may be interactions of stimulation intensity and frequency on outcomes.

In two studies described above [41,42], rTMS was associated with changes in RSFC or EEG variables. Further, Davis and colleagues observed that RSFC changes were associated with better cognitive performance, including increased similarity in brain activation patterns during encoding and retrieval during a memory task [42,44].

Intriguingly, prior rTMS studies targeting dlPFC also suggest that brain state during rTMS may influence the brain's response and related cognitive effects. That is, the same rTMS protocol may yield different effects when administered during task performance or at rest. In one study, Bakulin and colleagues applied rTMS to left dlPFC during different phases of a modified Sternberg task and observed differences in n-back performance were associated with phase of stimulation [39]. Specifically, the authors found that when rTMS was applied in absence of the modified task, 10 Hz rTMS to the left dlPFC was associated with significantly increased scores on the n-back task. Conversely, when rTMS was applied during any phase of the modified task, no significant benefit was observed. Other authors have speculated that rTMS during a task may invert the responses putatively associated with high-frequency and low-frequency rTMS [62,73,74]. While there is some evidence of efficacy differences between rTMS during task and rest, further study will be required to rigorously evaluate whether effects are truly inverted and if the same inversion is present for other stimulation targets.

Box 1. Parameters and approaches for repetitive transcranial magnetic stimulation

- TMS uses a powerful electromagnet to apply a focal, transient magnetic pulse to stimulate activity in the neurons of underlying gray matter [21]. When multiple TMS pulses are applied in series or in more complex temporal patterns, the procedure is called repetitive transcranial magnetic stimulation (rTMS). Initial research surrounding rTMS indicated transient effects associated with stimulation [21]. Critically, it has also been reported that rTMS can modify the brain's intrinsic functional networks over extended periods [49,60,75].
- rTMS approaches can apply stimulation in simple series or in more complex patterns such as theta-burst stimulation (TBS). rTMS frequencies are typically described as either "excitatory" or "inhibitory" [21] as a function of stimulation frequency (1 Hz vs. 1 Hz, respectively). Classification into either form of rTMS is determined by changes observed in the motor evoked potential following rTMS to the primary motor cortex. "Excitatory" frequencies are reported to be associated with increases in brain activity while "inhibitory" frequencies are reported to be related to increased long-term depression of synaptic transmission. In TBS, pulse sequences are applied at frequencies and in patterns which putatively mirror neural oscillatory patterns associated with cognition [76,77]. The response to theta-burst rTMS varies by the rest period between stimulation. Despite the differences in rest periods between forms of TBS, the 50 Hz TBS is applied at a repeated 5 Hz frequency [78–80]. In continuous theta-burst stimulation (cTBS), pulses are applied during a 40 second period of stimulation followed by a short rest period. Alternatively, stimulation can be applied in 10 shorter periods consisting of a triplet of pulses followed by a rest period called intermittent theta-burst stimulation (iTBS). Application of TBS rTMS in different patterns can produce divergent effects on brain activity, cognition, and behavior [21]. iTBS has been hypothesized to be associated with promoting brain activity, while cTBS has been putatively associated with increased long-term depression of synaptic transmission [78–80].
- rTMS protocols can also apply different intensities of simulation often tailored to each individual subject. Stimulation intensity is often individualized by first gauging an individual's *motor threshold* [21]. This involves measuring the elecotromyographic (EMG) response to single-pulse TMS of primary motor cortex in a distal muscle either at rest (resting motor threshold, RMT) or in flexion (active motor threshold, AMT) [21]. The TMS pulse causes the targeted corticospinal tract to fire and trigger an overt response in the target muscle. After the cortical area associated with the predetermined muscle of interest, frequently is the abductor pollicis brevis of the right hand, is located, an adaptive stepwise procedure is used determine the individual's RMT/AMT. This procedure is a guided titration of TMS intensities near the strength that caused the initial EMG response. For RMT, the target intensity is the minimum stimulation strength required to generate a 50 µV or greater peak-to-peak intensity in five of ten stimulations as measured by EMG. The active motor threshold is similar but employs a higher threshold, 200 µV. This higher threshold is required to determine the measured response is due to stimulation and not flexion-related noise in the EMG. Following the motor thresholding procedure, the intensity of the rTMS protocol can then be individualized so that, for example, all participants receive rTMS at 110% of their unique RMT.

2.2.2. rTMS of dlPFC: MCI and AD

Although AD and MCI (especially aMCI) are often associated with clinical memory deficits, rTMS studies in these populations have frequently assessed general cognitive outcomes rather than memory-specific outcomes. Still, studies of rTMS in individuals with MCI and AD have yielded relatively consistent results. Much of this consistency may be derived from the greater homogeneity of rTMS parameters selected for studies of these populations versus studies of healthy individuals.

For example, in our survey of this literature, high-frequency rTMS was frequently used in studies of patients with MCI/AD. In several studies that applied high-frequency rTMS to left dlPFC, stimulation was associated with improved scores on one or more common cognitive assessments, including the MoCA, MMSE, and/or ADAS-Cog [14,34,36,45]. In a smaller number of studies, significant improvements were also reported on domain-specific assessments of memory abilities such as associative memory and relational memory [38,45,74,81]. Notably, all studies targeting dlPFC in clinically impaired individuals observed cognitive improvement in at least one domain. Specifically, improvements in AVLT, paired associate learning, MMSE, ADAS-cog, Rivermead Behavioral Memory

Test, letter-number sequencing, association memory, recognition, logical memory, MoCA, and word/image association were observed following ten or more sessions of rTMS for cognitively impaired individuals [11,14,34–36,38,45].

In another study, Rutherford and colleagues recruited patients with AD and applied 20 Hz rTMS at 100% RMT to bilateral dlPFC (serially, one hemisphere at a time) across 13 sessions [11]. Of special note, longitudinal follow-up with participants revealed they had significantly attenuated rates of decline compared to participants randomly assigned to a control condition. Replication of this promising finding would be an important step toward generalization to clinical treatment.

Finally, it has also been reported that low-frequency rTMS of right dlPFC was associated with cognitive improvement [33,82–84]. This finding may be interesting in the context of both healthy and pathological aging, as there is some evidence that right dlPFC exhibits hyperactivity associated with diminished cognitive performance [85,86].

2.2.3. rTMS of Other Frontal Lobe Areas

While the dlPFC has been the most common target for rTMS in the frontal lobe, several other sites in frontal regions have also been targeted. Among these sites are precentral gyrus, middle frontal gyrus, and right medial frontopolar cortex. Jung and colleagues have explored the effects of 1 Hz rTMS to left and right precentral gyrus, two additional non-association areas [49]. They observed decreased connectivity between the DMN and the right motor network, the insular network, and the visual network attributable to rTMS. Additionally, rTMS during task engagement resulted in decreased connectivity between the DMN and the dorsal attention network and increased connectivity between the DMN and the frontoparietal network.

Regarding right medial frontopolar cortex as a target, one study investigated the effects of single-session 20 Hz or 1 Hz rTMS [50]. In this instance, the authors reported RSFC changes associated with improved cognition for the 20 Hz stimulation group and changes associated with poorer cognition following 1 Hz rTMS in the low-frequency stimulation group.

rTMS of middle frontal gyrus (MFG) has also been explored by Wang and colleagues [48]. This group administerd 10 Hz rTMS to either a left or right middle frontal gyrus target for two consecutive days. While improvements in hippocampal-dependent relational memory were found following stimulation of the right hemisphere target, no such changes were present following rTMS to the left site. Different effects of rTMS applied to left and right MFG could be attributable to laterality, but replication would be an important step to aid the interpretation of these findings.

Interestingly, a second study of rTMS applied to MFG used RSFC to determine an rTMS target. Lynch and colleagues applied a connectome-based approach to identify independent targets for each subject within right MFG based on within-network RSFC [47]. The authors applied a single session of cTBS rTMS to right MFG, and observed reduced working memory performance associated with stimulation.

2.3. rTMS of Parietal Lobe Sites

2.3.1. rTMS of AG: Healthy Young and Healthy Old

Outside of the frontal lobe, much of the association cortex accessible to typical TMS approaches lies in lateral portions of the parietal lobe. In the context of memory-related rTMS studies, locations in the inferior parietal lobule have been targeted most frequently. This is likely due to associations with memory task performance based on neuropsychological and neuroimaging studies [87–90]. In particular, left AG has been a popular choice for rTMS-based modulation of memory function.

rTMS of AG has proved fruitful for memory researchers, illustrated most clearly by the work of Voss and colleagues [12,22,54,61]. Angular gyrus is a cortical area within the effective range of rTMS that exhibits strong RSFC with hippocampus. By targeting a DMN component functionally connected to hippocampus, many researchers have applied rTMS measured the effects of rTMS on hippocampal-dependent memory function. In particular, Voss and colleagues have frequently demonstrated success in modulation of memory performance, brain activity, or both using a paradigm involving 20 Hz rTMS to left AG at 100% RMT [12,13,52–55,58,60]. The only significant source of heterogeneity within the studies using this paradigm was the number of rTMS sessions.

The bulk of the 20 Hz rTMS studies from Voss and colleagues targeting left AG used five rTMS sessions [12,13,54–56,58]. Perhaps unsurprisingly, rTMS studies using five rTMS sessions with similar parameters frequently observed similar outcomes. These studies reported improvement in both measures of memory and RSFC. More specifically, RSFC changes associated with improved cognition are observed in the DMN. These consisted of strengthened RSFC between left AG and left hippocampus. In addition to these primary findings, it has also been reported that rTMS promotes hippocampal RSFC with DMN components beyond AG [12]. Consistent with a mechanistic explanation for rTMS effects on memory, these changes in RSFC were also accompanied by significant cognitive changes [12,13,54–56,58]. The aforementioned improvements in relational memory performance following rTMS were significantly greater compared to participants in the placebo-sham conditions. Similar increases in cognitive performance and changes in RSFC have also been observed under several rTMS protocols originating from the same group [55,91,92].

Dosage, operationalized as number of stimulation sessions, may be a key factor in determining the efficacy of AG rTMS as a memory-enhancing therapy. Several studies have varied rTMS dosing to investigate this relationship. In one study, Freedberg and colleagues observed that three, four, or five sessions of rTMS to left AG resulted in similar RSFC changes, but those authors did not assess changes in memory [60]. In a study with similar design but inconsistent findings, Hendrikse and colleagues reported finding no significant cognitive benefit following four rTMS sessions [52]. To explore a potential minimum threshold, a dose-finding study was carried out by Freedberg and colleagues. Those authors reported that a minimum of 5 rTMS sessions was required for reliable, statistically significant change in hippocampal-AG RSFC [51]. While these studies indicate that a minimum number of rTMS sessions may be necessary to obtain reliable effects, Hermiller and colleagues also reported a single session of cTBS rTMS was adequate to induce comparable changes in RSFC [53]. While 20 Hz rTMS studies suggest that three to five sessions may be required to generate reliable RSFC changes, the recent cTBS study reportedly requiring only a single session is intriguing. It seems possible that different stimulation frequencies or different sets of stimulation parameters may require a unique number of minimum rTMS sessions for significant changes to be observed. Future research exploring this possibility is warranted.

Right AG has also been targeted with rTMS. In a single study, Tambini and colleagues applied cTBS rTMS to right AG [61]. Following this, significant cognitive improvement was observed and coupled with RSFC changes. Unfortunately, this was the only study targeting right AG, and additional research into right AG rTMS in healthy individuals is warranted.

2.3.2. rTMS of AG: MCI and AD

Although results from healthy young and old adults demonstrate the potential for rTMS to improve memory abilities, similar findings have not been reported for AD and (a)MCI. While new clinical trials are proceeding at the time of this writing [93], only one recent study was identified applying rTMS to AG in individuals with mild to severe AD. Velioglu and colleagues administered ten sessions of 20 Hz rTMS at 100% MT to left AG [57]. Visual recognition memory performance and the clock drawing test improved after stimulation. Notably, the cognitive improvements were associated with changes in RSFC and, somewhat surprisingly, significant changes in other blood-derived, neurally-relevent biomarkers. Following rTMS, individuals were reported to have elevated blood brain derived neurotrophic factor measures and lower oxidative status measures. While intriguing, caution is warranted when interpreting these findings because biomarker measures derived from peripheral blood and CSF do not always exhibit strong correlation [94].

2.3.3. rTMS of Other Parietal Lobe Sites

Several parietal regions beyond AG have been targeted with rTMS. The next most common stimulation site was precuneus [62–64]. As precuneus lies at the core of the default mode network [95], several studies have identified significant cognitive or brain changes following rTMS targeting precuneus.

One such study by Chen and colleagues applied ten sessions of 10 Hz rTMS to precuneus in individuals with subjective cognitive decline [63]. Following stimulation, these researchers observed significantly improved episodic memory and RSFC between precuneus and posterior hippocampus. Improvement in these domains is reminiscent of AG stimulation, mainly due to the notable hippocampal RSFC changes. A similar outcome was also reported by Koch and colleagues [64]. Here again, ten sessions of rTMS were administered but with 20 Hz stimulation. Following stimulation, the authors noted significant improvement in episodic memory coupled with changes in RSFC and EEG profiles. Several studies also targeted precuneus with low-frequency or cTBS rTMS and found transient impairments in memory or metacognition [62,96–98].

Two studies reported applying rTMS to superior parietal regions, and both reported cognitive changes in healthy young adults. Both studies reported outcomes consistent with expectations for high-frequency and low-frequency rTMS stimulation. Specifically, Wang and colleagues observed significant improvement in recalling face/word pairs following two sessions of 10 Hz rTMS of their target in superior parietal cortex [48]. Alternatively, Ribeiro and colleagues observed acute cognitive impairment following one session of 1 Hz of rTMS to superior parietal cortex [65]. Beyond association regions in the parietal cortex, post-central gyrus has also been targeted due to its functional connections with the insula [66]. Following five sessions of 10 Hz rTMS, Addicott and colleagues reported increased RSFC between the target and left insula. The directionality of these findings is consistent with the putative associations between high- and low-frequency stimulation and cognitive enhancement/impairment and in some instances required fewer than five stimulation sessions.

3. Multitarget Stimulation

While rTMS studies have most frequently targeted a single cortical region, some investigators have also tested the effect of multitarget rTMS. As the name suggests, multitarget rTMS involves targeting multiple, distal brain regions for stimulation within the same paradigm either serially or, less often, simultaneously. The potential benefits of multitarget stimulation include modulation of brain activity in locations in one or more functional brain networks, and this approach could provide additive or interactive cognitive enhancement [99].

For example, one study employing multitarget stimulation serially targeted several temporal and parietal stimulation locations [100]. Here, the researchers used 20 Hz stimulation over frontal and parietal targets every weekday for six weeks. Following stimulation, adults with AD exhibited a significant increase in ADAS-cog performance, and there was evidence that this effect endured for up to 12 weeks. The reported durability of this improvement is unusual in the literature and could reflect lasting modulation of underlying functional brain networks.

The "neuroAD protocol" is another line of research using a multitarget rTMS approach [68,69,101,102]. The protocol involves stimulation of six distinct targets regions: left and right dlPFC, left and right somatosensory association cortex, Broca's area, and Wernicke's area. Targeting these areas, the authors sought to improve multiple behaviorally relevant functional networks impacted by AD [70]. During each stimulation session, three of the six targets were serially stimulated.. Three different brain regions were selected for stimulation every session, with each site being stimulated in 15 sessions [101]. Stimulating at 100–110% RMT was associated with significant improvement in ADAS-Cog performance following rTMS [70]. Meanwhile, stimulation at 90% RMT reported observed increases in MMSE scores [103]. Unique among rTMS therapies for memory, the neuroAD protocol was recently submitted to the FDA for consideration as an intervention for patients with MCI or AD. At the time of writing, the most recent FDA determination was that the cognitive benefits were not substantial enough to warrant approval due to their modest efficacy (less than a three-point improvement on ADAS-Cog) [104]. Regarding concerns about the protocol's efficacy, it is possible that the limited magnitude of cognitive benefit associated with the protocol could be due to the inclusion of individuals with substantial AD-related cognitive impairment. For individuals with more mild impairment, evidence of greater cognitive improvement was present: nearly a third of these individuals improved by four or more points on the ADAS-Cog [70]. If upheld, this finding would suggest that the neuroAD intervention is more effective in earlier disease stages, such as MCI, rather than AD.

Where the neuroAD protocol targeted several locations serially during a session, the development of new TMS coils has also allowed stimulation of multiple cortical areas simultaneously. The ability to broadly stimulate bilateral frontal, temporal, and parietal areas has been explored with "H"-style TMS coils. Specifically, 10 Hz rTMS has been applied using an H coil for twelve consecutive sessions in individuals with AD [67]. Improvements were noted in ADAS-Cog scores but not in several other measures (MMSE, depression, or caregiver ratings of subjective improvement).

4. Developments Relevant to Treating Memory loss with rTMS

Approaches using rTMS to treat memory loss have evolved substantially over the last two decades, as have insights from neuroscience regarding functional brain organization, neurodegenerative diseases, and brain mechanisms supporting memory processes. These developments are important considerations for investigators designing new rTMS interventions for memory loss. Furthermore, the integration of key concepts into new paradigms could improve the efficacy and reproducibility of future rTMS research. Here, we review some key developments including acknowledgment of the brain's large-scale functional networks, computational modeling of rTMS stimulation fields, and frequency-specific effects of rTMS.

4.1. Functional Brain Networks

The last decade has seen a tremendous expansion of the field's understanding of the brain's intrinsic functional organization. Readily identifiable, large-scale functional brain networks have been reliably observed both in group studies and at the level of individual participants [105,106]. This development may offer benefits for rTMS approaches similar to those provided by stereotactic alignment of structural MRI data with the physical brain: improved rigor and reproducibility through precision alignment to previously identified stimulation targets. Here, a key concept is the identification of stimulation targets using individualized maps of functional networks overlaid onto the physical brain. Similar targeting has already been applied with success in rTMS studies seeking to treat depression [107,108]. If implemented, this approach could supplement and refine earlier approaches that identify targets based on physical distance, gross neuroanatomical landmarks, or coordinate-based targets derived from brain atlases.

Acknowledging functional network architecture in the design of rTMS interventions will help to ensure that the same functional network is being stimulated across different participants. For example, while dlPFC has shown promise as an rTMS target for treating memory loss [35,41,74], dlPFC is a large region of association cortex which includes several distinct functional networks [105,106]. Furthermore, the territory of these networks varies between individuals [109,110]. Stimulation of the same dlPFC location based on neuroanatomy or template-derived coordinates could therefore affect a different selection of functional networks between subjects unless targets are selected for each participant according to their brain's unique functional organization.

A related consideration is that stimulation of different functional networks would be expected to affect different cognitive processes. A strong implication of rTMS not guided by functional network consideration is that cognitive benefits of rTMS interventions could vary between individuals as a function of the stimulated networks rather than stimulation efficacy per se. Alternatively, otherwise similar cognitive benefits might be attributable to changes in different cognitive processes between individuals. Taking memory performance as an example, deficits in executive functions [111,112] or depressed mood [113] have been associated with memory impairments, so by inference, rTMS-associated improvements in executive functions or mood might be expected to enhance apparent memory performance, but without affecting underlying memory processes. While positive outcomes for patients are always welcome, interpretation of this type of finding could be confounded if superficially similar outcomes are attributable to different mechanisms. Integration of functional neuroimaging data into new TMS protocols to support network-specific targeting could help to avoid this specific confound.

While integration of functional neuroimaging data in rTMS intervention design is expected to enhance rigor, approaches to processing neuroimaging data can vary greatly and affect interpretation. It has been well documented that even when using the same dataset, different groups can generate significantly different findings [114]. This is not surprising because the number of possible analysis paths available to investigators is enormous; one recent report estimated that a typical fMRI dataset might afford nearly 7,000 unique analysis pipelines [115]. Thorough documentation of all steps of functional neuroimaging analysis is therefore essential, and widely-used workflows for analysis might be considered. For example, the Human Connectome Project [116] provides a standardized "minimal preprocessing pipeline" for structural and functional MRI data that appears to deliver reliable results [117]. This and similar pipelines can provide investigators with a predetermined workflow for MRI data processing, ensuring that all groups perform the same steps in the same order. Also, adoption of a common approach to analyzing neuroimaging data could reduce a significant source of heterogeneity for rTMS interventions that include neuroimaging outcomes.

4.2. Modeling of TMS Field Locale/Stimulation Strength

Selection of TMS stimulation sites can be refined by anatomical and functional considerations as described above, and recent advances in computational modeling of electrical fields induced by non-invasive brain stimulation techniques (including TMS but also transcranial electrical stimulation) may support still further enhancement. Tools such as the SimNIBS toolkit [118] allow researchers to model the induced magnetic and electrical fields for an individual brain based on structural imaging data. The models then estimate the spatial extent of brain tissue affected by each TMS pulse. These estimates are important when considering the anatomical focality of the stimulation produced by a set of TMS parameters.

Model estimates of stimulation extent may also help investigators to understand which functional brain networks are most likely to be affected by TMS at a specific location. In combination with processed functional neuroimaging data, stimulation models can highlight functional networks that are most likely to be affected by TMS at a specific location. New studies could clearly benefit from this approach, and previous studies might benefit retroactively if the necessary data (structural MRI, resting-state fMRI, stimulation coordinates, and stimulation intensity) were collected.

4.3. Stimulation Frequency and Patterning

Historically, rTMS frequencies have sometimes been dichotomized into either "excitatory" or "inhibitory" stimulation [21] as a function of stimulation frequency (1 Hz vs. 1 Hz, respectively). Classification as excitatory or inhibitory has been driven by changes observed in the motor evoked potential following rTMS to the primary motor cortex. Unfortunately, this simple scheme for classification may be overly reductionist, not addressing potentially important complexities while limiting exploration of new rTMS protocols.

We respectfully suggest that the current "excitatory vs. inhibitory" dichotomy might benefit from a different characterization: high-frequency vs. low-frequency stimulation. Our suggestion for revised terminology arises from the neurophysiology of TMS. Crucially, it is not the case that "excitatory" stimulation causes an overt response at the rTMS target while "inhibitory" stimulation suppresses this response. Rather, irrespective of stimulation frequency, some neurons at the target location depolarize, making "inhibitory" a mischaracterization of the stimulatory effect from the standpoint of a cellular response. Findings from active rTMS, or rTMS performed during task performance, also weigh against the historical labeling of rTMS protocols. Active rTMS has been reported to invert the expected rTMS response [21,40,119,120]. That is, during active rTMS, typical "inhibitory" rTMS protocols have been associated with improvements in cognitive performance in some cases, whereas the same protocol at rest would be associated with reduced performance. "Excitatory" protocols similarly have been reported to swap responses in active conditions further supporting that such classification may be improper. Finally, evidence from studies applying physiological considerations in rTMS protocol determination also suggests that these classifications may be unfitting. One example of the importance of physiological considerations is "inhibitory" rTMS to the right dlPFC. In this instance, it has been observed that following rTMS, episodic memory performance is reported to significantly increase despite the "inhibitory" classification of stimulation [33,84]. It is important to note that right dlPFC does exhibit increased connectivity associated with reduced cognition [85,86]. In this way, although the "inhibitory" protocol improved cognition, it may have also acted to reduce the associated increase in connectivity. From a RSFC standpoint, "inhibitory" rTMS may be properly named in this instance, but the opposing cognitive outcomes add unnecessary confusion to the rTMS field. In this way, the classification of rTMS frequencies into "excitatory" or "inhibitory" addresses few specific instances and may inaccurately map onto neurophysiological (or other) outcomes.

As recent studies have enriched our understanding of how brain tissues and brain networks respond to rTMS frequencies and patterns, investigators now have a larger menu of frequencies from which to choose along with a better understanding of likely effects on underlying brain activity. For example, high-frequency rTMS protocols have been associated with increased within-network connectivity of a targeted functional network [42,50]. This may be an important consideration for efficacy because in other work, stronger within-network connectivity has been associated with better cognitive outcomes in neurological disease such as stroke [121]. Meanwhile, low-frequency rTMS has sometimes been associated with decreases in within-network connectivity accompanied by increases in between-network connectivity [42,84]. While this association may not be as robust as the association of high-frequency rTMS with stronger within-network connectivity, the potential for frequency-dependent effects on connectomic measures presents exciting possibilities for basic and clinical research.

Regarding the effects of different frequencies within the "high" or "low" categories, little is known. Very few published studies have measured whether different rTMS frequencies with the same expected activation valence (e.g., high-frequency, 10 Hz vs. 20 Hz) produce different effects. Instead, published work has more often contrasted high and low frequencies or the same stimulation frequency at one stimulation location versus another [42,44,50]. This gap in the literature may be important because the few publications on the topic suggest that varying stimulation frequency can affect cognitive outcomes. In one important demonstration, rTMS at 20 Hz and iTBS were associated with different cognitive outcomes following one session of rTMS targeting left AG [53]. Future research on rTMS methods may help to titrate stimulation frequencies and patterns that combine continued safety with greater efficacy. For the immediate future, new rTMS interventions may benefit by simply acknowledging the expected strengthening of within-network connectivity associated with typical high-frequency rTMS.

5. Suggestions for Studies Using rTMS to Treat Memory Loss

While rTMS shows promise as a potential intervention to enhance declarative/relational memory abilities or to treat memory loss (age-related or pathological), substantial between-study heterogeneity in design has made direct comparisons difficult. Here, we will close our review by discussing study design features and rTMS parameters that we expect will enhance the rigor, reproducibility, and efficacy of new investigations. These include, but are not limited to, selecting a functional network to target, finding suitable stimulation locations within that network, thoughts on TMS coil placement, selection of rTMS frequency to utilize, numbers of rTMS sessions, and the importance of longitudinal follow-ups.

5.1. Stimulation Site Selection

Any rTMS study must select one or more stimulation sites. Predictably, stimulation at different sites has been associated with different cognitive and behavioral outcomes. Acknowledging this, studies focused on memory enhancement or treatment of memory loss should select one (or more than one) site previously associated with memory abilities. Based on prior work and insights from the normative functional organization of the brain, we offer two broad insights and several more specific recommendations.

Perhaps our strongest recommendation is that investigators should consider selecting targets based on functional network locations in addition to structural features or coordinates. The parallel, interdigitated nature of the brain's functional networks [122] makes reliably targeting a specific network through structural features impractical. Conversely, functional targeting is a relatively simple enhancement that can be readily implemented [107,108]. Regarding which networks to target, two may be especially important for normal memory function [24,123]: the default mode network, which is often described as including the medial temporal lobes and hippocampus, structures essential for normal memory; and the frontoparietal network [90], which has been frequently implicated in fMRI studies observing "subsequent memory effects" (increases in activation related to remembered versus forgotten items). Importantly, functionally determined rTMS targets could potentially be derived from resting-state or task-based neuroimaging data (or both); each offers advantages. Resting-state fMRI is relatively easy to collect from most populations and affords the opportunity to readily identify intrinsic networks [124–126]. Alternatively, task-based fMRI, perhaps collected during memory task performance, might offer even more refined targets because of the direct association with memory performance [127]. In either case, individualized stimulation targets derived from analysis of functional neuroimaging data are strongly predicted to provide more consistent results than other approaches.

Turning to specific cortical locations, one possibility is the left posterior lateral parietal lobule, or more specifically, left AG. Left AG is a region of association cortex that has well-characterized structural connections with the medial temporal lobe and RSFC with the hippocampus [22]. This connectivity and the necessity of hippocampus for normal memory functions [26,128] make left AG an appealing target. As reviewed here, significant prior work has demonstrated that rTMS of left AG can improve declarative/relational memory in healthy young and healthy older participants [12,13,51,58]. Additionally, stimulation of left AG does not have any known association with relief from depressive symptoms or executive functions, potential confounds related to stimulating other sites (e.g., dlPFC).

rTMS of left dlPFC has also been previously associated with improved memory performance. However, the above concerns regarding potential confounds related to mood and executive functions may apply to stimulation of this region. Irrespective of which location is selected, we strongly recommend individual targeting of a specific functional network rather than a location guided by simple distance, neuroanatomical features, or transformed atlas coordinates.

5.2. Stimulation Site Targeting

Less complex but no less important than selection of a stimulation site is targeting of the stimulation site during an rTMS session. Earlier methods using EEG or scalp landmarks [37,65,100] can be substantially improved upon by TMS instruments that support real-time stereotactic alignment of structural MRI data and the participant's physical brain [53,56,58,129]. Extending the same stereotactic coordinates to the TMS coil allows accurate, reproducible targeting of a specific brain region during one or more TMS sessions. Recently, stereotactic localization of a target brain region has been further enhanced by robotic systems that can maintain precise head-coil positioning to account for head motion during rTMS sessions [129]. Whether automated or manual, stereotactic alignment systems substantially enhance experimental rigor for rTMS studies..

5.3. Frequency Selection

rTMS frequencies and protocols are dichotomized into "excitatory" (high-frequency and iTBS) or "inhibitory" (low-frequency and cTBS) frequencies [21]. While this dichotomy captures some important differences, factors beyond rTMS frequency also contribute the excitatory or inhibitory influence of rTMS. One such factor is the underlying physiology of the rTMS target and the functional network to which it belongs. rTMS of right dlPFC is a prime example of the role target physiology can play. Multiple reports suggest that 1 Hz rTMS of right dlPFC caused significant improvement in cognitive abilities [33,82–84]. That might be consistent with an "excitatory" influence of an "inhibitory" frequency. Whatever the underlying mechanism, this outcome exemplifies the complex relationship between rTMS parameters and cognitive outcomes.

Neurophysiological considerations may also provide insight into what rTMS frequencies may generate potent responses. For example, Chung and colleagues investigated whether iTBS at a frequency matched to an individual's brain activity would outperform the "excitatory" 50 Hz iTBS rTMS [130]. While both the individual and 50 Hz iTBS were reported to significantly improve cognition, individualized iTBS was also associated with significant changes in EEG measures. These reports illustrate the potential impact of neurophysiological considerations on rTMS outcomes. Stimulation frequency is an rTMS parameter that could benefit from more study, including refinement of methods for determining individualized stimulation frequencies based on observed neurodynamics of a given brain.

5.4. Number of Sessions

Perhaps the greatest degree of consensus in the rTMS literature lies in the number of rTMS sessions necessary for reliable memory enhancement. Specifically, multiple consecutive days of rTMS appear to be necessary to reliably observe improvements in memory performance that endure for one or more days after stimulation. Regarding the absolute number of sessions required, some research has been conducted with the explicit goal of dose estimation. Following up on prior work that tested the effects of rTMS applied to left AG, one studied estimated that a minimum of five sessions was required for benefits to memory performance [51], while a similar study by the same group estimated that as few as three simulation sessions was adequate to observe significant changes in RSFC between the stimulation site in left AG and the hippocampus [60]. To the best of our knowledge, these two studies are the only published works examining the effects of different numbers of rTMS sessions for left AG rTMS. More research on dosing of rTMS to treat memory impairment would be helpful. However, based on these dose-finding studies and other studies reporting significant changes after left AG stimulation, a minimum of five stimulation sessions appears to be a reasonable criterion [51,60]. Notably, ongoing clinical trials in patients with MCI or AD may incorporate even more sessions, such as the "20 weekday sessions during a period of 2 to 4 weeks" in a trial by Taylor and colleagues [93].

5.5. Longitudinal Follow-Up

rTMS therapies for memory would be most beneficial if the effects endured for some prolonged period after stimulation. Unfortunately, many rTMS publications do not report longitudinal measures. Without longitudinal follow-up, the durability and dose-response curves of rTMS therapies are impossible to determine, and this creates challenges for future efforts to translate rTMS research to clinical applications. Collection of longitudinal follow-up measures, perhaps one, three, and six months after completion of an rTMS protocol, would be a welcome addition to the design of future studies.

5.6. Methodological Heterogeneity Versus Discovery Science

We have noted the heterogeneous methodologies of rTMS interventions for memory, and we have suggested that this creates challenges for interpretation and generalization. In that context, the suggestions we offer in this section of our review are intended to highlight opportunities for investigators to enhance their study designs based on recent advances and best practices. However, we do not wish to promote a rigidly proscriptive methodological homogeneity; the field of rTMS for memory (or other cognitive) enhancement is much too young to suggest that any single approach is optimal. Discovery science and exploratory research remain essential to progress in rTMS interventions for memory. So, while departures from typical rTMS protocols should be well-justified, as long as they are conducted with great scientific rigor, such efforts may well prove effective, informative, or both. Standard approaches for rTMS will only be enhanced by novel efforts, and we fully expect that a review of best practices written a decade from now would differ significantly from our current work largely due to new basic science findings.

6. Conclusions

The brain systems that support declarative/relational memory are imperfect recorders that are negatively impacted by age and disease. Potential treatments for memory loss (or interventions to enhance memory performance) would be beneficial, and published work describing rTMS interventions offer preliminary evidence that non-invasive brain stimulation may offers symptom-modifying therapies. Our review of the current literature highlights many published examples of rTMS interventions that successfully modulated memory, often through multi-day high-frequency stimulation of regions in frontal or parietal association cortex. Unfortunately, the current rTMS literature suffers from significant heterogeneity which creates challenges for interpretation and comparison. To address this, we have offered suggestions for the design of future rTMS investigations with the goal of enhancing rigor and reproducibility. Our intent is not proscriptive; rather, we hope to encourage best practices that will speed the transition of rTMS-based memory modulation from laboratories to memory clinics where new therapies are sorely needed. By reducing methodological heterogeneity, introducing neuroimaging measures, and incorporating longitudinal follow-up, forthcoming memory-related rTMS studies have the opportunity to prove the method's validity, generalizability, and translational potential to treat clinical memory loss.

Author Contributions: Conceptualization, C.J.P., D.L.M., and D.E.W.; Methodology, C.J.P. and D.E.W.; Validation, C.J.P., D.L.M., and D.E.W.; data curation, C.J.P.; writing—original draft preparation, C.J.P.; writing—review and editing, C.J.P., D.L.M., and D.E.W.; supervision, D.E.W. All authors have read and agreed to the published version of the manuscript.

Funding: Authors CJP, DLM, and DEW received support from NIH/NIA award R01AG064247.

Conflicts of Interest: The authors declare no conflict of interest.

References

1. Alzheimer's Association. 2021 Alzheimer's disease facts and figures. *Alzheimers Dement.* **2021**, *17*, 327–406. [CrossRef]
2. Russ, T.C.; Morling, J.R. Cholinesterase inhibitors for mild cognitive impairment. *Cochrane Database Syst. Rev.* **2012**, CD009132. [CrossRef]
3. Cummings, J.L.; Tong, G.; Ballard, C. Treatment Combinations for Alzheimer's Disease: Current and Future Pharmacotherapy Options. *J. Alzheimers Dis. JAD* **2019**, *67*, 779–794. [CrossRef] [PubMed]
4. Moll van Charante, E.P.; Richard, E.; Eurelings, L.S.; van Dalen, J.-W.; Ligthart, S.A.; van Bussel, E.F.; Hoevenaar-Blom, M.P.; Vermeulen, M.; van Gool, W.A. Effectiveness of a 6-year multidomain vascular care intervention to prevent dementia (preDIVA): A cluster-randomised controlled trial. *Lancet* **2016**, *388*, 797–805. [CrossRef]
5. Vellas, B.; Carrie, I.; Gillette-Guyonnet, S.; Touchon, J.; Dantoine, T.; Dartigues, J.F.; Cuffi, M.N.; Bordes, S.; Gasnier, Y.; Robert, P.; et al. Mapt study: A multidomain approach for preventing Alzheimer's disease: Design and baseline data. *J. Prev. Alzheimers Dis.* **2014**, *1*, 13–22. [PubMed]
6. Hampel, H.; Vergallo, A.; Aguilar, L.F.; Benda, N.; Broich, K.; Cuello, A.C.; Cummings, J.; Dubois, B.; Federoff, H.J.; Fiandaca, M.; et al. Precision pharmacology for Alzheimer's disease. *Pharmacol. Res.* **2018**, *130*, 331–365. [CrossRef] [PubMed]
7. Albert, M.S.; DeKosky, S.T.; Dickson, D.; Dubois, B.; Feldman, H.H.; Fox, N.C.; Gamst, A.; Holtzman, D.M.; Jagust, W.J.; Petersen, R.C.; et al. The diagnosis of mild cognitive impairment due to Alzheimer's disease: Recommendations from the National Institute on Aging-Alzheimer's Association workgroups on diagnostic guidelines for Alzheimer's disease. *Alzheimers Dement.* **2011**, *7*, 270–279. [CrossRef] [PubMed]
8. McKhann, G.M.; Knopman, D.S.; Chertkow, H.; Hyman, B.T.; Jack, C.R.; Kawas, C.H.; Klunk, W.E.; Koroshetz, W.J.; Manly, J.J.; Mayeux, R.; et al. The diagnosis of dementia due to Alzheimer's disease: Recommendations from the National Institute on Aging-Alzheimer's Association workgroups on diagnostic guidelines for Alzheimer's disease. *Alzheimers Dement.* **2011**, *7*, 263–269. [CrossRef]
9. Mol, M.E.M.; van Boxtel, M.P.J.; Willems, D.; Jolles, J. Do subjective memory complaints predict cognitive dysfunction over time? A six-year follow-up of the Maastricht Aging Study. *Int. J. Geriatr. Psychiatry* **2006**, *21*, 432–441. [CrossRef]
10. Freitas, C.; Mondragón-Llorca, H.; Pascual-Leone, A. Noninvasive brain stimulation in Alzheimer's disease: Systematic review and perspectives for the future. *Exp. Gerontol.* **2011**, *46*, 611–627. [CrossRef]
11. Rutherford, G.; Lithgow, B.; Moussavi, Z. Short and Long-term Effects of rTMS Treatment on Alzheimer's Disease at Different Stages: A Pilot Study. *J. Exp. Neurosci.* **2015**, *9*, 43–51. [CrossRef]
12. Wang, J.X.; Voss, J.L. Long-lasting enhancements of memory and hippocampal-cortical functional connectivity following multiple-day targeted noninvasive stimulation. *Hippocampus* **2015**, *25*, 877–883. [CrossRef]
13. Nilakantan, A.S.; Mesulam, M.-M.; Weintraub, S.; Karp, E.L.; VanHaerents, S.; Voss, J.L. Network-targeted stimulation engages neurobehavioral hallmarks of age-related memory decline. *Neurology* **2019**, *92*, e2349–e2354. [CrossRef] [PubMed]
14. Li, X.; Qi, G.; Yu, C.; Lian, G.; Zheng, H.; Wu, S.; Yuan, T.-F.; Zhou, D. Cortical plasticity is correlated with cognitive improvement in Alzheimer's disease patients after rTMS treatment. *Brain Stimulat.* **2021**, *14*, 503–510. [CrossRef] [PubMed]
15. Deng, Z.D.; Lisanby, S.H.; Peterchev, A.V. Electric field depth–focality tradeoff in transcranial magnetic stimulation: Simulation comparison of 50 coil designs. *Brain Stimulat.* **2012**, *6*, 1–13. [CrossRef] [PubMed]
16. Velanova, K.; Jacoby, L.L.; Wheeler, M.E.; McAvoy, M.P.; Petersen, S.E.; Buckner, R.L. Functional-anatomic correlates of sustained and transient processing components engaged during controlled retrieval. *J. Neurosci. Off. J. Soc. Neurosci.* **2003**, *23*, 8460–8470. [CrossRef]
17. Blumenfeld, R.S.; Ranganath, C. Dorsolateral Prefrontal Cortex Promotes Long-Term Memory Formation through Its Role in Working Memory Organization. *J. Neurosci.* **2006**, *26*, 916–925. [CrossRef] [PubMed]
18. Mars, R.B.; Grol, M.J. Dorsolateral Prefrontal Cortex, Working Memory, and Prospective Coding for Action. *J. Neurosci.* **2007**, *27*, 1801–1802. [CrossRef] [PubMed]
19. Barbey, A.K.; Koenigs, M.; Grafman, J. Dorsolateral Prefrontal Contributions to Human Working Memory. *Cortex J. Devoted Study Nerv. Syst. Behav.* **2013**, *49*, 1195–1205. [CrossRef]
20. Center for Devices and Radiological Health. *Repetitive Transcranial Magnetic Stimulation (rTMS) Systems—Class II Special Controls Guidance for Industry and FDA Staff*; FDA: Silver Spring, MD, USA, 2021.
21. Rotenberg, A.; Horvath, J.C.; Pascual-Leone, A. (Eds.) *Transcranial Magnetic Stimulation*; Springer: New York, NY, USA, 2014; Volume 89, pp. 69–142. ISBN 978-1-4939-0878-3.
22. Thakral, P.P.; Madore, K.P.; Schacter, D.L. A Role for the Left Angular Gyrus in Episodic Simulation and Memory. *J. Neurosci.* **2017**, *37*, 8142–8149. [CrossRef]
23. Amodio, D.M.; Frith, C.D. Meeting of minds: The medial frontal cortex and social cognition. *Nat. Rev. Neurosci.* **2006**, *7*, 268–277. [CrossRef]
24. Buckner, R.L.; Carroll, D.C. Self-projection and the brain. *Trends Cogn. Sci.* **2007**, *11*, 49–57. [CrossRef]
25. Greicius, M.D.; Srivastava, G.; Reiss, A.L.; Menon, V. Default-mode network activity distinguishes Alzheimer's disease from healthy aging: Evidence from functional MRI. *Proc. Natl. Acad. Sci. USA* **2004**, *101*, 4637–4642. [CrossRef]
26. Milner, B. The medial temporal-lobe amnesic syndrome. *Psychiatr. Clin. N. Am.* **2005**, *28*, 599–611. [CrossRef] [PubMed]
27. Spreng, R.N.; Mar, R.A.; Kim, A.S.N. The Common Neural Basis of Autobiographical Memory, Prospection, Navigation, Theory of Mind, and the Default Mode: A Quantitative Meta-analysis. *J. Cogn. Neurosci.* **2009**, *21*, 489–510. [CrossRef] [PubMed]

28. Fjell, A.M.; McEvoy, L.; Holland, D.; Dale, A.M.; Walhovd, K.B. What is normal in normal aging? Effects of Aging, Amyloid and Alzheimer's Disease on the Cerebral Cortex and the Hippocampus. *Prog. Neurobiol.* **2014**, *117*, 20–40. [CrossRef] [PubMed]
29. Seeley, W.W.; Crawford, R.K.; Zhou, J.; Miller, B.L.; Greicius, M.D. Neurodegenerative diseases target large-scale human brain networks. *Neuron* **2009**, *62*, 42–52. [CrossRef] [PubMed]
30. Weiler, M.; Stieger, K.C.; Long, J.M.; Rapp, P.R. Transcranial Magnetic Stimulation in Alzheimer's Disease: Are We Ready? *eNeuro* **2020**, *7*. [CrossRef] [PubMed]
31. Chou, Y.; Ton That, V.; Sundman, M. A systematic review and meta-analysis of rTMS effects on cognitive enhancement in mild cognitive impairment and Alzheimer's disease. *Neurobiol. Aging* **2020**, *86*, 1–10. [CrossRef]
32. Heath, A.; Taylor, J.; McNerney, M.W. rTMS for the treatment of Alzheimer's disease: Where should we be stimulating? *Expert Rev. Neurother.* **2018**, *18*, 903–905. [CrossRef]
33. Turriziani, P. Enhancing memory performance with rTMS in healthy subjects and individuals with Mild Cognitive Impairment: The role of the right dorsolateral prefrontal cortex. *Front. Hum. Neurosci.* **2012**, *6*, 62. [CrossRef]
34. Yuan, L.-Q.; Zeng, Q.; Wang, D.; Wen, X.-Y.; Shi, Y.; Zhu, F.; Chen, S.-J.; Huang, G.-Z. Neuroimaging mechanisms of high-frequency repetitive transcranial magnetic stimulation for treatment of amnestic mild cognitive impairment: A double-blind randomized sham-controlled trial. *Neural Regen. Res.* **2021**, *16*, 707–713. [CrossRef]
35. Drumond Marra, H.L.; Myczkowski, M.L.; Maia Memória, C.; Arnaut, D.; Leite Ribeiro, P.; Sardinha Mansur, C.G.; Lancelote Alberto, R.; Boura Bellini, B.; Alves Fernandes da Silva, A.; Tortella, G.; et al. Transcranial Magnetic Stimulation to Address Mild Cognitive Impairment in the Elderly: A Randomized Controlled Study. *Behav. Neurol.* **2015**, *2015*, 287843. [CrossRef]
36. Cui, H.; Ren, R.; Lin, G.; Zou, Y.; Jiang, L.; Wei, Z.; Li, C.; Wang, G. Repetitive Transcranial Magnetic Stimulation Induced Hypoconnectivity Within the Default Mode Network Yields Cognitive Improvements in Amnestic Mild Cognitive Impairment: A Randomized Controlled Study. *J. Alzheimers Dis. JAD* **2019**, *69*, 1137–1151. [CrossRef]
37. Schluter, R.S.; Jansen, J.M.; van Holst, R.J.; van den Brink, W.; Goudriaan, A.E. Differential Effects of Left and Right Prefrontal High-Frequency Repetitive Transcranial Magnetic Stimulation on Resting-State Functional Magnetic Resonance Imaging in Healthy Individuals. *Brain Connect.* **2018**, *8*, 60–67. [CrossRef]
38. Bagattini, C.; Zanni, M.; Barocco, F.; Caffarra, P.; Brignani, D.; Miniussi, C.; Defanti, C.A. Enhancing cognitive training effects in Alzheimer's disease: rTMS as an add-on treatment. *Brain Stimulat.* **2020**, *13*, 1655–1664. [CrossRef]
39. Bakulin, I.; Zabirova, A.; Lagoda, D.; Poydasheva, A.; Cherkasova, A.; Pavlov, N.; Kopnin, P.; Sinitsyn, D.; Kremneva, E.; Fedorov, M.; et al. Combining HF rTMS over the Left DLPFC with Concurrent Cognitive Activity for the Offline Modulation of Working Memory in Healthy Volunteers: A Proof-of-Concept Study. *Brain Sci.* **2020**, *10*, 83. [CrossRef] [PubMed]
40. Beynel, L.; Davis, S.W.; Crowell, C.A.; Hilbig, S.A.; Lim, W.; Nguyen, D.; Palmer, H.; Brito, A.; Peterchev, A.V.; Luber, B.; et al. Online repetitive transcranial magnetic stimulation during working memory in younger and older adults: A randomized within-subject comparison. *PLoS ONE* **2019**, *14*, e0213707. [CrossRef]
41. Chung, S.W.; Rogasch, N.C.; Hoy, K.E.; Sullivan, C.M.; Cash, R.F.H.; Fitzgerald, P.B. Impact of different intensities of intermittent theta burst stimulation on the cortical properties during TMS-EEG and working memory performance. *Hum. Brain Mapp.* **2018**, *39*, 783–802. [CrossRef] [PubMed]
42. Davis, S.W.; Luber, B.; Murphy, D.L.K.; Lisanby, S.H.; Cabeza, R. Frequency-specific neuromodulation of local and distant connectivity in aging and episodic memory function. *Hum. Brain Mapp.* **2017**, *38*, 5987–6004. [CrossRef] [PubMed]
43. Fitzsimmons, S.M.D.D.; Douw, L.; van den Heuvel, O.A.; van der Werf, Y.D.; Vriend, C. Resting-state and task-based centrality of dorsolateral prefrontal cortex predict resilience to 1 Hz repetitive transcranial magnetic stimulation. *Hum. Brain Mapp.* **2020**, *41*, 3161–3171. [CrossRef]
44. Wang, W.-C.; Wing, E.A.; Murphy, D.L.K.; Luber, B.M.; Lisanby, S.H.; Cabeza, R.; Davis, S.W. Excitatory TMS Modulates Memory Representations. *Cogn. Neurosci.* **2018**, *9*, 151–166. [CrossRef]
45. Wu, X.; Ji, G.-J.; Geng, Z.; Zhou, S.; Yan, Y.; Wei, L.; Qiu, B.; Tian, Y.; Wang, K. Strengthened theta-burst transcranial magnetic stimulation as an adjunctive treatment for Alzheimer's disease: An open-label pilot study. *Brain Stimul. Basic Transl. Clin. Res. Neuromodul.* **2020**, *13*, 484–486. [CrossRef]
46. Xue, S.-W.; Guo, Y.; Peng, W.; Zhang, J.; Chang, D.; Zang, Y.-F.; Wang, Z. Increased Low-Frequency Resting-State Brain Activity by High-Frequency Repetitive TMS on the Left Dorsolateral Prefrontal Cortex. *Front. Psychol.* **2017**, *8*, 2266. [CrossRef]
47. Lynch, C.J.; Breeden, A.L.; Gordon, E.M.; Cherry, J.B.C.; Turkeltaub, P.E.; Vaidya, C.J. Precision Inhibitory Stimulation of Individual-Specific Cortical Hubs Disrupts Information Processing in Humans. *Cereb. Cortex* **2019**, *29*, 3912–3921. [CrossRef] [PubMed]
48. Wang, H.; Jin, J.; Cui, D.; Wang, X.; Li, Y.; Liu, Z.; Yin, T. Cortico-Hippocampal Brain Connectivity-Guided Repetitive Transcranial Magnetic Stimulation Enhances Face-Cued Word-Based Associative Memory in the Short Term. *Front. Hum. Neurosci.* **2020**, *14*, 541791. [CrossRef]
49. Jung, J.; Bungert, A.; Bowtell, R.; Jackson, S.R. Modulating Brain Networks with Transcranial Magnetic Stimulation Over the Primary Motor Cortex: A Concurrent TMS/fMRI Study. *Front. Hum. Neurosci.* **2020**, *14*, 31. [CrossRef]
50. Riedel, P.; Heil, M.; Bender, S.; Dippel, G.; Korb, F.M.; Smolka, M.N.; Marxen, M. Modulating functional connectivity between medial frontopolar cortex and amygdala by inhibitory and excitatory transcranial magnetic stimulation. *Hum. Brain Mapp.* **2019**, *40*, 4301–4315. [CrossRef] [PubMed]

51. Freedberg, M.; Reeves, J.A.; Toader, A.C.; Hermiller, M.S.; Kim, E.; Haubenberger, D.; Cheung, Y.K.; Voss, J.L.; Wassermann, E.M. Optimizing Hippocampal-Cortical Network Modulation via Repetitive Transcranial Magnetic Stimulation: A Dose-Finding Study Using the Continual Reassessment Method. *Neuromodul. J. Int. Neuromodul. Soc.* **2020**, *23*, 366–372. [CrossRef] [PubMed]
52. Hendrikse, J.; Coxon, J.P.; Thompson, S.; Suo, C.; Fornito, A.; Yücel, M.; Rogasch, N.C. Multi-day rTMS exerts site-specific effects on functional connectivity but does not influence associative memory performance. *Cortex J. Devoted Study Nerv. Syst. Behav.* **2020**, *132*, 423–440. [CrossRef]
53. Hermiller, M.S.; VanHaerents, S.; Raij, T.; Voss, J.L. Frequency-specific noninvasive modulation of memory retrieval and its relationship with hippocampal network connectivity. *Hippocampus* **2019**, *29*, 595–609. [CrossRef] [PubMed]
54. Hermiller, M.S.; Karp, E.; Nilakantan, A.S.; Voss, J.L. Episodic memory improvements due to noninvasive stimulation targeting the cortical–hippocampal network: A replication and extension experiment. *Brain Behav.* **2019**, *9*, e01393. [CrossRef]
55. Kim, S.; Nilakantan, A.S.; Hermiller, M.S.; Palumbo, R.T.; VanHaerents, S.; Voss, J.L. Selective and coherent activity increases due to stimulation indicate functional distinctions between episodic memory networks. *Sci. Adv.* **2018**, *4*, eaar2768. [CrossRef]
56. Nilakantan, A.S.; Bridge, D.J.; Gagnon, E.P.; VanHaerents, S.A.; Voss, J.L. Stimulation of the Posterior Cortical-Hippocampal Network Enhances Precision of Memory Recollection. *Curr. Biol. CB* **2017**, *27*, 465–470. [CrossRef]
57. Velioglu, H.A.; Hanoglu, L.; Bayraktaroglu, Z.; Toprak, G.; Guler, E.M.; Bektay, M.Y.; Mutlu-Burnaz, O.; Yulug, B. Left lateral parietal rTMS improves cognition and modulates resting brain connectivity in patients with Alzheimer's disease: Possible role of BDNF and oxidative stress. *Neurobiol. Learn. Mem.* **2021**, *180*, 107410. [CrossRef] [PubMed]
58. Wang, J.X.; Rogers, L.M.; Gross, E.Z.; Ryals, A.J.; Dokucu, M.E.; Brandstatt, K.L.; Hermiller, M.S.; Voss, J.L. Targeted Enhancement of Cortical-Hippocampal Brain Networks and Associative Memory. *Science* **2014**, *345*, 1054–1057. [CrossRef] [PubMed]
59. Wynn, S.C.; Hendriks, M.P.H.; Daselaar, S.M.; Kessels, R.P.C.; Schutter, D.J.L.G. The posterior parietal cortex and subjectively perceived confidence during memory retrieval. *Learn. Mem.* **2018**, *25*, 382–389. [CrossRef] [PubMed]
60. Freedberg, M.; Reeves, J.A.; Toader, A.C.; Hermiller, M.S.; Voss, J.L.; Wassermann, E.M. Persistent Enhancement of Hippocampal Network Connectivity by Parietal rTMS Is Reproducible. *eNeuro* **2019**, *6*. [CrossRef]
61. Tambini, A.; Nee, D.E.; D'Esposito, M. Hippocampal-targeted Theta-burst Stimulation Enhances Associative Memory Formation. *J. Cogn. Neurosci.* **2018**, *30*, 1452–1472. [CrossRef]
62. Bonnì, S.; Veniero, D.; Mastropasqua, C.; Ponzo, V.; Caltagirone, C.; Bozzali, M.; Koch, G. TMS evidence for a selective role of the precuneus in source memory retrieval. *Behav. Brain Res.* **2015**, *282*, 70–75. [CrossRef]
63. Chen, J.; Ma, N.; Hu, G.; Nousayhah, A.; Xue, C.; Qi, W.; Xu, W.; Chen, S.; Rao, J.; Liu, W.; et al. rTMS modulates precuneus-hippocampal subregion circuit in patients with subjective cognitive decline. *Aging* **2020**, *13*, 1314–1331. [CrossRef] [PubMed]
64. Koch, G.; Bonnì, S.; Pellicciari, M.C.; Casula, E.P.; Mancini, M.; Esposito, R.; Ponzo, V.; Picazio, S.; Di Lorenzo, F.; Serra, L.; et al. Transcranial magnetic stimulation of the precuneus enhances memory and neural activity in prodromal Alzheimer's disease. *NeuroImage* **2018**, *169*, 302–311. [CrossRef]
65. Ribeiro, J.A.; Marinho, F.V.C.; Rocha, K.; Magalhães, F.; Baptista, A.F.; Velasques, B.; Ribeiro, P.; Cagy, M.; Bastos, V.H.; Gupta, D.; et al. Low-frequency rTMS in the superior parietal cortex affects the working memory in horizontal axis during the spatial task performance. *Neurol. Sci. Off. J. Ital. Neurol. Soc. Ital. Soc. Clin. Neurophysiol.* **2018**, *39*, 527–532. [CrossRef]
66. Addicott, M.A.; Luber, B.; Nguyen, D.; Palmer, H.; Lisanby, S.H.; Appelbaum, L.G. Low- and High-Frequency Repetitive Transcranial Magnetic Stimulation Effects on Resting-State Functional Connectivity between the Postcentral Gyrus and the Insula. *Brain Connect.* **2019**, *9*, 322–328. [CrossRef]
67. Leocani, L.; Dalla Costa, G.; Coppi, E.; Santangelo, R.; Pisa, M.; Ferrari, L.; Bernasconi, M.P.; Falautano, M.; Zangen, A.; Magnani, G.; et al. Repetitive Transcranial Magnetic Stimulation With H-Coil in Alzheimer's Disease: A Double-Blind, Placebo-Controlled Pilot Study. *Front. Neurol.* **2020**, *11*, 614351. [CrossRef]
68. Rabey, J.M.; Dobronevsky, E.; Aichenbaum, S.; Gonen, O.; Marton, R.G.; Khaigrekht, M. Repetitive transcranial magnetic stimulation combined with cognitive training is a safe and effective modality for the treatment of Alzheimer's disease: A randomized, double-blind study. *J. Neural Transm.* **2013**, *120*, 813–819. [CrossRef]
69. Nguyen, J.-P.; Suarez, A.; Kemoun, G.; Meignier, M.; Le Saout, E.; Damier, P.; Nizard, J.; Lefaucheur, J.-P. Repetitive magnetic stimulation combined with cognitive training for the treatment of Alzheimer's disease. *Neurophysiol. Clin. Clin. Neurophysiol.* **2017**, *47*, 47–53. [CrossRef]
70. Sabbagh, M.; Sadowsky, C.; Tousi, B.; Agronin, M.E.; Alva, G.; Armon, C.; Bernick, C.; Keegan, A.P.; Karantzoulis, S.; Baror, E.; et al. Effects of a combined transcranial magnetic stimulation (TMS) and cognitive training intervention in patients with Alzheimer's disease. *Alzheimers Dement. J. Alzheimers Assoc.* **2020**, *16*, 641–650. [CrossRef]
71. Beynel, L.; Davis, S.W.; Crowell, C.A.; Dannhauer, M.; Lim, W.; Palmer, H.; Hilbig, S.A.; Brito, A.; Hile, C.; Luber, B.; et al. Site-specific effects of online rTMS during a working memory task in healthy older adults. *bioRxiv* **2019**, 642983. [CrossRef] [PubMed]
72. Singh, A.; Erwin-Grabner, T.; Sutcliffe, G.; Paulus, W.; Dechent, P.; Antal, A.; Goya-Maldonado, R. Default mode network alterations after intermittent theta burst stimulation in healthy subjects. *Transl. Psychiatry* **2020**, *10*, 75. [CrossRef] [PubMed]
73. Bashir, S.; Al-Hussain, F.; Hamza, A.; Shareefi, G.F.; Abualait, T.; Yoo, W.-K. Role of Single Low Pulse Intensity of Transcranial Magnetic Stimulation over the Frontal Cortex for Cognitive Function. *Front. Hum. Neurosci.* **2020**, *14*, 205. [CrossRef] [PubMed]
74. Cotelli, M.; Manenti, R.; Cappa, S.F.; Zanetti, O.; Miniussi, C. Transcranial magnetic stimulation improves naming in Alzheimer disease patients at different stages of cognitive decline. *Eur. J. Neurol.* **2008**, *15*, 1286–1292. [CrossRef] [PubMed]

75. Hawco, C.; Voineskos, A.N.; Steeves, J.K.E.; Dickie, E.W.; Viviano, J.D.; Downar, J.; Blumberger, D.M.; Daskalakis, Z.J. Spread of activity following TMS is related to intrinsic resting connectivity to the salience network: A concurrent TMS-fMRI study. *Cortex J. Devoted Study Nerv. Syst. Behav.* **2018**, *108*, 160–172. [CrossRef] [PubMed]
76. Buzsáki, G. Theta oscillations in the hippocampus. *Neuron* **2002**, *33*, 325–340. [CrossRef]
77. Buzsáki, G.; Draguhn, A. Neuronal oscillations in cortical networks. *Science* **2004**, *304*, 1926–1929. [CrossRef]
78. Huang, Y.-Z.; Edwards, M.J.; Rounis, E.; Bhatia, K.P.; Rothwell, J.C. Theta burst stimulation of the human motor cortex. *Neuron* **2005**, *45*, 201–206. [CrossRef]
79. Oberman, L.; Eldaief, M.; Fecteau, S.; Ifert-Miller, F.; Tormos, J.M.; Pascual-Leone, A. Abnormal modulation of corticospinal excitability in adults with Asperger's syndrome. *Eur. J. Neurosci.* **2012**, *36*, 2782–2788. [CrossRef]
80. Stagg, C.J.; Wylezinska, M.; Matthews, P.M.; Johansen-Berg, H.; Jezzard, P.; Rothwell, J.C.; Bestmann, S. Neurochemical Effects of Theta Burst Stimulation as Assessed by Magnetic Resonance Spectroscopy. *J. Neurophysiol.* **2009**, *101*, 2872–2877. [CrossRef] [PubMed]
81. Cotelli, M.; Manenti, R.; Cappa, S.F.; Geroldi, C.; Zanetti, O.; Rossini, P.M.; Miniussi, C. Effect of Transcranial Magnetic Stimulation on Action Naming in Patients With Alzheimer Disease. *Arch. Neurol.* **2006**, *63*, 1602–1604. [CrossRef]
82. Rami, L.; Gironell, A.; Kulisevsky, J.; Garcia-Sánchez, C.; Berthier, M.; Estévez-González, A. Effects of repetitive transcranial magnetic stimulation on memory subtypes: A controlled study. *Neuropsychologia* **2003**, *41*, 1877–1883. [CrossRef]
83. Sandrini, M.; Censor, N.; Mishoe, J.; Cohen, L.G. Causal Role of Prefrontal Cortex in Strengthening of Episodic Memories through Reconsolidation. *Curr. Biol.* **2013**, *23*, 2181–2184. [CrossRef] [PubMed]
84. Turriziani, P.; Smirni, D.; Mangano, G.R.; Zappalà, G.; Giustiniani, A.; Cipolotti, L.; Oliveri, M. Low-Frequency Repetitive Transcranial Magnetic Stimulation of the Right Dorsolateral Prefrontal Cortex Enhances Recognition Memory in Alzheimer's Disease. *J. Alzheimers Dis. JAD* **2019**, *72*, 613–622. [CrossRef] [PubMed]
85. Kumar, S.; Zomorrodi, R.; Ghazala, Z.; Goodman, M.S.; Blumberger, D.M.; Cheam, A.; Fischer, C.; Daskalakis, Z.J.; Mulsant, B.H.; Pollock, B.G.; et al. Extent of Dorsolateral Prefrontal Cortex Plasticity and Its Association With Working Memory in Patients With Alzheimer Disease. *JAMA Psychiatry* **2017**, *74*, 1266–1274. [CrossRef] [PubMed]
86. Joseph, S.; Zomorrodi, R.; Ghazala, Z.; Knezevic, D.; Blumberger, D.M.; Daskalakis, Z.J.; Mulsant, B.H.; Pollock, B.G.; Rajji, T.K.; Kumar, S. Dorsolateral prefrontal cortex excitability assessed using TMS-EEG and its relationship with neuropsychiatric symptoms in Alzheimer's dementia. *Alzheimers Dement.* **2020**, *16*, e042956. [CrossRef]
87. Berryhill, M. Insights from neuropsychology: Pinpointing the role of the posterior parietal cortex in episodic and working memory. *Front. Integr. Neurosci.* **2012**, *6*, 31. [CrossRef]
88. Berryhill, M.E.; Chein, J.; Olson, I.R. At the intersection of attention and memory: The mechanistic role of the posterior parietal lobe in working memory. *Neuropsychologia* **2011**, *49*, 1306–1315. [CrossRef]
89. Gilmore, A.W.; Nelson, S.M.; McDermott, K.B. A parietal memory network revealed by multiple MRI methods. *Trends Cogn. Sci.* **2015**, *19*, 534–543. [CrossRef] [PubMed]
90. Uncapher, M.R.; Wagner, A.D. Posterior parietal cortex and episodic encoding: Insights from fMRI subsequent memory effects and dual-attention theory. *Neurobiol. Learn. Mem.* **2009**, *91*, 139–154. [CrossRef] [PubMed]
91. Hermiller, M.S.; Chen, Y.F.; Parrish, T.B.; Voss, J.L. Evidence for Immediate Enhancement of Hippocampal Memory Encoding by Network-Targeted Theta-Burst Stimulation during Concurrent fMRI. *J. Neurosci.* **2020**, *40*, 7155–7168. [CrossRef]
92. Warren, K.N.; Hermiller, M.S.; Nilakantan, A.S.; Voss, J.L. Stimulating the hippocampal posterior-medial network enhances task-dependent connectivity and memory. *eLife* **2019**, *8*, e49458. [CrossRef] [PubMed]
93. Taylor, J.L.; Hambro, B.C.; Strossman, N.D.; Bhatt, P.; Hernandez, B.; Ashford, J.W.; Cheng, J.J.; Iv, M.; Adamson, M.M.; Lazzeroni, L.C.; et al. The effects of repetitive transcranial magnetic stimulation in older adults with mild cognitive impairment: A protocol for a randomized, controlled three-arm trial. *BMC Neurol.* **2019**, *19*, 326. [CrossRef] [PubMed]
94. Laske, C.; Stransky, E.; Leyhe, T.; Eschweiler, G.W.; Maetzler, W.; Wittorf, A.; Soekadar, S.; Richartz, E.; Koehler, N.; Bartels, M.; et al. BDNF serum and CSF concentrations in Alzheimer's disease, normal pressure hydrocephalus and healthy controls. *J. Psychiatr. Res.* **2007**, *41*, 387–394. [CrossRef] [PubMed]
95. Andrews-Hanna, J.R.; Reidler, J.S.; Sepulcre, J.; Poulin, R.; Buckner, R.L. Functional-Anatomic Fractionation of the Brain's Default Network. *Neuron* **2010**, *65*, 550–562. [CrossRef] [PubMed]
96. Hebscher, M.; Meltzer, J.A.; Gilboa, A. A causal role for the precuneus in network-wide theta and gamma oscillatory activity during complex memory retrieval. *eLife* **2019**, *8*, e43114. [CrossRef] [PubMed]
97. Hebscher, M.; Ibrahim, C.; Gilboa, A. Precuneus stimulation alters the neural dynamics of autobiographical memory retrieval. *NeuroImage* **2020**, *210*, 116575. [CrossRef]
98. Ye, Q.; Zou, F.; Lau, H.; Hu, Y.; Kwok, S.C. Causal Evidence for Mnemonic Metacognition in Human Precuneus. *J. Neurosci.* **2018**, *38*, 6379–6387. [CrossRef]
99. Gu, S.; Pasqualetti, F.; Cieslak, M.; Telesford, Q.K.; Yu, A.B.; Kahn, A.E.; Medaglia, J.D.; Vettel, J.M.; Miller, M.B.; Grafton, S.T.; et al. Controllability of structural brain networks. *Nat. Commun.* **2015**, *6*, 8414. [CrossRef]
100. Zhao, J.; Li, Z.; Cong, Y.; Zhang, J.; Tan, M.; Zhang, H.; Geng, N.; Li, M.; Yu, W.; Shan, P. Repetitive transcranial magnetic stimulation improves cognitive function of Alzheimer's disease patients. *Oncotarget* **2017**, *8*, 33864–33871. [CrossRef]

101. Andrade, S.M.; de Oliveira, E.A.; Alves, N.T.; Dos Santos, A.C.G.; de Mendonça, C.T.P.L.; Sampaio, D.D.A.; da Silva, E.E.Q.C.; da Fonsêca, É.K.G.; de Almeida Rodrigues, E.T.; de Lima, G.N.S.; et al. Neurostimulation Combined With Cognitive Intervention in Alzheimer's Disease (NeuroAD): Study Protocol of Double-Blind, Randomized, Factorial Clinical Trial. *Front. Aging Neurosci.* **2018**, *10*, 334. [CrossRef]
102. Rabey, J.M.; Dobronevsky, E. Repetitive transcranial magnetic stimulation (rTMS) combined with cognitive training is a safe and effective modality for the treatment of Alzheimer's disease: Clinical experience. *J. Neural Transm.* **2016**, *123*, 1449–1455. [CrossRef]
103. Gandelman-Marton, R.; Aichenbaum, S.; Dobronevsky, E.; Khaigrekht, M.; Rabey, J.M. Quantitative EEG after Brain Stimulation and Cognitive Training in Alzheimer Disease. *J. Clin. Neurophysiol. Off. Publ. Am. Electroencephalogr. Soc.* **2017**, *34*, 49–54. [CrossRef] [PubMed]
104. Gaithersburg, H. FDA Executive Summary Prepared for the March 21, 2019 Meeting of the Neurological Devices Panel 2019. Available online: https://www.fda.gov/advisory-committees/advisory-committee-calendar/june-3-4-2021-neurological-devices-panel-medical-devices-advisory-committee-meeting-announcement (accessed on 9 August 2021).
105. Power, J.D.; Cohen, A.L.; Nelson, S.M.; Wig, G.S.; Barnes, K.A.; Church, J.A.; Vogel, A.C.; Laumann, T.O.; Miezin, F.M.; Schlaggar, B.L.; et al. Functional network organization of the human brain. *Neuron* **2011**, *72*, 665–678. [CrossRef] [PubMed]
106. Yeo, B.T.T.; Krienen, F.M.; Sepulcre, J.; Sabuncu, M.R.; Lashkari, D.; Hollinshead, M.; Roffman, J.L.; Smoller, J.W.; Zöllei, L.; Polimeni, J.R.; et al. The organization of the human cerebral cortex estimated by intrinsic functional connectivity. *J. Neurophysiol.* **2011**, *106*, 1125–1165. [CrossRef]
107. Fox, M.D.; Buckner, R.L.; White, M.P.; Greicius, M.D.; Pascual-Leone, A. Efficacy of transcranial magnetic stimulation targets for depression is related to intrinsic functional connectivity with the subgenual cingulate. *Biol. Psychiatry* **2012**, *72*, 595–603. [CrossRef] [PubMed]
108. Fox, M.D.; Liu, H.; Pascual-Leone, A. Identification of reproducible individualized targets for treatment of depression with TMS based on intrinsic connectivity. *NeuroImage* **2013**, *66*, 151–160. [CrossRef]
109. Gordon, E.M.; Laumann, T.O.; Adeyemo, B.; Gilmore, A.W.; Nelson, S.M.; Dosenbach, N.U.F.; Petersen, S.E. Individual-specific features of brain systems identified with resting state functional correlations. *NeuroImage* **2017**, *146*, 918–939. [CrossRef] [PubMed]
110. Gordon, E.M.; Laumann, T.O.; Adeyemo, B.; Petersen, S.E. Individual Variability of the System-Level Organization of the Human Brain. *Cereb. Cortex* **2017**, *27*, 386–399. [CrossRef]
111. Malhotra, P.A. Impairments of attention in Alzheimer's disease. *Curr. Opin. Psychol.* **2019**, *29*, 41–48. [CrossRef]
112. Rizzo, M.; Anderson, S.W.; Dawson, J.; Myers, R.; Ball, K. Visual attention impairments in Alzheimer's disease. *Neurology* **2000**, *54*, 1954–1959. [CrossRef]
113. Strömgren, L.S. The influence of depression on memory. *Acta Psychiatr. Scand.* **1977**, *56*, 109–128. [CrossRef]
114. Botvinik-Nezer, R.; Holzmeister, F.; Camerer, C.F.; Dreber, A.; Huber, J.; Johannesson, M.; Kirchler, M.; Iwanir, R.; Mumford, J.A.; Adcock, R.A.; et al. Variability in the analysis of a single neuroimaging dataset by many teams. *Nature* **2020**, *582*, 84–88. [CrossRef] [PubMed]
115. Carp, J. On the Plurality of (Methodological) Worlds: Estimating the Analytic Flexibility of fMRI Experiments. *Front. Neurosci.* **2012**, *6*, 149. [CrossRef] [PubMed]
116. Van Essen, D.C.; Ugurbil, K.; Auerbach, E.; Barch, D.; Behrens, T.E.J.; Bucholz, R.; Chang, A.; Chen, L.; Corbetta, M.; Curtiss, S.W.; et al. The Human Connectome Project: A data acquisition perspective. *NeuroImage* **2012**, *62*, 2222–2231. [CrossRef]
117. Glasser, M.F.; Sotiropoulos, S.N.; Wilson, J.A.; Coalson, T.S.; Fischl, B.; Andersson, J.L.; Xu, J.; Jbabdi, S.; Webster, M.; Polimeni, J.R.; et al. The minimal preprocessing pipelines for the Human Connectome Project. *NeuroImage* **2013**, *80*, 105–124. [CrossRef]
118. Thielscher, A.; Antunes, A.; Saturnino, G.B. Field modeling for transcranial magnetic stimulation: A useful tool to understand the physiological effects of TMS? In Proceedings of the 2015 37th Annual International Conference of the IEEE Engineering in Medicine and Biology Society (EMBC), Milano, Italy, 25–29 August 2015; pp. 222–225.
119. Cohen, L.G.; Celnik, P.; Pascual-Leone, A.; Corwell, B.; Falz, L.; Dambrosia, J.; Honda, M.; Sadato, N.; Gerloff, C.; Catalá, M.D.; et al. Functional relevance of cross-modal plasticity in blind humans. *Nature* **1997**, *389*, 180–183. [CrossRef] [PubMed]
120. Cohen, L.G.; Weeks, R.A.; Sadato, N.; Celnik, P.; Ishii, K.; Hallett, M. Period of susceptibility for cross-modal plasticity in the blind. *Ann. Neurol.* **1999**, *45*, 451–460. [CrossRef]
121. Siegel, J.S.; Seitzman, B.A.; Ramsey, L.E.; Ortega, M.; Gordon, E.M.; Dosenbach, N.U.F.; Petersen, S.E.; Shulman, G.L.; Corbetta, M. Re-emergence of modular brain networks in stroke recovery. *Cortex J. Devoted Study Nerv. Syst. Behav.* **2018**, *101*, 44–59. [CrossRef] [PubMed]
122. Braga, R.M.; Buckner, R.L. Parallel Interdigitated Distributed Networks within the Individual Estimated by Intrinsic Functional Connectivity. *Neuron* **2017**, *95*, 457–471. [CrossRef]
123. Buckner, R.L.; Andrews-Hanna, J.R.; Schacter, D.L. The brain's default network: Anatomy, function, and relevance to disease. *Ann. N. Y. Acad. Sci.* **2008**, *1124*, 1–38. [CrossRef]
124. Lindquist, M.A.; Geuter, S.; Wager, T.D.; Caffo, B.S. Modular preprocessing pipelines can reintroduce artifacts into fMRI data. *Hum. Brain Mapp.* **2019**, *40*, 2358–2376. [CrossRef]
125. Smith, S.M.; Beckmann, C.; Andersson, J.; Auerbach, E.J.; Bijsterbosch, J.; Douaud, G.; Duff, E.; Feinberg, D.A.; Griffanti, L.; Harms, M.P.; et al. Resting state fMRI in the Human Connectome Project. *NeuroImage* **2013**, *80*, 144–168. [CrossRef]
126. Wig, G.S.; Schlaggar, B.L.; Petersen, S.E. Concepts and principles in the analysis of brain networks. *Ann. N. Y. Acad. Sci.* **2011**, *1224*, 126–146. [CrossRef] [PubMed]

127. Davis, S.W.; Kragel, J.E.; Madden, D.J.; Cabeza, R. The Architecture of Cross-Hemispheric Communication in the Aging Brain: Linking Behavior to Functional and Structural Connectivity. *Cereb. Cortex* **2012**, *22*, 232–242. [CrossRef] [PubMed]
128. Cohen, N.J.; Squire, L.R. Preserved learning and retention of pattern-analyzing skill in amnesia: Dissociation of knowing how and knowing that. *Science* **1980**, *210*, 207–210. [CrossRef]
129. Neggers, S.F.W.; Langerak, T.R.; Schutter, D.J.L.G.; Mandl, R.C.W.; Ramsey, N.F.; Lemmens, P.J.J.; Postma, A. A stereotactic method for image-guided transcranial magnetic stimulation validated with fMRI and motor-evoked potentials. *NeuroImage* **2004**, *21*, 1805–1817. [CrossRef] [PubMed]
130. Chung, S.W.; Sullivan, C.M.; Rogasch, N.C.; Hoy, K.E.; Bailey, N.W.; Cash, R.F.H.; Fitzgerald, P.B. The effects of individualised intermittent theta burst stimulation in the prefrontal cortex: A TMS-EEG study. *Hum. Brain Mapp.* **2019**, *40*, 608–627. [CrossRef]

MDPI
St. Alban-Anlage 66
4052 Basel
Switzerland
Tel. +41 61 683 77 34
Fax +41 61 302 89 18
www.mdpi.com

Brain Sciences Editorial Office
E-mail: brainsci@mdpi.com
www.mdpi.com/journal/brainsci

www.ingramcontent.com/pod-product-compliance
Lightning Source LLC
LaVergne TN
LVHW070143100526
838202LV00015B/1880